Dr. King Osborne, P.E.

THEORY OF COMPUTATION

Harper & Row
Computer Science and Technology Series

THEORY OF COMPUTATION

Derick Wood

University of Waterloo

HARPER & ROW, PUBLISHERS, New York
Cambridge, Philadelphia, San Francisco, Washington,
London, Mexico City, São Paulo, Singapore, Sydney

To Ma Li

Sponsoring Editor: John Willig
Project Editor: Thomas R. Farrell
Cover Design: 20/20 Services, Inc.
Text Art: RDL Artset, Ltd.
Production: Willie Lane
Printer and Binder: R. R. Donnelley & Sons Company

Theory of Computation

Library of Congress Cataloging-in-Publication Data

Wood, Derick, 1940–
Theory of computation.

Bibliography: p.
Includes index.
1. Computational complexity. 2. Machine theory.
3. Electronic data processing—Mathematics. I. Title.
QA267.W66 1987 511 86-11976
ISBN 0-06-047208-1

86 87 88 89 9 8 7 6 5 4 3 2 1

Contents

Preface xiii

Course Outlines xvii

PART I INTRODUCTION 1

Chapter 0 PRELIMINARIES

0.1 Sequences, Sets, and Tuples 3
  0.1.1 Sets and Their Specification 3
  0.1.2 Set Operations 6
  0.1.3 Multisets 8
  0.1.4 Sequences 8
  0.1.5 Tuples 9
0.2 Directed Graphs 9
  0.2.1 Digraphs 10
  0.2.2 Multiset Digraphs and Trees 14
0.3 Functions, Relations, and Closure 17
  0.3.1 Functions 17
  0.3.2 Big-O Notation 19
  0.3.3 Relations 20
  0.3.4 Closure 25
0.4 Cardinality, Countability, and Enumerability 27
0.5 Proof Methods and Techniques 28
  0.5.1 Direct Proof 29
  0.5.2 Proof by Contradiction 30
  0.5.3 Proof by Induction 31
  0.5.4 The Pigeonhole Principle 34
  0.5.5 Counting 35

0.5.6 Diagonalization 36
0.6 Additional Remarks 38
0.6.1 Summary 38
0.6.2 Further Reading 38
Exercises 38
Programming Projects 49

Chapter 1 LANGUAGES AND COMPUTATION
1.1 Programming Problems and Computation 50
1.2 Alphabets, Words, and Languages 63
1.2.1 Alphabets and Words 64
1.2.2 Languages and Operations 68
1.2.3 Defining Languages 71
1.2.4 Definition Summaries 76
1.3 Languages, Problems, and Computation 77
1.4 Additional Remarks 80
1.4.1 Summary 80
1.4.2 History 80
1.5 Springboard 81
1.5.1 The λ-calculus 81
1.5.2 Markov Algorithms 81
1.5.3 Post Systems and Production Systems 82
1.5.4 Recursive Functions 83
1.5.5 Formal Semantics 84
1.5.6 Program Correctness 85
1.5.7 Program Construction 86
Exercises 86
Programming Projects 90

PART II MODELS 93

Chapter 2 FINITE AUTOMATA
2.1 States, State Diagrams, and Transitions 95
2.2 Deterministic Finite Automata 98
2.3 Nondeterministic Finite Automata 111
2.4 Minimization and Simplification 127
2.5 *DFAs* and Tries 133
2.6 λ-*FAs* and Lazy *FAs* 135
2.6.1 λ-*FAs* and λ-Transitions 135
2.6.2 Lazy Finite Automata 140
2.7 Additional Remarks 142
2.7.1 Summary 142
2.7.2 History 143
2.8 Springboard 143
2.8.1 The Syntactic Monoid 143

2.8.2 Life and Cellular Automata 144
2.8.3 Communication Protocols 144
2.8.4 Security 144
2.8.5 Tree Automata 145
2.8.6 Avoiding Spaghetti Programming 145
Exercises 147
Programming Projects 153

Chapter 3 REGULAR EXPRESSIONS
3.1 Motivation and Definition 155
3.2 Regular Expressions into Finite Automata 161
3.3 Extended Finite Automata into Regular Expressions 166
3.4 Extended Regular Expressions 175
3.5 Additional Remarks 179
3.5.1 Summary 179
3.5.2 History 179
3.6 Springboard 180
3.6.1 A Differential Calculus 180
3.6.2 Descriptional Complexity 180
3.6.3 VLSI 181
3.6.4 Path Expressions 181
Exercises 182
Programming Projects 185

Chapter 4 CONTEXT-FREE GRAMMARS
4.1 Basics of Context-Free Grammars 187
4.2 Simplifications 211
4.2.1 Redundant Symbols 212
4.2.2 Empty Productions 215
4.2.3 Unit Productions 220
4.2.4 Binary Form and Chomsky Normal Form 221
4.2.5 Greibach Normal Form and Two-Standard Normal Form 224
4.3 Linear and Regular Grammars 229
4.4 Extended Context-Free Grammars 232
4.5 Additional Remarks 236
4.5.1 Summary 236
4.5.2 History 237
4.6 Springboard 237
4.6.1 Recognition and Parsing 237
4.6.2 Ambiguity 238
4.6.3 *W*-grammars 238
4.6.4 Simplifications and Efficiency 239
4.6.5 Context-Free Expressions 239
4.6.6 E0L Grammars and Languages 239
4.6.7 ALGOL-Like Languages 240

Exercises 240
Programming Projects 246

Chapter 5 PUSHDOWN AUTOMATA
5.1 Deterministic Pushdown Automata 248
5.2 Nondeterministic Pushdown Automata 259
5.3* Counter Automata 270
5.4* Two-Way Deterministic Pushdown Automata 271
5.5 Additional Remarks 273
5.5.1 Summary 273
5.5.2 History 274
5.6 Springboard 274
5.6.1 Data Structure Automata 274
5.6.2 Preset Pushdown Automata 274
5.6.3 Multihead and Multipushdown Automata 274
5.6.4 Finite-Buffer *PDAs* and Lazy *PDAs* 274
5.6.5 Small *PDAs* 275
Exercises 276
Programming Projects 279

Chapter 6 TURING MACHINES
6.1 The Turing Machine 280
6.2 Turing Machine Programming 295
6.3* Simplifications 301
6.4 Extensions 311
6.4.1 Two-Way Infinite Tape Turing Machines 312
6.4.2 Nondeterministic Turing Machines 316
6.4.3 Multihead Turing Machines 318
6.4.4 Multitape Turing Machines 320
6.5 Universal Turing Machines 320
6.6 Resource-Bounded Turing Machines 326
6.7 Additional Remarks 331
6.7.1 Summary 331
6.7.2 History 331
6.8 Springboard 331
6.8.1 The Busy Beaver Game 331
6.8.2 The Turing Micros 332
6.8.3 Computable Functions 333
6.8.4 *NP*-Completeness 333
6.8.5 Simulation Efficiency 333
6.8.6 Recursively Enumerable Languages 334
6.8.7 Context-Sensitive Languages 334
Exercises 334
Programming Projects 337

Chapter 7 FUNCTIONS, RELATIONS, AND TRANSLATIONS

7.1 Introduction 338
7.2 Finite Transducers 339
7.3 Translation Grammars 348
7.4 Pushdown Transducers 352
7.5 Turing Machines and Computable Functions 355
7.6 Recursive Functions 359
7.7 Additional Remarks 364
 7.7.1 Summary 364
 7.7.2 History 364
7.8 Springboard 364
 7.8.1 Lexical Analysis 364
 7.8.2 Syntax-Directed Translations 364
 7.8.3 File Compaction 365
 7.8.4 Translation Expressions 365
 7.8.5 String Similarity 365
 7.8.6 Recursive Function Theory 365
Exercises 366
Programming Projects 368

PART III PROPERTIES 371

Chapter 8 FAMILY RELATIONSHIPS

8.1 Finite, Regular, and Context-Free Languages 373
 8.1.1 Introduction 373
 8.1.2 Regular Languages 375
 8.1.3* Unary Regular Languages 379
8.2 Context-Free and Tractable Languages 382
 8.2.1 Deterministic Context-Free Languages 382
 8.2.2* Tractable Languages 383
 8.2.3 Context-Free Pumping Lemma 387
 8.2.4* Parikh's Theorem 394
8.3* Tractable and Recursive Languages 397
8.4 Recursive and *DTM* Languages 403
8.5 Additional Remarks 406
 8.5.1 Summary 406
 8.5.2 History 406
8.6 Springboard 407
 8.6.1 Hierarchy Results 407
 8.6.2 Pumping Lemmas 407
 8.6.3 Nonlanguages and Closure Properties 408
Exercises 408

Chapter 9 CLOSURE PROPERTIES

9.1 Boolean and Reversal Operations 412

9.2 Mappings 420
9.2.1 Morphism and Substitution 420
9.2.2* Inverse Morphism and Finite Transduction 429
9.3 Catenation, Quotient, and Star 434
9.4 Composition 435
9.5 Cylinders, Cones, and *AFLs* 438
9.6 Additional Remarks 440
9.6.1 Summary 440
9.6.2 History 440
9.7 Springboard 441
9.7.1 Closure Properties 441
9.7.2 Cones 441
Exercises 441

Chapter 10 DECISION PROBLEMS
10.1 Decidability and Membership 446
10.1.1 Introduction 446
10.1.2 Finite Automata 447
10.1.3 Context-Free Grammars 447
10.1.4 Deterministic Turing Machines 448
10.2 Emptiness and Finiteness 450
10.2.1 Finite Automata 450
10.2.2 Context-Free Grammars 451
10.2.3 Deterministic Turing Machines 452
10.3 Containment, Equivalence, and Intersection 454
10.3.1 Finite Automata 454
10.3.2 Context-Free Grammars 455
10.4 Context-Free Ambiguity 462
10.5 Additional Remarks 462
10.5.1 Summary 462
10.5.2 History 462
10.6 Springboard 463
10.6.1 Post's Correspondence Problem 463
Exercises 464

PART IV ONWARD 469

Chapter 11 FURTHER TOPICS
11.1 *NP*-Complete Problems 472
11.1.1 The Ordering $\propto$ 472
11.1.2 An *NP*-Complete Problem 475
11.1.3 Satisfiability 476
11.1.4 Traveling Salesman 481
11.1.5 Subset Sum 488
11.1.6 Hierarchical Grammars 492

11.2 Rewriting Systems and Church's Thesis 497
11.2.1 Phrase Structure Grammars 497
11.2.2 Interactive Lindenmayer Grammars 500
11.3 Parsing 504
11.3.1 General *CFG* Parsing 504
11.3.2 Deterministic Top-Down Parsing 507
11.3.3 Deterministic Bottom-Up Parsing 515
11.4 Additional Remarks 523
11.4.1 Summary 523
11.4.2 History 523
11.5 Springboard 524
11.5.1 Dealing with *NP*-Completeness 524
11.5.2 *LALR* Grammars 525
11.5.3 Other Kinds of Completeness 525
Exercises 526

BIBLIOGRAPHY 531

INDEX 547

Preface

This book is intended to be used as the basis of a one- or two-term introductory course in the theory of computation at the junior and senior levels. It concentrates on the fundamental models for languages and computation together with their properties. The models are finite automata, regular expressions, context-free grammars, pushdown automata, Turing machines, and transducers. The properties that are presented are family relationships (the relative power of the models), closure properties, decidability properties, and complexity properties. The book contains simple and elegant proofs of many results that are usually considered difficult. For example the proof that we give to show that every finite automaton has an equivalent regular expression is little known.

Throughout the text, algorithms are given in a PASCAL-like notation, since there is an emphasis on constructions and programming. In particular each of Chapters 0-7 has programming projects in addition to the more usual exercises. There is also an emphasis on practical applications throughout the text; for example, finite automata and pattern matching, regular expressions and text editing, extended context-free grammars and syntax diagrams, finite transducers and data compression. The text contains an abundance of worked examples to help the student understand the concepts as they are introduced.

Each chapter terminates with a summary of the material, its history, and a springboard; the last two items include references to the current literature. The springboard introduces a few topics for further investigation that can be used as projects for high-caliber students.

The text consists of four parts. Preliminaries are dealt with in Part I. Chapter 0 reviews sets, relations, graphs, functions, enumerability, and proof techniques. Chapter 1 introduces decision problems, formal languages, and their relationship.

Part II presents five models for languages and computation in Chapters 2-6: finite automata, regular expressions, context-free grammars, pushdown automata, and Turing machines. Chapters 2-6 are written to be order-independent to a large extent. In Chapter 7 these models are adapted for relations and functions. It presents finite transducers (from finite automata), translation grammars (from context-free grammars), pushdown transducers (from pushdown automata), and Turing transducers (from Turing machines). Turing transducers give rise to Turing-computable functions, the largest known class of computable functions. Finally, primitive recursive functions and recursive functions are introduced as an alternative method of specifying functions.

Part III covers the fundamental properties of the models. Chapter 8 explores the relative expressive power of the models, that is, family relationships. Chapter 9 investigates how instances of each model can be formed using a building block approach, that is, the closure properties of the associated language families. Decision problems for the models are studied in Chapter 10. Both basic decidable and undecidable problems are considered.

Part IV explores some further topics in a single chapter, Chapter 11. The topics are NP-completeness, general grammars, and parsing.

Lemmas and theorems are numbered sequentially within sections, in the same sequence. Examples are numbered sequentially within sections and continuations of the same example are indicated by using uppercase letters in addition. Exercises and programming projects are also numbered by section; programming projects are distinguished by an initial letter "P". Finally, an asterisk "*" and a double asterisk "**" indicate that a section or exercise is "difficult" and "more difficult", respectively; hence, the corresponding sections can be omitted on a first reading.

Many people have helped me to develop the present version of the text by giving their encouragement, their comments, or their methods of presenting material. These include Ray Bagley who reviewed the manuscript and made a large number of excellent suggestions; John Brzozowski, my colleague, who not only read and commented on an earlier version, but also by his teaching methods suggested many improvements; Patrick Dymond, who discussed the content of Chapters 6, 7, and 8; Nathan Friedman, a reviewer, who gave some excellent suggestions; Ed Gurari, another reviewer, who made many useful comments; Heikki Mannila, who was willing to participate in a discussion about a suggested approach at the drop of a hat; Areski Nait-Abdallah, who encouraged the germination of this text; Thomas Ottmann, who helped me in the development of Chapters 1 and 7; Jan van Leeuwen, who gave me some much needed positive response in the early days of the text; and Sheng Yu, who was willing to listen and comment on my mutterings. The many anonymous reviewers who waded through preliminary drafts have my sympathy as well as my thanks.

However, without the foundation of friendship and collaboration that I have enjoyed over many years with Hermann Maurer and Arto Salomaa I would not have been able to contemplate this book, let alone write it.

I am very grateful to Mary Chen, who typeset this book, for her concern, patience, and excellent work. She designed the layout, suggested type styles, and much more. I wish to thank Ian Allen, who was responsible for the macros that implemented her suggested layout. The excitement, continuous encouragement, and warmth of my editor, John Willig, my project editor, Thomas Farrell, and other staff members of Harper & Row, even when I fell behind schedule, was much appreciated.

Last, but not least, I tender thanks to the Lord of the Universe who enabled me to meet criticism and deadlines, to persevere, and to aim for excellence. Dei Gloria.

I am sure that errors remain to be found; however, they are mine and mine alone. I will be glad to hear of any errors that you find and any corrections or suggestions that you have.

Derick Wood

Course Outlines

This text contains much more than one term's worth of material; therefore, I give some suggested course outlines below. These reflect different emphases that an instructor may wish to give or that a syllabus requires. First, however, I wish to give some general guidelines to using this book.

The text has an unusual arrangement of material, in that a variety of models are presented in Part II and their properties are discussed mostly in Part III. This provides the basis of a course in which only models are emphasized; perhaps in a first course. It also allows the student to see similar results, concerning properties of models, side by side, making it easier for the student to compare and contrast proof techniques. One effect of this separation is that complete coverage of a standard topic such as finite automata and regular expressions will need not only material from Chapters 2 and 3, but also snippets from Chapters 8, 9, and 10 and also, possibly, from Chapter 7.

The text also contains an extensive review of discrete mathematics in Chapter 0. Experience has shown that students have little prior exposure to or understanding of graphs, relations, closure, and proof techniques. This will vary tremendously from university to university, therefore this is omitted from the course outlines given below. It is anticipated that students begin with Chapter 1, which introduces the basic questions of the theory of computaion, the basics of language theory, and also sketches an important connection between languages and decision problems.

In a thirteen-week course at the University of Waterloo we have been able to cover finite automata, regular expressions, context-free grammars, Turing machines, closure properties, family relationships, and decidability properties. More recently we covered only finite automata, regular expressions, Turing machines, and their properties together with an introduction to context-free grammars. A sample thirteen-week schedule might be as follows:

Problems, decision problems, computation, and languages: 1.1-1.4

(includes a review of 0.4 and 0.5)

Finite automata and regular expressions:

2.1-2.3 (excluding 2.3.1), 2.6.1-2.6.2, 3.1, 3.2

8.1 (for nonregular languages)

9.1, 9.3 (for regular languages), 10.1.1

Turing machines:

6.1-6.5, 10.1.3, 10.2.3

Context-free grammars and pushdown automata:

4.1, 4.2 (excluding 4.2.5), 4.3, 5.1, 5.2

8.1 (for non-context-free languages)

9.1, 9.2, 9.3 (excluding *CF* substitution)

10.1.2, 10.2.2, 10.3.2

In a quartermester system a restricted version of this coverage could be given. However, it is probably better to concentrate in depth on either finite automata, regular expressions, and Turing machines, or finite automata, regular expressions, context-free grammars, and pushdown automata. It is important to realize that the second approach does not allow for undecidability results to be presented.

If two quartermesters are available, then the first could be devoted to finite automata, regular expressions, Turing machines, and undecidability. The second could then cover context-free grammars, pushdown automata, transductions, and their properties.

Part I

INTRODUCTION

Chapter 0

Preliminaries

In this chapter we review the mathematical notions that are a foundation for this text, namely, the notions of discrete mathematics. The review includes sets, sequences, directed graphs, functions, relations, enumerability, and proof techniques.

0.1 SEQUENCES, SETS, AND TUPLES

Throughout the text sets of objects are met and used. These are typically, but not exclusively, sets of words (see Section 1.2). For this reason it is crucial that you be comfortable with sets, their specification, and operations upon them. Hand in hand with sets we will also consider multisets, sequences, and tuples.

0.1.1 Sets and Their Specification

A *set* is a collection of *elements* taken from some prespecified *universe*, for example, the universe of integers or the universe of English words. A set may contain the whole universe, or nothing at all — that is, it could be the *empty set*, denoted by $\varnothing$. A set is *finite* if it contains a finite number of elements and is *infinite* otherwise. The empty set is a finite set, whereas the set of all integers, denoted by $\mathbb{Z}$, is infinite, as is the set of all positive integers $\mathbb{N}$, the set of all nonnegative integers $\mathbb{N}_0$, and the set of all reals $\mathbb{R}$.

A finite set can be specified by simply listing its *members* or elements in braces; for example,

$$\{2,3,5,7\}$$
$$\{the, a, of, for, to\}$$

$$\{\ \}$$

are finite sets of four, five, and zero elements, respectively. Thus, $\{\ \} = \varnothing$. The order in which the members are listed is unimportant, so $\{7,2,5,3\}$ specifies the same set as $\{2,3,5,7\}$. Moreover, the number of times a member is listed is unimportant, so $\{7,2,7,7,5,2,3,7\}$ specifies, once more, the same set as $\{2,3,5,7\}$.

The *cardinality* of a finite set A is the number of elements in A and is denoted by $\#A$. For example, for $A=\{2,3,5,7\}$, $\#A=4$, and for $\varnothing$, $\#\varnothing=0$. We return to this issue of cardinality in Section 0.4, where we discuss the cardinality of infinite sets as well.

Clearly infinite sets cannot be specified by listing their members! We must specify them indirectly. One method of specifying them is to provide a property that all members of the required set must satisfy. For example,

$$A = \{x : x \text{ is a positive integer and } x \text{ is divisible by 3 or 7}\}$$

$$B = \{x : x = 2^i, \text{ for some integer } i \geq 0\}$$

Of course, finite sets can also be specified in this manner; for example,

$$C = \{x : x \text{ is an English word containing the vowels a, e, and u }\}$$

$$D = \{x : x \text{ is an English word and } x \text{ is a palindrome}\}$$

$$E = \{x : x \text{ is a positive integer, } x \text{ is a prime, and } x \text{ is less than } 8\}$$

The general format of such specifications is

$$\{x : P(x)\}$$

meaning the set of all elements x, from the given universe, satisfying the property P.

Given a set A and an element e from its universe we write e *is in* A if, indeed, e is a member of A, and e *is not in* A if e is not a member of A.

Given two sets A and B, we write $A \subseteq B$, which is read as A is *contained in* B, if every member of A is also a member of B. We say A is a *subset* of B and, equivalently, B is a *superset* of A. We also write $B \supseteq A$ which is read as B *contains* A. If $A \subseteq B$ and $B \subseteq A$ we write $A = B$, that is, A *equals* B. We write $A \neq B$, that is, A *is not equal to* B, if there is a member of A which is not a member of B, or vice versa. We write $A \subset B$, which is read as A is *properly contained* in B, if A is contained in B and A is not equal to B, that is, $A \subseteq B$ and $A \neq B$. We also write $B \supset A$ which, in this case, is read as B *properly contains* A. If A is *not contained* in B, then we write $A \not\subseteq B$ and, finally, we write A *inc* B, when A is incomparable with B, that is, $A \not\subseteq B$ and $B \not\subseteq A$.

Two further methods of specifying (infinite) sets are based on the notion of closure. Let the set we wish to define have some operation defined upon its elements. For simplicity we deal only with binary operations, that is, operations that require two operands — the more general situation is dealt with in Section 0.3. Addition, multiplication, and division are examples of binary operations. We first introduce the *recursive* or *inductive definitional method*. Let $\circ$ be a binary operation; then a set A is *recursively defined*, with respect to $\circ$, by the following:

(i) Certain *initial* or *base elements* belong to A.
(ii) If a and b are in A, then $a \circ b$ belongs to A.
(iii) No other elements belong to A.

For example, let A be defined by

(i) 7 belongs to A.
(ii) If a, b are in A, then $a + b$ is in A.
(iii) No other elements belong to A.

We argue that $A = \{7i : i \geq 1\}$. First, 7 is in A by (i) and $7+7 = 14$ is in A by (ii). Now $7 + 14$ and $14 + 14$ are in A by (ii), and so on. On the one hand, we cannot omit any positive multiple of 7 since it can be found as a sequence of applications of rule (ii) with 7. On the other hand, we cannot obtain any other integers. Clearly, negative integers and zero cannot be obtained. Integers without 7 as divisor cannot be obtained because (ii) preserves divisibility by 7, that is, if $a \textbf{ mod } 7 = 0$ and $b \textbf{ mod } 7 = 0$, then $(a+b) \textbf{ mod } 7 = 0$. (Recall that $i \textbf{ mod } j$ is the remainder on dividing integer i by integer j.)

We say a set A is *closed* under a binary operation $\circ$ if, for all a, b in A, $a \circ b$ is in A. In general, a set A defined recursively is always closed under the operations introduced in rule (ii). Our example has only one *constructor*, as it is sometimes called, namely, addition. In the Exercises at the end of this chapter we consider more complex examples.

If a set A is not closed under a given binary operation $\circ$, then there exist at least two elements a and b in A for which $a \circ b$ is not in A. Let B be a superset of A, that is, $B \supseteq A$ and let B be closed under $\circ$. We say B is a *closure of* A with respect to $\circ$, if there is no B' such that $A \subseteq B' \subset B$ and B' is closed under $\circ$. B is a smallest $\circ$-closed superset of A. Indeed, as we show in Section 0.3, such a B always exists, and it is unique. This gives rise to the *closure definitional method*. Let $\circ$ be a binary operation and let a set A be *closure defined* by

> A is the smallest set containing some initial or base elements which is closed under $\circ$.

Again we may introduce finitely many operations in the definition as well as finitely many initial elements. For example, let the set B be defined by: B is the smallest set containing 7 which is closed under $+$. It is not difficult to see that $B = \{7i : i \geq 1\}$ — the same set as the one above.

These two definitional methods are useful in many areas of computer science and they are used in various places in the text. The first method provides a direct means for constructing members of the defined set, whereas the second method does not. However, the methods are interchangeable in that if a set can be defined in one way it can also be defined in the other way.

0.1.2 Set Operations

Given two sets A and B, we define their *union*, denoted by $A \cup B$, by

$$A \cup B = \{x : x \text{ is in } A \text{ or } x \text{ is in } B\}$$

Their *intersection*, denoted by $A \cap B$, is defined by

$$A \cap B = \{x : x \text{ is in } A \text{ and } x \text{ is in } B\}$$

Their *difference*, denoted by $A - B$, is defined by

$$A - B = \{x : x \text{ is in } A \text{ and } x \text{ is not in } B\}$$

$A - B$ consists of those members of A which are left when those also in B are removed. It is straightforward to see, from the definitions that the containments $A \subseteq A \cup B$, $B \subseteq A \cup B$, $A \cap B \subseteq A$, $A \cap B \subseteq B$, $A \cap B \subseteq A \cup B$, and $A - B \subseteq A$ always hold.

For a given set A over some universe U, we denote the *complement* of A by either $\overline{A}$ or $\neg A$, and we define it by

$$\overline{A} = \neg A = U - A$$

The notation $\overline{A}$ is the traditional choice, while the notation $\neg A$ is often more convenient, as we shall see, in Chapter 3, particularly. In the Exercises we consider the properties of union, intersection, and complement, for example

Distributivity	$A \cap (B \cup C) = (A \cap B) \cup (A \cap C)$
	$A \cup (B \cap C) = (A \cup B) \cap (A \cup C)$
Idempotence	$A \cup A = A;\quad A \cap A = A$
Involution	$\neg(\neg A) = A$
Commutativity	$A \cap B = B \cap A;\quad A \cup B = B \cup A$
Associativity	$A \cup (B \cup C) = (A \cup B) \cup C$
	$A \cap (B \cap C) = (A \cap B) \cap C$

Observe that the difference operation is noncommutative, unlike union and intersection. In other words, $A \cup B = B \cup A$ and $A \cap B = B \cap A$, but, in general, $A - B \neq B - A$. Indeed, $A - B = B - A$ if and only if $A = B$.

A useful tool for visualizing the effect of the various operations is the *Venn diagram*. Given the two sets A and B displayed in Figure 0.1.1, their union and intersection are displayed in Figures 0.1.2 and 0.1.3, respectively. Shaded areas represent the indicated sets.

We extend union and intersection to an arbitrary number of sets as follows. Let $A_1, \ldots, A_n$ be n sets, $n \geq 1$; then

$$\bigcup_{i=1}^{n} A_i = \{x : x \text{ is in } A_i, \text{ for some } i, 1 \leq i \leq n\}$$

and

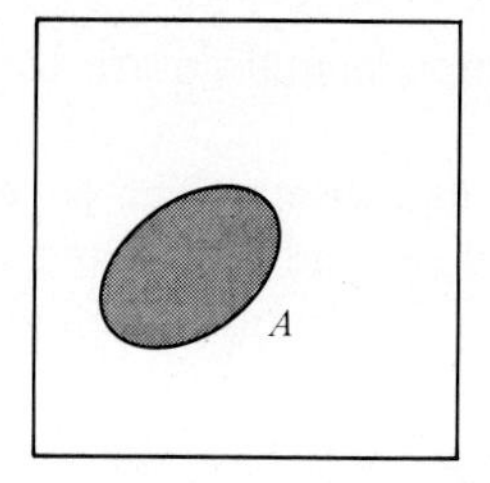

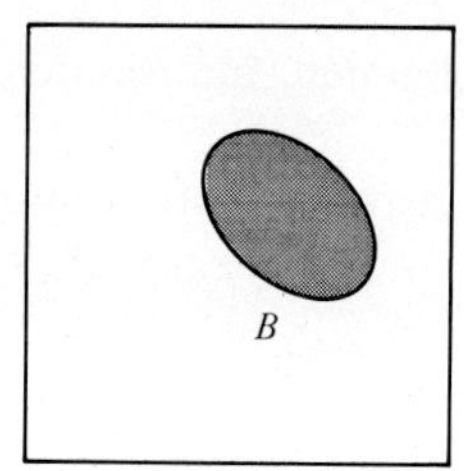

Figure 0.1.1

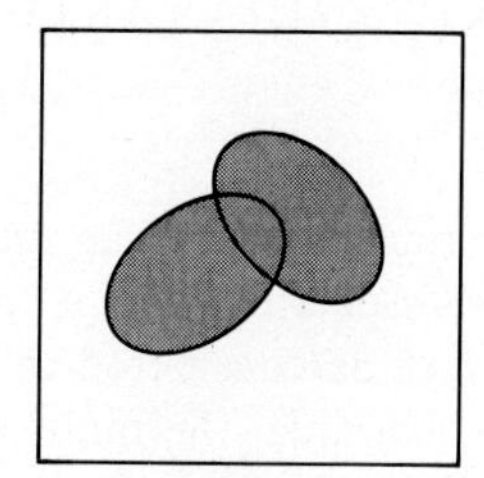

Figure 0.1.2

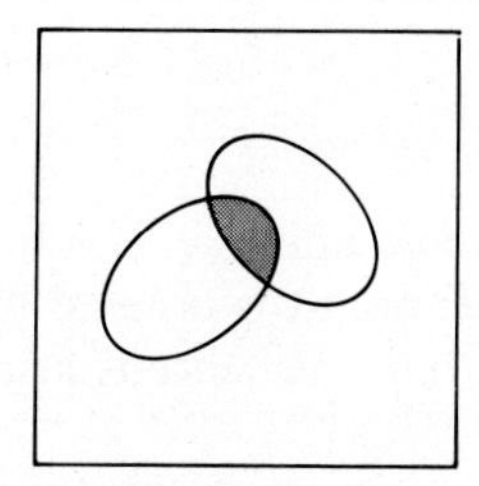

Figure 0.1.3

$$\bigcap_{i=1}^{n} A_i = \{x : x \text{ is in } A_i, 1 \le i \le n\}$$

Often we wish to take an infinite union or intersection, which we denote by $\cup_{i=1}^{\infty} A_i$, $\cap_{i=1}^{\infty} A_i$, where we have sets $A_1, A_2, \ldots, A_i$.

The operations of union, intersection, and complementation are not independent. They are related by De Morgan's rules

$$A \cup B = \overline{\overline{A} \cap \overline{B}} \quad \text{and} \quad A \cap B = \overline{\overline{A} \cup \overline{B}}$$

The operations of union, intersection, and complementation are called the *Boolean operations.*

Given a set A we are often interested in its subsets. The collection of all subsets of A, called the *power set* of A, is also a set and is denoted by 2^A. In other words $2^A = \{B : B \subseteq A\}$. Observe that $\emptyset$ is always an element of 2^A. For example, if

$$A = \{2,3,5,7\}$$

then

$$B = \{\emptyset,\{2\},\{3\},\{5\},\{7\},\{2,3\},\{2,5\},\{2,7\},\{3,5\},\{3,7\},\{5,7\},$$
$$\{2,3,5\},\{2,3,7\},\{2,5,7\},\{3,5,7\},\{2,3,5,7\}\}$$

0.1.3 Multisets

A *multiset* is a collection of elements, from some universe, in which repetitions are not ignored. Each member has associated with it the number of times it appears, namely, its *multiplicity*. A set can be considered to be a multiset in which all multiplicities are equal to one. For example, the multiset $S = \{2,3,3,5,7,5,7,5\}$ is a multiset in which 3 has multiplicity 2, while 2, 5, and 7 have multiplicities 1, 3, and 2, respectively. Given a multiset we can refer to its underlying set; for example, the underlying set of S is the set $\{2,3,5,7\}$. If we wish to refer to a member a of a multiset and its multiplicity m_a explicitly, we write $m_a : a$. The Boolean operations for sets can be extended to multisets in a natural way. We leave this extension to the Exercises.

0.1.4 Sequences

A *sequence* of elements, from some universe, is a list of elements possibly, but not necessarily, with repetitions. Because it is ordered by position, we can speak of the 1st, 2nd, and ith elements. A *finite sequence* is usually specified by listing its elements, for example

$$0,1,1,2,3,5,8,13$$
$$the, cat, sat, on, the, mat$$

while an *infinite sequence* is specified either by using ellipses — for example

$$0,1,1,2,3,5,8,13,\ldots$$

when the specification of the remaining elements is clear, as in this case — or by giving a property that the ith element (the *generic* element) must satisfy. For example,

$$F_0, F_1, F_2, \ldots$$

where $F_0 = 0$, $F_1 = 1$, and for all $i > 1$, $F_i = F_{i-1} + F_{i-2}$. This particular sequence is the Fibonacci sequence.

A *finite sequence* can be *empty*, that is, it may consist of no elements. Since the *empty sequence* would be represented by , that is, nothing at all, in our representation, we prefer to indicate the empty sequence by using a special symbol for it, namely, λ. It is important to realize the difference between a sequence with one element, for example, 3 or *cat*, and λ, which represents a sequence of *no* elements! This choice of representation implies λ should not belong to any universe we are using as a basis for sequences.

The *length* or *size* of a finite sequence S is the number of positions it contains. It is denoted by $|S|$. For example $|\lambda| = 0, |3| = 1$, $|cat| = 1$, and $|0,1,1,2,3,5,8,13| = 8$.

0.1.5 Tuples

Before leaving this section we introduce (*ordered*) *tuples*, or *vectors* as they are often known. Let $n \geq 0$ be an integer. Then an *n-tuple* over some given universe is a sequence of length n over the universe. However, it is usually represented by enclosing the sequence in parentheses. For example,

$$(0,1,1,2,3,5,8)$$
$$(the, cat)$$
$$(\lambda)$$

are a 7-tuple of integers, a 2-tuple of English words, and a 0-tuple, respectively. A 0-tuple is usually written as ( ), since there is no difficulty in recognizing the empty sequence within the parentheses. An n-tuple is also called an *n-vector*. A *tuple* is an n-tuple for some $n \geq 0$. If $n = 2,3,4,5,6,7$, and 8, we prefer the words (*ordered*) pair, triple, quadruple, quintuple, sextuple, septuple, and octuple. 1-tuples are usually identified with their elements, that is, (x) is treated the same as x.

Given sets $A_1, \ldots, A_n$, $n \geq 1$, the *cartesian product* of $A_1, \ldots, A_n$, denoted by $A_1 \times A_2 \times \cdots \times A_n$, is defined by

$$A_1 \times \cdots \times A_n = \{(x_1, \ldots, x_n) : x_i \text{ is in } A_i,\ 1 \leq i \leq n\}$$

that is, it is the set of all n-tuples formed from $A_1, \ldots, A_n$ so that the ith position in each n-tuple contains only members of A_i, for each i, $1 \leq i \leq n$. If the A_i are all equal to some set A, then we use A^n to denote $A_1 \times \cdots \times A_n$.

0.2 DIRECTED GRAPHS

Directed graphs are a basic tool in presenting the structure of machines and the grammatical structure of sentences, as we shall see. This is because they are ideal for representing relations. For these reasons, in this section, we review the basic definitions associated with directed graphs.

0.2.1 Digraphs

A *directed graph*, or *digraph* G, is an ordered pair $G = (V,E)$ of *vertices* or *nodes* V and *directed edges* $E \subseteq V \times V$, where V and E are finite sets. Figure 0.2.1 displays a digraph of 7 vertices and 6 edges.

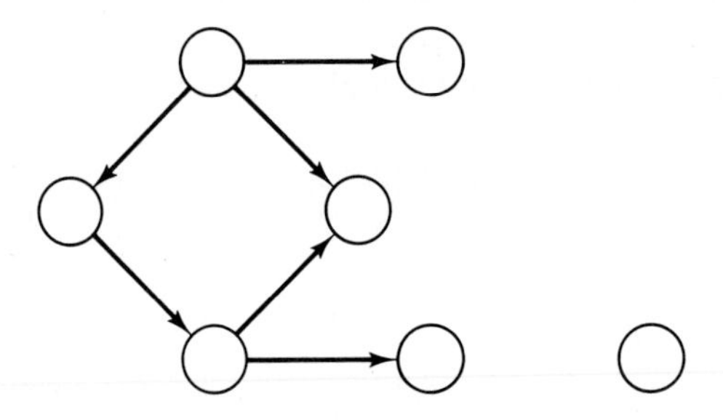

Figure 0.2.1

Given two vertices u and v in a digraph $G = (V,E)$, then u is a *predecessor* of v and v is a *successor* of u if there is an edge (u,v) in E. A *path* from u to v is a sequence of edges $(u_0,u_1),(u_1,u_2), \ldots ,(u_{n-1},u_n)$ from E, for some $n \geq 0$, where $u_0 = u$ and $u_n = v$. Clearly for every edge (u,v) in E there is a path from u to v. The *length* of a path is the number of edges in the path or, alternatively, it is the length of the sequence. It is convenient to allow the *empty path*, that is, $n = 0$ in the above sequence, which has length 0. The sequence of edges in a path also specifies a sequence of vertices, that is,

$$(u_0,u_1),(u_1,u_2), \ldots ,(u_{n-1},u_n)$$

determines the vertex sequence

$$u_0,u_1,u_2, \ldots ,u_{n-1},u_n$$

We say a path is *minimal* if no vertex appears twice in its vertex sequence. If a path is nonempty and $u_0 = u_n$, then the path is a *cycle*. It is a *minimal cycle* if its vertex sequence, ignoring u_n, contains no vertex twice.

The *in-degree* of a vertex v is the number of edges leading into v, the *out-degree* of v is the number of edges leading out of v, and the *degree* of v is the number of edges in which v appears.

Example 0.2.1 Given a PASCAL program P, define a digraph $G = (V,E)$ in which V is the set of subprograms in P and, for all u,v in V, E contains edge (u,v) if and only if subprogram u contains an invocation or call of subprogram v. We say that G is the *call digraph* of P.

Then a subprogram u is recursive if and only if there is a cycle from u to itself.

Strictly speaking, u is only *putatively* recursive, since the necessary sequence of invocations may never occur in any execution of P. An example PASCAL program P is given below, while in Figure 0.2.2 we have its call digraph G.

```
program P;
var a,b : integer;
function p1(i : integer): integer;
begin   if i = 0 then p1 := 0 else
        if i > 0 then p1 := i * i else
        {i < 0} p1 := p1(-i)
end {p1};
function p3(i : integer): integer; forward;
function p2(i : integer): integer;
begin
        if i = 0 then p2 := i else
          p2 := p3(i - 2) * i
end {p2};
function p3;
begin
        if i < 0 then   p3 := abs(i) * p2(i-1) else
                        p3 := i
end {p3}
begin read(a,b);
      writeln(p1(abs(a)));
      writeln(p2(b)))
end.
```

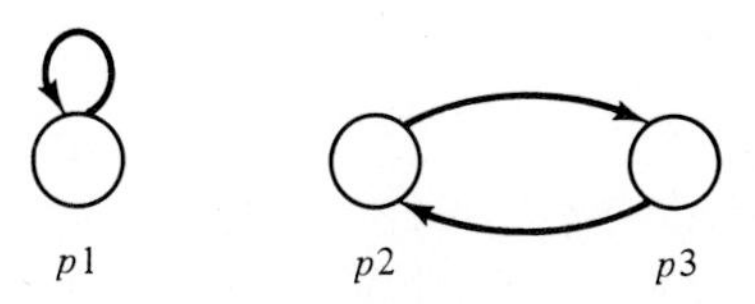

Figure 0.2.2

Observe that $p1$ is recursive, indeed *directly recursive*, but because the only call of $p1$ is with a nonnegative argument, recursion can never occur in this program. On the other hand, $p2$ and $p3$ are *mutually* recursive. This is indicated by a cycle of length 2 in G, and this recursion can occur.

Given a digraph $G = (V,E)$, we say two vertices u and v in V are *adjacent* in G if either (u,v) or (v,u) is in E. The adjacency relationships of a digraph can be represented by a *bit matrix*, that is, a matrix whose entries are either 0 or 1. Let $\#V = n$, A be an $n \times n$ bit matrix, and let the vertices be numbered from 1 to n. More precisely, let $f : \{1, \ldots, n\} \rightarrow V$ be a bijection that defines the numbering. We say A is the *adjacency matrix of G* if for all i, j, $1 \leq i,j \leq n$, $A_{ij} = 1$ if and only if $(f(i), f(j))$ is in E. The adjacency matrix of the call digraph of Figure 0.2.2 is displayed in Figure 0.2.3, where $f(i) = pi$, $1 \leq i \leq 3$. We say a digraph $G = (V,E)$ is *reflexive* if (u,u) is in E for all u in V. It is *symmetric* if for all (u,v) in E, (v,u) is also in E. These correspond in the adjacency matrix to all ones in the diagonal and to a symmetric matrix, respectively. We say G is *transitive* if for all (u,v), (v,w) in E implies (u,w) is in E. Transitivity corresponds to the presence of short cuts. We return to these concepts at the end of the next section, for now let us consider them in the context of call digraphs.

	1	2	3
1	1	0	0
2	0	0	1
3	0	1	0

Figure 0.2.3

If a call digraph is reflexive, then all procedures are directly recursive. If a call digraph is symmetric, then whenever procedure a calls procedure b, for some a and b, procedure b also calls procedure a. In other words a and b are mutually recursive. Finally, if a call digraph is transitive, then whenever there is a sequence of calls of the form

procedure a calls procedure b
procedure b calls procedure c

. . .

procedure p calls procedure q

it is also possible to obtain

procedure a calls procedure q

When compiling programs written in some high-level programming language it is useful to determine those procedures that are nonrecursive since this knowledge can be helpful in the optimization phase. Now recursion may be so indirect that it is detected only after a long chain of procedure calls. For example, in the above sequence,

procedure a calls procedure b

. . .

procedure p calls procedure q

with $q = a$. However, if the call digraph is transitive it is only necessary to check whether or not (a,a) is in E to determine if a is recursive. Call digraphs are not usually transitive, but, fortunately, we can extend them so that they are transitive. This extension is called the *transitive closure*. It is defined using the closure definitional method of Section 0.1.

Let $G = (V,E)$ be a digraph and let $\circ$ be a binary operation defined on E by: for all (u,v), (v,w) in E, $(u,v)\circ(v,w) = (u,w)$. Then $G^+ = (V,E^+)$ is the *transitive closure* of G if

E^+ is the smallest set containing E and closed under $\circ$.

That E^+ is well-defined and unique is demonstrated at the end of Section 0.3. For now, observe that G^+ can be obtained by adding edges to G as follows

If $(u,v),(v,w)$ are in E, for some u,v,w in V, then add (u,w) to E.

Continue to do this until no new edges are added. In Figure 0.2.4 the transitive closure of the call digraph of Figure 0.2.2 is displayed. By the definition of transitive closure, if (a,a) is in E^+, then there is a cycle in E and, hence, a is recursive. On the other hand, if a is recursive, then there is a cycle beginning and ending at a; therefore, (a,a) is in E^+.

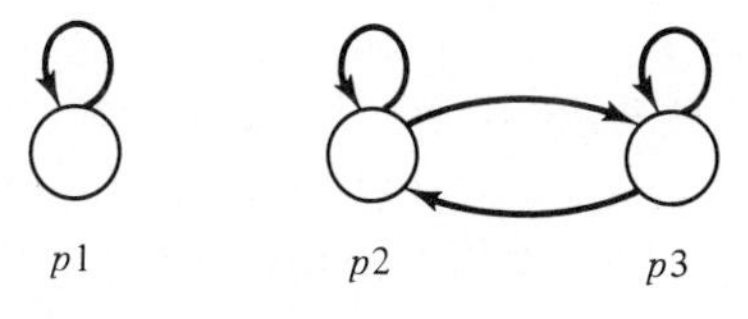

Figure 0.2.4

Letting A^+ be the adjacency matrix of $G^+ = (V,E^+)$, there is a close relationship between A, the adjacency matrix of G, Boolean matrix multiplication, and A^+. This relationship is explored in the Exercises.

For documentation purposes it is useful to know which procedures may be invoked as a result of invoking some procedure a. This leads to the following general notion.

Given a digraph $G = (V,E)$ and nodes u and v we say v is *reachable* from u, or u*-reachable*, if there is a path from u to v. We say G is u*-connected* if for all nodes v in V, v is u-reachable. In other words, G is u*-connected* if in G^+, u is adjacent to every node. We also say G is *node connected* in this case. G is *strongly connected* if G is u-connected for all u in V. In this case, with one exception, the adjacency matrix A^+ of G^+ consists of all ones. The exception is when G consists of a single vertex without an edge. When G is not strongly connected it can be partitioned into strongly connected components; see the Exercises. Since reachability and node connectedness are basic concepts which are useful in Chapters 2 and 4, for example, we give an algorithm to determine all nodes which are u-reachable for a given node u. The algorithm proceeds by finding all nodes adjacent to u, then all nodes adjacent to these, and so on. It terminates when no new u-reachable nodes are discovered. G is u-connected if and only if the set of u-reachable nodes equals V.

Algorithm *Reachability.*
On entry: A digraph $G = (V,E)$ and a node u in V.
On exit: The set R of all u-reachable nodes in G.

```
begin R := {u}; N := {u};
    repeat T := ∅
        for all v in N do T := T ∪ {w : (v,w) is in E};
        N := T − R; {The new u-reachable nodes}
        R := R ∪ N
    until N = ∅;
end
```

Observe that N, the set of new nodes, is always disjoint from the previous value of R. This implies that a node appears at most once as the value of the control variable v in the for loop. The repeat loop must, therefore, terminate, since there are only a finite number of nodes. Clearly each node added to R during this loop is u-reachable; hence, it remains only to demonstrate that every u-reachable node is in R on termination. A formal proof requires induction (see Section 0.5) so we argue only informally here, leaving the formal proof for the Exercises. If a node v is u-reachable, then there is a path from u to v in G. Indeed, there is a shortest such path

$$(u_0,u_1), \ldots, (u_{n-1},u_n)$$

where $u_0 = u$ and $u_n = v$. Since it is a shortest path the u_i, $0 \le i \le n$, are all distinct. This implies that $u_1, \ldots, u_{n-1}$ are also u-reachable and must be in R on termination. Otherwise $u_1, \ldots, u_{n-1}$ are not in R and this cannot hold. This gives the central idea of a formal proof.

The final concept we wish to introduce is that of the *dual* of a digraph. Formally, we have

> Let $G = (V,E)$ be a digraph; then the *dual digraph* $G_D = (V_D,E_D)$ of G satisfies $V_D = E$ and $E_D = \{((u,v),(v,w)) : (u,v),(v,w) \text{ are in } E\}$.

The edges in a digraph become nodes in its dual and, essentially, nodes become edges. In Figure 0.2.5 we give three examples of digraphs and their duals; we have labeled the nodes and edges to clarify the correspondence. A syntax diagram (Chapter 4) is the dual of a state diagram (Chapter 2), and vice versa — this is the reason for our interest in dual digraphs.

After introducing the basic notions associated with digraphs we consider multiset digraphs and trees.

0.2.2 Multiset Digraphs and Trees

As we shall see, particularly in Chapter 2, we need to allow a digraph to have not only a set of edges, but also a *multiset* of edges. In other words, allow more than

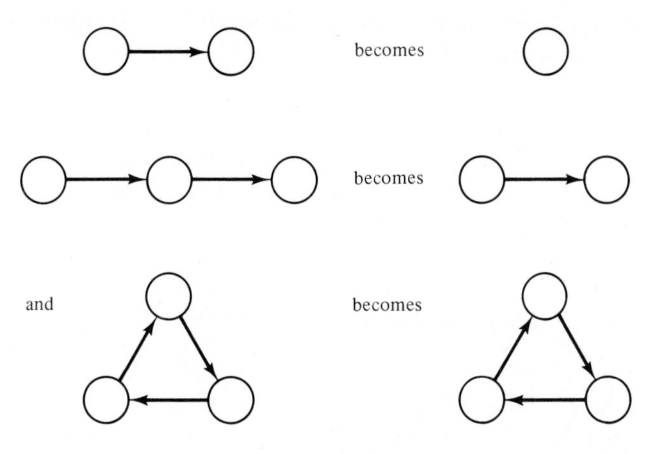

Figure 0.2.5

one edge (u,v) from u to v. To distinguish these multiple edges we introduce *edge labels* from some given domain. The notion of a path is easily extended to this case, when the edge labels are also used to determine the path uniquely. In this setting a *labeled edge* can be considered to be a triple (u,a,v) which specifies an edge from u to v with label a. Thus, such a *multiset digraph* G can be considered to be specified by a triple (V,E,L), where L is a set of *labels* and $E \subseteq V \times L \times V$.

Trees could be viewed as restricted digraphs. However, the trees we usually meet in computer science invariably require the successors of each node to be ordered. By this we simply mean that we want to refer to the first, second, . . . , rth successor of a node. For this reason we use tuples to define trees.

A (*directed*,*ordered*) *tree* T of n *internal nodes*, for some $n \geq 0$, is

(a) the *empty tree* if $n = 0$;
(b) specified by an r-tuple $(u, T_1, \ldots, T_r)$ otherwise, for some $r \geq 1$, where u with out-degree r is the *root* of T and $T_1, \ldots, T_r$ are the r *subtrees* of u with $n_1, \ldots, n_r$ internal nodes, respectively, where $n = 1 + n_1 + \cdots + n_r$.

Since each node in a tree has at most one predecessor, we speak of the *parent* of a node and the *children* of a parent. Similarly we can speak of a child's siblings to its left and right. When displaying trees we represent an internal node as a node in a digraph, while empty subtrees or *external nodes* are represented by □. Figure 0.2.6 is a tree with 8 internal nodes; the edges are directed down the page; and the children of a node are ordered from left to right, by convention.

Since the children of a node are *totally ordered* (see Section 0.3.2), the external nodes of a tree are also totally ordered; see the Exercises. The totally-ordered sequence of external nodes is called the *frontier* of a tree. We often wish to label the nodes of a tree — for example, in Figure 0.2.7 we have a binary search tree, while in Figure 0.2.8 we have a syntax tree (see Chapter 4). These are both examples of *node-labeling* rather than the edge-labeling we required for multiset digraphs. Various properties of trees are explored in the Exercises.

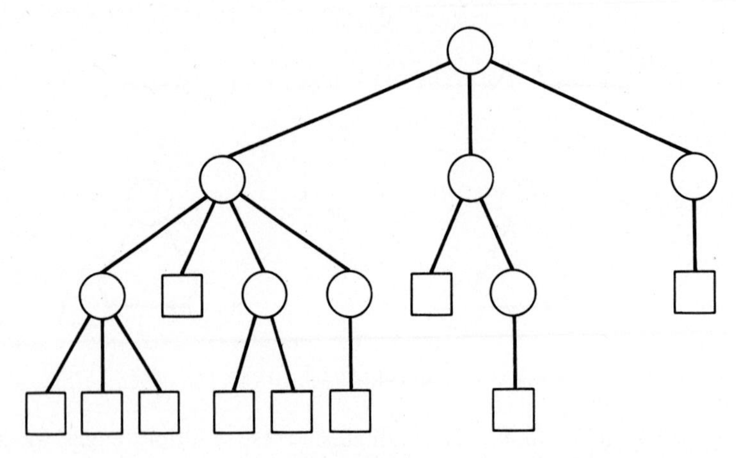

Figure 0.2.6

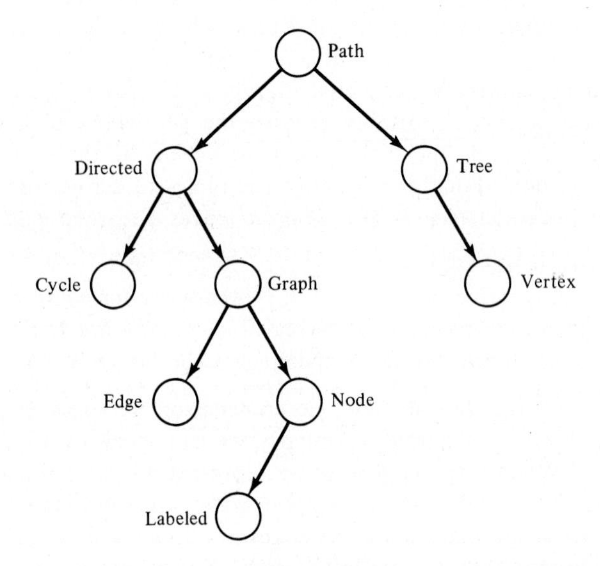

Figure 0.2.7

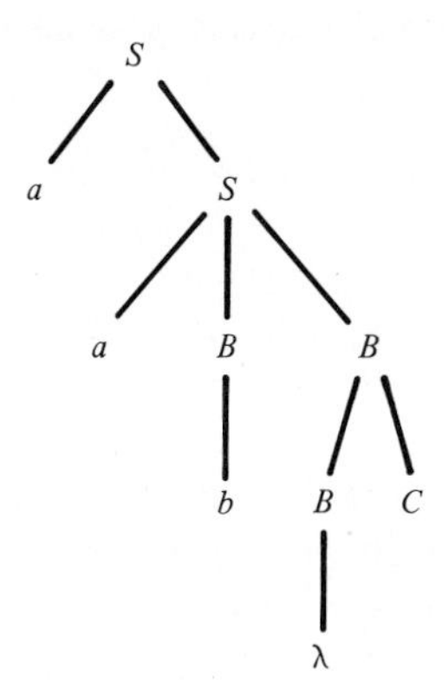

Figure 0.2.8

0.3 FUNCTIONS, RELATIONS, AND CLOSURE

Functions and relations of one type or another permeate this text, so it is imperative that you be fluent in their definition and use. They both capture relationships between sets. We begin with functions.

0.3.1 Functions

Given two sets A and B, a *(partial) function* f *from* A *into* B associates with each a in A at most one element b of B. This association is denoted by writing $f(a) = b$, and the function is denoted by writing $f : A \rightarrow B$. For a given a in A, if f does not associate any element of B with a, then $f(a)$ is said to be *undefined*, denoted by $f(a) =$ ***undef***. The value ***undef*** is usually not in any of the sets considered in this text. It is convenient, however, to define $f($***undef***$) =$ ***undef***, for all functions.

There are two kinds of function deserving special mention. The first are the *identity functions*. A function $f : A \rightarrow A$, for some set A, is an identity function if $f(a) = a$, for all a in A. The second are the *characteristic functions*. Let A and B be two sets with $B \subseteq A$; then the function $f : A \rightarrow \{$**false**,**true**$\}$ is the *characteristic function for* B if for all a in A, if a is in B, then $f(a) =$ **true**; otherwise $f(a) =$ **false**.

Example 0.3.1 Examples of functions abound in everyday student life. Given the set of students S enrolled at your university and the set of possible student IDs I, then $f : S \rightarrow I$ represents the relationship of students and their IDs. Although most students have IDs, $f(s) =$ ***undef*** is possible — for example, transfer students who haven't completed registration yet. However, if we represent the set of students by the set of their names N, then, in general, $g : N \rightarrow I$ is not well-defined because there may be two students with the same name. But we may write

$g : I \rightarrow N$, since we expect each ID to belong to at most one student.

We say $f : A \rightarrow B$ is a *total function* if $f(a)$ is in B, for all a in A, otherwise it is said to be *nontotal.* f is said to be *surjective* or *onto* if for all b in B there is an a in A with $f(a) = b$. f is said to be *injective* or *one-to-one into*, if it is total and for all a and a' in A, $a \neq a'$ implies $f(a) \neq f(a')$. In this case f can be considered to specify an inverse function f^{-1} defined by: for all b in B, $f^{-1}(b) = a$ if $f(a) = b$, for some a, and $f^{-1}(b) = \textbf{\textit{undef}}$, otherwise. If f is both injective and surjective, then f is said to be a *bijection*, or *one-to-one onto.* It is convenient to speak of f^{-1} even when f is not injective. In this case f^{-1} is the function $f^{-1} : B \rightarrow 2^A$ defined by $f^{-1}(b) = \{a : f(a) = b\}$.

Given two functions f and g from A into B, we say f and g are *equal* if $f(a) = g(a)$, for all a in A.

Example 0.3.2 In the United States there are the set G of governors and the set S of states. Similarly, in Canada, there are the set G' of lieutenant-governors and the set S' of provinces. Then $f : S \rightarrow G$ associates with every state its governor and $f' : S' \rightarrow G'$ associates with every province its lieutenant-governor. f and f' are total and one-to-one onto, that is, are bijections. So $f^{-1} : G \rightarrow S$ associates with every governor the state the governor represents and similarly for f'^{-1}.

We can extend the notion of functions of one variable with one result to functions of many variables with many results in a straightforward manner.

Given $m+n$ sets $A_1, \ldots, A_m, B_1, \ldots, B_n$ for $m,n \geq 1$, an *m,n-place (partial) function* f from $A_1 \times \cdots \times A_m$ into $B_1 \times \cdots \times B_n$ associates with each m-tuple $(a_1, \ldots, a_m)$ in $A_1 \times \cdots \times A_m$ either a unique element $(b_1, \ldots, b_n)$ of $B_1 \times \cdots \times B_n$, or no elements of $B_1 \times \cdots \times B_n$, in other words, it is undefined. In the first case we write $f(a_1, \ldots, a_m) = (b_1, \ldots, b_n)$, and in the second case we write $f(a_1, \ldots, a_m) = \textbf{\textit{undef}}$. The functional relationship is denoted by writing $f : A_1 \times \cdots \times A_m \rightarrow B_1 \times \cdots \times B_n$. To ensure consistency we define $f(a_1, \ldots, a_m) = \textbf{\textit{undef}}$ if $a_i = \textbf{\textit{undef}}$, for some i, $1 \leq i \leq m$.

Given an m,n-place function $f : A_1 \times \cdots \times A_m \rightarrow B_1 \times \cdots \times B_n$ and an n,p-place function $g : B_1 \times \cdots \times B_n \rightarrow C_1 \times \cdots \times C_p$; then their *composition* is denoted by $gf : A_1 \times \cdots \times A_m \rightarrow C_1 \times \cdots \times C_p$ and is defined by the following:

For all $(a_1, \ldots, a_m)$ in $A_1 \times \cdots \times A_m$,
$$gf(a_1, \ldots, a_m) = g(f(a_1, \ldots, a_m))$$

Observe that $gf(a_1, \ldots, a_m)$ is undefined if either $f(a_1, \ldots, a_m) = \textbf{\textit{undef}}$ or $f(a_1, \ldots, a_m) = (b_1, \ldots, b_n)$, for some $b_1, \ldots, b_n$ in $B_1 \times \cdots \times B_n$, and $g(b_1, \ldots, b_n) = \textbf{\textit{undef}}$.

Given an m,n-place function f for which $n = 1$ we simply refer to it as an *m-place function.* The word "function" is used in a generic sense, although most of the time it will refer to 1-place functions, that is, functions with only one argument and only one result. We will usually deal with only *m-place* functions $f : A_1 \times \cdots \times A_m \to B$, although functions with many results do occur in a natural manner. For example, consider $f : A \times F \to O \times T$, where A is a set of airlines, F a set of flight numbers, O a set of originating airports, and T a set of originating departure times. On a given day, for a given airline and flight number the originating airport and originating departure time are unique.

We say a function $f : A_1 \times \cdots \times A_m \to B$ is *Boolean* if $B = \{\mathbf{false}, \mathbf{true}\}$, the set of Boolean values. The characteristic functions are one example of such functions.

0.3.2 Big-O Notation

We measure the time and space complexity of a problem or program by total functions from $\mathbb{N}$ to $\mathbb{N}$, since time and space are measured in positive integral units as is the size of the input data. In order to compare time or space complexities of problems and programs we are usually interested only in their order, that is, multiplicative constants and lower-order terms are ignored. The big-O notation is used for this purpose.

Given two total functions $f, g : \mathbb{N} \to \mathbb{N}$, we write $f(n) = O(g(n))$, if there are positive integers c and d such that, for all $n \geq d$,

$$f(n) \leq cg(n)$$

In this case f *is said to be big-O of* g.

Similarly, we write $f(n) = \Omega(g(n))$, if there are positive integers c and d such that, for all $n \geq d$,

$$cf(n) \geq g(n)$$

In this case we say f *is big-omega of* g.

If $f(n) = O(g(n))$ and $f(n) = \Omega(g(n))$, then we write $f(n) = \Theta(g(n))$, that is, f *is big-theta of* g.

Example 0.3.3 Let $f(n) = n$ and $g(n) = 1{,}000{,}000n$; then $f(n) \leq cg(n)$ for $c = 1$ and $d = 1$, so $f(n) = O(g(n))$. Moreover, $cf(n) \geq g(n)$ for $c = 1{,}000{,}000$ and $d = 1$, so $f(n) = \Omega(g(n))$. Hence, $f(n) = \Theta(g(n))$.

If $f(n) = n^2 + 3n + 50$ and $g(n) = 17n \log n + 20n$, then there are no c and d with $f(n) \leq cg(n)$ for all $n \geq d$. However, since $n \geq \log n$ for all $n \geq 1$ we have $cf(n) \geq g(n)$ for $c = 17$ and $d = 1$ or $c = 8$ and $d = 8$, so $f(n) = \Omega(g(n))$.

Whenever $f(n) = O(g(n))$, then $g(n)$ is an *upper bound* for $f(n)$ and whenever $f(n) = \Omega(g(n))$, $g(n)$ is a *lower bound* for $f(n)$. The big-O notation compares the rate of growth of functions rather than their values, so when $f(n) = \Theta(g(n))$, $f(n)$ and $g(n)$ have the same rates of growth, but can be very different in their values.

0.3.3 Relations

Despite the above examples of functions there are many relationships that cannot be captured functionally. For example, students who participate in a common course, the relationship of students and courses they take, the association of professors with courses they teach, etc. For these situations we need the more general notion of relation.

An *n-ary relation* R, $n \geq 1$, with respect to sets $A_1, \ldots, A_n$ is any subset R of $A_1 \times \cdots \times A_n$. Equivalently, the relation is specified by the Boolean function $f : A_1 \times \cdots \times A_n \rightarrow \{\textbf{false}, \textbf{true}\}$, that satisfies $f(a_1, \ldots, a_n) = \textbf{true}$ if and only if $(a_1, \ldots, a_n)$ is in R, that is, f is the characteristic function of R. The values $n = 1,2,3,4, \ldots$ gives rise to *unary*, *binary*, *ternary*, *quaternary*, $\ldots$, *n-ary* relations, respectively. A relation R is finite if and only if R is a finite set. Observe that a function $f : A_1 \times \cdots \times A_m \rightarrow B_1 \times \cdots \times B_n$ can be viewed as a relation $R \subseteq A_1 \times \cdots \times A_m \times B_1 \times \cdots \times B_n$, defined by

$$R = \{(a_1, \ldots, a_m, b_1, \ldots, b_n) : f(a_1, \ldots, a_m) = (b_1, \ldots, b_n)\}$$

For this reason, we will usually identify the relation a function defines with the function itself.

We are particularly interested in binary relations, in which case we prefer to write aRb rather than (a,b) is in R. A finite binary relation is best viewed as a digraph.

Example 0.3.4 Assume we have a set, $C = \{c_1, c_2, c_3, c_4\}$, of four courses and a set, $P = \{p_1, p_2\}$, of two professors who teach the four courses. Then we might have the digraph of Figure 0.3.1, which represents the teaching assignment as the binary relation $T \subseteq P \times C$, where $T = \{(p_1,c_1),(p_1,c_2),(p_1,c_3),(p_2,c_1),(p_2,c_4)\}$.

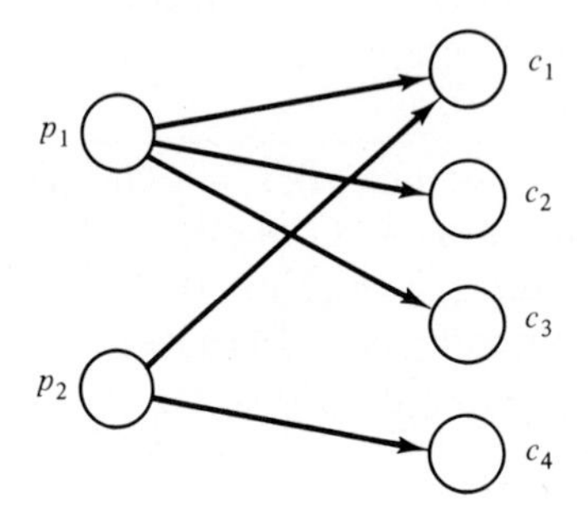

Figure 0.3.1

The digraph corresponding to a finite binary relation R has a node corresponding to each element of the two sets and for every pair of nodes u and v there is a directed edge from u to v if and only if uRv.

Observe that T is not functional. There does not exist any function $f : P \rightarrow C$ with $T = \{(p, f(p)) : p \text{ in } P\}$ and there does not exist any function $g : C \rightarrow P$ with $T = \{(f(c), c) : c \text{ in } C\}$. One might, initially, expect there to be a function $g : C \rightarrow P$; however, as in this case, it is possible for more than one professor to be assigned the same course. This might be because it has two sections, or they intend to do team teaching.

Not only do binary relations $R \subseteq A \times B$ often occur, but in the majority of cases they are relations over a single base set, that is, $A = B$. We call such a relation a *binary relation over* A.

Example 0.3.5 Consider the set S of students on your campus together with the binary relation $E \subseteq S \times S$, defined by

> For all students s_1, s_2 in S, $s_1 E s_2$ if and only if s_1 and s_2 are both enrolled in some common course c.

Consider the set C of courses currently being taught on your campus together with the binary relation $K \subseteq C \times C$ defined by

> For all courses c_1 and c_2 in C with $c_1 \neq c_2$, $c_1 K c_2$ if and only if there is some student enrolled in both c_1 and c_2.

The K relation is necessary when scheduling final examinations, since courses that are *K-related* cannot have examinations at the same time.

Just as we can compose functions we can also compose binary relations. Let $R \subseteq A \times B$ and $S \subseteq B \times C$ be two relations. Then the *composition* of R with S, denoted by $R \circ S$, is the relation $R \circ S \subseteq A \times C$, defined by

$$R \circ S = \{(a, c) : (a, b) \text{ is in } R \text{ and } (b, c) \text{ is in } S \text{ for some } b \text{ in } B\}$$

Example 0.3.6A Let A be the set of people in the world. Consider the two relations *child* and *sibling* defined as follows.

> For all a and b in A, (a, b) is in *child* if a is a child of b.
> For all a and b in A, (a, b) is in *sibling* if a and b are siblings.

Then *child* $\circ$ *child* corresponds to being a grandchild, that is (a, b) is in *child* $\circ$ *child* exactly when a is a grandchild of b.

In contrast *sibling* $\circ$ *sibling* $=$ *sibling* if we assume everyone is a sibling of themselves.

Finally,

$$child \circ sibling = \{(a,c) : \text{there is some } b \text{ in } A \text{ such that}(a,b) \text{ is in } child \text{ and } (b,c) \text{ is in } sibling\}.$$

What relation is this? If (a,c) is in *child* $\circ$ *sibling*, then a is either a child of c or a is a nephew or niece of c.

Clearly, the composition of two relations can be extended to the composition of k relations, for $k \geq 2$. However, we are more interested in the composition of a relation with itself, since this notion will appear throughout this text.

Let R be a relation over A. Then by R^0 we denote the relation $\{(a,a) : a \text{ is in } A\}$ and by R^i, $i \geq 1$, we denote the relation $R \circ R^{i-1}$. We call this the *ith-power of R*. The *transitive closure of R*, denoted by R^+, is the relation $\cup_{i=1}^{\infty} R^i$ and the *reflexive transitive closure of R*, denoted by R^*, is the relation $\cup_{i=0}^{\infty} R^i$. We consider these concepts in more detail after we introduce additional properties that a binary relation may satisfy.

Example 0.3.6B We have already introduced $child^2$. What is $child^i$, $i \geq 3$? (a,b) is in $child^3$ if there are b and c such that $(a,b),(b,c)$, and (c,d) are in *child*. Alternatively, (a,d) is in $child^3$ if there is b such that (a,b) is in *child* and (b,d) is in $child^2$. In other words (a,d) is in $child^3$ if a is a great-grandchild of d. Therefore, $child^i$, $i \geq 3$, corresponds to being a great^{i-2}-grandchild.

This means that (a,b) is in $child^+$ exactly when a is a descendant of b and (a,b) is in $child^*$ when either a is a descendant of b or $a = b$.

A binary relation R over A is *reflexive* if aRa, for all a in A, otherwise it is *irreflexive* or not reflexive.

The relation E of Example 0.3.5 is reflexive since all students are enrolled in the same course as themselves. However, K of Example 0.3.5 is irreflexive or not reflexive, since by definition for any course c in C, (c,c) is not in K. Moreover, *sibling* of Example 0.3.6 is also reflexive.

Graphically, reflexivity corresponds to edges from nodes to themselves; see Figure 0.3.2.

A binary relation R over A is *symmetric* if whenever aRb, then bRa.

E of Example 0.3.5 is symmetric since if a student s_1 shares a course with student s_2, then s_2 shares it with s_1. The relation *sibling* of Example 0.3.6 is symmetric, while *child* is not.

A symmetric relation corresponds graphically to a digraph in which between every two nodes either there is no edge or there are two edges in opposite directions. A binary relation is *asymmetric* if it is not symmetric. Figure 0.3.2 displays an asymmetric relation and Figure 0.3.3 a symmetric relation.

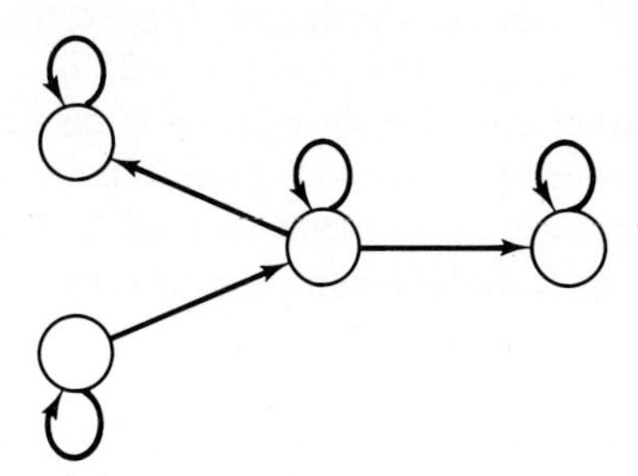

Figure 0.3.2

A binary relation R is *antisymmetric* if whenever aRb and bRa, then $a = b$. The relation displayed in Figure 0.3.4 is antisymmetric.

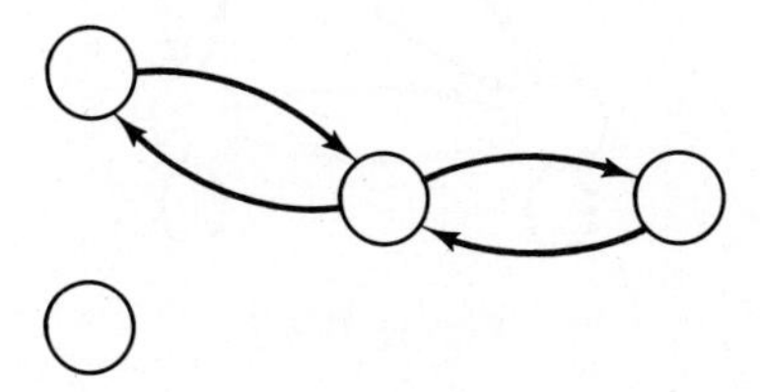

Figure 0.3.3

A binary relation R over A is *transitive* if whenever aRb and bRc, then aRc; otherwise, it is *intransitive* or *nontransitive*. Both E and K of Example 0.3.5 are, in general, nontransitive, whereas *sibling* of Example 0.3.6 is transitive. In terms of digraphs a relation being transitive corresponds to the existence of short cuts. In other words, whenever there is a nonempty path from a node a to a node b there must be an edge from a to b. See Figure 0.3.4 for a digraph that is transitive.

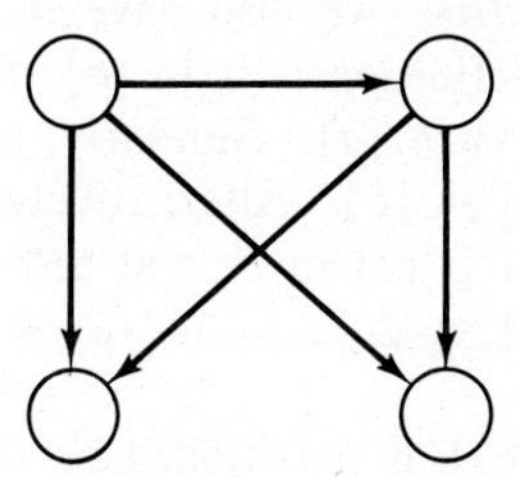

Figure 0.3.4

If a binary relation R over A is reflexive, symmetric, and transitive, then it is called an *equivalence relation.* The relation *sibling* of Example 0.3.6 is an equivalence relation, another equivalence relation is displayed in Figure 0.3.5. An equivalence relation R over A defines disjoint subsets of A, called *equivalence classes.* The equivalence class of an element a of A with respect to R, denoted by $[a]_R$ or simply $[a]$ if R is understood, is defined by

$$[a]_R = \{b : b \text{ is in } A \text{ and } aRb\}$$

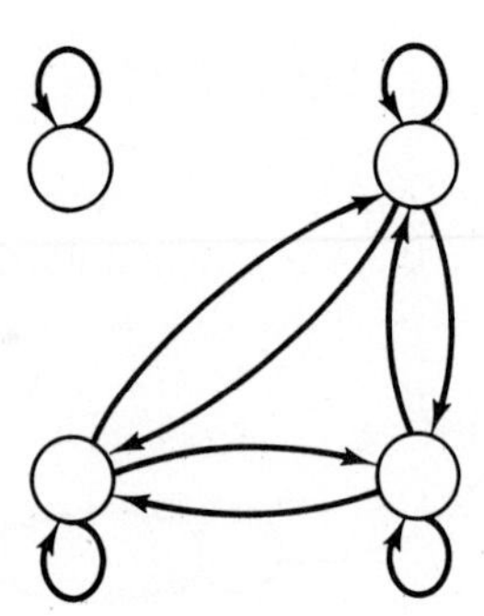

Figure 0.3.5

Since R is reflexive a is in $[a]$, and if aRb then $[a] = [b]$, because of symmetry and transitivity. Indeed, we have the following lemma, which we prove with the direct method. Consult Section 0.5.1 before reading the proof.

Lemma 0.3.1 *Let R be an equivalence relation over A. Then for all a,b in A either $[a] = [b]$ or $[a] \cap [b] = \varnothing$.*

Proof: (direct)
If $a = b$, then clearly $[a] = [b]$. Therefore, assume $a \neq b$. If aRb, then b is in $[a]$ and since by symmetry bRa, we also have a is in $[b]$. The first situation implies $[b] \subseteq [a]$, since by definition c is in $[b]$ if and only if bRc. Now, aRc by transitivity and, hence, c is in $[a]$. Similarly, a is in $[b]$ implies $[a] \subseteq [b]$, therefore we conclude that $[a] = [b]$. Alternatively, if (a,b) is not in R, then for all c in A, aRc implies (b,c) is not in R and bRc implies (a,c) is not in R, by transitivity. Hence, $[a] \cap [b] = \varnothing$. ■

Lemma 0.3.1 implies that A is partitioned by R into equivalence classes, that is, $A = \cup_{a \text{ in } A} [a]$. We say that R has *finite index* if there are only a finite number of equivalence classes, and has *infinite index* otherwise. Figure 0.3.5 displays an equivalence relation which has two equivalence classes.

A binary relation R over A is a *partial order* if it is reflexive, transitive, and antisymmetric. The relation of Figure 0.3.4 is not a partial order, but it is easily modified to give one, see Figure 0.3.6. Another more general example of a partial order is given by the set inclusion relation $\subseteq$ over a collection of sets (see Chapter 8). Let $S = \{i : 1 \leq i \leq n\}$, for some $n \geq 1$, and consider the collection $C = 2^S$. Then for all A,B in 2^S, $A \subseteq B$, $B \subseteq A$, or they are incomparable. Clearly $\subseteq$ is reflexive and transitive; moreover, it is antisymmetric because $A \subseteq B$ and $B \subseteq A$ implies $A = B$.

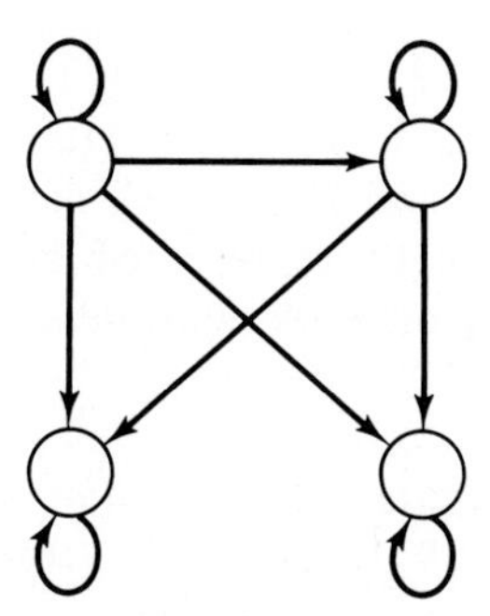

Figure 0.3.6

A binary relation R over A is a *total order* if it is a partial order and for all a,b in A, either aRb or bRa. A total order is often called a *linear order* because the elements of A can be laid out on a straight line such that a is to the left of b if and only if aRb. The integers under the relation $\leq$ form a total order, as do the reals.

Finally, before leaving this section, we consider the notion of *closure*.

0.3.4 Closure

In Section 0.3.3 we defined, for a binary relation R over a set A, the notions of R^+ and R^*. These were called the transitive and the reflexive, transitive closure of R, respectively. To begin our investigation of closure in general we demonstrate that the names of R^+ and R^* were well chosen.

First, R^+ is transitive and R^* is both reflexive and transitive. That R^* is reflexive follows immediately from its definition, since $R^0 \subseteq R^*$ and $R^0 = \{(a,a) : a \text{ is in } A\}$. It is transitive because whenever (a,b) is in R^*, then (a,b) is in R^i, for some $i \geq 0$. Similarly, whenever (b,c) is in R^*, (b,c) is in R^j, for some $j \geq 0$. Thus, by the definition of R^{i+j}, (a,c) is in R^{i+j} and also in R^*. A similar argument shows R^+ is also transitive.

Second, let us define a binary operation $\cdot$ on the elements of such a relation R by

For all (a,b), (c,d) in R, $(a,b)\cdot(c,d) = (a,d)$, if $b = c$ and it is undefined otherwise.

We say a relation R is closed under $\cdot$ if for all (a,b), (b,c) in R we have (a,c) in R. Now, using the closure definitional method of Section 0.1, we can define: A relation R' which is a smallest relation containing R and closed under $\cdot$.

It should be clear that such an R' is transitive. We show that R' equals R^+ and is, therefore, unique. Because R' is a smallest transitive relation containing R, it follows immediately that $R' \subseteq R^+$ and $R \subseteq R'$. Therefore, we need show only that $R^+ - R \subseteq R'$. Consider an element (a,b) in $R^+ - R$. Then (a,b) is in R^i, for some $i \geq 2$. This implies that there exist $c_1, \ldots, c_{i-1}$ such that

$$(a,c_1),(c_1,c_2), \ldots, (c_{i-1},b)$$

are in R, from the definition of R^i. This implies that $(a,c_1)\cdot(c_1,c_2) = (a,c_2)$ is in R', and this in turn implies (a,c_3), $(a,c_4), \ldots, (a,c_{i-1})$ and, finally, (a,b) are in R'. In other words, $R^+ = R'$ and R^+ is well named as the transitive closure of R. If R is finite, then it can be represented by a digraph G and, moreover, obtaining the transitive closure of R is equivalent to obtaining the transitive closure of G.

Theorem 0.3.2 *Let R be a finite binary relation and $G = (V,E)$ be its associated digraph. Then $G^+ = (V,E^+)$ is the transitive closure of G, where $E^+ = \{(u,v) :$ there is a nonempty path from u to v in $G\}$, and G^+ represents the transitive closure of R.*

Proof: This has been left to the Exercises. ■

Furthermore, we may also define the *ith approximation* $G_i = (V,E_i)$ to G^+, where $E_i = \cup_{j=1}^{i} E^j$, and, in this case, be assured that $E^+ = \cup_{i=1}^{m} E^i$ for some finite m. Again, see the Exercises.

We have demonstrated that the closure-definitional method yields a unique result in the specific case of transitive closure. We now prove this in general.

Let R be a binary relation over A and B be a subset of A. We say B is *R-closed* if for all a in B, aRb implies b is in B. For example, let A be the set of all integers and R be the relation "is the square of"; then $B = \{9,4,3,2\}$ is R-closed, but $\{9,4,2\}$ is not R-closed, since $9R3$.

We extend this notion to an n-ary relation R over a set A, $n \geq 2$, by: $B \subseteq A$ is R-closed if, for all $b_1, \ldots, b_{n-1}$ in B, $(b_1, \ldots, b_n)$ is in R implies b_n is in B. For example, let A be the set of all integers and R the ternary relation over A defined by

$$(a_1,a_2,a_3) \text{ is in } R \text{ if and only if } a_1 \times a_2 = a_3.$$

Then A is R-closed as is the subset $B = \{ia : i \geq 1\}$ for some integer a. We more usually say that A and B are *closed under multiplication.*

Given an n-ary relation R over some set A and given a subset $B \subseteq A$ if B is not R-closed, then we often wish to find the smallest superset of B that is R-closed. Clearly, A *is* R-closed, so there always exists at least one R-closed superset. It is potentially possible that there may be more than one smallest

R-closed superset of B. Before proving that there is a unique smallest R-closed superset of B, we must define "smallest" more carefully. We say $B' \supseteq B$ is a *smallest R-closed superset of B* if there is no R-closed superset B'' satisfying $B' \supset B'' \supseteq B$. Observe that if B is, fortuitously, R-closed, then $B' = B$, as required. Section 0.5.1 on the direct proof method should be consulted before reading the following proof.

Theorem 0.3.3 *Let A be a set, R be an n-ary relation over A, $n \geq 2$, and B be a subset of A. Then B has a smallest R-closed superset B' and, moreover, B' is unique.*

Proof: (direct) B has at least one R-closed superset, namely, A itself. Define B' to be the common intersection of all R-closed supersets of B.

Clearly, $B \subseteq B' \subseteq A$ and $B' \subseteq B''$ for all R-closed supersets B'' of B. We claim it is the required smallest R-closed superset of B. We first argue that B' is R-closed. Consider any $b_1, \ldots, b_{n-1}$ in B with $(b_1, \ldots, b_n)$ in R for some b_n. Now for all R-closed supersets B'' of B, we have $B \subseteq B''$; therefore, $b_1, \ldots, b_n$ are in B''. But this implies, by the definition of B', that $b_1, \ldots, b_n$ are in B'. In other words, we have demonstrated that B' is R-closed.

Now B' is the smallest R-closed superset of B by its definition as the common intersection of all R-closed supersets of B and it is, therefore, unique. ■

Because of this theorem we may now write that B' is the R-closure of B. But how do we find the R-closure of a given set? In most cases the R-closure of B is an infinite set, as is B itself, but we can see its finer structure with the following computational schema.

Let $B_0 = B$, and define the *ith approximation* B_i, $i \geq 1$, to the R-closure of B by

$$B_i = B_{i-1} \cup \{b : \text{there are } b_1, \ldots, b_{n-1} \text{ in } B_{i-1} \text{ and } b \text{ in } A \text{ such that } (b_1, \ldots, b_{n-1}, b) \text{ is in } R\}.$$

Then the B_i satisfy:

$$B = B_0 \subseteq B_1 \subseteq B_2 \subseteq \cdots \subseteq B'.$$

Indeed, $B' = \cup_{i=0}^{\infty} B_i$; the proof of which is left to the Exercises.

0.4 CARDINALITY, COUNTABILITY, AND ENUMERABILITY

What does it mean for two sets to be the same size? If they are finite we mean they have the same number of elements. But what if they are infinite? Can one infinite set be larger than another infinite set? The standard approach to this issue is to use bijections as the the basis of comparing sets.

Given two sets A and B, we say that they have *equal cardinality* if there is a bijection $f : A \rightarrow B$. Each member a of A is married to a member b of B and b is married to a.

This definition is consistent with the usual one for finite sets, since two finite sets A and B have equal cardinality, under this definition, if and only if $\#A = \#B$. For infinite sets we consider three examples. First, the set of all integers $\mathbb{Z}$ has equal cardinality with the set of nonnegative integers $\mathbb{N}_0$. Define $f : \mathbb{Z} \rightarrow \mathbb{N}_0$ by $f(i) = -2i-1$, for all $i < 0$ and $f(i) = 2i$, for all $i \geq 0$. Then $f(0) = 0$, $f(-1) = 1$, $f(1) = 2$, $f(-2) = 3$, $f(2) = 4$, $f(-3) = 5$, etc. f is clearly a total function. The proof that it is a bijection is left to the Exercises.

Second, let P denote the set of prime integers. We prove in Section 0.5.2 that there are an infinity of primes. Assuming this result we show that P has equal cardinality with $\mathbb{N}$, the set of positive integers. Define $g : P \rightarrow \mathbb{N}$ by $g(i)$ = the integer j such that i is the jth smallest prime. So $g(2) = 1$, $g(3) = 2$, $g(5) = 3$, $g(7) = 4$, etc. Again the proof that g is a bijection is left to the Exercises.

The primes have a natural ordering associated with them that the function g mimics. Since g is bijective, $g^{-1} : \mathbb{N} \rightarrow P$ is also total. We obtain the natural ordering of the primes with the following *counting* or *enumeration* loop

$$\textbf{for } i := 1 \textbf{ to } \infty \textbf{ do } \mathit{write}(g^{-1}(i))$$

This concept is used in Section 1.1 and is important in its own right.

Third, $\mathbb{N}$ and $\mathbb{N}_0$ have equal cardinality. Simply define $h : \mathbb{N} \rightarrow \mathbb{N}_0$ by $h(i) = i - 1$, for all $i > 0$. It is easy to show that h is a bijection, and hence, $\mathbb{N}$ and $\mathbb{N}_0$ have equal cardinality.

We say that a set is *countable* or *enumerable* if either it is finite or it is infinite and has equal cardinality with $\mathbb{N}$.

Since $\mathbb{Z}, \mathbb{N}_0$, and P have equal cardinality with $\mathbb{N}$ we can enumerate or count their elements. The bijection that demonstrates equal cardinality provides the enumeration sequence.

We say a set is *uncountable* or *nonenumerable* if it is not countable. As we shall demonstrate in Section 0.5 the set of all reals between 0 and 1 is nonenumerable. Therefore, informally, there are more reals than there are integers. Fortunately, most of the time we will deal only with enumerable sets.

0.5 PROOF METHODS AND TECHNIQUES

The workhorse of any mathematical investigation is proof, that is, mathematical proof. Intuition or creativity, call it what you will, enables us to formulate claims or hypotheses which may, if we can prove them, become facts. We establish a new fact by inferring, deducing, or proving its truth from previously known facts. Any course in theoretical computer science will apply proof techniques and also develop problem-solving skills. It is the intent of this section to review the three basic methods of proof and to introduce three specific proof techniques. The methods are direct proof, proof by contradiction, and proof by (mathematical) induction.

The techniques are the pigeonhole principle, counting, and diagonalization.

0.5.1 Direct Proof

In a direct proof we enumerate all cases to show that something does hold. An example of direct proof is in the solution of the following problem.

Consider the set S of all distinct 2×2 matrices whose entries are only 0 or 1. Such a matrix is called a *bit matrix*. Is there a matrix

$$M = \begin{bmatrix} M_{11} & M_{12} \\ M_{21} & M_{22} \end{bmatrix}$$

in S which satisfies the following condition

$$\{(M_{11}, M_{12}), (M_{11}, M_{21}), (M_{21}, M_{22}), (M_{12}, M_{22})\} = \{(0,0), (0,1), (1,0), (1,1)\}$$

Informally, can the four distinct bit 2-vectors be found when reading rows from left to right and columns from top to bottom. Do not be fooled into thinking that this is a useless problem. It points the way to a basic uncountability result, as we shall see. For example, with

$$M' = \begin{bmatrix} 0 & 1 \\ 1 & 0 \end{bmatrix}$$

we find $(0,1), (1,0)$ from the rows and $(0,1), (1,0)$ from the columns. So M' is not such a matrix. The restriction to reading in only one direction is crucial, since the following M'' gives all four possibilities if we are allowed to read in both directions

$$M'' = \begin{bmatrix} 1 & 1 \\ 0 & 0 \end{bmatrix}$$

$(1,1), (0,0)$ are obtained from the rows and $(1,0), (0,1)$ from the columns. At this point try to find an M in S. Do you think one exists? Spend some minutes to make up your mind before reading on.

We prove that there is no such M.

Lemma 0.5.1 *Let S be the set of matrices defined above. Then each M in S realizes at most three of the 2-vectors* $(0,0), (0,1), (1,0)$ *and* $(1,1)$.

Proof: (direct) Consider an arbitrary M in S. First, assume M contains $(0,0)$. This implies M is one of the four forms

$$\begin{bmatrix} 0 & 0 \\ ? & ? \end{bmatrix} \begin{bmatrix} ? & ? \\ 0 & 0 \end{bmatrix} \begin{bmatrix} 0 & ? \\ 0 & ? \end{bmatrix} \begin{bmatrix} ? & 0 \\ ? & 0 \end{bmatrix}$$

If any of the $?$ is replaced by 0, then we have $(0,0)$ in both a row and column. If both of the $?$ are replaced by 1, then we have two appearances of either $(0,1)$ or $(1,0)$. Hence, M contains at most three of the 2-vectors in this case.

Second, if M contains $(1,1)$, then we conclude, in a similar way, that M contains at most three of the 2-vectors.

Third, and finally, if M contains neither $(0,0)$ nor $(1,1)$, then it can only contain two 2-vectors, namely, $(0,1)$ and $(1,0)$.

We have enumerated all the cases and, therefore, we have established the result. ■

Because this text is concerned with languages, which are simply special kinds of sets, we are often faced with the problem of proving that two sets A and B are equal, that is, $A = B$ (or we might have to prove that $A \subset B$ or A and B are incomparable). Typically, the two sets are not self-evidently equal, since they are specified indirectly. A proof that $A = B$ is usually broken down into two proofs. First, we prove $A \subseteq B$ and, second, that $B \subseteq A$. To prove $A \subseteq B$ we have three basic approaches, direct proof, proof by contradiction, and proof by induction. From Chapter 1 onward we will demonstrate a number of proofs of $A \subseteq B$ by contradiction and induction. Hence, we provide one example of a direct proof here, based on the *generic* approach.

Lemma 0.5.2 *For all sets A and B, $A \cap (A \cup B) = A$.*

Proof: (direct) $A \cap (A \cup B) = A$ if and only if $A \cap (A \cup B) \subseteq A$ and $A \subseteq A \cap (A \cup B)$.

Claim 1: $A \cap (A \cup B) \subseteq A$.
Let x be a *generic element* of $A \cap (A \cup B)$, that is, x is not chosen because of its value but simply because it is an arbitrary element of $A \cap (A \cup B)$. If we can prove that x is in A, then because x could have equally well been any element of $A \cap (A \cup B)$, we can deduce that: For all x in $A \cap (A \cup B)$, x is in A. In other words, $A \cap (A \cup B) \subseteq A$.

Now, by the definition of intersection, x is in A and x is in $A \cup B$. But this implies immediately that x is in A as required; therefore, $A \cap (A \cup B) \subseteq A$.

Claim 2: $A \subseteq A \cap (A \cup B)$.
Let x be a generic element of A this time! If we can show that x is also in $A \cap (A \cup B)$, then we will have shown that $A \subseteq A \cap (A \cup B)$. But x is in $A \cup B$, so x is also in $A \cap (A \cup B)$, by the definition of intersection. This implies $A \subseteq A \cap (A \cup B)$. ■

0.5.2 Proof by Contradiction

In a proof by contradiction we establish the validity of a claim by assuming it does not hold and inferring a contradiction. For example, we could prove Lemma 0.5.1 by assuming there is an M which realizes all four bit 2-vectors, and then deducing that this produces a contradiction. More formally, if our assumptions imply the stated conclusion, then the negation of the conclusion implies the negation of our assumptions, and vice versa. An elegant proof by contradiction, which is also an important proof in the development of mathematics, is found in the following theorem.

Theorem 0.5.3 *There are an infinite number of primes.*

Proof: (by contradiction) Assume there are only a finite number of primes, k say, where $k \geq 1$. Observe that k is at least 1 since 2 is a prime. Denote the k primes by $p_1, \ldots, p_k$ and consider the integer

$$N = p_1 p_2 \cdots p_k + 1$$

Clearly, N is not divisible by any of $p_1, \ldots, p_k$, since each gives a remainder of 1. Therefore, N is either a new prime or it is a product of new primes. In both cases we have shown that there exists a prime greater than p_i, $1 \leq i \leq k$. But this contradicts the assumption that only k primes exist. ■

A second proof by contradiction concerns the infinite sequence of integers $s_0, s_1, s_2, \ldots, s_i$, defined by

$$s_i = 2^i \ , \ i \geq 0.$$

This is a typical geometric series. We wish to prove

Theorem 0.5.4 *The set $S = \{s_{i+1} - s_i : i \geq 0\}$ is infinite.*

Proof: (by contradiction) Assume that S is finite. This implies that there is an integer $m > 0$ such that $m \geq s_{i+1} - s_i$, for all $i \geq 0$. Thus, we deduce that $m \geq s_{m+1} - s_m = 2^{m+1} - 2^m$, that is, $m \geq 2^m$. But $m < 2^m$ for all $m \geq 0$ (this is either self-evident or could be proved by induction, see below). Hence, we have obtained a contradiction. Therefore, our assumption is false and S is infinite. ■

Similar proofs to this one are to be found in Section 8.1.

0.5.3 Proof by Induction

We now turn to the third method of proof — proof by induction. There are two forms of induction. Although they are equivalent, we will find that one is always more appropriate than the other in a given situation.

The basic method of induction, called *induction*, has the following two steps. For simplicity assume we are proving a claim about nonnegative integers.

> *First, establish that the claim holds for a basis value b. Second, establish that the claim holds for value $n+1$ when it holds for the value n, for all $n \geq b$.*

Having completed these two steps we conclude that the claim holds for all integers $n \geq b$.

The second method of induction, called *strong induction*, has the same first step but a "stronger" second step.

> *First, establish that the claim holds for a basis value b. Second, establish that the claim holds for value $n+1$ when it holds for all values $b, b+1, \ldots, n$.*

Just as recursion is the most frequently used method of specifying infinite series and sets (and languages, as we shall see), so (mathematical) induction is the most frequently used method of proof associated with such series and sets (and languages). It is bootstrapping par excellence. For example, consider the sequence $s_1, s_2, \ldots, s_i$, where $s_i = 2i - 1$, $i \geq 1$. Then the sum of the first n terms $S_n = s_1 + \cdots + s_n$ is well known to be n^2. This can be proved by induction.

Theorem 0.5.5 *Let $S_n = 1+3+ \cdots +2n-1$, for all $n \geq 1$. Then $S_n = n^2$, for all $n \geq 1$.*

Proof: (by induction on n)

Basis: The case $n = 1$. Immediately it follows that

$$S_n = s_1 = 1 = 1.1 = n^2$$

Induction Hypothesis: Assume $S_n = n^2$ for some $n \geq 1$. We establish that it holds for $n+1$. Since n is arbitrary, this implies it holds for all $n \geq 1$. Using induction we will have proved the theorem.

Induction Step: We extend the initial portion by one position. Consider S_{n+1}. By definition $S_{n+1} = S_n + s_{n+1}$ and by the induction hypothesis $S_n = n^2$. Therefore, $S_{n+1} = n^2 + s_{n+1} = n^2 + 2n + 1$, from the definition of S_i. This implies, $n^2 + 2n + 1 = (n+1)^2$. Hence, $S_{n+1} = (n+1)^2$.

We have assumed the theorem to hold for S_n, for some $n \geq 1$; we have shown it to hold for S_1 (the Basis); and we have demonstrated, under these conditions, that it holds for S_{n+1}. Therefore, it holds for all S_n. ■

Proofs by induction are subtle, indeed so subtle that at first sight they appear to assume what is to be proved. That this is not so is evident after a careful examination of a number of such proofs has been made. Constructing inductive proofs requires painstaking attention to details. Throughout the text ample opportunities are provided for the reader to attempt such proofs.

A second example is geometrical in flavor. Consider $n \geq 0$ infinite straight lines arranged in the plane so that they satisfy the following two conditions:

(i) no two lines are parallel. This implies each pair of lines intersect;
(ii) no three lines have a common intersection point.

The lines in such an arrangement partition the plane into disjoint regions. See Figure 0.5.1 for an arrangement of four lines. Remarkably, the number of regions is uniquely determined by the number of lines and it equals $1 + n(n+1)/2$.

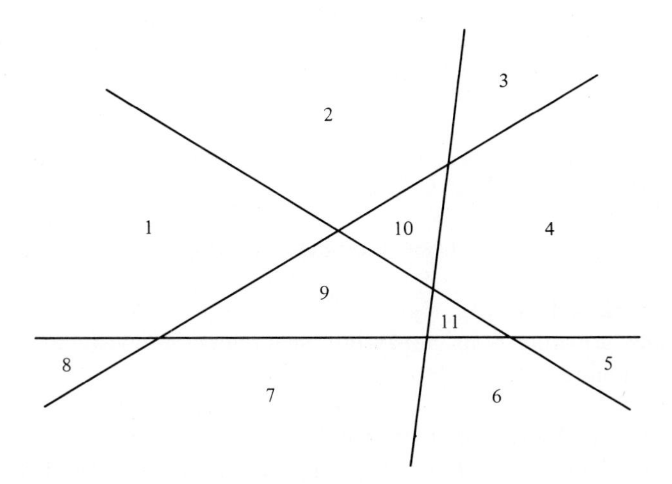

Figure 0.5.1

Theorem 0.5.6 *The number of regions in an arrangement of n lines, as defined above, is* $1 + n(n+1)/2$.

Proof: (by induction on the number of lines)

Basis: $n = 0$ and $n = 1$. Every arrangement satisfies the two conditions in both cases. If $n = 0$, then there is one region and $1 + 0(0+1)/2 = 1$. If $n = 1$, then there are two regions and $1 + 1(2)/2 = 2$.

Induction Hypothesis: Assume the theorem holds for all arrangements of n lines that satisfy conditions (i) and (ii), for some $n \geq 1$. Note that the basis of the induction is actually $n = 1$.

Induction Step: Consider an arrangement A of $n+1$ lines that satisfy conditions (i) and (ii). Since $n \geq 1$, $n+1 \geq 2$ and there are at least two lines in the arrangement. Remove one line L from the arrangement. This leaves an allowable arrangement B, since removing lines cannot affect conditions (i) and (ii). Moreover, B has n lines. By the induction hypothesis the number of regions in B is uniquely determined by n and equals $1 + n(n+1)/2$. Now reconsider A and L. The line L, by conditions (i) and (ii), must intersect each line B at a unique intersection point. Therefore, there are n distinct intersection points of L with the lines in B. Because of condition (ii) the lines in B can be ordered, with respect to their intersection with L, as $L_1, L_2, \ldots, L_n$. L partitions the unbounded region to the left of L_1 into two regions. L partitions the region between L_1 and L_2, that is, intersected by L, into two regions, and so on. Therefore, L divides each of $n+1$

regions in B into two regions. Therefore, A has

$$1+n(n+1)/2+n+1$$

regions, that is,

$$1+(n+1)(n+2)/2$$

regions. Observe that we required there be at least one intersection point of L with the lines in B in the above argument. This is the reason we assumed $n \geq 1$.

Since no assumption other than conditions (i) and (ii) was made about A, the number of regions is always $1+n(n+1)/2$. ■

Having covered the three basic approaches we turn to the three techniques.

0.5.4 The Pigeonhole Principle

The pigeonhole principle is an apparently self-evident proposition yet its applications are far reaching.

Theorem 0.5.7 The Pigeonhole Principle
Let $n \geq 1$ be an integer, let there be n pigeonholes, and let there be $n+1$ items of mail to be placed in the pigeonholes. Then, however the items are placed, at least one pigeonhole will contain at least two items.

Proof: (by induction on the number of pigeonholes)

Basis: $n = 1$. Since there is only one pigeonhole, both items must be placed in it, and the theorem holds in this case.

Induction Hypothesis: Assume the theorem holds for some $n \geq 1$.

Induction Step: Consider the case of $n+1$ pigeonholes. Consider what can happen to one particular pigeonhole P as a result of placing $n+2$ items. There are three possible outcomes. The first outcome is that P contains at least two items and, in this case, P is the pigeonhole of the theorem statement. The second is that P contains one item. This implies that $n+1$ items have been distributed among the remaining n pigeonholes and by the induction hypothesis at least one of them contains two items. The third is that P contains no items of mail. This implies that after distributing $n+1$ items one of the remaining n pigeonholes contained at least two items, by the induction hypothesis. Clearly, the distribution of the $(n+2)$nd item does not effect this result.

In all three cases we have shown that one pigeonhole contains at least two items; hence, the induction step has been established. This implies, by the principle of induction, that the theorem holds. ■

The pigeonhole principle can be generalized for m items of mail; see the Exercises. One application of this result is in characterizing whether or not a given digraph has a cycle.

Lemma 0.5.8 *Let* $G = (V,E)$ *be a digraph with* $\#V = n \geq 1$. *Then* G *has a cycle if and only if* G *has a path of length at least* n.

Proof:

if: Assume G has a path

$$(u_0,u_1), \ldots, (u_{m-1},u_m)$$

with $m \geq n$. We consider only the initial portion of this path, that is a path of length exactly n. The vertex sequence of this path is

$$u_0,u_1, \ldots, u_n$$

which contains $n+1$ vertex appearances. We have $n+1$ vertices appearing in the vertex sequence and only n distinct vertices in G. Let the vertex appearances in the vertex sequence correspond to the items of mail and the vertices in G be the mail boxes; then at least two vertex appearances must correspond to the same vertex in G by invoking the pigeonhole principle. In other words, there are appearances u_i and u_j, $0 \leq i < j \leq n$ with $u_i = u_j$. This implies that the portion of the path from u_i to u_j, namely,

$$(u_i,u_{i+1}), \ldots, (u_{j-1},u_j)$$

contains at least one edge and is, therefore, a cycle. We have demonstrated that if G has a path of length at least n it has a cycle; we now consider the converse.

only if: If G has a cycle

$$(u_0,u_1), \ldots, (u_{m-1},u_m)$$

with $u_0 = u_m$ and $m \geq 1$, then either $m \geq n$ in which case there is nothing to prove or $m < n$. In this latter case repeat the cycle n times giving the path

$$(u_0,u_1), \ldots, (u_{m-1},u_m),(u_m,u_{m+1}), \ldots, (u_{mn-1},u_{mn})$$

which has length $mn \geq n$ since $m \geq 1$. ■

0.5.5 Counting

The second technique, *counting*, gives a method of demonstrating that particular sets are larger than others.

Assume you are given an $n \times n$ bit matrix M, for some $n \geq 1$. Each row of M determines a bit n-vector and each column likewise. You are asked whether or not the set of all bit vectors appearing in the rows and columns of M exhausts the set of all possible bit n-vectors. This is a generalization of the 2×2 problem discussed earlier. An example of a 4×4 matrix is displayed in Figure 0.5.2. We prove this is not the case by counting how many distinct bit n-vectors there are.

Theorem 0.5.9 *For all* $n \geq 1$, *the rows and columns of an* $n \times n$ *bit matrix do not exhaust all possible bit* n*-vectors.*

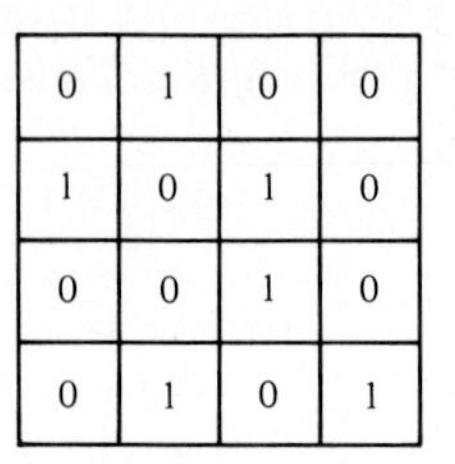

0	1	0	0
1	0	1	0
0	0	1	0
0	1	0	1

Figure 0.5.2

Proof: (by counting) Since each position may be 0 or 1, there are 2^n distinct bit n-vectors, but only $2n$ rows and columns in an $n \times n$ bit matrix. Now $2n < 2^n$ whenever $n < 2^{n-1}$ and this holds for all $n > 2$. Therefore, the theorem holds for all $n > 2$. The cases $n = 1$ and $n = 2$ are proved directly. The case $n = 2$ was proved as Lemma 0.5.1, while the case $n = 1$ is immediate. Thus, the theorem holds for all $n \geq 1$. ■

0.5.6 Diagonalization

But what if we ask for a counterexample to the matrix problem discussed above? A bit n-vector which does not appear in the given $n \times n$ bit matrix? The proof of the cases $n = 1$ and $n = 2$ provide such counterexamples, but for $n > 2$ this is not so obvious. For this purpose we introduce our third technique. This is called *diagonalization*, which can also be used to prove Theorem 0.5.9.

For the matrix M of Figure 0.5.2 construct the 4-vector $(\overline{M}_{11}, \overline{M}_{22}, \overline{M}_{33}, \overline{M}_{44})$, where $\overline{M}_{ii}$ denotes the complement of M_{ii}, giving $(1,1,0,0)$. We claim that $(1,1,0,0)$ does not appear in any column or row. Simply observe that no row or column vector begins with $(1,1)$. However, we prefer to give an argument which generalizes for any n and any $n \times n$ bit matrix.

Theorem 0.5.10 *For all $n \geq 1$, and for an arbitrary $n \times n$ bit matrix M, the n-vector $V = (\overline{M}_{11}, \overline{M}_{22}, \ldots, \overline{M}_{nn})$ appears as neither a row nor a column vector.*

Proof: (by contradiction) This beautiful proof uses the diagonal entries of the matrix to construct a counterexample. Assume V appears in the ith row (the column case is similar and is omitted), that is, $V = (M_{i1}, M_{i2}, \ldots, M_{in})$. This implies $(M_{i1}, \ldots, M_{in}) = (\overline{M}_{11}, \overline{M}_{22}, \ldots, \overline{M}_{nn})$. In other words

$$M_{ij} = \overline{M}_{jj}, \text{ for all } j, \ 1 \leq j \leq n.$$

Specifically this holds for $j = i$, since $1 \leq i \leq n$. Thus, $M_{ii} = \overline{M}_{ii}$, a contradiction. Hence, V does not appear in any row or column of M. ■

This technique has much wider application than is demonstrated by this finite example. For our final result we use it to demonstrate that there are more real numbers than there are integers.

Theorem 0.5.11 *The set of real numbers is not enumerable.*

Proof: (by contradiction using diagonalization)
Consider the real numbers in the open interval $(0,1)$. It is sufficient to show that this set of real numbers is not enumerable. Assume that $(0,1)$ is enumerable. Then there is a bijection $f : \mathbb{N} \longrightarrow (0,1)$ so that every positive integer is associated under f with a unique real, and vice versa. Representing reals in $(0,1)$ by their infinite binary expansions we can picture f as an infinite matrix M so that M_{ij} is the jth bit in the binary expansion of $f(i)$; such an M is illustrated in Figure 0.5.3. Note that each real has at most two binary expansions. For example, 0.5 in decimal can be represented by 0.1 and 0.0111 . . . , which has 1's in every position apart from the first. To avoid this difficulty we choose as representations those that do not terminate with infinitely recurring 1's.

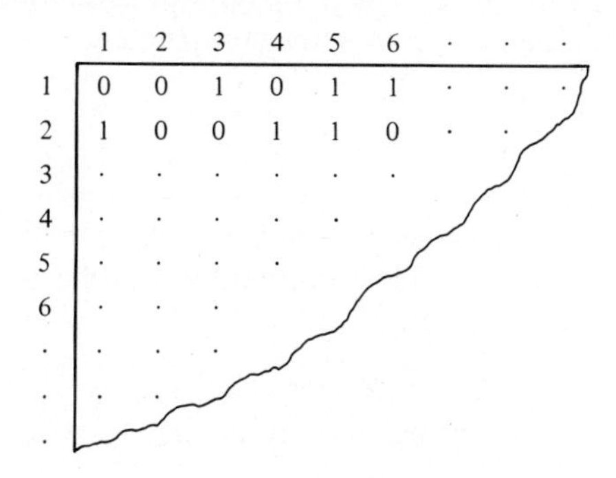

Figure 0.5.3

Now take, as in our finite example, the real r whose binary expansion is

$$r = .\overline{M}_{11}\overline{M}_{22}\overline{M}_{33} \cdots$$

and, as before, if r appears in some row j, then we deduce that

$$M_{jj} = \overline{M}_{jj}$$

a contradiction. Thus, we have demonstrated a real r which is not equal to $f(i)$ for any integer i, contradicting the existence of such an f in the first place. ∎

The reader may argue that the r that we have chosen may terminate with infinitely recurring 1's. In other words, r corresponds to some real in the enumeration, but it doesn't appear as a row because we have excluded such expansions. The avoidance of this difficulty is left to the Exercises.

0.6 ADDITIONAL REMARKS

0.6.1 Summary

This chapter has reviewed the basics of discrete mathematics required to read the present text with understanding. The required topics include: sequences, sets, tuples, digraphs, trees, functions, relations, closure, cardinality, countability, enumerability, and proof methods and techniques. The Exercises which follow are to be used to test your fluency with these topics.

0.6.2 Further Reading

A number of good introductory texts for discrete mathematics have appeared in the last ten years. Any one of Bobrow and Arbib (1974), Prather (1976), Stanat and McAllister (1977), Stone (1973), and Tremblay and Manohar (1975) is worth dipping into if the abbreviated review presented above causes difficulties. Larson (1983) and Solow (1982) provide excellent reviews of induction and the pigeonhole principle as well as examining how proofs are constructed.

EXERCISES

Note that Exercises are numbered as $i.j$, meaning the jth Exercise related to the material appearing in Section 0.i. This numbering scheme is used throughout the text.

1.1 You are given the set A of integers defined by $A = \{x : x = pqr$, for three primes p, q, and r not necessarily distinct$\}$. Are the following integers members of A?

(i) -30; (ii) 0; (iii) 1; (iv) 30; (v) 125; (vi) 35.

Justify your answers.

1.2 You are given five sets of English words, denoted by S_a, S_e, S_i, S_o, S_u. They are defined by

$$S_\alpha = \{y : y \text{ is a word and } y \text{ contains } \alpha\}$$

where α ranges over a, e, i, o, and u. Express the following sets of English words in terms of these five sets and the set operations.

(i) all words containing either a, e, or i;
(ii) all words containing at least e, o, and u;
(iii) all words containing exactly e, o, and u;
(iv) all words not containing e or i;
(v) all words containing a and e but not i.

1.3 Prove the following identities

(i) $A \cup A = A$.
(ii) $A \cap A = A$.
(iii) $(A \cup B) \cup C = A \cup (B \cup C)$.
(iv) $A \cap (B \cap C) = (A \cap B) \cap C$.
(v) $A \cup (B \cap C) = (A \cup B) \cap (A \cup C)$.
(vi) $A \cap (B \cup C) = (A \cap B) \cup (A \cap C)$.
(vii) $A \cap \varnothing = \varnothing = \varnothing \cap A$.
(viii) $A \cup (A \cap B) = A$.
(ix) $\overline{\overline{A}} = A$.
(x) $\overline{A \cup B} = \overline{A} \cap \overline{B}$.
(xi) $\overline{A \cap B} = \overline{A} \cup \overline{B}$.

1.4 Given three sets A, B, and C with $A \cup B = A \cup C$, does $B = C$? Justify your answer.

1.5 For any two sets A and B prove that $2^A \cup 2^B \subseteq 2^{A \cup B}$ and demonstrate that equality need not hold. Can you characterize when equality is obtained? How are $2^A \cap 2^B$ and $2^{A \cap B}$ related, if at all?

1.6 Prove that $A \times (B \cap C) = (A \times B) \cap (A \times C)$.

1.7 Is it possible that $\#A = \# 2^A$, for some finite set A? Justify your answer and if it is positive characterize when this can occur.

1.8 Is it possible for two sets A and B that $A \times B = B \times A$? Justify your answer and if it is positive characterize when this can happen.

1.9 Prove that $(A \cap B) \times (C \cap D) = (A \times C) \cap (B \times D)$.

1.10 Consider the following recursive definition of a set of integers. The set A is defined by the following specification: (i) it includes the initial element 7; (ii) if a and b are in A, then $a + b$ and $a \times b$ are in A; and (iii) no other elements are in A.

What are the elements of A? If $a \times b$ is replaced by (a) $a - b$ or (b) a **div** b, that is, integer division, then what effect if any is there on the set A. Rephrase the above definition as a closure definition.

1.11 Consider the following recursive definition of a set of integers. The set A is defined by the following: (i) it includes the initial elements 5 and 7; (ii) if a and b are in A, then $a+b$ and $a - b$ are in A; and (iii) no other elements are in A.

What are the elements of A? If 5 and 7 are replaced by two arbitrary distinct integers, i and j, under what conditions on i and j is the set A changed.

1.12 Extend the definitions of the Boolean operations to multisets and define and prove identities similar to those in Exercise 1.3.

1.13 Prove that for all finite sets A and B,

$$\#(A \times B) = \#A \times \#B$$

1.14 Let $A_1, A_2, \ldots, A_i$, be an (countable) infinity of sets. Prove

$$\neg(\bigcup_{i=1}^{\infty} A_i) = \bigcap_{i=1}^{\infty} \neg A_i \text{ and } \neg(\bigcap_{i=1}^{\infty} A_i) = \bigcup_{i=1}^{\infty} \neg A_i,$$

that is, De Morgan's rules can be extended to the infinite case.

2.1 Let $G = (V,E)$ be a digraph and v a vertex of G. Prove that the sum of the in-degrees of all vertices in V equals the sum of the out-degrees of all vertices in V and that these are equal to the number of edges in G.

2.2 Given two $n \times n$ bit matrices, A and B, define their *Boolean sum* to be the $n \times n$ matrix C, where $C_{ij} = 0$, if $A_{ij} = 0$ and $B_{ij} = 0$, and $C_{ij} = 1$, otherwise. We denote this by $C = A$ **or** B. Similarly their *Boolean product* is the matrix C, where $C_{ij} = 0$ if $\Sigma_{k=1}^{n} A_{ik} B_{kj} = 0$, and $C_{ij} = 1$, otherwise. We denote this by $C = A$ **and** B.

Let $G = (V,E)$ be a directed graph, where $\#V = n \geq 1$ and $V = \{v_1, \ldots, v_n\}$. An $n \times n$ bit matrix R is said to be the *reachability matrix* of G if $R_{ij} = 1$ when there is a path from v_i to v_j in G and $R_{ij} = 0$, otherwise. Prove that $R = I$ **or** A^n, where I is the $n \times n$ identity matrix, A is the $n \times n$ adjacency matrix, and A^n is the Boolean product of A with itself n times. Relate G, A, and R to G^* and A^*, where G^* is the reflexive transitive closure of G and A^* is the adjacency matrix of G^*.

2.3 Let $G = (V,E)$ be a digraph, $\#V = n$, $V = \{v_1, \ldots, v_n\}$, and A be the $n \times n$ adjacency matrix of G. Prove that the i,jth entry of A^m, $m \geq 1$, is the number of paths from v_i to v_j of length m, where A^m is the matrix product of A with itself m times.

2.4 Let $G = (V,E)$ be a digraph, $\#V = n$, $V = \{v_1, \ldots, v_n\}$ and A be the $n \times n$ adjacency matrix of G. Then an $n \times n$ matrix D is said to be the *distance matrix* of G if $D_{ij} =$ is the length of the minimal path from v_i to v_j, if there is a path from v_i to v_j, and is ∞ otherwise. What relation if any is there between D and powers of A? Justify your answer.

2.5 The call graph of a collection of procedures provides the basis for answering a number of questions. Let $G = (V,E)$ be such a graph, where $\#V = n$, $V = \{p_1, \ldots, p_n\}$ and A be the $n \times n$ adjacency matrix of G. Prove that

(i) $A_{ii}^n = 1$ if and only if p_i is recursive.
(ii) $A_{ij}^n = 1$ if and only if a call of p_i can lead, eventually, to a call of p_j. Thus, the set $\{p_j : A_{ij}^n = 1\}$ is the *subordinate set* of p_i, that is, it delineates those procedures in the collection which must be included with p_i in any program. Similarly, the set $\{p_i : A_{ij}^n = 1\}$ is the *dependency*

set of p_j, that is, it details which procedures in the collection depend on p_j.

2.6 A path in a digraph is a *spanning* or *Hamiltonian cycle* if every vertex in the digraph appears in the cycle. Prove that a digraph is strongly connected if and only if it has a Hamiltonian cycle. Prove that a digraph G of $n > 1$ vertices with adjacency matrix A is strongly connected if and only if $A_{ij}^n = 1$ for all i,j, $1 \leq i,j \leq n$.

2.7 Let $G = (V,E)$ be a digraph; then $G' = (V',E')$ is said to be a *subgraph* of G if $V' \subseteq V$ and $E' = E \cap V' \times V'$. We say G' is a *strongly connected component* of G if G' is strongly connected and, further, we say it is a *strong component* of G if there is no strongly-connected component $G'' = (V'',E'')$ of G with $V' \subset V''$, that is, it is maximal!

(i) Prove that every vertex of G lies in exactly one strong component.
(ii) Prove that $\overline{G} = (\overline{V},\overline{E})$, where $\overline{V} = \{V' : G' = (V',E')$ is a strong component of $G\}$ and $\overline{E} = \{(V',V'') :$ there are vertices v' in V', v'' in V'' with (v',v'') in $E\}$, is acyclic. $\overline{G}$ is called the *condensation* of G.

2.8 Let $P = \{p_1, \ldots, p_m\}$ represent a set of programs in a computer system at time t (strictly speaking, we mean processes). Let $R = \{r_1, \ldots, r_n\}$ represent the set of resources in the system at time t. Then the *allocation digraph* $G_t = (R,E)$ at time t, is defined by

(r_i,r_j) is in E if and only if there is a program p_k which has been allocated resource r_i and is waiting for resource r_j, at time t.

We say that the computer system is in a state of *deadlock* at time t if there are programs $p_{i_1}, \ldots, p_{i_k}$ such that p_{i_1} holds resource r_{i_1} and is waiting for resource r_{i_2}, p_{i_2} holds r_{i_2} and is waiting for $r_{i_3}, \ldots$, and p_{i_k} holds resource r_{i_k} but is waiting for resource r_{i_1}.

Prove that the system is in deadlock at time t if and only if G_t contains a strong component with at least two vertices. If deadlock is present, what do the vertices and edges in such a strong component inform you about the system?

2.9 Let $G = (V,E)$ be an acyclic digraph. Prove that there is at least one vertex with in-degree 0 and one vertex with out-degree 0.

2.10 We can define a constructor or binary operation $\circ$ on binary trees as follows

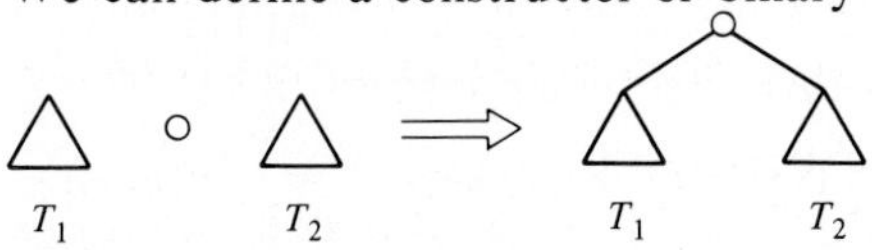

Based on this operation give a recursive definition of the set of binary trees.

2.11 Prove that *Reachability* is correct, that is it always terminates and always produces the set of all u-reachable nodes in a given digraph (use induction).

2.12 Based on the ideas in Exercise 2.10 give a recursive definition of the set of digraphs.

2.13 Prove that the external nodes of a tree are totally ordered.

2.14 The *height* of a tree T is the length of the longest path from the root to an external node. Prove that it equals $ht(T)$ given by the recursive definition

> For a tree T of n internal nodes, $n \geq 0$, $ht(T) = 0$ if $n = 0$ and $ht(T) = 1 + \max(\{ht(T_1), \ldots, ht(T_r)\})$, otherwise, and in this case $T = (u, T_1, \ldots, T_r)$.

2.15 The *weight* of a tree T is the number of external nodes in T. Prove that it equals $wt(T)$ defined as

> For a tree T of n internal nodes, $n \geq 0$, $wt(T) = 1$ if $n = 0$ and $wt(T) = wt(T_1) + \cdots + wt(T_r)$, where $T = (u, T_1, \ldots, T_r)$, otherwise.

3.1 For a function $f : A \rightarrow B$ and subsets A_1 and A_2 of A

(i) Prove $f(A_1 \cup A_2) = f(A_1) \cup f(A_2)$.
(ii) Prove $f(A_1 \cap A_2) \subseteq f(A_1) \cap f(A_2)$.
(iii) Characterize when equality holds in (ii).

3.2 Let P_n be the set of the first n primes, $n \geq 0$, for example, $P_3 = \{2,3,5\}$. Give examples of functions $f : P_m \rightarrow P_n$ which are

(i) not total;
(ii) injective but not bijective;
(iii) surjective but not bijective;
(iv) bijective.

3.3 Give examples of functions $f : \mathbb{Z} \rightarrow \mathbb{Z}$ which are:

(i) injective and not bijective;
(ii) surjective but not injective.

3.4 Let $f : \mathbb{R} \rightarrow \mathbb{R}$ be defined by $f(x) = x^3$ and $g : \mathbb{R} \rightarrow \mathbb{R}$ be defined by $g(x) = x/(x-5)$, if $x \neq 5$. Are f and g total? Are fg and gf total? Justify your answers. If f and g are defined on $\mathbb{Z}$ rather than $\mathbb{R}$ with $g(x) = x \textbf{ div } (x-5)$, how does this effect your answers, if at all?

3.5 Define $f : \mathbb{N} \rightarrow \mathbb{N}_0 \times \mathbb{N}_0$ by $f(i) = (j,k)$, where $i = 2^j(2k+1)$. Prove that f is bijective.

3.6 Let $A_n = \{0, \ldots, n-1\}$, for $n \geq 1$, and let $g_n : A_n \rightarrow A_n$ be defined by $g_n(i) = (i+1)^2 \textbf{ mod } n$, where $p \textbf{ mod } q$ is the remainder when p is divided by q. Prove that g_n is bijective if and only if n is prime.

3.7 Let $A_n = \{0, \ldots, n-1\}$ and $f : A_n \rightarrow A_n$. Prove that f is onto if and only if it is total. Prove that if f is onto it is bijective, and vice versa.

3.8 Let $A_n = \{0, \ldots, n\}$, $n \geq 0$, and $F_n = \{f : f$ is a total function, $f : A_n \rightarrow \{0,1\}\}$, that is, the set of all total functions from $A_n \rightarrow \{0,1\}$. How many such functions are there? How does this change if we allow f to be partial.

3.9 Let $A_m = \{0, \ldots, m\}$, $m \geq 0$ and $F_{m,n} = \{f : A_m \rightarrow A_n\}$, $m,n \geq 0$. How many functions are there in $F_{m,n}$?

[*Hint*: Each f defines an m-tuple $(f(0), \ldots, f(m))$ over $A_n \cup \{$***undef***$\}$.]

3.10 Classify the following functions as total or partial, surjective, injective, bijective, or none of these

(i) $f : \mathbb{N}_0 \times \mathbb{N}_0 \rightarrow \mathbb{N}_0$ by $f(i,j) = i - j$ if $i \geq j$.
(ii) $f : \mathbb{N}_0 \times \mathbb{N}_0 \rightarrow \mathbb{N}_0$ by $f(i,j) = i - j$ if $i \geq j$ and $f(i,j) = 0$ otherwise.
(iii) $f : \mathbb{N}_0 \times \mathbb{N}_0 \rightarrow \mathbb{N}_0 \times \mathbb{N}_0$ by $f(i,j) = (i \textbf{ mod } j, i \textbf{ div } j)$ if $j > 0$.
(iv) $f : \mathbb{N}_0 \times \mathbb{N}_0 \rightarrow \mathbb{N}$ by $f(i,j) = (i+1) \times (j+1)$.
(v) $f : \mathbb{Z} \times \mathbb{Z} \rightarrow \mathbb{Z}$ by $f(i,j) = i + j$.

3.11 We say $f : A \rightarrow B$ is one-to-one if f has an inverse $g : B \rightarrow A$. That is, $gf(a) = a$, for all a in A and $fg(b) = b$ for all b in B. Prove that f is one-to-one if and only if it is a bijection.

3.12 Let $A_n = \{0, 1, \ldots, n-1\}$, for $n \geq 1$, and $f_i : A_n \rightarrow A_n$ be defined by $f_i(j) = i \times j \textbf{ mod } n$. Prove that for a given n, f_i is bijective, for some i, $0 \leq i \leq n-1$, if and only if n is prime.

3.13 Prove that $f : A \rightarrow B$ is injective if and only if for all $A_1, A_2 \subseteq A$, $A_1 \subset A_2$ implies $f(A_1) \subset f(A_2) \subseteq B$.

3.14 For which of the following pairs of functions, f and g, does $f(n) = O(g(n))$, $f(n) = \Omega(g(n))$, and $f(n) = \Theta(g(n))$?

(i) $f(n) = [\sqrt{n}\,]$, where [ ] denotes "integer part of"; $g(n) = 1000n$.
(ii) $f(n) = [\log_{10} n]$; $g(n) = [\log_2 n]$.
(iii) $f(n) = [\sqrt[3]{n}\,]$; $g(n) = [\sqrt{n}\,]$.
(iv) $f(n) = n^2$; $g(n) = [n \log n]$.
(v) $f(n) = 176n^2 - 36n + 17$; $g(n) = n^2$.
(vi) $f(n) = [n \log n] + [\sqrt{n}\,]$; $g(n) = [n \log^2 n]$.

3.15 Let C be the set of cities in the United States. Define aRb to hold for two cities a and b in C when there is a nonstop bus that runs from a to b.

(i) Is R reflexive?
(ii) Is R transitive?

(iii) Is R a function?
(iv) What do R^0, R^2, and R^+ mean?

3.16 We define a ternary relation R over $\mathbb{N}$ by: (a,b,c) is in R exactly when c **mod** $(a * b) = 0$.

(i) We can obtain two "functions" from R, namely: $f : \mathbb{N} \rightarrow \mathbb{N} \times \mathbb{N}$ and $g : \mathbb{N} \times \mathbb{N} \rightarrow \mathbb{N}$ defined by $f(a) = (b,c)$ and $g(a,b) = c$ when (a,b,c) is in R. Are either f or g well defined?
(ii) Is $\mathbb{N}$ closed under R?

3.17 Let A be any set and $A_1, \ldots, A_m$, $m \geq 1$ be disjoint nonempty subsets of A satisfying $A = \cup_{i=1}^{m} A_i$, that is, $A_1, \ldots, A_m$ form a partition of A. Define a relation R over A by: aRb exactly when a and b are in the same subset A_i, $1 \leq i \leq m$.

(i) Prove that R is an equivalence relation over A.
(ii) Prove that the equivalence classes of R are exactly the $A_1, \ldots, A_m$.

3.18 Let A be any set and define a relation R over A by aRb exactly when $a = b$.

(i) Prove that R is an equivalence relation. We call such an R the identity relation.
(ii) Define $f : A \rightarrow A$ by $f(a) = b$ exactly when aRb. Is f well defined? What is f usually called?

3.19 Let R be a relation over some set A and define the inverse relation R^{-1} over A by $aR^{-1}b$ if and only if bRa. Prove that R is an equivalence relation if and only if

(i) $R = R^{-1}$,
(ii) $RR \subseteq R$, and
(iii) R contains the identity relation over A.

3.20 Given an ordered binary tree $G = (V,E)$ we define the relation "shares a common ancestor," denoted by C, recursively by

(i) uCu for all u in V;
(ii) uCv, $u \neq v$, if $parent(u)Cv$, $uCparent(v)$, or $parent(u)Cparent(v)$.

Prove that

(i) C is an equivalence relation.
(ii) It has only one equivalence class.

3.21 As in Exercise 3.20, but define the relation "is an ancestor of," denoted by A, as follows

(i) for all u in V, uAu;
(ii) for all u,v in V, $u \neq v$, uAv if $uAparent(v)$.

Prove that A is transitive and antisymmetric. Is A a partial or total order?

3.22 Given an ordered binary tree $G = (V,E)$ we denote the relation "is to the left of" by L and we define it recursively by

(i) for all vertices u, v in V, if $u \neq v$, $parent(u) = parent(v)$, and $u = leftchild(parent(u))$, then we have uLv;
(ii) for all u, v in V, if $u \neq v$, and $parent(u)Lv$, $uLparent(v)$, or $parent(u)Lparent(v)$, then we have uLv.

Which properties does L satisfy? Justify your answers.

3.23 Prove that for $G = (V,E)$ and $G^+ = (V,E^+)$, $E^+ = \cup_{i=0}^{m} E_i$ for some finite m.

4.1 Given that two sets A and B are enumerable, prove that (i) $A \cup B$, (ii) $A \cap B$, (iii) $A \times B$ are enumerable. (iv) Is $A - B$ enumerable? Justify your answer. (v) As an application of these results prove that the set of rational numbers is enumerable.

4.2 Given that A is enumerable and infinite, prove that 2^A is non-enumerable (see Section 0.5).

4.3 Prove that the following sets are enumerable:

(i) $\mathbb{N}_0$.
(ii) The set of all primes.
(iii) $\{x : x$ is in $\mathbb{N}$ and $x = i^2$, some i in $\mathbb{Z}\}$.
(iv) $\{x : x$ is in $\mathbb{N}$ and $x = p + q$, for two primes p and $q\}$.
(v) $\{x : x$ is in $\mathbb{N}$ and $x = p * q + r$, for three primes p,q, and $r\}$.

4.4 The examples in Exercise 4.3 point to the possibility of a general result, namely,

> Let A be an enumerable set and $B \subseteq A$ be any subset of A. Then B is enumerable.

Is this result valid? Justify your answer.

4.5 Given countably many enumerable sets $A_1, A_2, \ldots$, is $\cup_{i=1}^{\infty} A_i$ enumerable? Is $\cap_{i=0}^{\infty} A_i$ enumerable? Is, abusing our notation somewhat, $\times_{i=1}^{\infty} A_i$ enumerable? Justify your answers in all cases.

4.6 Given a nonenumerable set A, prove that it has an infinite enumerable subset B.

4.7 Given a nonenumerable set A and an arbitrary partition of A into countably many disjoint subsets $A_1, A_2, \ldots$, that is, $\cup_{i=1}^{\infty} A_i = A$ and for all $i,j \geq 1$, $A_i \cap A_j \neq \varnothing$ implies $i = j$, prove that at least one of the A_i

must be nonenumerable. (This can be considered to be the extension of the pigeonhole principle (see Section 0.5) to countably many pigeonholes and uncountably many pieces of mail.)

4.8 Prove that the set of all functions from $I\!N$ to $I\!N$ is nonenumerable.

[*Hint:* Relate this set to $2^{I\!N}$.]

4.9 Let A be a set satisfying $A = \cup_{i=0}^{\infty} A_i$, for some A_i, where $A_i \cap A_j = \emptyset$, for all i, j, $i \neq j$. Given that each A_i is finite, prove that A is enumerable.

4.10 Let F be the set of all functions $f : I\!N_0 \rightarrow I\!N_0$ satisfying $\{i : f(i) \neq 0\}$ is finite.

(i) Prove that F is enumerable.
(ii) If we change the condition on f to $\{i : f(i) \neq 25\}$ is finite, is F still enumerable?
(iii) If we change the condition on f to $\{i : f(i) \neq c_f\}$ is finite, where c_f is in $I\!N_0$, is F still enumerable?

Justify your answers in all cases.

4.11 Let (x_0, y_0) be some given point in $I\!R^2$, that is, in the plane. Prove that the set of all straight lines through (x_0, y_0) has equal cardinality with $I\!R$. Does this hold for the set of all rays beginning at (x_0, y_0)? (A ray or half-line is "half a line" so that it has one endpoint in the plane.)

4.12 Given the real line, that is, $I\!R$, consider the set of all closed intervals on the line. Does this set have equal cardinality with $I\!R$? Assume we fix one endpoint of the intervals to be some given value x_0. Does this affect your answer?

4.13 Given two sets A and B of equal cardinality, are 2^A and 2^B of equal cardinality? Justify your answer.

5.1 In analogy to Theorem 0.5.1 consider 2×2 matrices whose entries may be 0, 1, or 2. Allow row vectors also to be formed by reading rows right to left and column vectors also by reading from bottom to top. Prove, by direct proof, that all possible ordered pairs with entries 0, 1, or 2 cannot be achieved by any 2×2 matrix.

5.2 Give a simpler proof of the result claimed in Exercise 5.1.

5.3 Consider 4×4 bit matrices. Again quadruples can be formed by reading from left to right and vice versa, and by reading from the top to bottom and vice versa. Give a simple proof that no 4×4 bit matrix realizes all quadruples in this way.

5.4 Prove Theorem 0.5.9 by induction.

5.5 Extend Theorem 0.5.6 to a simple arrangement of planes in three-dimensional space. Prove the resulting claim by induction. Can you extend this result to d-dimensions?

5.6 Prove that three chess queens cannot be placed in nonattacking positions on a 3×3 chess board. We say that two chess queens attack each other if they share a diagonal, row, or column.

5.7 Prove directly that in any binary tree there is exactly one more external node than there are internal nodes. Extend this result to m-ary trees, $m \geq 2$, that is, trees in which all internal nodes have out-degree m.

5.8 Given a binary tree, prove that no node can be its own predecessor or successor.

5.9 Prove, by induction, that there are $\binom{2n}{n}/(n+1)$ distinct binary trees having n internal nodes.

5.10 Given a binary tree, assign values to its nodes using the rules

(i) all external nodes have value 1;
(ii) all internal nodes have a value which is the sum of the values of their successors.

Prove that the value of an internal node is the number of external nodes descended from it, that is, in its subtree. This value is usually called the *weight* of a node.

5.11 Given a binary tree prove that for all internal nodes u in the tree

$$0 < \frac{\text{the weight of the lighter successor of } u}{\text{the weight of } u} \leq 1/2$$

where the lighter successor is chosen arbitrarily if they have equal weight.

5.12 Prove, by induction, that

(i) $\Sigma_{i=1}^{n}(3i^2 - 3i + 1) = n^3$, for all $n \geq 1$.
(ii) $\Sigma_{i=1}^{n} i^2 = \dfrac{n(n+1)(2n+1)}{6}$.

5.13 Consider the following "inductive proof" that all people are the same sex.

Claim: *Given any group of n people, n* ≥ 1, *they are all of the same sex.*

Proof of Claim: (by induction on *n*)

Basis: $n = 1$. Clearly every person is the same sex as themselves.

Induction Hypothesis: Assume the claim holds for all k, $1 \leq k \leq n$, for some $n \geq 1$.

one person from the group, person A, say. The remaining group G consists of n members. Therefore, by the induction hypothesis, they are all of the same sex. Now let person A return to the group and remove another person, person B from the group. Again we are left with a group G' of n members and, therefore, of the same sex. But now since A is in G' A is the same sex as those in G' and, similarly, B is the same sex as those in G. Clearly, this implies A is the same sex as B and therefore all members of G are the same sex. Hence, by the principle of induction, we conclude that the claim is valid. ■

What is wrong in the above "proof"? It is self-evident that it must be incorrect, but this is not a reason. We should not be able to prove that an untrue claim holds, otherwise the whole of mathematics is in danger.

5.14 Let A be any finite set. Prove that $\# 2^A = 2^{\# A}$.

5.15 Is your proof of Problem 5.14 dependent on the finiteness of A? Prove that A and 2^A never have equal cardinality, whether or not A is finite. Note that diagonalization cannot be used since it depends on enumerability. However, a similar technique can be applied.

5.16 Given an infinite set A we can define the sequence $S_0, S_1, \ldots, S_i$, of families of sets by

(i) $S_0 = A$;
(ii) $S_{i+1} = 2^{S_i}$, $i \geq 0$.

Are there then an i and j, $0 \leq i < j$ such that S_i and S_j have equal cardinality?

5.17 A (*undirected*) *graph* G is defined as an ordered pair (V,E), where V is a finite set of vertices and E is a set of *undirected edges*, that is, $E \subseteq \{\{u,v\} : u,v \text{ in } V\}$. Prove that if $\#V = n$ and $\#E \geq n$, then G has an undirected cycle.

[*Hint:* Use the pigeonhole principle with subsets of V as pigeonholes. Extend your proof to directed graphs of n vertices and $\geq 2n-1$ edges.]

5.18 Given a sequence $s = s_1, s_2, \ldots, s_k$, a sequence $s_{i_1}, s_{i_2}, \ldots, s_{i_r}$, where $1 \leq i_j < i_{j+1}$, for all j, $1 \leq j \leq r$, is said to be a *scattered subsequence of* s.

You are given a sequence $s = s_1, \ldots, s_k$, of distinct integers, where $k = mn+1$, for some $m \geq 1$, and some $n \geq 1$. Using the pigeonhole principle, prove that s contains a scattered subsequence $s_{i_1}, \ldots, s_{i_r}$ such that either $r > m$ and $s_{i_1} > s_{i_2} > \cdots > s_{i_r}$ or $r > n$ and $s_{i_1} < \cdots < s_{i_r}$.

[*Hint:* Use an $m \times n$ matrix of mailboxes and consider the longest decreasing and increasing scattered subsequences starting with each s_i.]

5.19 The pigeonhole principle can be restated as a property of total functions over a finite domain and range, namely,

> Let $f : D \to R$ be a total function for a finite domain D and finite range R, where $\#D > \#R$. Then there exist two elements, d_1 and d_2, in D with $f(d_1) = f(d_2)$.

How may this restatement be extended to functions over infinite domains and finite ranges, and infinite domains and infinite ranges?

5.20 How can the pigeonhole principle be applied to the following questions. How many students do you require in your course to ensure that

(i) at least two students were born on the same day of the week;
(ii) at least two students obtain the same final grade (grade points or letter grade) in the course;
(iii) at least two students were born on the same day of same month?

5.21 Another extension to the pigeonhole principle it to ensure that some mailbox has at least three items of mail in it (in general k items). State and prove such a generalization for three items. Can you then obtain a result for k items, $k \geq 0$? This is called the *general pigeonhole principle.*

5.22 Prove that if $n+1$ distinct integers are selected from the set $\{i : 1 \leq i \leq 2n\}$, then one of the selected integers will divide some other selected integer exactly.

5.23 Modify the proof of Theorem 0.5.11 so that the problem of each real having at most two binary expansions is resolved completely.

PROGRAMMING PROJECTS

P1.1 Implement a package for the manipulation of finite sets. It should contain the operations of union, intersection, and cartesian product. Also membership, containment, and equality testing should be possible. Investigate how to represent and manipulate infinite sets.

P1.2 Investigate methods for implementing a package for either the manipulation of sequences or the manipulation of tuples.

P2.1 Implement a package for the manipulation of digraphs which includes transitive closure.

Chapter 1

Languages and Computation

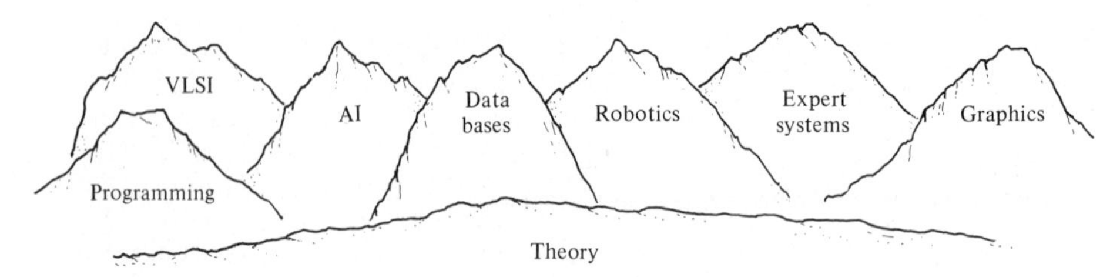

Sierras de la Computacion

1.1 PROGRAMMING PROBLEMS AND COMPUTATION

A major concern of computer science is software construction in all its aspects, while the theory of computation focuses on the theoretical underpinnings of software construction. It's direct impact is seen, for example, in the construction of compilers. The theoretical bases for the lexical and parsing stages of compiler construction are now well accepted and understood. Another example is to be found in text editors. Here the ability to specify a search for a string using a regular expression and be guaranteed fast searching are due directly to the underlying theory. In this text we study the fundamentals of the theory of computation. However we begin with some further motivation for its study. This is followed by an examination of the relationship between programs and formal languages which provides an additional example of the impact of theory on practice.

If you climb in the Himalayas without a mountain guide, then your chances of survival are drastically reduced. If you are building a jet aircraft and are unfamiliar with the effects of metal fatigue, then your plane will disintegrate. If you are constructing a suspension bridge and are unfamiliar with the effect of wind-induced vibration, then your bridge will probably collapse in a storm. If you are replanting a forest and are unfamiliar with effects of various kinds of forest on the ecology, then you will destroy the local ecology. If you build a skyscraper and are unfamiliar with the wind-tunnel effect, then the glass panes in the windows will fall onto the streets below. If you implement a software package for electronic mail and are unfamiliar with the effects of flow, then your package will never run.

In all of the above examples theory provides insight into what might happen if . . . . We consider three further examples, two fictitious morality tales about specific computing problems and one dream about what might have been.

Once in the recent past in the Ultimate Kingdom, its ruler King Zan Zah, a modern king, purchased many IPM 4000 machines to computerize the kingdom with the help of his wise men. Before too many months had passed they were infested by a plague of bugs — one aspect of the dreaded software crisis. Most bugs caused the programs to loop indefinitely. In an attempt to deal with the plague Zan Zah decided that his wise men should implement a termination testing program to save valuable time, effort, and tax dollars. A termination testing program determines for a given input program whether or not it terminates for all possible inputs; see Figure 1.1.1. The wise men toiled and toiled and toiled. And, one by one they admitted defeat. They were unable to write such a program. Zan Zah, not wishing to lose face in the world outside the Ultimate Kingdom, ordered the wise men to be executed in one month's time, and then Zan Zah decreed that any one in the U.K. who could solve this problem would become the heir to the kingdom. People everywhere in the kingdom attempted to solve the problem, but it was only toward the end of the month that Dr. Chu Ling in the Ivory Tower heard of the decree. Immediately, Chu Ling rushed to obtain a audience with Zan Zah. Zan Zah who was now wishing that IPM had never been incorporated, was exhausted with and disspirited by the many interviews he had granted, when claimed solutions were shown to be incorrect, etc. Thus it was a reluctant Zan Zah who met with Chu Ling. Zan Zah was amazed that Chu Ling had nothing with him apart from a writing block and pencils. "Where is your program?" he asked. "I have no program," replied Chu Ling, "because there can be no such program." To the astonishment of Zan Zah, she awakened increasing interest in the Ivory Tower, by elegantly proving that *no such program* could be written for the IPM 4000 or any other machine — real or abstract.

Zan Zah was delighted. Although the plague couldn't be removed as easily as expected, there was no loss of face. No one in the outer kingdoms could do it, either. He made Chu Ling the heir and initiated the annual Chu Ling Award for those in the Ivory Tower who made similar advances and to encourage the people of the kingdom to listen to those in the Ivory Tower.

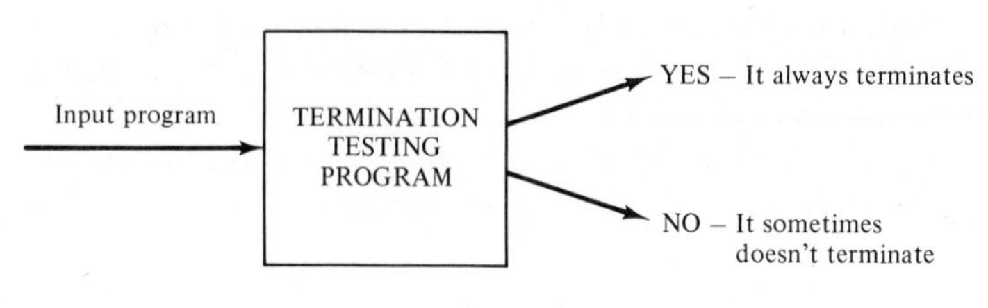

Figure 1.1.1

The second story has become part of the folklore of theoretical computer science. Indeed, there are those who believe it really happened!

Once upon a time at W. University Professor Bettergood was teaching Computer Science 140, Introductory PASCAL Programming, and she was fed up. It was not so much the class sessions — after all, they met only twice a week for an hour at a time. Well, to be precise 50 minutes. No, it wasn't the class sessions with 500 students — it was grading the assignments. In the past, in the good, rich old days, the department had employed a multitude of teaching assistants to grade the assignments. But now she had only one and although he worked hard, he worked according to union rules (even teaching assistants are part of organized labor these days). This left Professor Bettergood with about 400 programs to grade every two weeks. With this number issues of style and clarity had gone out the window — she graded only for correctness. Although it took time, she went through with it for a whole term. But when two terms later she found herself in the same situation she decided to do something about it. She took all the grading herself — after all, what's another 100 programs when you are already grading 400? At the same time she employed her teaching assistant to write a correctness checking program, a CCP. For each assignment Professor Bettergood would provide her solution program and the CCP would take this together with a student's attempt and check whether or not the student program had the same effect as the professor's program. Since the teaching assistant was interested in expert systems, he was delighted with the task. He began work immediately, rapidly implementing CCP1. CCP1 worked well with the professor's program against itself, but when given one of the student programs it ran and ran for hours and hours. In quick succession, at least initially, there followed CCP1.3, CCP1.8, CCP2, . . . , and CCP2.7.3, at which time the term was over. Professor Bettergood and the teaching assistant were both frustrated and exhausted, since *no* satisfactory CCP had been produced.

The project and the story might have continued for many more terms except that a few terms later the teaching assistant attended a theory of computation course. It was a course he had put off to the last possible moment before graduation, having tried to get out of it, but the department stood firm. Much to his surprise he discovered that the CCP project was doomed to failure — it couldn't be written with the generality they wanted! He had been trying to implement an equivalence testing program. Such a program when given two programs would either report they give the same answer for the same inputs — that is, they are equivalent — or report that there is some input data for which they give different answers. See Figure 1.1.2.

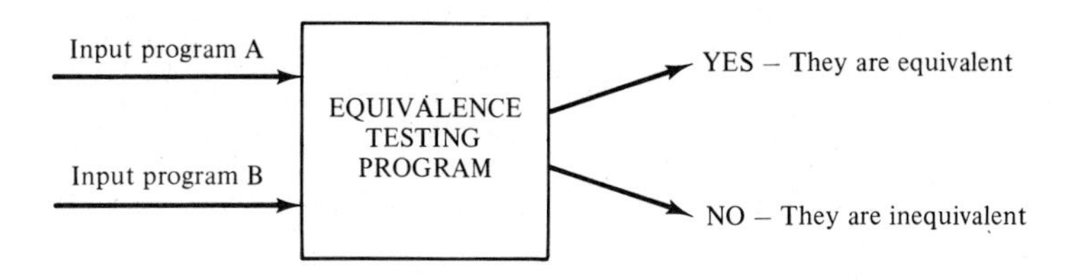

Figure 1.1.2

In the examples and stories theory acts as an early warning system of dangerous areas. It implies that you need a guide or guidebook on your journey through the Sierras de la Computacion. To go on without such guidance would be foolish and dangerous. But, you might say, "Just tell me the dangerous problems or, even better, give me file of them; I don't need to attend the theory course; I'll just memorize them for you." Well, unfortunately there is no such file, since the list of such problems is infinite. At most you can expect to be exposed to some of the better-known problems and, at the same time, be provided with enough theoretical background to begin to recognize new dangerous problems for yourself. Knowledge of some of the results is not enough; you need to be able to *derive* them for yourself and, more importantly, develop a *feel* for what kind of problems are dangerous. This is the aim of the text — to give you feel and fluency with some of the fundamental abstract models (in the theory of computation) and their properties as well as to develop your ability with proof techniques.

Theory, acting as an early warning system, provides a science of the impossible — what shouldn't be attempted because it cannot be done. However, a good theory should also provide guidelines for the possible: In our mountaineering metaphor a guide should not only tell us the dangerous areas, but also tell us of simple routes, difficult routes, and severe routes, for example. We find that the theory of computation does indeed do this at many different levels. At the coarsest level it partitions problems into three classes; (i) the impossible problems; (ii) the possible-with-unlimited-resources-but-impossible-with-limited-resources problems; and (iii) the possible-with-limited-resources problems. These are usually known as the undecidable, the intractable, and the tractable problems, respectively, when the resource is time. At a finer level the theory also considers problems which require, for example, time linearly proportional to their size. But the theory also faces and deals with questions as fundamental and diverse as

PROGRAM EXISTENCE
  Does there exist a program for a given problem?

PROGRAM SPECIFICATION
  How should programs be specified?

EXPRESSIVE POWER
  Does one method of specifying programs have greater expressive power than a second, that is, can more programs be specified using it?

PROGRAM TERMINATION
Does a given program always halt or terminate?

PROGRAM CORRECTNESS
Is a given program correct?

PROGRAM CONSTRUCTION
How is a correct program obtained?

PROGRAM SEMANTICS
What is the meaning of a given program?

PROGRAM EFFICIENCY
Is there a more efficient program for a given problem?

PROGRAM IMPLEMENTATION
Can a given program be implemented on a given machine?

MACHINE COMPARISON
Is one machine more powerful than a second? This is the machine version of EXPRESSIVE POWER.

This list is *not* exhaustive. It serves merely to emphasize the breadth and the scope any theory of computation should encompass. As will be seen we exclude the topics of program correctness, program construction, and semantics, while focusing on the remaining topics from the viewpoint of language and automata theory. In the remainder of this chapter we consider the relationship of problems and machines to language and automata theory. Specifically, we will demonstrate that the study of language and automata theory is, by proxy, the basis for the study of the theory of computation in its wider sense. Before beginning this demonstration we present a dream of what might have been. The dream looks at the EXPRESSIVE POWER question, a fundamental question that we raise for each of the methods discussed in this text.

From the beginning the Computing and Information Agency was infiltrated with dissension. The CIA had been set up as a federal agency with a budget beyond everyone's dreams and with wide-ranging power. They could co-opt any individual citizen who could contribute to their goals; they could arraign and hold, without due process of law, any individual who was working against them. The president had given an emotional, patriotic address at the establishment of the CIA. He had referred implicitly to the achievements of Markov and his followers. Boldly he announced that "before the decade was out, we, the citizens of this free country, would have obtained the solution of the fundamental issue in the theory of computing. Which theory has most expressive power?"

These were turbulent, exciting years. A handful of men, brilliant and courageous men, dared go where no man had gone before. Chomsky, Church, Kleene, Lindenmayer, Markov, Post, and Turing explored terrains beyond the imagining of us mortals. Yes, we felt like mortals and we viewed them as gods. As agents of the CIA they created new worlds and set out to explore them — the Columbuses of a new age.

What they found is now recorded in the annals of history. They not only arrived, not only saw, but they also conquered. They established that each of their theories had the same expressive power, although each was so different and, more importantly in the eyes of the populace, they showed that Markov's theory also had exactly the same expressive power as theirs; no more and no less. The political impact of this result was shattering — it gave back to a lost country its self-respect, and it shattered the superiority of Markov and his followers. This led to a thaw in the Cold War, when the two peoples could once more mix and interchange ideas freely.

We discuss the approaches of Church, Markov, and Post in Section 1.5, the approaches of Kleene and Turing in Chapter 7, and the approaches of Chomsky and Lindenmayer in Chapter 11. These seven approaches have, indeed, the same expressive power. We now return to our theme of programming problems and computation.

Given a programming problem it is easier to convince someone that there is a program which solves the problem (if one exists) than to convince someone there is no program for the problem (if one doesn't). This is because in the former case we need only outline a solution informally, but in the latter case we need to show that over *all* possible programs none solves the given problem. In order to deal with negative results of this kind we need to be precise about what constitutes a program. We require a precise description of allowable or legal programs. Observe that we do not need such a precise description when demonstrating a positive result (although if we wish to execute a program we do). For this reason we discuss below the basic conditions that a description of a program and a machine to execute it must satisfy. This approach is due to Alan Turing and is not the only one. Alonzo Church took a complete different approach to these problems, which turns out to be equivalent; see Section 1.5.1.

Before beginning to write a program for a problem we must be given, or arrive at, a problem specification. Consider the following simple example of a problem specification

SQUARE
INSTANCE: An integer n.
QUESTION: What is the value of $n \times n$?

SQUARE is a simple but illuminating example of a *generic* specification of a problem. Each value of n determines an *instance* of SQUARE. For example, $n = 7$ determines the problem instance: what is the value of 7×7? We are more concerned with *families* or *genera* of similar *problem instances*, rather than stand alone instances. To families of related problems, for which programs exist, there correspond, in a natural way, programs with input. For example, we have the PASCAL program *Square*

```
program Square;
var n : integer;
```

```
begin
    read(n);
    writeln(n * n)
end.
```

corresponding to the problem SQUARE. Unfortunately, PASCAL programs have some limitations which ensures that this correspondence isn't total, namely, it holds only for integers n in the range

$$-trunc(sqrt(maxint)) \leq n \leq trunc(sqrt(maxint))$$

where *maxint* is the predefined installation-dependent integer constant specifying the maximal integer of type **integer** which can be represented in PASCAL at a given installation, *sqrt* is the square root function, and *trunc* truncates a real value to yield an integer value. Similarly, all installations have a restriction on the size of PASCAL programs. Moreover, most installations also have a restriction on how long a program may be executed before it is timed-out.

Any theory of computation (or programming) must face these issues. Should a theory restrict the size of possible programs, restrict the size of input data, or restrict the length of a computation (the execution time)? But a more fundamental issue it should first settle is: what is a program? Do we mean a PASCAL (or ADA) program? No, is the standard answer to this question, since it is preferable to allow a more general notion of program, which we prefer to call *algorithm*, in this more abstract setting. On the one hand, the notion of algorithm should be general enough to encompass PASCAL and other high-level language programs, assembly language programs, machine language programs, as well as firmware, special-purpose computers, flowcharts, and other means of specifying computing processes — for example, numerical calculations specified in English. On the other hand, it should not be so general that the notion of algorithm is vague and useless to the development of a theory of computation. It is generally agreed that any algorithm worthy of the name should satisfy the following elementary condition

1. *It consists of a finite set of instructions.* Surely not an unreasonable requirement since we wish to write down our algorithms!

However this is insufficient as it stands. Just as programs do not stand alone, but require a machine to execute them, so it is for algorithms.

2. *There is a computing agent that can carry out the instructions of the algorithm.* When discussing problems and programs above, we ignored the necessity for a machine or computer to execute our program, although, in principle, it could be executed by a human being. So a computing agent may be considered to be a computer or a human (indeed in the twenties and thirties it was still common for computer to mean a person who carried out calculations — just as earlier typewriter meant the person who typed, not the machine!) Programs and machines belong together. A program requires a machine to execute it and a machine requires a program to instruct it. Having introduced the computing agent it is necessary to define the elementary requirements it should

satisfy.

3. *The computing agent must be able to store, retrieve, and make steps in a computation.* This requirement is satisfied by computers and humans; they have a memory.
4. *The computing agent must carry out the instructions in discrete steps.* This prohibits analog or continuous computation. For example, in pouring an after-dinner liqueur for someone I ask the person to tell me when, rather than asking do you want 50 milliliters, or a 100, or . . . ? The first question is analog, the second discrete. Again, computers are discrete.
5. *The computing agent must carry out the instructions deterministically or mechanistically.* This prohibits the computing agent from acting in a probabilistic manner, or in a nondeterministic manner. Essentially the choice of the next instruction to be carried out should be solely and uniquely determined by the initial values and the instructions carried out so far. It should not be determined by the toss of a die, nor should it be determined by whim. Determinism implies that the algorithm can be carried out again and again with the same initial values to give the same final result. This is called *repeatability*, and is a foundation stone of science. Programs that rely on random numbers during their execution are not probabilistic in the sense meant here. This is simply because the so-called random numbers are in reality pseudo-random and depend on some initial values. Nondeterminism is used in some of the models discussed in this text; however, we will see that they are deterministic if viewed in the right way.

These five conditions capture the essence of programs and computers by way of algorithms and computing agents. However, as we already observed there are a number of questions that should still be answered if we are to use these conditions as our basis for a theory of computation. The first and, perhaps, the most important is:

Q1. *Should an algorithm halt or terminate in a finite number of steps on each input?*

Our experience tells us that we should answer this question with a resounding yes. However, as we shall see, it is essential in the theory of computation to allow "algorithms" which do not terminate on all inputs. We call these *partial algorithms*.

Three more basic questions are:

Q2. *Should the size of a set of instructions be restricted?*
Q3. *Should the size and number of the initial values or input data be restricted?*
Q4. *Should the number of computation steps or execution time be restricted?*

When given a specific computer each of these parameters *is* restricted. However, in principle, there are no *a priori* bounds for them. So in a general theory of computation there is no reason, at least initially, to restrict them. So we answer Q2-Q4 negatively, but, having said this, Q4 is perhaps better answered with: no in a general theory, and yes in some circumstances. For example, if we restrict the length of a computation in some way, then which problems have algorithms? This leads into the area of computational complexity, one of the key areas in the theory of computation.

What other restrictions are possible?

Q5. *Should the memory of the computing agent be restricted?*

We should separate the memory required to store the program from the additional memory required to store the intermediate calculations and results, otherwise we are reconsidering Q2. Again it seems there is no good reason to *a priori* fix the amount of additional memory, so we answer no in general. However, just as restricting computation length or time makes sense in some circumstances, so does restricting memory or space.

As our computing agent may be human, should the human's creative and intellectual ability be allowed to effect the ability of the computing agent?

Q6. *Should the ability of the computing agent be restricted?*

Since we require a general theory of computation, one feels that it should not be based on whimsy, the intellectual and creative abilities of the arbitrarily chosen computing agent. We, therefore, answer Q6 with a firm yes.

Having arrived, in an informal framework, at some basic conditions that a general theory of computation must satisfy, we now begin to develop a formal approach. Every algorithm has initial values or input data and resulting values or output data. Each problem specifies, however implicitly, its initial and final values. For example, SQUARE has initial and final values consisting of a single integer. Therefore, every algorithm has, correspondingly, initial values or input data and final values or output data. We prescribed that the computing agent be deterministic and this implies a functional relationship between the input and output data — the same input yields the same output. This black box view of algorithms is captured by Figure 1.1.3, where f is the function computed by the black box. For example, if f is defined by *Square* then $f : \mathbb{Z} \rightarrow \mathbb{Z}$, where $f(i) = i * i$, for all i in $\mathbb{Z}$. In this case f is a total function, but in general it is partial. Recall from Section 0.3 that by function we mean partial function. If $f : \mathbb{Z} \rightarrow \mathbb{Z}$ is a (partial) function computed by an algorithm and $f(x)$ is undefined for some x in $\mathbb{Z}$, then this corresponds to the algorithm terminating with an undefined value.

We consider that every problem Π specifies a function $f_\Pi : D_\Pi \rightarrow R_\Pi$, where D_Π is the *domain* of the problem and function and R_Π is the *range*. The function f_Π is a *problem function*. Every value d in D_Π determines an *instance* of the problem Π. For example, the problem *SQUARE* specifies the function $f_{SQUARE} : \mathbb{Z} \rightarrow \mathbb{Z}$ by $f_{SQUARE}(i) = i * i$, for all i in $\mathbb{Z}$. Every value d in $D_{SQUARE} = \mathbb{Z}$ determines an instance of *SQUARE*, for example, 7. Although the restriction to problems specifying functions seems severe, it follows practice and tradition.

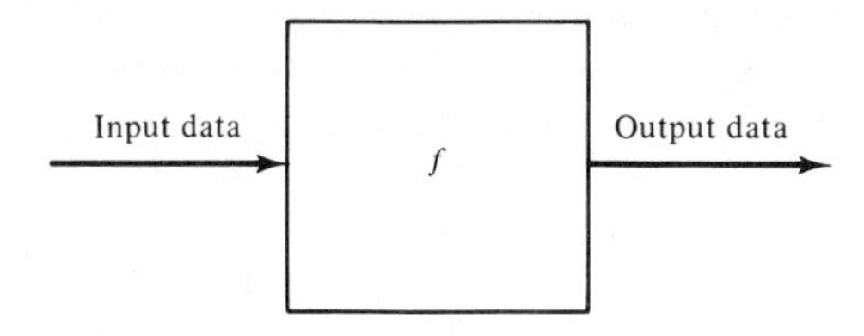

Figure 1.1.3

Similarly, we consider every algorithm A to define a function $f_A : D_A \to R_A$, where D_A is the *domain* of A and f_A, and R_A is the *range* of A and f_A. The function f_A is an *algorithm function.* In this case $f(d) =$ ***undef*** corresponds to the algorithm A terminating with the value ***undef*** or terminating with no value. These possibilities are illustrated with programs for the following problem.

INTEGER SQUARE ROOT
INSTANCE: An integer n
QUESTION: What is the nonnegative integer m that satisfies $n = m \times m$?

We can write the program segment A to solve this problem.

```
[A] Given n;
    for i := 0 to n do
        if n = i * i then return i;
    return undef
```

If n is negative or n is positive, but has no integer square root, then the statement **return undef** is, eventually, executed. This ensures $f_A(n) =$ ***undef*** exactly when either $n < 0$ **or** ($n > 0$ and $n \neq i \times i$ for any integer i). Alternatively, we might write "program" segment B.

```
[B] Given n;
    if n ≥ 0 then
    begin i := 0;
    1 : if n = i * i then return i
       else if i > n then return undef;
         i := i+1; goto 1
    end
```

For $n < 0$, $f_B(n) =$ ***undef*** corresponds to B terminating but not assigning a value to $f_B(n)$. If we omitted the test for $i > n$, then we would obtain a partial algorithm. This is because when $n > 0$ and $n \neq i \times i$, for any $i > 0$, the goto loop never terminates.

The domain of an algorithm or problem function A, is usually of the form $B_1 \times \cdots \times B_m$, for some $m \geq 0$ and sets $B_1, \ldots, B_m$, and the range is of the form $C_1 \times \cdots \times C_n$, for some $n \geq 0$ and sets $C_1, \ldots, C_n$. We say f_A is the m,n-place function computed by algorithm A. On the other hand, given an m,n-

place function f we say it is *algorithm-computable* if there is an algorithm A with $f = f_A$. We say that algorithm A is *correct* with respect to problem Π if $f_A = f_\Pi$. The functional view of algorithms provides a formal counterpart to the intuitive and informal notion that two algorithms are the same, that is equivalent. We say that two algorithms A and B are *equivalent* if $f_A = f_B$.

Finally, the functional view allows us to focus on an important subclass of problems and algorithms, namely, *decision problems* and *decision algorithms*. A problem Π is a *decision problem* if f_Π is a total Boolean function. A *decision algorithm* is an algorithm which computes a *total* Boolean function. For example,

SQUARE TESTING
INSTANCE: Two integers p and q.
QUESTION: Is $q = p^2$?

is a decision problem for squareness. This is because it defines a total function $f : \mathbb{Z}\times\mathbb{Z} \rightarrow \{\textbf{false},\textbf{true}\}$, where $f(p,q) = \textbf{true}$ if $q = p^2$ and $f(p,q) = \textbf{false}$ if $q \neq p^2$.

An algorithm A for SQUARE TESTING must define a function $f_A = f$; hence f_A is also total. Intuitively, the totality of f_A is necessary because A must decide for each input whether it corresponds to a **false** or **true** output. Therefore, A must always provide a result for every allowed input. The possibilities of silence or a don't know response are ruled out. However, it is conceivable that we don't know the Boolean function defined by a decision problem. For example, consider

FERMAT TESTING
INSTANCE: A positive integer n.
QUESTION: Are there nonzero integers x,y, and z
such that $x^n + y^n = z^n$?

Then $f_{FERMAT\ TESTING} : \mathbb{N} \rightarrow \{\textbf{false},\textbf{true}\}$ is a total Boolean function, but the values of n for which $f(n) = \textbf{true}$ are unknown. In fact, we do not even known if these values of n are bounded.

Decision problems, decision algorithms, and total Boolean functions play a central rôle in a theory of computation. The reasons for this are two-fold. First, every problem can be transformed into a closely related decision problem and, second, the algorithm of the closely related decision problem can often be used to solve the original problem. We have illustrated the first part of this reasoning with SQUARE and SQUARE TESTING, we now define a generic transformation of problems to decision problems. For simplicity we only consider 1,1-place problem functions.

Given a problem Π and its problem function $f_\Pi : B \rightarrow C$, we obtain a decision problem D from Π as follows. D specifies a total Boolean function $f_D : B \times C \rightarrow \{\textbf{false},\textbf{true}\}$ which is defined by

$$f_D(b,c) = \begin{cases} \textbf{true} & \text{if } f_\Pi(b) = (c) \\ \textbf{false} & \text{if } f_\Pi(b) \neq (c) \end{cases}$$

D has as input data both possible input and possible output data of Π and it answers the question: Is this the output expected from this input? So an algorithm for D can be viewed as a *checker*, *validator*, or *verifier* for the problem Π.

Let A be an algorithm for Π, that is $f_A = f_\Pi$. It appears that we can use A as a subprogram to obtain an algorithm $\bar{A}$ for D; see Figure 1.1.4. If f_A is a total function, then $\bar{A}$ is an algorithm computing f_D, but if, as is more usual, f_A is not total, then for input b, $f_A(b)$ may be undefined. Since $\bar{A}$ is purported to be a decision algorithm this implies A cannot be used, in general, as an integral part of $\bar{A}$.

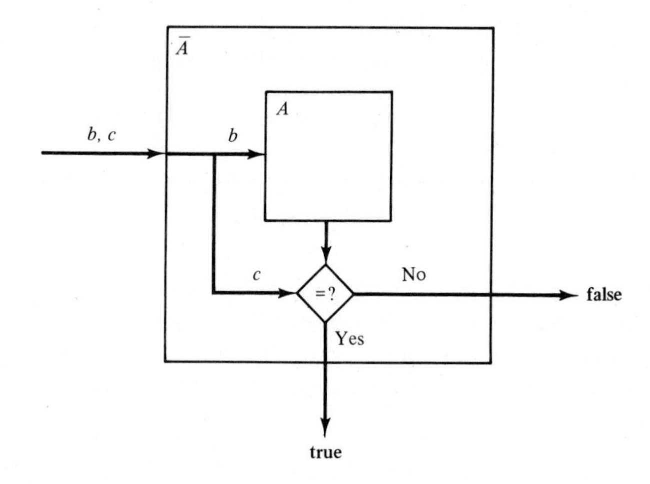

Figure 1.1.4

Now consider the converse problem of obtaining an algorithm from a decision algorithm. For example, given a decision algorithm T for SQUARE TESTING we can obtain an algorithm S for SQUARE from T, in a PASCAL-like notation, as follows

[S] Given p; **for** $q := 0$ **to** ∞ **do**
  if $f_T(p,q)$ **then return** q

Since both f_{SQUARE} and f_T are total, this algorithm will eventually discover the q satisfying $p = q^2$ if one exists. We prefer to write the algorithm using a set-oriented loop as

[S] Given p; **for** q **in** $\mathbb{Z}_0$ **do**
  if $f_T(p,q)$ **then return** q

But how do we generalize this? Let $f : B \times C \rightarrow \{\mathbf{false}, \mathbf{true}\}$ be the total Boolean function computed by some decision algorithm $\overline{A}$, for a decision problem D, where B is considered to be the input and C the output. If Π is the original problem with $f_\Pi : B \rightarrow C$, then we might try to obtain an algorithm A for Π from $\overline{A}$ as shown in Figure 1.1.5. There are two difficulties, at least, with this algorithm.

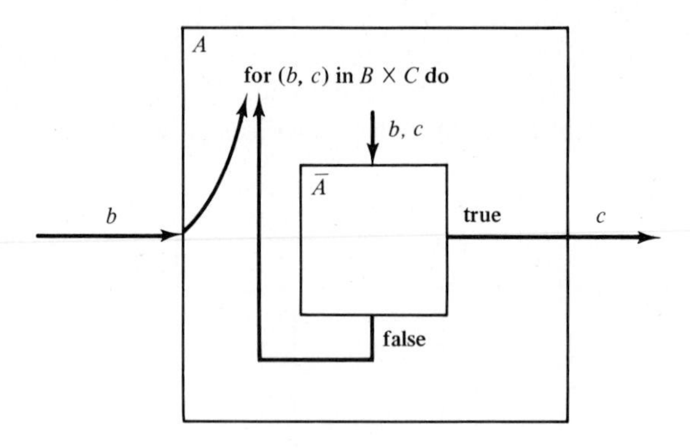

Figure 1.1.5

First, for a given b, there may be more than one value c with $f(b,c) = \mathbf{true}$. In this case which value of c does A choose? The classical method of overcoming this difficulty is to take the minimal c with $f(b,c) = \mathbf{true}$, if there are any at all. Hand in hand with this redefinition a minimalization operator is provided which allows us to determine the minimal value. See Section 7.5 for more details of this approach.

Second, the loop provides an enumeration of the elements of C if $f(b,c) = \mathbf{false}$ for all c. On the one hand, such enumerations do not always exist. For example, the set of reals cannot be enumerated and neither can the set of all subsets of the integers. On the other hand, we are usually dealing with the set of integers and other enumerable sets, so in most cases the above tentative algorithm is indeed an algorithm. For example, in the algorithm S above we need to enumerate the set of all nonnegative integers, but this is simple to do: $0, 1, 2, \ldots,$ is the obvious enumeration that we used above.

The above examples and discussion demonstrates something of the close relationship existing between problems and decision problems, and algorithms and decision algorithms. Before discussing the issues that should be addressed by any theory of computation we emphasize one aspect of decision algorithms that we return to not only in Section 1.3 but also as a major theme throughout the text.

Given a decision problem, D, that specifies a total Boolean function $f_D : A_1 \times \cdots \times A_m \rightarrow \{\mathbf{false}, \mathbf{true}\}$ an instance of D that is determined by $(a_1, \ldots, a_m)$ is said to be a *yes* or *true instance* if $f_D(a_1, \ldots, a_m) = \mathbf{true}$ and is a *no* or *false instance* otherwise. f_D and, hence, D determine a subset D_Y of $A_1 \times \cdots \times A_m$ of yes instances

$$D_Y = \{(a_1, \ldots, a_m) : (a_1, \ldots, a_m) \text{ is in } A_1 \times \cdots \times A_m \text{ and } f_D(a_1, \ldots, a_m) = \mathbf{true}\}$$

For example, with SQUARE TESTING we have the total Boolean function $f_{SQUARE\ TESTING} : \mathbb{Z} \times \mathbb{Z} \rightarrow \{\mathbf{false}, \mathbf{true}\}$ and this function is defined by $f_{SQUARE\ TESTING}(b,c) = \mathbf{true}$ if and only if $c = b^2$. So (2,4) and (7,49) are yes instances of SQUARE TESTING. Moreover, the set of yes instances is

$$SQUARE\ TESTING_Y = \{(b,c) : b \text{ and } c \text{ are integers and } c = b^2\}$$

that is, it equals

$$\{(b,b^2) : b \text{ is an integer}\}$$

As we shall in Section 1.3 this notion provides us with the bridge between language theory and a theory of computation. However, we first need to introduce the basics of language theory.

1.2 ALPHABETS, WORDS, AND LANGUAGES

Unsigned decimal integers are written as a sequence of digits, for example, 2745. *English words* are written as a sequence of letters, for example, catholic. *English sentences* are written as a sequence of words, for example, this sentence. *PASCAL programs* are written as a sequence of symbols.

These examples have two things, at least, in common.

First, the underlying set or universe whose elements are composed to give the sequences is finite. Decimal integers are formed from the ten decimal or denary digits $0, \ldots, 9$; English words are formed from the 26 letters of the English alphabet (or 52 if we distinguish between lower- and uppercase); English sentences are formed from the finite, but large number, of English words; PASCAL programs are formed from a finite number of characters. Finiteness is fundamental and prevents some natural sequences from being included, for example, sequences of positive integers, since the underlying set of positive integers is infinite.

Second, the sequences are not allowed to be infinite sequences. This is simply because we wish to write them down, so even Joyce's *Ulysses* contains only finite sequences of words, although it may not appear to be the case when reading it! However, although each sequence is finite in length, there is no bound on the length of allowable sequences. So although most sentences contain fewer than 100 words, it is possible to write longer sentences and such sentences should not be prohibited, *in advance*.

We introduce a notation for finitely generated sequences of elements from a finite underlying set. A finite underlying set is called an *alphabet*, a finite sequence of its elements is called a *word*, and a set of finite sequences is called a *language*. We define the basic notation and terminology for these concepts in detail.

1.2.1 Alphabets and Words

An *alphabet* is a finite, nonempty set of elements. The elements of the alphabet are usually called *symbols* or *letters*. For example,

$$D = \{0,1,2,3,4,5,6,7,8,9\}$$

and

$$A = \{a,b,c,d,e,f,g,h,i,j,k,l,m,n,o,p,q,r,s,t,u,v,w,x,y,z\}$$

A *word over an alphabet* Σ is a finite sequence of symbols from Σ, usually written without any separating commas. For example,

0102
656670
0

are words over D, while

aaba
cdxjol
streetcar

are words over A.

Note the technical use of the words alphabet and word. With this terminology we speak of the alphabet V of English words rather than a vocabulary, in which case a word over V means a sequence of English words. Throughout we use alphabet and word in this technical sense. If we need to use word in its nontechnical sense we will prefix it with the natural language to which it belongs, for example Finnish word and English word.

The *empty word* is the empty sequence and is denoted by λ. The empty word is a word over every alphabet. The *universal set* or *universal language* of all words over an alphabet Σ is denoted by Σ^*. Σ^* is always an infinite set and, as we shall see, it is enumerable.

We now consider operations on words and the relationships between them. The *length* of a word x is denoted by $|x|$ and is simply the number of symbols in the given word, so $|0102| = 4$, $|cdxjol| = 6$, and $|\lambda| = 0$. Let x be a word over an alphabet Σ and let a be in Σ. Then the *a-length* of x is denoted by $|x|_a$ and is the number of times a occurs in x. For example, $|abbaa|_a = 3$, $|abbaa|_b = 2$, $|abbaa|_c = 0$. Clearly $|\lambda|_a = 0$, for all a in Σ and $|x| = \Sigma_{a \text{ in } \Sigma} |x|_a$.

Given two words x and y over an alphabet Σ, the *catenation of* x *with* y, denoted by xy, is the word obtained by appending the word y to x. For example, if $x = 0102$ and $y = 656670$, then $xy = 0102656670$ and $yx = 6566700102$. Note that $\lambda\lambda = \lambda$, since appending no symbols to no symbols gives no symbols and, moreover, $x\lambda = \lambda x = x$, for all words x. This, apparently trivial, point is often misunderstood. Recall that λ is not a symbol of the alphabet, but is rather a denotation for the empty word. λ denotes   and $\lambda\lambda$ denotes   and these are clearly the same. Since $\lambda x = x\lambda = x$, for all words x, we say that λ is the *identity* with respect to catenation.

Let Σ be an alphabet. For all words, x, y, and z over Σ,

$$(xy)z = x(yz) = xyz$$

where parentheses are used to indicate which catenation operation is carried out first. In other words, catenation is associative. (Hence, Σ^* is a monoid with respect to catenation — the *free monoid generated by* Σ.)

Catenation of words is analogous to multiplication of reals, particularly in their relation to length and logarithm, respectively. Given two positive reals x and y we have $\log(x \times y) = \log x + \log y$ and, similarly, given two words x and y we have $|xy| = |x| + |y|$. We have $\log(x \times 1) = \log x + \log 1 = \log x$, for a positive real x, and we have $|x\lambda| = |x| + |\lambda| = |x|$, for a word x. This analogy can be carried a step further with powers of words.

Given a word x over an alphabet Σ we define x^0 to be λ and for all $i \geq 1$, x^i to be xx^{i-1}. So $(0102)^0 = \lambda$, $(0102)^1 = 0102$, and $(0102)^3 = 010201020102$. Now $|x^i| = i \cdot |x|$ in analogy with $\log(x^i) = i \cdot \log x$ for reals. We say x^i is the *ith power of* x. For all words x, and all $i,j \geq 0$, $x^i x^j = x^{i+j} = x^j x^i$.

Example 1.2.1 Given the alphabet $\{a,b,c\}$, the words

$$\lambda$$
$$cab$$
$$aaaa$$
$$c$$
$$ba$$

are words over $\{a,b,c\}$. They have lengths

$$|\lambda| = 0$$
$$|cab| = 3$$
$$|aaaa| = 4$$
$$|c| = 1$$
$$|ba| = 2$$

Note that the catenation of cab with $aaaa$, that is,

$$cabaaaa$$

has length

$$|cab| + |aaaa| = 3+4 = 7$$

as expected. Similarly,

$$| aaaa | = | a^4 | = 4 | a | = 4$$

Again

$$(cab)^3(ba)^4 = cabcabcabbabababa$$

and its length is

$$3 | cab | + 4 | ba | = 9 + 8 = 17.$$

Finally, observe that

$$(cab)^3(cab)^5 = (cab)^5(cab)^3 = (cab)^8$$
$$= cabcabcabcabcabcabcabcab$$

Given two words x and y over an alphabet Σ, we say that x *equals* y, written $x = y$, if they have the same length and the same symbols at the same positions. We say x is a *prefix* of y if there is a word z over Σ such that $xz = y$. Intuitively, we have partial equality from the left. For example, if $x = can$ and $y = canadian$, then x is a prefix of y, since $xz = y$, where $z = adian$. On the other hand, if $x = can$ and $y = scan$ then x is not a prefix of y and y is not a prefix of x.

We say x is a *proper prefix* of y if $x \neq \lambda$, $x \neq y$, and x is a prefix of y. Clearly, λ is always a prefix of any word x including itself and a word x is always a prefix of itself.

In an analogous manner we may define *suffix* and *proper suffix*. For example, if $x = well$ and $y = unwell$, then x is a suffix of y and is, in fact, a proper suffix of y.

Finally, we say that x is a *subword* of y if there are words w and z with $wxz = y$. For example, $x = in$ is a subword of *dinner* and it is a *proper subword* since $x \neq dinner$ and $x \neq \lambda$.

Example 1.2.2 Consider the word $x = through$. Then

$$\lambda, t, th, thr, thro, throu, throug, through$$

are all the prefixes of x,

$$\lambda, h, gh, ugh, ough, rough, hrough, through$$

are all the suffixes of x, and

$$\lambda, t, h, r, o, u, g, th, hr, ro, ou, ug, gh$$

$$thr, hro, rou, ugh, thro, hrou, roug, ough,$$

$$throu, hroug, rough, throug, hrough, through$$

are all the subwords of x.

Letting x be a word over some alphabet Σ we define the *reversal*, or *mirror image*, of x as $a_n \cdots a_1$ if $x = a_1 \cdots a_n$, where a_i is in Σ, $1 \leq i \leq n$. We denote the reversal of x by x^R. $\lambda^R = \lambda$ and $a^R = a$, for all a in Σ. We simply write the symbols of x in reverse order.

An alternative recursive definition of x^R can also be given. For all words x over some alphabet Σ, x^R is defined recursively as

(i) x, if $x = \lambda$;
(ii) $y^R a$, if $x = ay$ for some a in Σ and y in Σ^*.

Example 1.2.3 Let $x = able$, over the alphabet A. Then x^R is, using the recursive definition

$$\begin{aligned}(able)^R &= (ble)^R a \\ &= (le)^R ba \\ &= e^R lba \\ &= \lambda^R elba \\ &= elba\end{aligned}$$

We prove that x^R defined recursively does indeed give what we expect. This is good example of a proof by induction. If you are unfamiliar with this technique read Section 0.5 first.

Lemma 1.2.1 *Let Σ be an alphabet and x be a word over Σ. Then x^R, using the recursive definition, equals $a_n \cdots a_1$, where $n = |x|$ and $x = a_1 \cdots a_n$, for some a_i in Σ, $1 \leq i \leq n$.*

Proof: (by induction on the length of x)

Basis: $|x| = 0$ implies $x = \lambda$. In this case the recursive definition implies $x^R = \lambda$ as expected.

Induction Hypothesis: Assume the lemma holds for all words x in Σ^* with $|x| \leq n$, for some $n \geq 0$.

Induction Step: Let x be a word over Σ such that $|x| = n+1$. Then $|x| \geq 1$ and, therefore, x contains at least one symbol from Σ. Let $x = a_1 \cdots a_{n+1}$, for some a_i in Σ, $1 \leq i \leq n+1$. Then by the recursive definition $x^R = (a_2 \cdots a_{n+1})^R a_1$, since $x \neq \lambda$.

Now, we obtain $|a_2 \cdots a_{n+1}| = n$ and by the induction hypothesis $(a_2 \cdots a_{n+1})^R = a_{n+1} \cdots a_2$. Combining these observations we conclude that $x^R = a_{n+1} \cdots a_2 a_1$ as desired. Since x is an arbitrary word in Σ^* with $|x| = n+1$, we have shown that x^R gives what we expect for all such words.

The lemma now follows by the principle of induction. ■

It can be proved in a similar manner that for all words x and y over some alphabet $(xy)^R = y^R x^R$; see the Exercises. Indeed, this leads to a programming project as well.

1.2.2 Languages and Operations

The universal language Σ^* given by an alphabet Σ is an infinite set but it is worth emphasizing that every word in Σ^* is finite, that is,it has finite length. We can define Σ^* recursively as follows.

Given an alphabet Σ, then Σ^* consists solely of words obtained from (i) and (ii)

(i) λ is in Σ^*;
(ii) if x is in Σ^*, then ax is in Σ^*, for all a in Σ.

A *language* L over an alphabet Σ is a set of words over Σ, that is $L \subseteq \Sigma^*$. Hence, $\varnothing$ is a language over every alphabet, as is $\{\lambda\}$. Observe that $\varnothing \neq \{\lambda\}$ since $\varnothing$ contains no words but $\{\lambda\}$ contains a single word, the empty word. We call $\{\lambda\}$ the *empty word set* to distinguish it from $\varnothing$, the empty set. For example, $\{abba, aba, a\}$ is a finite language over the alphabet $\{a, b\}$ and $\{a^i b^i : i \geq 1\}$ is an infinite language over $\{a, b\}$. We denote by $alph(x)$, the set of symbols in the word x. This definition implies $alph(\lambda) = \varnothing$, $alph(abba) = \{a, b\}$, and $alph(2001) = \{0, 1, 2\}$. Let $alph(L) = \cup_{x \text{ in } L} alph(x)$, for all languages L. If $L \subseteq \Sigma^*$, then $alph(L) \subseteq \Sigma$, but equality doesn't necessarily hold, for example, with $\Sigma = \{a, b\}$ and $L = \{a\}$, $alph(L) = \{a\}$.

Since languages are sets the Boolean operations can be applied to them. However, there is a cautionary note about complementation to be observed. Consider a language L, over the digit alphabet D, defined as the set of all words containing only 0's and 1's, that is binary sequences. Then $alph(L) = \{0, 1\}$. But what is $\overline{L}$? Is it $D^* - L$ or is it $\{0, 1\}^* - L$? Both definitions are reasonable; we choose the former.

Apart from the Boolean operations there are operations that are specific to languages. Given two languages $L_1 \subseteq \Sigma_1^*$ and $L_2 \subseteq \Sigma_2^*$ their *catenation* (*product*), denoted by $L_1 L_2$, is defined by

$$L_1 L_2 = \{x_1 x_2 : x_1 \text{ is in } L_1 \text{ and } x_2 \text{ is in } L_2\}.$$

Product is similar to but different from cartesian product. For example, let $L_1 = \{a, ab\}$ and $L_2 = \{b, bb\}$, then $L_1 L_2 = \{ab, abb, abbb\}$ contains only three words, whereas $L_1 \times L_2 = \{(a, b), (a, bb), (ab, b), (ab, bb)\}$ contains four pairs of words.

Catenation satisfies a number of properties. It is *associative*, that is, for all languages A, B, and C

$$A(BC) = (AB)C = ABC$$

It has an *identity*, since

$$\{\lambda\}A = A\{\lambda\} = A$$

for all languages A. This implies that the set or *family* of all languages over some alphabet Σ, namely, 2^{Σ^*}, is a monoid with respect to the catenation product. Catenation also has a *zero*, since

$$\emptyset A = A\emptyset = \emptyset$$

for all languages A. The interaction of catenation with union and intersection is studied in the Exercises. Catenation is distributive over union, so

$$A(B \cup C) = AB \cup AC \text{ and } (B \cup C)A = BA \cup CA$$

but not over intersection, so, in general,

$$A(B \cap C) \neq AB \cap AC \text{ and } (B \cap C)A \neq BA \cap CA$$

We also extend powers to apply to languages.

Given a language $L \subseteq \Sigma^*$ and a nonnegative integer i, define the *ith power of* L, denoted by L^i, by

$$L^i = \begin{cases} \{\lambda\}, & \text{if } i = 0 \\ LL^{i-1}, & \text{if } i \geq 1 \end{cases}$$

L^i consists of all words formed by catenating i words from L. We define the *star* or *closure* of L, denoted by L^*, by

$$L^* = \bigcup_{i=0}^{\infty} L^i$$

L^* consists of all words formed by catenating a finite number of, possibly zero, words from L. The *plus* or *positive closure* of L, denoted by L^+, is defined by

$$L^+ = \bigcup_{i=1}^{\infty} L^i$$

L^+ consists of all words formed by catenating a finite number, never zero, of words from L.

Example 1.2.4 Consider the language $L \subseteq D^*$ defined as all those words over D which do not contain any of the digits $2, 3, \ldots, 9$. Then L contains λ, 0, 1, and 01000, for example. Furthermore, L^i, for all $i \geq 1$, equals L. $L \subseteq L^i$ since λ is in L and every word x in L may be written as $\lambda^{i-1}x$ which is in L^i, by definition. Conversely, $L^i \subseteq L$ since every word x in L^i does not contain any digits $2, \ldots, 9$, and, therefore, is in L, by definition. Thus, $L = L^i$. Finally, since $L^0 \subseteq L$ we have $L^* = L$ and $L^+ = L$. Indeed, it is not difficult to see that $L = \{0,1\}^*$.

On the other hand, consider $L = \{a, ba\}$. Then, for $i > 1$, $L^i \neq L^{i-1}$, since L^i contains a^i which is not in L^{i-1}. Hence, $L \subset L^*$ and $L \subset L^+$. Also we have $L^+ \subset L^*$ since λ is not in L.

Star and plus are related since $L^+ \subseteq L^*$, for all languages L. Moreover, $L^* = \{\lambda\} \cup L^+$ by definition. But $L^+ \neq L^* - \{\lambda\}$ as we might expect, since λ is in L^+ if λ is in L. For example, let $L = \{\lambda, a\}$. Then $L^+ = \{a^i : i \geq 0\}$; hence, $L^* = L^+$ and $L^+ \neq L^* - \{\lambda\}$. It can be proved that $L^+ = L^* - \{\lambda\}$ if and only if λ is not in L.

Let L be a language. Then both L^+ and L^* are closed under catenation of words. Indeed, L^+ is the catenation closure of L. This implies, in other words, that L^+ is a semigroup with respect to catenation, and L^* is a monoid with respect to catenation. This implies $L^+ = L$ if and only if L is closed under catenation. A useful formula that the student should know is $L^+ = LL^* = L^*L$, for all L.

Star and plus satisfy a number of properties, a remarkable one being

$$\varnothing^+ = \varnothing \quad \text{but} \quad \varnothing^* = \{\lambda\}$$

For $\{\lambda\}$ we obtain

$$\{\lambda\}^+ = \{\lambda\} \quad \text{and} \quad \{\lambda\}^* = \{\lambda\}$$

All four identities follow directly from the definitions. We also have

$$AA^+ = A^+A$$

$$(A^+)^+ = A^+ \qquad (A^+)^* = A^*$$

$$(A^*)^+ = A^* \qquad (A^*)^* = A^*$$

and

$$A \subseteq B \quad \textit{implies} \quad A^+ \subseteq B^+$$

$$A \subseteq B \quad \textit{implies} \quad A^* \subseteq B^*$$

We now extend reversal to languages. Let L be an arbitrary language over some alphabet; then the *reversal*, L^R, of L is defined as

$$L^R = \{x^R : x \text{ is in } L\}$$

In other words, it is the reversal of all words in L.

Some properties of reversal are

$$(A \cup B)^R = A^R \cup B^R$$

$$(A \cap B)^R = A^R \cap B^R$$

$$(A^R)^R = A$$
$$\overline{A}^R = \overline{A^R}$$
$$(AB)^R = B^R A^R$$
$$(A^+)^R = (A^R)^+$$
$$(A^*)^R = (A^R)^*$$

where A and B are arbitrary languages.

We consider functions or mappings of languages. This is discussed in more detail in Section 9.2. Let Σ and Δ be two alphabets; then a function $f : \Sigma^* \rightarrow \Delta^*$ is a (*word*) *morphism* (or *homomorphism*) if

(i) $f(\lambda) = \lambda$; and
(ii) $f(xy) = f(x)f(y)$, for all x,y in Σ^*.

Strictly speaking, condition (i) is not needed since it is implied by condition (ii). However, we keep it for emphasis. A morphism can always be specified by the images of the symbols in the domain alphabet, since condition (ii) implies for all x in Σ^*, $x \neq \lambda$, $f(x) = f(a_1 \cdots a_m) = f(a_1) \cdots f(a_m)$, where $x = a_1 \cdots a_m$, for some a_i in Σ, $1 \leq i \leq m$.

For example, let $\Sigma = \{a,b,c,d\}$ and $\Delta = \{0,1\}$ and f be the morphism specified by $f(a) = 0, f(b) = 1, f(c) = 00, f(d) = \lambda$. Then

$$f(dbcaad) = f(d)f(b)f(c)f(a)f(a)f(d) = 10000$$

However, if $g : \Sigma^* \rightarrow \Delta^*$ satisfies $g(ab) = 111$ and $g(a) = 0$, then g is not a morphism.

If $f(a) \neq \lambda$, for all a in Σ, f is a *λ-free morphism.* One well-known example of a morphism, indeed a λ-free morphism, is Morse code (see Section 9.2); another is the ASCII internal code for characters (see Table 1.2.1), where the morphism is $f : Char \rightarrow OctalCode$.

1.2.3 Defining Languages

We have introduced the formal notion of alphabets, words, and languages, but we have not discussed how languages should be defined or specified. This text considers a number of different techniques in Chapters 2-6 and in Chapter 11. In the remainder of this section we examine some possible techniques and introduce a fundamental classification of them.

The first technique is, simply, to list the words in the language. Of course, this can only be used for finite languages and, in practice, "small" ones. For example the set of all FORTRAN identifiers is a finite set, but it is impractical to define it by providing a list of its elements. We discuss the issue of what is small in the Exercises. Despite this the set of finite languages, which we denote by $\boldsymbol{L}_{FIN}$, is an important set. Sets of languages are usually called *families of languages*, so $\boldsymbol{L}_{FIN}$ is usually called the *family of finite languages*.

Table 1.2.1 Seven-Bit ASCII Code.

Octal Code	Char.	Octal Code	Char.	Octal Code	Char.	Octal Code	Char.
000	NUL	040	SP	100	@	140	`
001	SOH	041	!	101	A	141	a
002	STX	042	″	102	B	142	b
003	ETX	043	#	103	C	143	c
004	EOT	044	$	104	D	144	d
005	ENQ	045	%	105	E	145	e
006	ACK	046	&	106	F	146	f
007	BEL	047	′	107	G	147	g
010	BS	050	(	110	H	150	h
011	HT	051	)	111	I	151	i
012	LF	052	*	112	J	152	j
013	VT	053	+	113	K	153	k
014	FF	054	,	114	L	154	l
015	CR	055	-	115	M	155	m
016	SO	056	.	116	N	156	n
017	SI	057	/	117	O	157	o
020	DLE	060	0	120	P	160	p
021	DCI	061	1	121	Q	161	q
022	DC2	062	2	122	R	162	r
023	DC3	063	3	123	S	163	s
024	DC4	064	4	124	T	164	t
025	NAK	065	5	125	U	165	u
026	SYN	066	6	126	V	166	v
027	ETB	067	7	127	W	167	w
030	CAN	070	8	130	X	170	x
031	EM	071	9	131	Y	171	y
032	SUB	072	:	132	Z	172	z
033	ESC	073	;	133	[	173	{
034	FS	074	<	134	\	174	|
035	GS	075	=	135	]	175	}
036	RS	076	>	136	^	176	~
037	US	077	?	137	—	177	DEL

A second technique, also borrowed from the specification of sets, is to define languages inductively. For example, let $L \subseteq \{(,)\}^*$ be defined by: (i) λ is in L; (ii) if x is in L, then (x) is in L; and (iii) if x and y are in L, then xy are in L. It is not difficult to see that L contains λ, ( ), ( )( ), and (( ),( )), for example, and that it is the set of all well-formed parenthesis expressions. If in an arithmetic expression we ignore all symbols apart from (and), then we are left with a word in L. We explore this particular example in more detail in the Exercises.

Since a language is not an arbitrary set, but is a subset of Σ^*, for some alphabet Σ, we might expect that there are definitional methods peculiar to languages. This is, indeed, the case and broadly speaking these methods can be split into three classes: *generators*, *recognizers*, and *parsers*.

A generator defines a language by giving a scheme for generating words in the language. It can be viewed as the black box of Figure 1.2.1; it has no input, only output. Language expressions, treated below, are one example of a generator. Regular expressions (see Chapter 3) and context-free grammars (see Chapter 4) are also examples of generators.

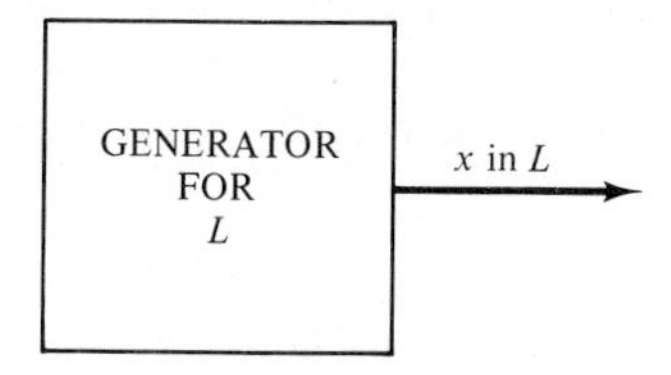

Figure 1.2.1

A recognizer defines a language by providing an algorithm or partial algorithm to recognize words in the language. A black box view of a recognizer is given in Figure 1.2.2. Finite automata (see Chapter 2), pushdown automata (see Chapter 5), and Turing machines (see Chapter 6) are examples of recognizers.

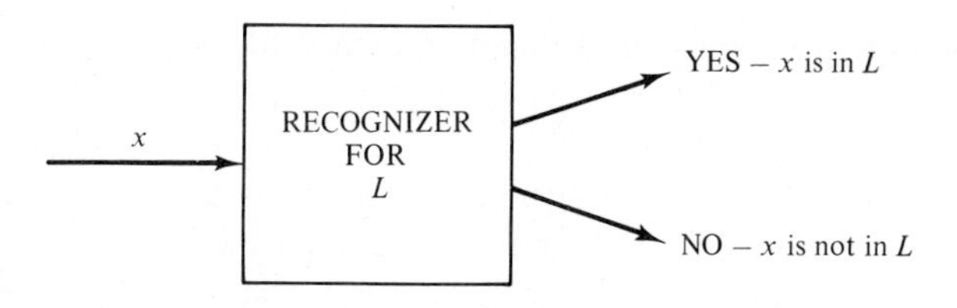

Figure 1.2.2

A parser is a hybrid consisting of a generator and a recognizer. It not only recognizes words as in a recognizer, it also recognizes only words defined by the given generator, see Figure 1.2.3. Furthermore, if a word is recognized, the parser also says how the word is generated by the generator. An important example of such a hybrid is a context-free-grammar parser, see Section 11.3 and also Sections 4.1 and 7.4.

Before we leave this section we consider one generative method of defining languages based on the operations that we have introduced for languages in Section 1.2.2. Given a language L we can obtain its complement, star, reversal, and plus. Similarly, given two languages K and L we can combine them with union, intersection, and catenation. In both cases we obtain, using a single operation, a new

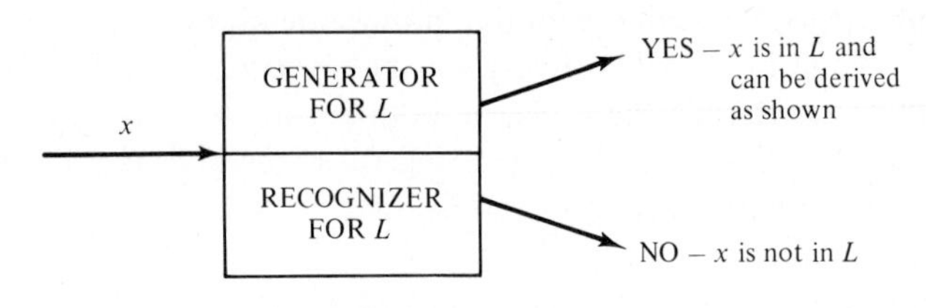

Figure 1.2.3

language from given ones. This leads to the idea of a language construction kit.

If we are given some initial languages we can combine them in this way to obtain new languages. It seems reasonable to begin with all subsets of some given alphabet Σ, since these are all finite, which means they can be written down. We then form finite combinations of these initial languages using the operations mentioned above.

Example 1.2.5 Let $\Sigma = \{a,b,c\}$. The initial languages are

$$\varnothing, \{a\}, \{b\}, \{c\}, \{a,b\}, \{a,c\}, \{b,c\}, \{a,b,c\}$$

and we can build

$$L_1 = \{a,b\}^*\{c\} \cup \{b,c\}$$

$$L_2 = \{a,b,c\}^*\{a\}\{a\}$$

$$L_3 = \neg(\{a,b,c\}^*\{a\}\{a\}),$$

where $\neg$ is the alternative method of denoting complementation, and

$$L_4 = \varnothing^*$$

L_1 is the set consisting of all words over $\{a,b\}$ followed by c and the letters b, L_2 is the set of all words over $\{a,b,c\}$ ending with a^2, L_3 is the set of all words over $\{a,b,c\}$ which do not end with a^2, and L_4 has only one word, the empty word.

A finite combination of languages over Σ using the operations of catenation, complement, intersection, plus, reversal, star, and union is called a *language expression over* Σ. Each language expression E over Σ specifies a language $L(E) \subseteq \Sigma^*$. We leave to the Exercises the formal recursive definition of language expressions, while in the remainder of this section we pose and answer one question: How many languages over Σ can be defined by language expressions over Σ?

To answer this we first prove the following,

Theorem 1.2.2 *For all alphabets Σ, Σ^* is enumerable.*

Proof: It is sufficient to prove that there is a bijection $f : \mathbb{N}_0 \longrightarrow \Sigma^*$, since $\mathbb{N}_0$ is known to be enumerable. We demonstrate how to construct such an f for

$\Sigma = \{a,b\}$ and then indicate how the construction can be generalized for arbitrary Σ.

Since $\{a,b\}$ has two symbols we let each word over Σ denote a number in binary using digits 1 and 2 rather than 0 and 1. So we obtain, letting a be 1 and b be 2

λ	0
a	1
b	2
aa	$11 = 3$
ab	$12 = 4$
ba	$21 = 5$
bb	$22 = 6$
aaa	$111 = 7$

With this representation each word x in Σ^* represents a unique integer. This would not hold with the usual representation, for letting a be 0 and b be 1 a, aa, aaa, . . . , all represent 0. Now $f : \mathbb{N}_0 \rightarrow \{a,b\}^*$ is defined by $f(i)$ = the word x over $\{a,b\}$ representing i. That f is injective follows from the representation of numbers as discussed above. That it is also surjective and, therefore, a bijection follows because every word represents some integer. So $\{a,b\}^*$ is enumerable.

To generalize this to $\Sigma = \{a_1, . . . ,a_m\}$, $m \geq 1$, let a_i denote the digit i, $1 \leq i \leq m$, in an m-adic number system. ■

One implication of this result is that although adding a new symbol to an alphabet adds new words, there are only as many words as before!

Returning to language expressions observe that each language expression over Σ is itself a word over some alphabet Δ. Indeed, $\Delta = \Sigma \cup \{,,\neg,\cup,\cap,*,+,^R,\{,\},(,),\}$! Letting $\boldsymbol{E}_\Sigma$ denote the set of all language expressions over Σ, $\boldsymbol{E}_\Sigma$ is itself a language, since $\boldsymbol{E}_\Sigma \subseteq \Delta^*$. We obtain:

Corollary 1.2.3 *$\boldsymbol{E}_\Sigma$ is enumerable.*

This follows since every infinite subset of an enumerable set is enumerable; see the Exercises for Section 0.4.

Let $\boldsymbol{L}_\Sigma$ be the set of languages over Σ given by language expressions over Σ. $\boldsymbol{L}_\Sigma$ is more usually called a *family* of languages.

Corollary 1.2.4 *$\boldsymbol{L}_\Sigma$ is enumerable.*

Proof: Define a function $f : \boldsymbol{L}_\Sigma \rightarrow \boldsymbol{E}_\Sigma$ by, for all L in $\boldsymbol{L}_\Sigma$, $f(L) = E$, for some E with $L(E) = L$. f is not injective, since two different language expressions can define the same language, for example, $\varnothing$ and $(\neg(\Sigma^*))^+$. But this implies $\boldsymbol{E}'_\Sigma = \{E : f(L) = E, \text{ for } L \text{ in } \boldsymbol{L}_\Sigma\} \subseteq \boldsymbol{E}_\Sigma$ and, moreover, by definition the range of f is $\boldsymbol{E}'_\Sigma$ Thus, $f : \boldsymbol{L}_\Sigma \rightarrow \boldsymbol{E}'_\Sigma$ is injective. Now a subset of an enumerable set is also enumerable; therefore, $\boldsymbol{E}'_\Sigma$ is enumerable and, hence, $\boldsymbol{L}_\Sigma$ is enumerable. ■

This result implies that there are languages over Σ not obtainable from any language expression, because the family of all languages over Σ equals 2^{Σ^*}, which is not enumerable; see Exercise 3.2.

Corollary 1.2.5 $\boldsymbol{L}_\Sigma \subset 2^{\Sigma^*}$

As we shall see this is not an isolated result. An analogous result can be proved for each language specification method introduced in Chapters 2-6. Furthermore, such a result can be proved for every specification method, those already existing and those to be defined in the future. On the one hand, every method fails the ultimate test of power — can it define all languages? On the other hand, as we shall see, different methods can have different expressive power — some define more languages than others.

1.2.4 Definition Summaries

To close this section we summarize the new definitions.

Definition Summary: Words

An *alphabet* is a finite, nonempty set of *symbols* or *letters*. A *word* over an alphabet Σ is a finite sequence of symbols from Σ. The *empty word* is denoted by λ and is the empty sequence of symbols. Σ^* denotes the set of all words over Σ. The *length* of a word x is denoted by $|x|$ and is defined as the number of symbols in x. For a in Σ, the *a-length* of x in Σ^* is the number of times a appears in x and is denoted by $|x|_a$.

Given two words x and y over Σ, their *catenation* is xy. $\lambda x = x\lambda = x$, for all x in Σ^*. By x^i we denote the *ith power* of x, where $x^0 = \lambda$ and $x^{i+1} = xx^i$.

For two words x and y over Σ, x is a *prefix* of y if $y = xz$, for some z in Σ^*. $x = y$ if x is a prefix of y and $|x| = |y|$. x is a *proper prefix* of y if it is a prefix, $x \neq y$, and $x \neq \lambda$. *Suffix* and *proper suffix* are defined similarly. We say x is a *subword* of y if $y = wxz$, for some w and z in Σ^*. x is a *proper subword* if $x \neq y$ and $x \neq \lambda$.

For a word x in Σ^*, x^R is the *reversal* of x, that is, x written backward.

Given a word x in Σ^*, $alph(x) \subseteq \Sigma$ denotes the *alphabet of* x, that is, the set of symbols appearing in x.

Definition Summary: Languages

Given an alphabet Σ, a *language* over Σ is a subset of Σ^*.

Since languages are sets, the Boolean operations of *union*, *intersection*, and *complement* are applicable.

Given two languages L_1 and L_2, possibly over different alphabets, a number of operations can be defined. The *catenation* (*product*) of L_1 and L_2 is denoted by L_1L_2 and equals the set $\{x_1x_2 : x_1 \text{ in } L_1 \text{ and } x_2 \text{ in } L_2\}$. The *ith power* of L_1 is defined by $L_1^i = \{\lambda\}$, if $i = 0$, and LL^{i-1}, otherwise. The *star* of $L_1 = \cup_{i=0}^{\infty} L_1^i$ and the *plus* of $L_1 = \cup_{i=1}^{\infty} L_1^i$, are denoted by L_1^* and L_1^+, respectively.

The *reversal* of L_1 is defined as $L_1^R = \{x^R : x \text{ is in } L_1\}$. Let Σ and Δ be two alphabets and $f : \Sigma^* \rightarrow \Delta^*$ be a function that satisfies $f(xy) = f(x)f(y)$, for all x, y in Σ^*. Then f is a *morphism.*

1.3 LANGUAGES, PROBLEMS, AND COMPUTATION

In Section 1.1 we have shown how decision problems and problems are closely related via decision algorithms and have abstracted decision problems further to obtain yes instances of these problems. In order to advance these ideas further we need to recognize that a computing agent necessarily deals with representations of input data and with representations of output data. For example, in a PASCAL program the input data is a sequence of characters as is the output data. We make this more precise using the basics of words and languages.

Given a decision problem Π, it specifies a function $f_\Pi : A_1 \times \cdots \times A_m \rightarrow$ {**false, true**}. Assume for now that each A_i can be represented as a set of words L_i over some alphabet Σ_i. For example, the nonnegative integers can be represented by $\{1\}^*$, that is, unary notation, or by $\{0,1\}^*$, that is, in binary notation. We assume that the Σ_i are not necessarily distinct. This representation must be functional from L_i to A_i, that is, to each representative in L_i there must correspond a unique element in A_i. However, the converse need not hold. There may be many representatives corresponding to the same element in A_i. For example, the number six can be represented as 110,0110,00110, etc. We call such a representation an *encoding.*

To begin to understand how we can encode more complex problems we look at one.

Example 1.3.1 A ubiquitous variant of a well-known optimization problem (see Sections 6.5 and 11.1 for further discussion) is

TRAVELING SALESMAN CHECKING(*TSC*)

INSTANCE: An integer $n \geq 1$, n distinct cities $C_1, \ldots, C_n$, the positive integral distances between every pair C_i, C_j, denoted by $d(C_i, C_j)$, and a positive integral bound B.

QUESTION: Does there exist a tour $C_{i_1}, \ldots, C_{i_n}$ of the cities such that

$$COST(i_1, \ldots, i_n) = \sum_{j=1}^{n-1} d(C_{i_j}, C_{i_{j+1}}) + d(C_{i_n}, C_{i_1}) \leq B.$$

A *tour* is, simply, a permutation of the n cities; each city is visited once and only once. This problem appears under many guises. A plotter is to display a set of given points. What is the best ordering of the points to minimize the total movement of the pen? A pop group has to give concerts at a number of North American cities. Which ordering of the concerts minimizes total travel time? A manufacturer of washing machines produces many different versions. Each involves changing the setup of the assembly shop. Which order of production of the different versions minimizes setup-change time?

Given an instance of *TSC* the number of cities is determined implicitly by the domain of the distance function. Let A be the set of all distance functions. Then the function f_{TSC} specified by *TSC* is defined by $f_{TSC} : A \times I\!N \rightarrow \{\textbf{false}, \textbf{true}\}$, where $f_{TSC}(d,B) = \textbf{true}$ if and only if the instance of *TSC* determined by d and B has a tour with cost at most B.

We encode pairs (d,B) as words as follows. We can assume that the cities $C_1, \ldots, C_n$ are represented by the integers $1..n$, since the names of the cities are not relevant to the problem. Now each $d : 1..n \times 1..n \rightarrow I\!N$ can be represented as a set of triples $(i,j,d(i,j))$, $1 \le i,j \le n$. This implies $A \subseteq 2^{I\!N^3}$. Thus, a pair (d,B) is represented as a pair

$$(\{(i,j,d(i,j)) : 1 \le i,j \le n\},B)$$

For example, with $n = 4$, d given by Table 1.3.1, and $B = 8$ we obtain

$$(\{(1,1,3),(1,2,2),(1,3,1),(1,4,2),(2,1,4),(2,2,1),(2,3,2),(2,4,3),(3,1,3),(3,2,1),$$
$$(3,3,4),(3,4,2),(4,1,1),(4,2,1),(4,3,1),(4,4,5)\},8)$$

So we are left with the problem of encoding integers. Let each nonnegative integer i be encoded as the word ab^ia. This gives

$$(\{(aba,aba,ab^3a),(aba,ab^2a,ab^2a), \ldots \},ab^8a)$$

as a possible encoding of this instance of *TSC*. Now either we include {, },,,(, and) as symbols in the representation or we replace them by words over $\{a,b\}$. For example,

$$\{ \equiv baab \qquad (\equiv baaab \qquad , \equiv bab$$
$$\} \equiv baaaab \qquad) \equiv baaaaab$$

We have had to make many choices, even when the representation of integers has been decided, since the same d can be written in many different ways. In other words, when a set is written we *order* its elements in some arbitrary way. So the triples in d could be written in $n^2!$ different orders, each giving rise to a different representation of the same set of triples. This is the reason that there may be many representatives of an element of A, but each representative determines a unique distance function.

We are now able to represent or *encode* $A_1 \times \cdots \times A_m$ by $L_1\{:\} L_2\{:\} \cdots \{:\} L_m$ where : is a new symbol not appearing in any Σ_i — the *separation symbol*. An instance determined by $(a_1, \ldots, a_m)$ is represented by $r(a_1) : \cdots : r(a_m)$, where $r(a_i)$ is a representative corresponding to a_i; in other words, it is a word. In this setting the representation of the set of yes instances is denoted by $L(r,\Pi)$ and is defined by

Table 1.3.1

d		j			
		1	2	3	4
i	1	3	2	1	2
	2	4	1	2	3
	3	3	1	4	2
	4	1	1	1	5

$$L(r,\Pi) = \{x_1 : \cdots : x_m : (a_1, \ldots, a_m) \text{ determines a yes instance of } \Pi \text{ and } x_i \text{ is a representation of } a_i,\ 1 \le i \le m\}$$

This is called *the language of yes instances of* Π *under* r. This, apparently trivial, exercise does have an important consequence, namely,

> For an instance $(a_1, \ldots, a_m)$ of a decision problem Π which has a representation r, $f_\Pi(a_1, \ldots, a_m) =$ **true** iff $x_1 : \cdots : x_m$ is in $L(r,\Pi)$, where x_i is a representation of a_i under r, $1 \le i \le m$.

Hence, we have replaced the evaluation of a function by a membership test; a computational problem has been replaced with a language-theoretical problem. This change of emphasis is one of the reasons that this text concerns itself with language- and automata-theoretical issues. Another reason is that the models discussed in Chapters 2-5 are useful in a number of practical situations, for example, pattern matching, syntax analysis, and compiler construction.

Let us summarize our journey so far. Beginning with the notions of problem, algorithm, and computing agent we have sketched how we can restrict our attention to decision problems and decision algorithms. In the final step, outlined above, we have transformed a computational situation into a language-theoretical one. In particular, this implies that ALGORITHM EXISTENCE

> *Does there exist an algorithm for a problem* Π*?*

has been replaced by DECISION ALGORITHM EXISTENCE

> Does there exist a decision algorithm for the decision problem D obtained from Π?

which has in turn been transformed into MEMBERSHIP TESTING

> Does there exist a decision algorithm for the language problem obtained from Π?

We shall see that not only is this transformation of importance for the existence of algorithms, but it is also important when considering the existence of efficient algorithms.

Having transformed our concerns into language issues we must turn our attention to the *finite* specification of languages or, equivalently, the *finite* specification of Boolean functions. Finite specifications are as essential to the specification of a language as a finite set of instructions is to the specification of an algorithm.

As we have seen we cannot hope to specify all languages over a fixed alphabet by means of finite specifications, because finite specifications can themselves be considered to be words over some fixed alphabet. Thus, if our goal is to be able to finitely specify all languages, then we are doomed to failure. Fortunately our goal is less ambitious, namely, *what can be expressed or specified*? This corresponds to the general question: *what functions can be computed*? Which is essentially ALGORITHM EXISTENCE once more. To this latter question Church's thesis claims that whatever new method of computing or specifying functions is invented then it will give, at most, those functions computable by Turing machines (see Section 7.4) or, equivalently, by the recursive functions (see Section 7.5). In language-theoretical terminology Church's thesis implies that all methods of finitely specifying languages will give at most the languages specified by Turing machines. We return to this topic in Section 11.2.

1.4 ADDITIONAL REMARKS

1.4.1 Summary

We have used Turing's approach to a theory of computation based on a program-machine dichotomy. This leads to the twin notions of ALGORITHM and COMPUTING AGENT for PROBLEMS and PROBLEM INSTANCES which give rise in a natural manner to ALGORITHM COMPUTABLE FUNCTIONS. We have considered the relationship between PROBLEMS and DECISION PROBLEMS and ALGORITHMS and DECISION ALGORITHMS which gives rise to YES INSTANCES and YES INSTANCE LANGUAGES translating FUNCTION COMPUTATION into MEMBERSHIP TESTING and translating problems about COMPUTABLE FUNCTIONS into problems about LANGUAGES.

1.4.2 History

The approach taken here is based on that of Alan Turing (1936). The functional view of programs and machines is persuasively argued in Scott (1967). The attempt to use decision algorithms as a basis for solving the corresponding problem has a classical counterpart in Kleene's theorem which involves primitive recursive predicates; see Kleene (1943,1952). Some introductory texts for the theory of computation are Arbib et al. (1981), Beckman (1980), Brainerd and Landweber (1974), Davis (1958), Davis and Weyuker (1983), Hennie (1977), Hermes (1969), Kfoury et al. (1982), Mallozi and De Lillo (1984), McNaughton (1982), Minsky (1967), Peter (1967), and Salomaa (1985).

The study of words and languages is now a well-established discipline and numerous introductory texts, apart from the present one, are available. Some of these are Aho and Ullman (1972), Beckman (1980), Berstel (1979), Bücher and

Maurer (1984), Davis and Weyuker (1983), Denning et al. (1978), Ginsburg (1966), Harrison (1965,1978), Hopcroft and Ullman (1979), Lewis and Papadimitriou (1981), McNaughton (1982), Revesz (1983), Salomaa (1973), and Savitch (1982). Of these Lewis and Papadimitriou (1981), and Salomaa (1973) are the most accessible.

1.5 SPRINGBOARD

In this and subsequent chapters the springboard subsection follows through some of the ideas into new areas or new applications. A springboard item may suggest further reading of books or papers and/or may contain specific problems. We do not attempt to provide an exhaustive list of topics or even to give all interesting topics. The lists merely explore some possibilities.

1.5.1 The λ-Calculus

Alonzo Church (1936) introduced a method of specifying functions that is distinct from the Turing dichotomy of program and machine. Before Church's work there was no rigorous and uniform notation for specifying functions. For example, with the usual mathematical notation for a function $f(x)$ of a single variable it is unclear whether $f(x)$ means the function or the result of applying it to the value x. Church distinguishes between them by letting $\lambda x.f(x)$ denote the function and $f(x)$ denote its value for the argument x. Church (1936, 1941) explores the consequences of this simple notion. This approach led him to the notion of effective computability and to the thesis that has been named for him.

The value of the λ-calculus for computer science is its rigorous treatment of functions with a simple uniform notation. This led to: the development of LISP (see McCarthy, 1960); the development of functional programming (see Burge, 1975, and Henderson, 1980); and an approach to semantics (see Landin, 1965), particularly in the way we view procedures and functions in programming languages.

1.5.2 Markov Algorithms

A.A. Markov (1954) developed another abstraction of algorithm which can be viewed as string processing. Each Markov algorithm consists of a sequence of rules or productions

$$\alpha_1 \rightarrow \beta_1, \ldots, \alpha_m \rightarrow \beta_m$$

over some given alphabet Σ. The algorithm is given as input data some word x over Σ. The rules are interpreted as a looping conditional construct, that is,

while x can be transformed **do**
if α_1 is a subword of x **then** replace its leftmost
occurrence in x with β_1 **else**

. . .

if α_m is a subword of x **then** replace its leftmost occurrence in x with β_m

Some of the rules can be designated as terminal, that is, when one of them is applied x cannot be further transformed. The loop may also terminate because none of the α_i appear in it. If an algorithm terminates for a given input, then the resulting word, the output, is uniquely determined.

Consider the Markov algorithm over $\Sigma = \{a, \#\}$ given by

$$a\#a \to aa\# \; ; \; \# \to \lambda$$

For any input word $a^i \# a^j$ it produces a^{i+j}, that is, it performs addition. Similarly,

$$a\#a \to \# \; ; \; \#a \to \# \; ; \; \# \to \lambda$$

performs nonnegative subtraction or monus. That is, $a^i \# a^j$ results in a^{i-j} if $i \geq j$; otherwise it results in λ.

Surprisingly, in view of their simple structure, Markov algorithms are as powerful as Turing machines and hence as powerful as the λ-calculus, recursive functions, and Post systems. The book of Galler and Perlis (1970) uses Markov algorithms as its fundamental approach.

1.5.3 Post Systems and Production Systems

Emil Post studied the formal systems of logic, as did most of the pioneers in the area of computability. He abstracted formal systems as rewriting systems, where rewriting corresponds to forming a conclusion from a set of premises. A *Post system* consists of a finite set P of inference rules of the form

$$\alpha_1, \ldots, \alpha_m \to \alpha$$

together with a finite axiom set S. The α_i are the premises, from which we infer the conclusion α. The α_i and α are words over an alphabet consisting of *variables* and *constants*. The axiom set S is a set of words consisting solely of constants.

To apply an inference rule we need to replace variables uniformly by constant words. In other words, each appearance of a variable in $\alpha_1, \ldots, \alpha_m, \alpha$ is replaced by the same constant word. This gives a constant version of the rule

$$\bar{\alpha}_1, \ldots, \bar{\alpha}_m \to \bar{\alpha}$$

in which no variables appear.

If $\bar{\alpha}_1, \ldots, \bar{\alpha}_m$ are in S, then we add $\bar{\alpha}$ to S. We say that a constant word x is inferred by a given Post system (P, S) if either it is in S or it can be obtained from S by a finite number of applications of inference rules.

For example, let

$$A \to aAb$$

be the only production and ab be the only axiom. Then

$$ab \rightarrow aabb$$

$$a^2b^2 \rightarrow a^3b^3$$

and, in general,

$$a^ib^i \rightarrow a^{i+1}b^{i+1}$$

Therefore, this Post system produces $\{a^ib^i : i \geq 1\}$. Secondly, let

$$A \rightarrow (A)\ ,\ \ A,B \rightarrow AB$$

be the productions and λ be the axiom. We obtain

$$(\ ),\ ((\ )),\ (\ )(\ )$$

and it can be shown that it produces all well-formed parenthesis expressions.

Recently, production systems, a variant of Post systems closer to their logical origin, have attained an importance in the development of expert systems. Winston (1977) provides a good introduction to their use in this manner, while Post (1936) contains the original formulation.

1.5.4 Recursive Functions

Recursive functions, due to Kleene (1936), are a key development in the theory of computation. Each recursive function is defined by combining some initial functions using composition, recursion, and minimalization. As in the work of Church (see Section 1.5.1) the dichotomy of program and machine is once more absent. The initial functions are the zero constant function, the successor function and the projection functions. Each recursive function is assumed to be an m-place partial function from $\mathbb{N}_0^m$ to $\mathbb{N}_0$, for some $m \geq 1$.

For example, let $add(x,y)$ be defined as $x+y$. Then we can define it recursively in terms of initial functions as

$$add(x,y) = \begin{cases} \pi_1^1(x), & \text{if } y = 0 \\ succ(\pi_2^3(y-1, add(x,y-1), x)), & \text{otherwise} \end{cases}$$

where π_1^1 is the 1-place projection function, and π_2^3 is the 3-place projection function which gives its second argument as result.

Not only do recursive functions capture the basic methods of building functions from given functions, but also they define the same class as the class defined by Turing machines and other devices. The class of (Turing-) computable functions is assumed to contain all functions that are computable by some mechanism or definable by some mechanism. This is known as the Church(-Turing) thesis. We consider recursive functions in somewhat more detail in Section 7.6. The books of Rogers (1967) and Machtey and Young (1978) are excellent introductory texts.

1.5.5 Formal Semantics

It has been said that the first version of FORTRAN was defined as the language that the first compiler compiled correctly. Whether or not this was the case it introduces one method of specifying the meaning or semantics of programs, namely, providing a compiler for the language. However, which compiler should be used? On which machine? These problems, as well as other more fundamental ones, led to investigations of formal methods for specifying the formal semantics of programs.

Initially, programmers were more concerned about the syntax of programs, that is, are its statements formed correctly, rather than do these statements mean what I intend them to mean? However, the development of ALGOL 60 (see Naur et al.,1963), changed this viewpoint. Although ALGOL 60 broke new ground in that its syntax was rigorously defined, the attempt to define its semantics using English was a dismal failure. For example, what is the meaning of the following sequence of statements?

$i := 2$; $n := 100$; $k := 0$;
for $i := 1$ **step** i **until** n **do** $k := k+i$;

Is the value of the increment evaluated once and for all on entry? Is the value of the upper bound evaluated once and for all on entry? Depending on which interpretation of the ALGOL 60 report (and even the revised version) is chosen the final value of k can be very different. These are not the only problems with the ALGOL 60 for statement; see Knuth (1967) and Wood (1969c).

This example demonstrates the need for the formal semantics of a programming language to be specified as rigorously and ruthlessly as its syntax. Unfortunately, having said this, the methods available for the formal specification of semantics are more complex and opaque than those for syntax specification. The methods which have been used correspond to the traditional methods used in logic. Useful surveys are to be found in Pratt (1982), de Bakker (1969), Pagan (1981), Wegner (1972b), Marcotty et al. (1976), Hoare and Lauer (1974), and Rustin (1972).

Corresponding to the definition-by-compiler approach we have the abstract machine approach; see McCarthy (1963) and McCarthy and Painter (1967). The Vienna Definition Language developed to express the semantics of PL/I is the best known example of this approach; see Wegner (1972a). A second approach is the axiomatic approach of Hoare, see Hoare and Lauer (1974) and Hoare and Wirth (1973). In this approach the meaning of each statement is defined in terms of axioms and rules of inference. A third approach is to view a program as computing a function. Here the meaning of a program is expressed recursively as the application of the function corresponding to its first statement to the initial "state", followed by the application of the function corresponding to the remainder of the program. This is called the denotational approach. Tennent (1976) is a good introduction to this approach; see also Tennent (1981).

1.5.6 Program Correctness

Program verification is the task of proving that programs meet their specifications, that is, they are correct.

In order to verify that a given program, in a given programming language, is correct, the semantics of the language must be well defined. Furthermore, the method used to specify the computation of the program must itself be rigorous; it must be a formal specification.

The classical method is the inductive assertion method of Floyd (1967a). It uses the predicate calculus and mathematical notation. A proof of correctness for a given program consists of three stages. In the first stage the conditions on the domain of the input data and the relationship between the input data and output data are expressed. For example, with

```
function factorial(n : integer) : integer;
var i, j, product : integer;
begin product := 1; j := 1; i := n;
      while i > 1 do
      begin j := j+1; product := product * j;
            i := i-1
      end;
      factorial := product
end;
```

we obtain both the initial condition: $n \geq 1$, and the terminating condition: $factorial(n) = n!$, for all $n \geq 1$. In the second stage we attach to each loop an *assertion* or *loop invariant*. Here we might obtain

$i+j = n+1$
$product = j!$

In the third stage we argue that the initial conditions imply the loop invariant holds, if the loop is executed. Then we argue that subsequent execution of the loop maintains the invariant and, finally, that termination of the loop implies the terminating condition holds. Since the loop is only entered if $i > 1$ the loop invariant holds trivially. It is easy to show that each subsequent execution of the loop maintains the loop invariant and, finally, that on termination $i = 1$; $j = n$; $product = n!$, which imply $factorial(n) = n!$, as desired. It only remains to observe that if $n = 1$, then $factorial(n) = 1$, to complete the proof of correctness.

Apart from the assertion method, there is the axiomatic method of Hoare, and methods based on denotational semantics; see Loeckxx and Sieber (1984). A useful, though now dated, survey is Elspas et al. (1972). An introductory article is Hantler and King (1976), while Anderson (1979) is a useful introductory text. DeMillo et al. (1979) argue, in a readable and provocative paper, that verifying real programs cannot be done.

The task of verifying programs has been avoided in the language PROLOG. In this language the specification is the program; see Clocksin and Mullish (1981). A similar approach has been taken with LUCID; see Ashcroft and Wadge (1976).

1.5.7 Program Construction

Rather than attempting to verify that a given program is correct, another approach is to construct a correct program in the first place. The languages LUCID (see Ashcroft and Wadge, 1976), and PROLOG (see Clocksin and Mullish, 1981), do this by, essentially, letting the formal specification be the program. However, this implies that we construct our "program" as a statement in the predicate calculus or a similar logic. High-level program constructs and structures are nonexistent. If we wish to write a program in ADA or PASCAL how then should we do it? Top-down design, structured programming, and stepwise refinement provide a discipline in which correct programs can be obtained; see Wood (1984), for example. This approach alone has the disadvantage that a rigorous proof of correctness does not result from its application. For this reason Dijkstra (1976) introduced the notions of a predicate transformer and weakest precondition. These enable a correct program to be derived, step by step, from its termination condition. Moreover, the derivation yields a proof of correctness as well. This approach is demonstrated in detail in Dijkstra (1976) and Gries (1981). Alagic and Arbib (1978), Linger et al. (1979), and Reynolds (1981) take the top-down, structured programming viewpoint in the development of correct programs and proofs of correctness.

EXERCISES

1.1 Translate the following problems into the generic format for problem specification given in Section 1.1. Note that you may have to make the size of data explicit.

- (i) Given two $n \times n$ bit matrices, compute their Boolean product.
- (ii) Given a sequence of keys form a binary search tree using the usual insertion procedure.
- (iii) Given a binary search tree compute the keys sorted in ascending order.
- (iv) Given two single-linked lists L_1 and L_2 compute the catenation of L_1 with L_2.
- (v) Given the names, gender, marital status, and birth dates of two people decide whether or not they can be legally married today.
- (vi) Given a PASCAL program decide if it is syntactically valid.
- (vii) Given a PASCAL program translate it into a FORTRAN program.
- (viii) Given two trees form a new tree using the constructor of Section 0.2.
- (ix) Given a text file, divided into lines, and a pattern, find the lines which contain the given pattern.
- (x) Given two sorted sequences of reals merge them into one sorted sequence.

1.2 For each of the problems specified in Exercise 1.1 define their associated functions.

1.3 Consider the following problem and algorithm.

MINIMUM MAXIMUM
INSTANCE: A sequence of n integers, $n \geq 1$
QUESTION: What are the values of the minimum and maximum integers in the given sequences?

```
begin
    Given a sequence S of n integers;
    Min := S[1]; Max := S[1];
    {Use array subscripting in the obvious way}
    for i := 2 to n do
        if Min > S[i] then Min := S[i] else
        if Max < S[i] then Max := S[i];
    {Min and Max are the desired values}
end
```

Let f and g be the functions specified by the problem and algorithm, respectively. Is $f = g$? If so prove it and if not why not.

1.4 Convert each of the problems specified in Exercise 1.1 into decision problems.

1.5 Assume you are given decision algorithms for each of the decision problems of Exercise 1.4. Which of them can be used as a basis for obtaining an algorithm for the initial problem using the technique discussed in the text. Justify your answer.

2.1 Some naturally occurring examples of words were given in the text. Give ten other examples.

2.2 We say a *word is infinite* if it has infinite length. Given an alphabet Σ does Σ^* contain any infinite words? Justify your answer.

Relax the restriction than an alphabet must be finite, that is allow an alphabet Σ be infinite. How does this effect your answer to the first part of the question? Justify your answer.

2.3 Given an alphabet Σ and the empty word over Σ, denoted by λ, is λ in Σ?

2.4 Consider the empty word λ. Is $\lambda\lambda\lambda = \lambda$? Is $\lambda^i = \lambda$, $i \geq 2$? Justify your answer.

2.5 Given an arbitrary alphabet Σ and two arbitrary words x and y over Σ, is the catenation of x with y always the same as the catenation of y with x? If it is, then prove it and if it is not, then provide a counterexample.

2.6 Let x,y, and z be three arbitrary words over an alphabet Σ. Prove that xy catenated with z is always equal to x catenated with yz; in other words, catenation is associative.

2.7 Given an arbitrary alphabet Σ and arbitrary words x and y over Σ, characterize when $xy = yx$.

2.8 Given the word $x = aaababbabbb$ write down all proper prefixes of x, all proper suffixes of x, and all proper subwords of x.

2.9 Let $\Sigma = \{a,b,c,d\}$, $x = aabaa$, $y = abcabc$, and $z = badcad$. Write the values of $alph(x)$, $alph(y)$ and $alph(z)$.

2.10 Let $\Sigma = \{a,b,c,d\}$ and $x = abdac$. Write the values of x^0, x^1, x^3, and x^7.

2.11 Let $L_i \subseteq \Sigma_i^*$, $i = 1,2,3$ be three arbitrary languages. Prove that

(i) $(L_1L_2)L_3 = L_1(L_2L_3)$.
(ii) $(L_1 \cup L_2)L_3 = L_1L_3 \cup L_1L_3$.
(iii) $L_1(L_2 \cup L_3) = L_1L_2 \cup L_1L_3$.
(iv) $(L_1 \cap L_2)L_3 \neq L_1L_3 \cap L_2L_3$ in general.
(v) $L_1(L_2 \cap L_3) \neq L_1L_3 \cap L_1L_3$ in general.

2.12 Let $L \subseteq \Sigma^*$ be an arbitrary language over an arbitrary alphabet. (i) Is $\neg(\neg L) = L$? (ii) Is $(\neg L)^*$ *always* equal to $\neg(L^*)$?; *ever* equal to $\neg(L^*)$?; *never* equal to $\neg(L^*)$? Justify your answers.

2.13 Let $L \subseteq \Sigma^*$ be an arbitrary language over an arbitrary alphabet. Characterize when $L^* = L$.

2.14 Let $L \subseteq \Sigma^*$ be an arbitrary language over an arbitrary alphabet. Is L^* always, sometimes, or never enumerable? Justify your answer.

2.15 Given two morphisms $f : \Sigma^* \rightarrow \Delta^*$ and $g : \Sigma^* \rightarrow \Delta^*$ then we can say they are *equal* if $f(x) = g(x)$, for all x in Σ^*. How would you test two such morphisms for equality? This is an example of a decision problem. Does there exist a decision algorithm?

2.16 Given two morphisms $f : \Sigma^* \rightarrow \Delta^*$ and $g : \Sigma^* \rightarrow \Delta$ equality is usually defined as in Exercise 2.14. However, if we are also given a language $L \subseteq \Sigma^*$ we can introduce *equality on L*. We say *f is equal to g on L*, written $f \underset{L}{=} g$, if for all x in L, $f(x) = g(x)$. Clearly, if two morphisms are equal they are equal on L. Does the converse also hold? Prove your answer.

2.17 Let $f : \Sigma^* \rightarrow \Delta^*$ be a morphism and consider the *inverse morphism* $f^{-1} : \Delta^* \rightarrow 2^{\Sigma^*}$ defined by

$$\text{For all } y \text{ in } \Delta^*, f^{-1}(y) = \{x : f(x) = y\}$$

For each of the following questions justify your answers.

(i) Can $f^{-1}(x)$ be infinite for some x in Δ^*? If so under what conditions?
(ii) For two words x and y over Δ, can $f^{-1}(x)$ and $f^{-1}(y)$ have words in common? Must they have words in common? Must they never have words in common?
(iii) Can $f^{-1}(x)$ be the empty set?
(iv) Can $f^{-1}(x)$ be equal to $\{\lambda\}$? If so under what conditions?
(v) Is $\{x\} = f^{-1}f(x)$ for any x in Σ^*?
(vi) Let x be any word in Δ^*. Is $f^{-1}(x) = f^{-1}ff^{-1}(x)$?

2.18 Let $\Sigma = \{a,b,c\}$ and let $L = \{c^i x c^j : i,j \geq 0$, where $x = \lambda$, $x = aw$, or $x = wb$, for some w in $\Sigma^*\}$. Is $L = \Sigma^*$? Prove your answer. What can you prove about L^2?

2.19 Let $\Sigma = \{a,b\}$ and consider the following recursive definition of L. L is defined by

(i) λ belongs to L.
(ii) If x is in L, then axb and bxa are in L.
(iii) If x and y are in L, then xy is in L.
(iv) No other words are in L.

Prove that L consists of exactly those words over Σ containing an equal number of a's and b's. If b and λ are initially in L and (ii)-(iv) are unchanged, what words are in L?

2.20 Derive a recursive definition of $L \subseteq \{a,b\}^*$ containing all words with twice as many as a's as b's.

2.21 Prove that for two arbitrary languages L_1 and L_2 that

(i) $(L_1 \cup L_2)^* = (L_1^* L_2^*)^*$;
(ii) $(L_1 L_2 \cup L_1)^* = L_1(L_2 L_1 \cup L_1)^*$.

2.22 (Due to D. Forkes.)
Let $L \subseteq \Sigma^*$ be an arbitrary language. Let $C_0 = L$ and define the languages S_i and C_i, for all $i \geq 1$, by $S_i = C_{i-1}^+$ and $C_i = \overline{S_i}$.

(i) Does S_1 always equal, never equal, sometimes equal C_2? Justify your answer.
(ii) Prove that $S_2 = C_3$, whatever the choice of L.
[*Hint:* Prove that C_3 is closed under catenation.]

2.23 Prove that $(xy)^R = y^R x^R$, for all words x and y over some given alphabet.

2.24 Can you give language expressions over $\{a,b\}$ for the following languages? In each case explain your answer.

(i) $\{aba, abba, abbba, a\}$.
(ii) $\{x : x$ is in $\{a,b\}^*$ and x contains at least one a and at least one $b\}$.

(iii) $\{a^j b^j : 1 \leq j \leq 7\}$.
(iv) $\{a^j b^j : 1 \leq j\}$.

2.25 Give a formal recursive definition of language expressions over a given alphabet Σ.

2.26 Prove that for each alphabet Σ, the set of all finite languages over Σ is enumerable.

2.27 Prove that for each alphabet Σ, 2^{Σ^*} is not enumerable.

2.28 You are given a finite language of not more than 1000 words over some alphabet. Assuming that only searching, that is, membership testing, is required which data structures would you consider and why?

2.29 Assuming the same set up as in Exercise 4.1 except that the language contains about (i) 10,000 words, (ii) 100,000 words, (iii) 10,000,000 words. How would your choice be effected, if at all? Would you consider any of these languages to be small? Explain your answers.

2.30 You are given a finite language which changes over time as a result of insertion and deletion of words (this is usually called the DICTIONARY problem). How would this knowledge effect your choices in Exercises 2.28 and 2.29?

2.31 You are told that the nesting of parentheses in expressions in PASCAL is at most three deep. Give a formal description of parenthesis expressions satisfying this restriction.

3.1 Given the decision problems of Exercise 1.4, show how each of them may be represented as a language problem.

PROGRAMMING PROJECTS

P1.1 Assume that PASCAL programs are restricted so that

(i) no subprograms are allowed;
(ii) at most a single **while not** *eof* (input) **do** loop is allowed;
(iii) only integer variables are allowed; and
(iv) the input is one text file as is the output.

Recalling the two tales do you think you can construct a PASCAL program to check:

(i) If such a restricted PASCAL program terminates?
(ii) If two such restricted PASCAL programs are equivalent?

If you think you can, then attempt it. If you think you can't, then explain why not (a formal proof is not required here).

P2.1 (due to J. L. Bentley.)
Consider the following shifting problem. You are given a one-dimensional array $A[1..m]$ of characters, for some $m \geq 1$. Now given an integer $i \geq 0$, cyclically shift the elements of A i places to the left.

For example, with

$$A = RHUBARB$$

a shift of two places yields

$$UBARBRH$$

while a shift of five places yields

$$RBRHUBA$$

Write a procedure $CyclicShift(A,1,m,i)$ to perform such a cyclic shift subject to the following conditions

(i) It should use at most one temporary character variable.
(ii) It should be efficient in terms of the number of element movements.
(iii) It should be simple and elegant.
(iv) It should use some concepts from Section 1.2.

P2.2 In Exercise 2.11 some of rules for symbolically manipulating language expressions involving union, intersection, and catenation are given. First, extend these rules to include star, the empty set, and the set $\{\lambda\}$ and, second, design a PASCAL (or C, LISP, etc.) program to perform these manipulations automatically. By this we mean interactively under the direction of the user with the aim of simplifying such expressions.

P2.3 Implement one or more of the approaches you have suggested for Exercises 2.28-2.31.

Part II

MODELS

Chapter 2

Finite Automata

2.1 STATES, STATE DIAGRAMS, AND TRANSITIONS

We introduce, in this chapter, the simplest acceptor or recognizer for language specifications, the finite automaton. It is also the simplest model of a computer, a natural outgrowth of combinational circuits, and the basis of the simplest method of specifying translations (see Section 7.1). It is used for pattern matching in text editors and for lexical analysis in compilers. It has been used to specify communication protocols and to model protection mechanisms. Despite this universality of usage it does have an inherent defect, namely, it is, apart from finite language specifiers, the weakest method of specification. As we shall see in Section 8.1 some very simple languages cannot be specified by any finite automaton.

Fundamental to finite automata, and to other machine models, is the concept of a *state*. A state is the "condition with respect to structure, form, constitution, phase, or the like: *a gaseous state; the larval state*," says the Random House Dictionary. The phrases "state of the art," "state of the nation," and "state of play" indicate that the concept of state is not restricted to machines, but is applicable to any system, for example, to your TV. A TV is either on or it is off; this is a two-state system; see Figure 2.1.1 in which a *state diagram* is displayed. At a more detailed level we might wish to differentiate the channels, in which case we may have a hundred or so states — one for "off" and the remainder meaning "on in channel i". Similarly, a dish-washer can be either in an off state or in an on-with-some-program state. Typically there are about 25 different programs possible; therefore, there are about 26 different states. In each of these examples there are a *finite* number of states. We will be dealing with machines having only a finite number of states throughout this text. Given a TV, not only is it in one of two

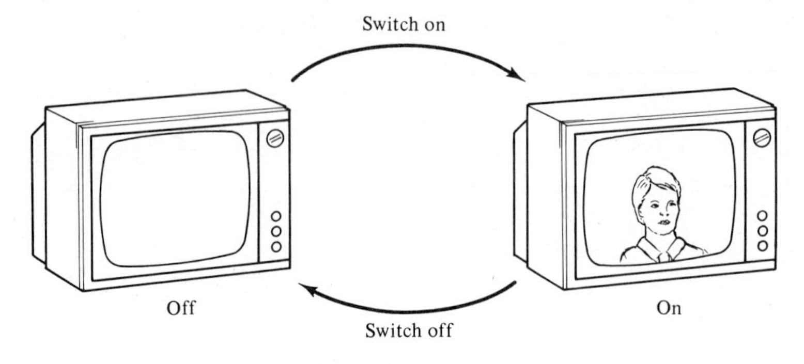

Figure 2.1.1

states, but we can switch a TV on or switch it off. In Figure 2.1.1 these options are indicated by the labeled directed edges. This enables us to change the state of a TV — surely a necessity if we ever want to switch it on or off! Switching a TV on or off is an example of a *state transition.* A TV, which is considered here to be a *finite state system*, indeed a two-state system, allows transitions from off to on, and vice versa. For this reason Figure 2.1.1 is also called a *transition diagram.* To abstract the TV system a step further notice that present day TVs have a pushbutton switch, which you press for off and press for on — the same action in each case. This gives the system of Figure 2.1.2.

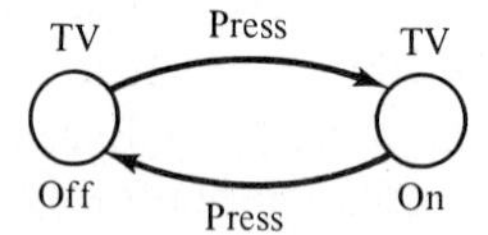

Figure 2.1.2

A sequence of "press" actions causes the TV to end up in the "On" state if either it began in the "On" state and there are an even number of presses, or it began in the "Off" state and there are an odd number of presses. If the "Off" state is the *start state* and the "On" state is the required *final state*, then all finite sequences with an odd number of presses cause this to happen. We now consider a second example, a coinchecker in a coffee vending machine. In this example we again find that there are only a finite number of states. This property is also shared with finite automata which are, therefore, sometimes called *finite state automata.* Hand in hand with the concept of state we have the concept of *system* and we speak of the current state of a system. The notions of state and system are so ubiquitous that there has developed a body of knowledge known as systems theory. Each of the machines we study in this text is a system in this sense.

Example 2.1.1 Assume we have a coffee vending machine which requires 30 cents for a cup of coffee. Such a machine usually accepts a variety of coins totaling at most 50 cents, makes change, and then produces the coffee; see Figure 2.1.3. We only study the coin checking aspect of the change maker. We assume only nickels, dimes, and quarters are legal tender. Now, the coin checker must accept legal coins, return illegal ones, and keep a running total. Once the running total is greater than or equal to 30 cents, the vending machine should return all further coins, make change, and go into coffee production. We may obtain the flowchart of the coin checker displayed in Figure 2.1.4

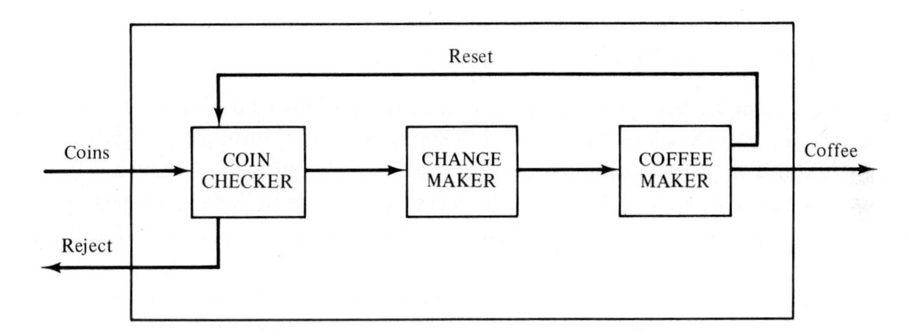

Figure 2.1.3 Coffee vending machine.

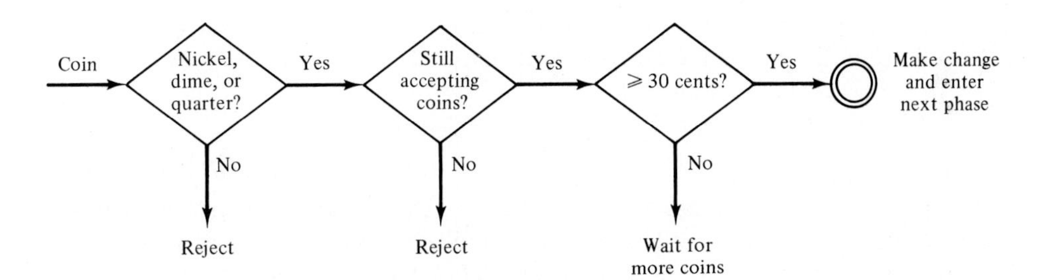

Figure 2.1.4 Coin checker.

To enable this flowchart to be converted into reality we need to keep track of the remaining change in nickels and dimes. Of course, a mechanism for checking the legality of coins is needed, but its details are not. The state of the coin checker at any time consists of the information crucial to its correct operation. It needs to know whether or not it should still accept coins, that is whether or not the current total is greater than or equal to 30 cents.

Associated with the coin checker is the one quantity *total*, which together with the current coin completely determines its actions, and the new value of this quantity; thus, we say that the coin checker is *deterministic*. The quantity *total* can be said to determine the state of the coin checker and a new coin causes a *transition* from one state to another state.

The state of the coin checker is an *abstraction* of the checking mechanism. There are a finite number of states since *total* is never more than 50 cents and never less than 0 cents.

We reconsider this example in more detail in the next section.

2.2 DETERMINISTIC FINITE AUTOMATA

In the two examples presented in Section 2.1, the notions of a *state, state* or *transition diagram*, a *state transition*, and *determinism* were introduced. Both example systems exhibit a *finite* number of states and a *finite* number of transitions. Such finite-state systems occur in many different settings: vending machines, puzzles, communications, computers, etc., and because of this the notion of a deterministic finite (state) automaton has become a central concept.

Without more ado we give the formal definition of such automata.

Definition A *deterministic finite automaton* (*DFA*) M is specified by a quintuple $M = (Q, \Sigma, \delta, s, F)$ where

Q is an alphabet of *state* symbols;
Σ is an alphabet of *input* symbols;
$\delta : Q \times \Sigma \rightarrow Q$ is a *transition function*;
s in Q is the *start state*; and
$F \subseteq Q$ is a set of *final states*.

In the TV example, $\Sigma = \{press\}$, $Q = \{Off, On\}$, $s = Off$, $F = \{On\}$, $\delta(Off, press) = On$ and $\delta(On, press) = Off$. Therefore, the state or transition diagram given in Figure 2.1.2 is simply a graph-theoretic representation of δ. We can also represent δ by a *state* or *transition table*, see Table 2.2.1.

Table 2.2.1

δ	press
On	Off
Off	On

For the coffee vending machine, Example 2.1.1, Σ is the set of possible coins, namely, nickels, dimes and quarters. The transition functions in both examples have been expressed pictorially by way of a state diagram.

Let's consider a formal example, which is slightly more difficult than the TV example, in that it has three states.

Example 2.2.1A Let $M = (Q,\Sigma,\delta,s,F)$ be defined by $Q = \{0,1,2\}$, $\Sigma = \{a\}$, $s = 0$, $F = \{2\}$, and $\delta(0,a) = 1$, $\delta(1,a) = 2$, $\delta(2,a) = 0$. The state diagram of M is displayed in Figure 2.2.1, where the start state is indicated by an ingoing wavy arrow [see Figure 2.2.2(a)] and each final state is drawn as a double circle [see Figure 2.2.2(b)].

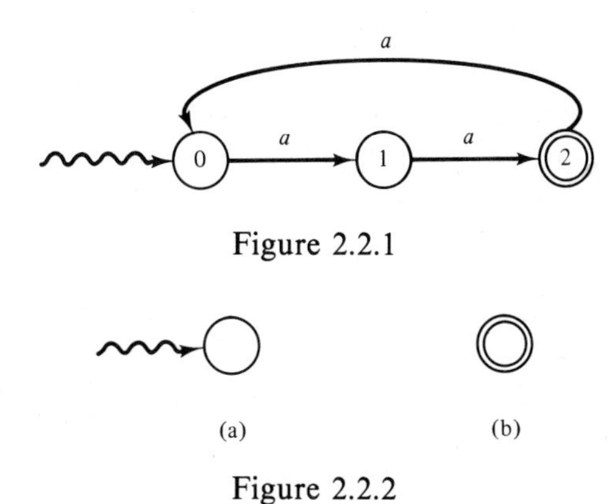

Figure 2.2.1

(a) (b)

Figure 2.2.2

Note that the example transition functions used above were both total. However, since a transition function is a (partial) function it may be nontotal. We discuss this further below.

DFAs are intended to read input words and accept or reject them, that is they are recognizers. To see how *DFAs* do this we treat them from a *machine-oriented viewpoint*.

Let $M = (Q,\Sigma,\delta,s,F)$ be a *DFA*, then we view it as a machine (a primitive computer) which has an *input tape* (or file) of *cells*, a reading head, and a finite control, see Figure 2.2.3. The finite control knows the transition function and the current state. The transition function and current state are analogous to a program and the current statement being executed (the instruction counter.) The input tape contains the input word, one symbol to each cell of the tape. This means that the tape has exactly the same number of cells as the input word has symbols. It can be thought of as a one-dimensional array of symbols or as a PASCAL text file.

The *DFA* is initialized with an input word x as follows

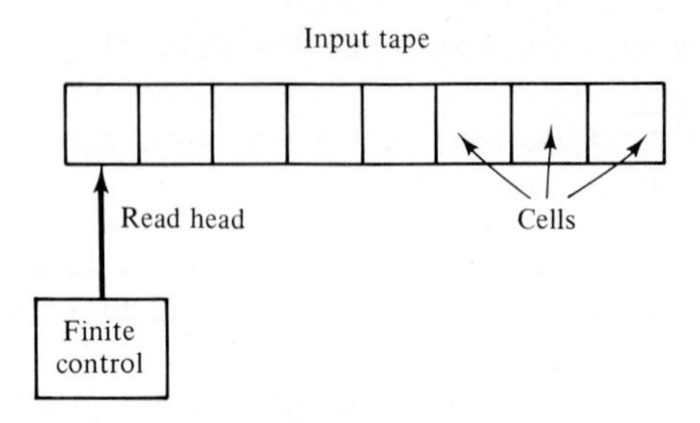

Figure 2.2.3

(i) x is placed on the input tape one symbol to a cell.
(ii) The reading head is positioned over the leftmost cell.
(iii) The *current state* is set to s.
(iv) The *DFA* is started.

The *DFA*, once started, begins its computation on the input word. As with any computer it has a *basic execute cycle.*

(i) The symbol under the reading head is read, that is, the *current symbol.* If there is no symbol under the reading head the *DFA* terminates. This occurs when the whole input word has been read.
(ii) The *next state* is computed from the current state and symbol using the transition function, that is, δ(current state, current symbol) = next state. If the current state is undefined the *DFA* aborts.
(iii) The reading head moves one cell to the right.
(iv) The next state becomes the current state and the execute cycle has been completed.

When there is no symbol under the reading head this is similar to having an end-of-file condition. Similarly, the current state being undefined is similar to setting an error condition in a program; typically, this would cause an abort.

Before giving an example to clarify this view, we introduce configurations.

Definition Let $M = (Q,\Sigma,\delta,s,F)$ be a *DFA*. We say that a word in $Q\Sigma^*$ is a *configuration of M*. It represents the current state of M and the remaining unread input of M.

A configuration of a *DFA*, M, contains all the information necessary to continue M's computation. In programming parlance it is equivalent to a dump of the current values of all variables of a program and the current position in the program. This information is sufficient to restart the program at the last dump point if the system has crashed, or the program has been timed out.

Initially, M is in configuration sx, where x is the input word. Finally, when M has read all its input, M is in a configuration q, for some state q.

Example 2.2.1B We are given the *DFA*, M, of Figure 2.2.1. For input aa we have initially Figure 2.2.4(a), that is, configuration $0aa$. Since $\delta(0,a) = 1$ we obtain Figure 2.2.4(b) or configuration $1a$. Finally, we obtain Figure 2.2.4(c) or configuration 2, as $\delta(1,a) = 2$, meaning $0aa$ gives 2 after a finite number of cycles of M.

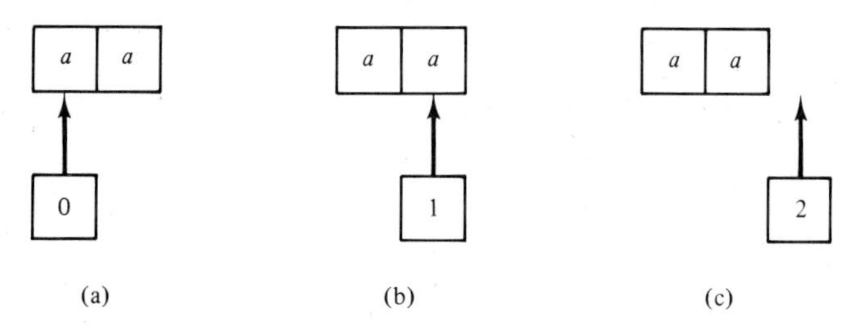

Figure 2.2.4

We write $0aa \vdash 1a \vdash 2$. This sequence of moves of M is called a *configuration sequence.* Since M is deterministic its starting and finishing configuration are sufficient to define it, so we also write $0aa \vdash^* 2$, meaning $0aa$ gives 2 after a finite number of cycles or computation steps of M.

We have introduced some notation which we now define formally.

Definition Let $M = (Q,\Sigma,\delta,s,F)$ be a *DFA*. If px and qy are two configurations of M, then we write $px \vdash qy$ (one execute cycle of M on px), if $x = ay$, for some a in Σ, and $\delta(p,a) = q$. We say *px yields qy*. Observe that $\vdash$ is a binary relation over $Q\,\Sigma^*$.

For $k \geq 1$, we write $px \vdash^k qy$ (k-steps of M on px) if either $k = 1$ and $px \vdash qy$ or $k > 1$ and there exists a configuration rz such that $px \vdash rz$ and $rz \vdash^{k-1} qy$.

Since $\vdash$ is a binary relation over $Q\,\Sigma^*$, its transitive closure which we denote by $\vdash^+$ is, by Theorem 0.3.2, well defined. However, we prefer to define it from below (see Section 0.3) as follows

> We write $px \vdash^+ qy$, where $\vdash^+$ is the transitive closure of $\vdash$, if $px \vdash^k qy$, for some $k \geq 1$.

In a similar manner we define $\vdash^*$, the reflexive transitive closure of $\vdash$. We write $px \vdash^* qy$ if either $px = qy$ or $px \vdash^+ qy$. We say that the sequence of configurations given by $px \vdash^* qy$ is a *configuration sequence.*

We can now formally define acceptance by a *DFA* and the language of a *DFA* as follows.

Definition Let $M = (Q,\Sigma,\delta,s,F)$ be a *DFA*. We say a word x in Σ^* is *accepted* by M if $sx \vdash^* f$, for some f in F. We say that $sx \vdash^* f$ is an *accepting configuration sequence.* A word that is not accepted is said to be *rejected.* The set of words accepted by M, called *the language accepted, defined,* or *recognized* by M is denoted by $L(M)$ and is defined as

$$L(M) = \{x : x \text{ is in } \Sigma^* \text{ and } sx \vdash^* f, \text{ for some } f \text{ in } F\}.$$

The notion of acceptance has caused *DFA* to stand for *deterministic finite acceptor* in some circles. We say that $L \subseteq \Sigma^*$ is a *DFA language* if there is a *DFA*, M, with $L = L(M)$. The collection of all *DFA* languages is denoted by $\mathbf{L}_{DFA}$ and is called the *family of DFA languages.*

Example 2.2.1C The words aa and $aaaaa$ are accepted by M, since $0aa \vdash^* 2$ and $0aaaaa \vdash^* 2$ and 2 is in F. Whereas the words λ, a, and $aaaa$ are rejected by F, since

$$0 \vdash^* 0;\ \ 0a \vdash^* 1;\ \ 0aaaa \vdash^* 1$$

and 0 and 1 are not in F.

The set of words accepted by M is

$$L(M) = \{aa(aaa)^i : i \geq 0\}$$

as we argued above, but which we now prove rigorously.

As is usual we prove rather more. Observe that $ia^m \vdash^* j$, where $j = (m+i) \textbf{ mod } 3$, hence a^m is accepted if and only if $i = 0$ and $2 = m \textbf{ mod } 3$, that is, $m = 3k+2$, for some $k \geq 0$.

Claim: For all i, $0 \leq i \leq 2$, $ia^m \vdash^* j$, where $j = (m+i) \textbf{ mod } 3$.

Proof of Claim: (by induction on m**)**

Basis: $m = 0$. Then $i \textbf{ mod } 3 = i$, so $i \vdash^* i$, and this holds by definition.

Induction Hypothesis: Assume the claim holds for all m, $0 \leq m \leq k$, for some $k \geq 0$.

Induction Step: Let $m = k+1$ and consider a^m. Note that $m > 0$. Now $ia^m \vdash i'a^{m-1} \vdash^* j$, for some i' and j, by definition. But, by definition, $i' = (i+1) \textbf{ mod } 3$. By the induction hypothesis, we obtain $i'a^{m-1} \vdash^* j$, where $j = (m-1+i') \textbf{ mod } 3$. Substituting for the value of i' we then obtain $j = (m-1+(i+1) \textbf{ mod } 3) \textbf{ mod } 3$, which by the well-known properties of the modulo operation yields

$$j = (m+i) \textbf{ mod } 3$$

as desired.

Thus, the claim is established. ■

Note that we took all pairs of states (i,j) into account in this proof, not just the required pair $(0,j)$. As we shall see this is usual in inductive proofs of this type.

This example demonstrates for one specific *DFA* that membership (in its language) is decidable. Recall that a decision problem is decidable if there is a decision algorithm which solves it, that is, an algorithm which terminates for every input producing the answer **false** or **true**. The corresponding generic decision problem can be stated as

DFA MEMBERSHIP
INSTANCE: A *DFA*, $M = (Q,\Sigma,\delta,s,F)$, and a word x in Σ^*.
QUESTION: Is x in $L(M)$?

and it is not difficult to see that it is also decidable. This is simply because there is an algorithm to compute the terminating state of M for input x; see Section 10.1. However, for Turing machines (see Chapter 6) the corresponding problem is undecidable, as we shall see in Section 10.1.

If the transition function of a *DFA* is total, then every input word will be read completely before the *DFA* stops. If the transition function is nontotal this is not the case. This classification of transition functions leads to a corresponding classification of *DFAs*.

Definition If the transition function of a *DFA* is total we say that the *DFA* is *complete* and otherwise we say it is *incomplete*.

Example 2.2.2 Let $M = (Q,\Sigma,\delta,0,F)$ have states 0, 1, and 2, input symbols a, b, and c, and its state diagram be as shown in Figure 2.2.5. Then M is incomplete, since $\delta(1,b) =$ ***undef*** as is $\delta(2,a)$. Thus, for example $0aabc \vdash^* 1b$ and no further move is possible.

We shall prove later that every incomplete *DFA* can be transformed into a complete one while preserving its language.

We now return to the coin checker example of Section 2.1.

Example 2.2.3 The legal coins arrive at the change maker as a sequence, that is, a word, hence we can view its state changes as displayed in Figure 2.2.6. The states correspond to the total amount inserted so far. A final state i means make change of $i-30$ cents and start the coffeemaker. Observe that we really recognize

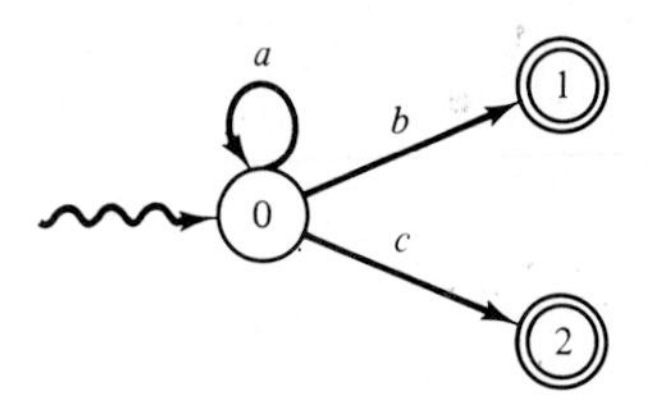

Figure 2.2.5

sequences rather than sets of coins, so both *dnq* and *ndq* appear, but not, of course, *qdn* or *qnd*, since the final coins are supernumerary here.

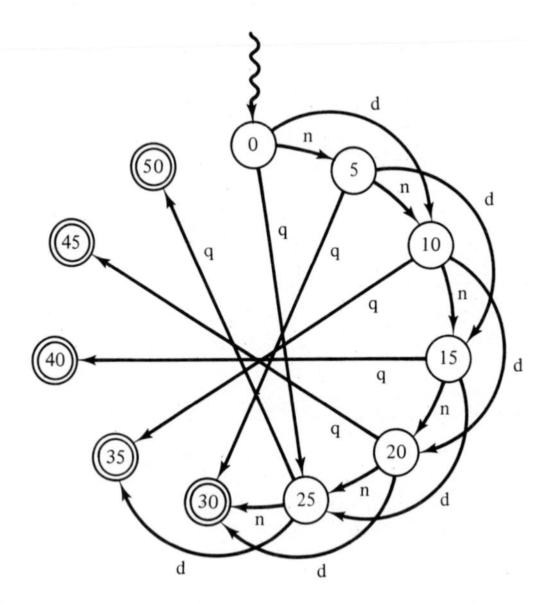

Figure 2.2.6 The state diagram of the coin checker.

Before considering renaming we look at one further example, which raises a number of questions.

Example 2.2.4 Let $L_i = \{a^i b^i\}$, $i \geq 1$. Construct *DFAs*, M_i, accepting L_i, for each $i \geq 1$.

M_1 is simply the state diagram of Figure 2.2.7, while M_2 is a slight modification of it as shown in Figure 2.2.8. M_3 follows a similar pattern as can be seen from Figure 2.2.9. What are M_4 and M_{10}? Are there any smaller *DFAs* accepting these languages? Letting $K_i = \cup_{j=1}^{i} L_j$, $i \geq 1$, construct a *DFA*, N_i, accepting K_i. Is $K = \cup_{j=1}^{\infty} L_i$ accepted by some *DFA*? Attempt to answer these questions before reading further.

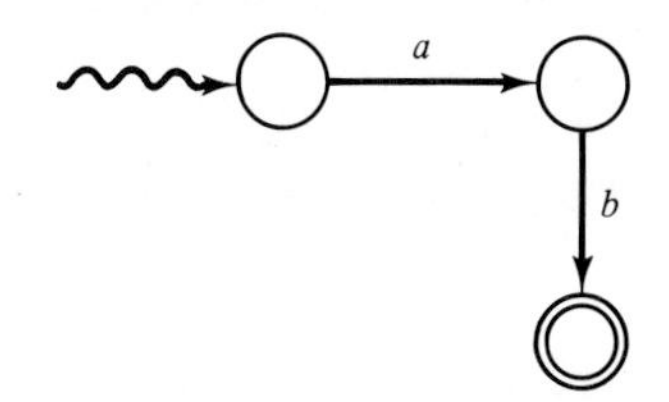

Figure 2.2.7

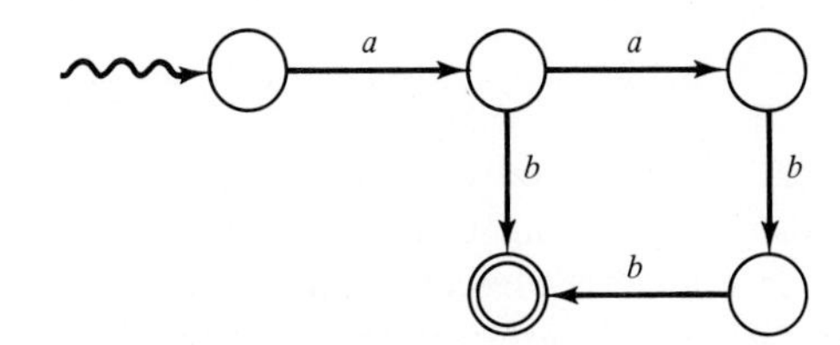

Figure 2.2.8

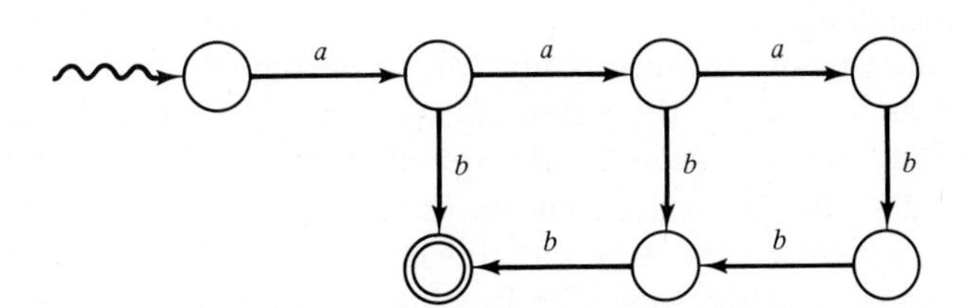

Figure 2.2.9

We prove directly that K is not a *DFA* language. The technique we use foreshadows a generally applicable technique that is established in Section 8.1.

Claim: $K = \{a^i b^i : i \geq 1\}$ *is not a DFA language.*

Proof of Claim: Assume K is a *DFA* language and obtain a contradiction. Under this assumption there is a *DFA*, $M = (Q, \{a, b\}, \delta, s, F)$, with $L(M) = K$. Letting $n = \#Q$ consider the accepting configuration sequence for $a^n b^n$, that is,

$$s_0 a^n b^n \vdash s_1 a^{n-1} b^n \vdash \cdots \vdash s_{2n-1} b \vdash s_{2n}$$

where $s_0 = s$ and s_{2n} is in F.

This sequence contains $2n+1$ states. Because $n+1 > n$, at least one state must appear more than once in the first n moves. This is an application of the pigeonhole principle. Let q be such a state, where $s_i = s_j = q$, for some i and j, $0 \leq i < j \leq n$.

This implies that

$$sa^n b^n \vdash^* qa^{n-i} b^n \vdash^+ qa^{n-j} b^n \vdash^+ s_{2n}$$

which in turn implies that

$$sa^{n-j+i} b^n \vdash^+ qa^{n-j} b^n \vdash^+ s_{2n}$$

if we omit the repeated appearance of q. In other words, $a^{n-j+i} b^n$ is in K. Because $j > i$ we have $n-j+i < n$ and we have shown that there is a word in K of the wrong form. This is the contradiction we sought, so $K \neq L(M)$ for any *DFA*, M. ■

In Example 2.2.4 we have proved that there is a language that is not a *DFA* language. We could also prove using counting that there are languages that are not *DFA* languages. For a given alphabet Σ, the family of all languages over Σ is 2^{Σ^*}. It can be shown (see the Exercises) that the family of all *DFA* languages over Σ is enumerable. This implies immediately that there exist languages that are not *DFA* languages.

A Remark on Renaming

Although we will typically use the symbol M to denote or stand for an automaton or machine and denote M by the quintuple (Q,Σ,δ,s,F), it is important to realize that these choices are not sacred. So we may speak of the *DFA*, X, where $X = (A,B,f,r,S)$. By the convention we have established on the order of items in the quintuple we know that A is the alphabet of state symbols, B is the alphabet of input symbols, f is the transition function, r is the initial state, and S is the set of final states. Moreover, the choice of a name for a state symbol is not restricted to be only a nonnegative integer. In particular, consider the *DFA*, $X = (e,B,f,r,S)$, where $A = \{r,s,t\}$, $B = \{a\}$, $S = \{s\}$, and f is defined as $f(r,a) = t$, $f(t,a) = s$, and $f(s,a) = r$. Then we have the state diagram shown in Figure 2.2.10.

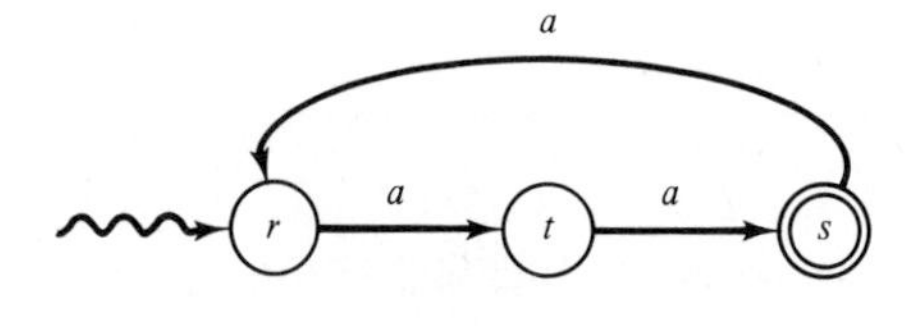

Figure 2.2.10

Since $L(X)$ only consists of words over $\{a\}$, it should be clear that $L(X) = L(M)$, for the *DFA*, M, of Example 2.2.1. However, $X \neq M$, since different state symbols, among other things, have been used. But X and M are closely related in that X is obtained from M by *renaming* the symbols of M appropriately. Conversely, M can be obtained from X by renaming the symbols of X appropriately. In both cases there is one and only one way to rename the symbols

0 becomes r;	Q becomes A;
1 becomes t;	Σ becomes B;
2 becomes s;	δ becomes f;
	F becomes S

to produce X from M. Its inverse gives M from X. A renaming is one example of an *isomorphism*.

Observe that if the symbol a in M is also renamed, say as b in X, that is, $B = \{b\}$, then $L(X) \neq L(M)$. For this reason whenever we produce a renaming of a *DFA* we always intend that the input symbols should not be renamed (or renamed identically), unless we *explicitly* include them. ■

The remark above introduced two different automata having the same language. This introduces the concept of equivalence, which we now define.

Definition Let M_1 and M_2 be two *DFAs*. If $L(M_1) = L(M_2)$ we say M_1 and M_2 are *equivalent*. Otherwise they are *inequivalent*.

If M_1 is a renaming of M_2, and vice versa, then they are equivalent. However, they may look quite different yet be equivalent.

Example 2.2.5 Let M_1 be as shown in Figure 2.2.11. Then M_1 and M of Figure 2.2.1 are equivalent, but are not renamings of each other.

At this stage a natural decision problem arises

DFA EQUIVALENCE
INSTANCE: Two *DFAs*, $M_1 = (Q_1, \Sigma_1, \delta_1, s_1, F_1)$ and $M_2 = (Q_2, \Sigma_2, \delta_2, s_2, F_2)$.
QUESTION: Are M_1 and M_2 equivalent?

We demonstrate in Chapter 10 that this problem is decidable.

Example 2.2.1 illustrates the ability of a *DFA* to perform *modulo counting*, since $L(M)$ consists of all words a^j for which $j \equiv 2 \textbf{ mod } 3$, that is, $j = 3i+2$, $i \geq 0$. As we shall see in Chapter 8 there are limits to the counting ability of *DFAs*. For example, there is no *DFA*, M, with

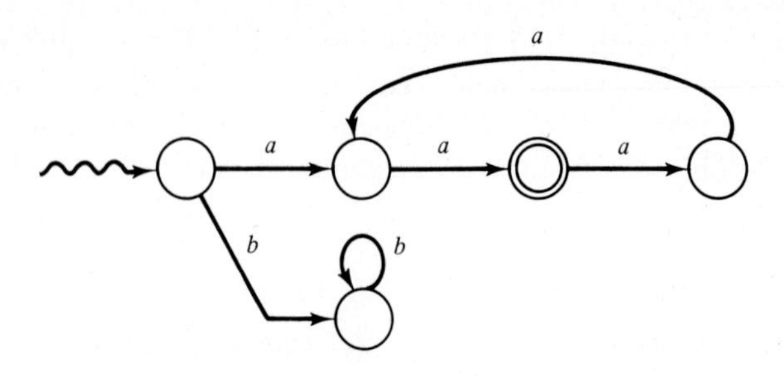

Figure 2.2.11

$$L(M) = \{a^p : \text{for all prime integers } p\}.$$

In our next example we consider a language in which certain combinations of letters are forbidden.

Example 2.2.6A Let $\Sigma = \{a,b\}$ and let $L \subseteq \Sigma^*$ be the set of all words which do not contain two consecutive a's. The following words are in L

$$\lambda, b, bb, \ldots, b^i, \ldots$$

Since they contain no a's at all the condition is satisfied vacuously. The word a is in L, but the words

$$aa, aaa, \ldots, a^i, \ldots$$

are certainly *not* in L. Finally, the following words are in L

$$ababab, abb, abba, abbbababbabb$$

A *DFA* to accept L must be able to detect the absence of two consecutive a's. Consider the state diagram of an incomplete *DFA*, $M = (\{0,1\},\Sigma,\delta,0,\{0,1\})$, shown in Figure 2.2.12.

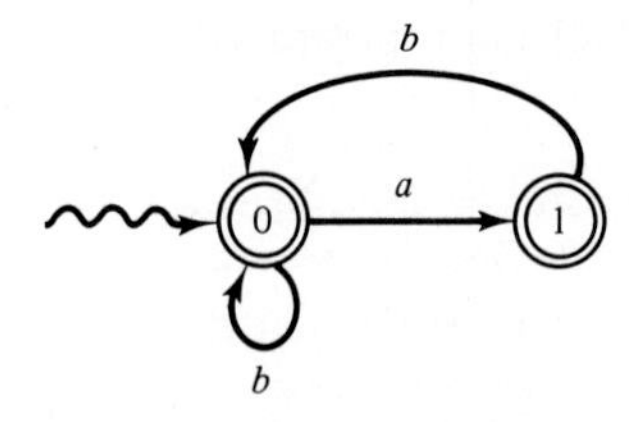

Figure 2.2.12

Now for every input word x which contains two consecutive a's M aborts, since the first a takes M to 1, and $\delta(1,a) = \textbf{\textit{undef}}$. Since M rejects all words which contain two consecutive a's we have $L(M) \subseteq L$. To prove $L = L(M)$ it suffices to prove $L \subseteq L(M)$. In other words, we need to prove that M excludes no words which do not contain two consecutive a's. We argue by induction on the length of input words. Here the only word of length 0, that is, λ, does not contain two consecutive a's, and it is accepted by M. Assume all words of length at most n, for some $n \geq 0$, which belong to L, are accepted by M. We show that all words in L of length $n+1$ are accepted by M.

Consider a word x in L, $|x| = n+1$. Clearly, x can be written as either ya or yb, where $|y| = n \geq 0$. We argue, by contradiction, that y is in L. For if y is not in L, then y contains two consecutive a's. This implies that x also contains two consecutive a's, which contradicts the starting assumption that x is in L. Hence, our second assumption that y is not in L must be false. Therefore, y is in L.

Returning to the induction step, we are now in a position to show that x is in $L(M)$. Since y is in L, then either $0y \vdash^* 0$ or $0y \vdash^* 1$. Consider them in turn.

Case 1: $0y \vdash^* 0$. Then either $y = \lambda$ or y ends with b. Immediately, $0ya \vdash^* 1$ and $0yb \vdash^* 0$ and x is in $L(M)$.

Case 2: $0y \vdash^* 1$. Then y ends with a. Now if $x = ya$, then it contains two consecutive a's, which contradicts the assumption that x is in L; therefore, $x = yb$ and $0yb \vdash^* 0$, and, hence, x is in $L(M)$.

We have shown that all words in L of length $n+1$ are in $L(M)$ given the induction hypothesis that all words in L of length at most n are in $L(M)$.

Thus, $L \subseteq L(M)$ by the principle of induction. ■

Hence, we have proved $L = L(M)$.

We have classified *DFAs* into ones with total transition functions and ones with nontotal transition functions, that is, complete and incomplete *DFAs*, respectively. We now demonstrate, first informally by example, that we can always complete an incomplete *DFA* by adding an additional state that corresponds to the value ***undef***. Such a state is called a *sink* or *completion* state.

Example 2.2.6B Add a new state 2 and appropriate transitions to M to give M'; see Figure 2.2.13. We have used the edge labeling "a,b" on the transition from 2 to itself to as an abbreviation for two transitions, one with a and one with b. Then M' is a complete *DFA* and since 2 is only reached for the first time when M' is in 1 and a second consecutive a is seen by M', M and M' are equivalent. From the programming viewpoint this corresponds, in a loose sense, to replacing in-line aborts or error terminations in a program by a single abort or error procedure.

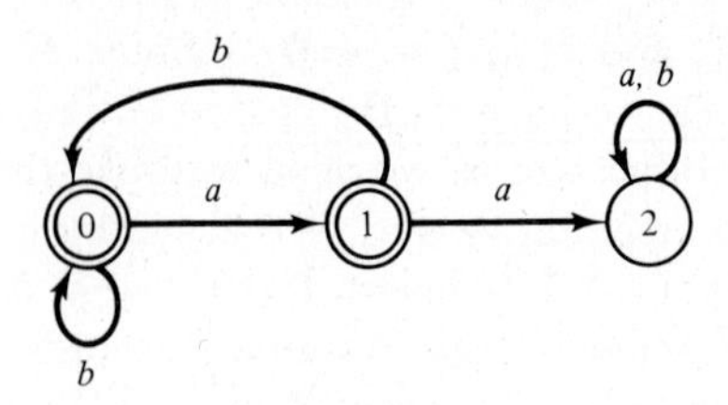

Figure 2.2.13

The above treatment of ***undef*** is generally applicable; thus, an equivalent complete *DFA* can always be obtained from an incomplete *DFA*.

Theorem 2.2.1 *Let $M = (Q, \Sigma, \delta, s, F)$ be an incomplete DFA, that is, there is a pair (q,a) in $Q \times \Sigma$ such that $\delta(q,a) =$* ***undef****. Then a DFA M' can be constructed such that M' is complete and $L(M') = L(M)$.*

Proof: Add a new state c to Q and extend δ to give δ', where, for all q in Q and for all a in Σ, $\delta'(q,a) = \delta(q,a)$ if $\delta(q,a) \neq$ ***undef***, and $\delta'(q,a) = c$ otherwise. To complete the transition function δ' add the transition $\delta'(c,a) = c$, for all a in Σ. $M' = (Q \cup \{c\}, \Sigma, \delta', s, F)$ is complete and $L(M') = L(M)$. ■

To terminate this section we consider a computer implementation of *DFAs*.

Example 2.2.7 *DFAs* lend themselves to computer implementation. The *DFA*, *M*, of Example 2.2.1 can be implemented in PASCAL as follows:

Let the input file be a file of type **char**, that is, a text file. Indeed let it be the standard file *input*. Then the transition function can be represented as a matrix: δ: **array**[0..2,'*a*'..'*a*']**of**−1..2, where 0, 1, and 2 represent the states 0, 1, and 2, respectively, and −1 represents ***undef***. Similarly, the set of final states can, indeed, be represented as a set of base type 0..2. So we obtain the following declarations

```
const startstate = 0;
type states = -1..2; symbols = 'a'..'a';
     transition = array[states,symbols] of states;
     stateset = set of states;

var state : states;
    symbol : symbols;
    delta : transition;
    finalstate : stateset;
```

In the body of the program we need to initialize *state, delta*, and *finalstate* as follows

```
state := startstate;
delta[0,'a'] := 1;
delta[1,'a'] := 2;
delta[2,'a'] := 0;
finalstate := [2];
```

Finally, to simulate M we have

```
while not eof (input) do                {input file exhausted?}
begin
    while not eoln (input) do           {skip end-of-line symbol}
    begin
        read(input, symbol);
        state coleq delta[state, symbol]
    end {while};
    readln(input)
end {while};
if state in finalstate then writeln(output, ' accepted ')
                       else writeln(output, ' rejected ' )
```

This simulation is a crude one; we have not echoed the input word; we have not displayed the configuration sequence; we have not shown any initial and final interrogation with the user (it is not user-friendly as written); we have not checked that the new state equals -1; and so on. However, it is a simulation of M via interpretation. We could also compile M, replacing

```
state := delta[state,symbol]
```

by

```
case state of
0 : state := 1;
1 : state := 2;
2 : state := 0;
end {case}
```

and removing the declarations of *transition* and *delta* and the initialization of the latter.

Rather than writing a new program for each new *DFA*, we could also write a *DFA* compiler or interpreter generator. We leave these as programming projects.

2.3 NONDETERMINISTIC FINITE AUTOMATA

In Section 1.1, when discussing the most basic properties of an algorithm and its associated computing agent, we specifically required the computing agent to be deterministic. Our reason for this was a simple one — it reflected reality: machines and programs *are* deterministic. The heading of this section abrogates this stance, thereby raising two natural questions. These are, first, why do we want to introduce nondeterministic machines and, second, how does such a machine operate. We approach the second question by responding to the first.

Nondeterministic automata are useful in proofs of various properties of finite automata. This is seen during our investigation of regular expressions in Chapter 3 and in other places in this text. From a more practical point of view we close this section by constructing a pattern-matching machine from a nondeterministic automaton. This turns out to be simpler than basing it directly on a *DFA*. Although these are excellent reasons for introducing nondeterminism, we consider one further reason. We look at the relationship between problems and decision problems as exemplified by one particular problem.

A well-known programming problem, often used to introduce the idea of backtracking, is

CHESS QUEENS
INSTANCE: An integer $n \geq 1$.
QUESTION: What is a solution to the n chess queens problem, that is, a placement of n chess queens on an $n \times n$ board such that no two queens attack each other?

If you have never written a program to solve this problem, then attempt to do so now, or alternatively, review a solution from some programming text. Figure 2.3.1 gives a solution for $n = 4$. Essentially programs to solve this problem carry out an exhaustive search of all possible placements. Letting the rows and columns of the board be indexed from 1 to n, then a placement can be represented as an n-tuple $(((1,c_1), \ldots, (n,c_n))$, where (i,c_i) denotes the placement of a queen on the square which is on row i and in column c_i. Usually this is represented as $(c_1, \ldots, c_n)$, since the ith position implicitly denotes the ith row. Furthermore since two queens in the same column attack each other we only need consider n-tuples $(c_1, \ldots, c_n)$ for which $\{c_1, \ldots, c_n\} = \{1, \ldots, n\}$, that is, $(c_1, \ldots, c_n)$ is a permutation of $(1, \ldots, n)$. Programs solving CHESS QUEENS enumerate, essentially, the permutations of $(1, \ldots, n)$, checking each permutation to determine if it is a solution. In other words, reducing CHESS QUEENS to the decision problem

CHESS QUEENS CHECKING
INSTANCE: An integer $n \geq 1$ and a permutation $(c_1, \ldots, c_n)$ of $(1, \ldots, n)$.
QUESTION: Is the given permutation a solution to the n chess queens problem?

Recall from Section 1.1 that if D is a decision algorithm for CHESS QUEENS CHECKING, then f_D is its associated total Boolean function. In other words, $f_D(n,(c_1, \ldots, c_n))$ is either **true** or **false**. The permutations of $(1, \ldots, n)$ can be enumerated, that is, there is a bijection $f_i : \{i : 1 \leq i \leq n!\} \rightarrow \{(c_1, \ldots, c_n)$ is a permutation of $(1, \ldots, n)\}$. Thus, we obtain, using the construction of Section 1.1, an algorithm for CHESS QUEENS from the decision algorithm D for CHESS QUEENS CHECKING as follows

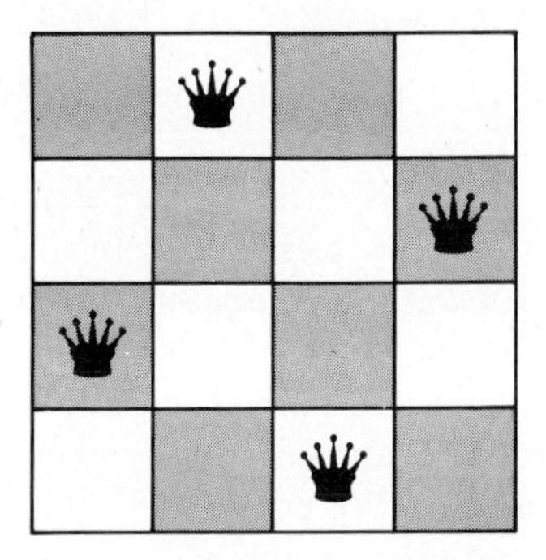

Figure 2.3.1 A nonattacking placement.

for $i := 1$ **to** $n!$ **do**
  if $f_D(n, f_n(i))$ **then return** i

The technique of backtracking speeds up the search immensely (see the Exercises), but that is not our purpose. Rather, we view this "generate a candidate, check the candidate" solution in a different way. Assume we have a new statement, the *choice statement*, which we may use in the following way

choose i **from** $1..n!$;
  if $f_D(n, f_n(i))$ **then return** i
end {choose}

The intent here is that **choose** introduces nondeterminism into deterministic programs. One meaning for it is that all possible choices are made; in other words, do all options. This corresponds to the multiple-universe view common in science fiction stories or to the many-worlds interpretation in quantum mechanics, whenever we make a choice the universe splits, so we exist concurrently in many universes corresponding to the different choices that have been made. The choice statement reflects the nonsequential nature of the set of permutations of $1..n$. Each execution of the for loop, in the first program segment above, is independent of the others; therefore, an enumeration of the $n!$ permutations is unnatural *in this context*. When the choice statement has a finite number of possible choices it can then be implemented directly on a nondeterministic machine simply by allowing the machine to reflect, by its transitions, the possible choices. The machine operates by considering all possible transitions at each step rather than the, at most, one transition possible in a deterministic machine.

Another meaning of the choice statement is that the best choice can be taken, that is, an "I don't care" selection of the best choice. Both interpretations have their place, but the second meaning is basic to the notion of acceptance in a nondeterministic machine, as we shall see.

Assume we are given a complete *DFA*, $M = (Q, \Sigma, \delta, s, F)$, and a word $x = a_1 \cdots a_n$ in Σ^*, for some $n \geq 0$ and some a_i in Σ, $1 \leq i \leq n$. Then x determines a sequence of states

$$s = p_0, p_1, \ldots, p_n$$

where $sa_1 \cdots a_i \vdash^* p_i$, for all i, $0 \le i \le n$. Because M is *deterministic* this sequence is unique. In each state p_i, $0 \le i < n$, one input symbol is read, and since δ is, in this case, a *function* there is at most one possible next state.

Although we are more comfortable with deterministic systems, for example, washing machines, telephones, calculators, computers, and computer programs, we live out our lives in a nondeterministic world, at least from our own perspective.[†] In other words, we choose one of several alternatives, usually attempting to choose the best. In simulation the use of pseudo-random number generators can be viewed as an attempt to handle nondeterminism. We consider, in this section, the addition of choice to *DFAs* to give nondeterministic machines. Since there are only a finite number of states there can only be a finite number of choices when given the current state and current input symbol, so such machines are still *finitely* specified. In a *DFA*, $M = (Q, \Sigma, \delta, s, F)$, the transition function δ can also be viewed as a relation over $Q \times \Sigma \times Q$. This alternative view of δ enables choice to be added easily.

Definition A *nondeterministic finite automaton* (*NFA*), M, is specified by a quintuple $M = (Q, \Sigma, \delta, s, F)$, where

- Q is an alphabet of *state symbols*;
- Σ is an alphabet of *input symbols*;
- $\delta \subseteq Q \times \Sigma \times Q$ is a *transition relation*;
- s in Q is a *start state*; and
- $F \subseteq Q$ is a set of *final states*.

For a triple (p, a, q) in δ, p corresponds to the current state, a to the current input symbol, and q to a next state. One frequently used definition of *NFAs* allows arbitrary words in transitions. However, for simplicity of presentation we delay the introduction of this version to Section 2.6. Clearly an *NFA*, $(Q, \Sigma, \delta, s, F)$, is a *DFA* if for all p in Q and all a in Σ, there is at most one triple (p, a, q) in δ, for some q. In this case we can treat δ as a function as before. When we are unconcerned about a *finite automaton* being deterministic or nondeterministic we refer to it simply as an *FA*.

Remark: It is sometimes convenient to consider the transition relation δ as a function, which, in general, gives state sets, that is $\delta : Q \times \Sigma \to 2^Q$ is defined by

For all p in Q and for all a in Σ: $\delta(p, a) = \{q : (p, a, q) \text{ is in } \delta\}$

[†] Some would argue that the universe is a deterministic system, however complex, and hence every action or "choice" is predetermined or predestined. However, from our own point of view this is certainly not the case, we face choices and appear to have freedom of choice

Example 2.3.1A Let M_1 be as shown in Figure 2.3.2 where the input alphabet is $\{a,b,c\}$. Then M_1 is nondeterministic, since $(0,a,1)$ and $(0,a,3)$ are in δ.

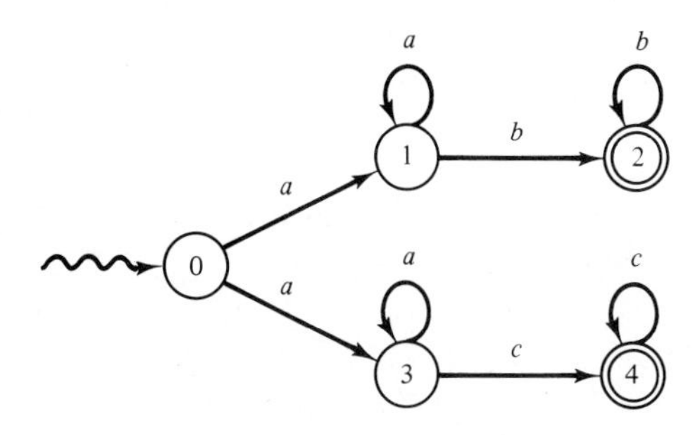

Figure 2.3.2

As in the case of *DFAs* we introduce the machine-oriented view of *NFAs*.

Definition Let $M = (Q,\Sigma,\delta,s,F)$ be an *NFA*. A *configuration* of M is a word in $Q\Sigma^*$, as for *DFAs*. Given two configurations px and qy of M, we write $px \vdash qy$ in M if $x = ay$ for some a in Σ and (p,a,q) is in δ. Similarly, we write $px \vdash^k qy$ in M if either $k = 1$, when $\vdash^1$ is identical to $\vdash$ or $x = az$, for some a in Σ and some z in Σ^+, and there is an r such that

$$px \vdash rz \text{ and } rz \vdash^{k-1} qy.$$

We write $px \vdash^+ qy$ if $px \vdash^k qy$ for some $k \geq 1$, and, finally, $px \vdash^* qy$ if either $px = qy$ or $px \vdash^+ qy$.

The notions $\vdash^+$ and $\vdash^*$ correspond to following one and only one alternative at each transition in an *NFA*. If we are given an *NFA*, $M = (Q,\Sigma,\delta,s,F)$, which happens to be a *DFA*, that is, $\#\,\delta(q,a) \leq 1$, for all q in Q and for all a in Σ, then acceptance of a word is well defined, namely,

x in Σ^* is accepted by M if and only if $sx \vdash^* q$, for some q in F.

Indeed, in Example 2.3.1 the same condition holds even though M_1 is *not* a *DFA* in disguise. Unfortunately, if we modify this example slightly we see that this is not always the case.

Example 2.3.2A Let M_2 be as displayed in Figure 2.3.3, where $\{a,b,c\}$ is the input alphabet. Note that for $A = a,b$, or c, $\delta(i,A) \neq \varnothing$, for all i, $0 \leq i \leq 5$. We have added a new state 5, the sink state, to M_1 of Example 2.3.1 to give M_2.

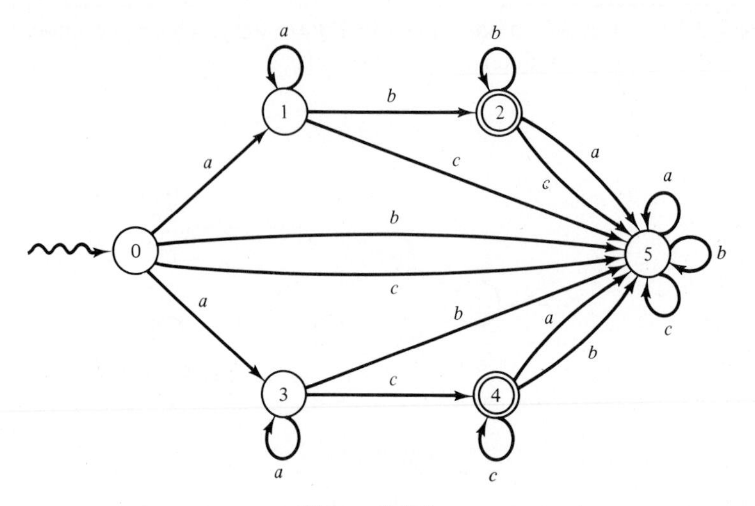

Figure 2.3.3

Now $0a \vdash 1$ and $0a \vdash 3$ as before, but $0ac \vdash^{+} 4$ and $0ac \vdash^{+} 5$, and 5 is not a final state. Indeed, no words are accepted by M_2 if we require that the set of obtained states must be a subset of F. This is simply because a word which reaches 2 is of the form $a^i b^j$, $i \geq 1$ and $j \geq 1$, and such a word also reaches 5, which is not a final state. Similarly, the only words that reach 4 are of the form $a^i c^j$, $i,j \geq 1$, and once more they also reach 5. But M_2 is only a completed version of M_1, and in M_1 words of these forms are accepted.

However, when introducing the choice statement the intent was that each chosen alternative should proceed independently and as the solution to CHESS QUEENS demonstrated, success of the whole corresponds to the success of a choice. Hence, in *NFA* we require that at least one sequence of choices leads to a final state for acceptance to occur.

Definition Given $M = (Q, \Sigma, \delta, s, F)$, an *NFA*, we say a word x in Σ^* is *accepted* if there is a configuration sequence $sx \vdash^* q$, for some q in F, and is *rejected* otherwise. The *language of M*, denoted by $L(M)$ as before, is defined by

$$L(M) = \{x : x \text{ is in } \Sigma^* \text{ and } sx \vdash^* q, \text{ for some } q \text{ in } F\}.$$

A language $L \subseteq \Sigma^*$ is said to be an *NFA language* if there exists an *NFA*, M, with $L = L(M)$ and we say two *NFAs*, M_1 and M_2, are *equivalent* if $L(M_1) = L(M_2)$ and *inequivalent* otherwise. Finally, the *family of NFA languages* is denoted by $\boldsymbol{L}_{NFA}$.

Example 2.3.2B It is straightforward to prove (see the Exercises) that

$0a^ib^j \vdash^* 2$, for $i,j \geq 1$, $0a^ic^j \vdash^* 4$, for $i,j \geq 1$, and that these are the only acceptable words. Thus, M_1 and M_2 accept the same set of words, that is,

$$L(M_1) = L(M_2) = \{a^ib^j : i,j \geq 1\} \cup \{a^ic^j : i,j \geq 1\}.$$

NFAs seem to be of no practical interest, since they do not correspond naturally to deterministic algorithms. However, as we shall see this is not the case. The reasons for this are that *NFAs* are simpler to define in many cases and, moreover, we can always construct a *DFA* from a given *NFA* which accepts exactly the same language. This latter result is a key theorem which we prove after one further example.

Example 2.3.3 Pattern Matching

A typical task in a text editor is searching for a given subword or pattern. Since we return to this topic below, for now we use it to show how an *NFA* can be defined simply, while an equivalent *DFA* is more difficult to define.

Assume texts are words over $\{a,b\}$ as are patterns. Now a pattern p occurs in a text t if and only if $t = \{a,b\}^*p\{a,b\}^*$. The *NFA*, M, of Figure 2.3.4 accepts all such texts for $p = ababa$. Observe that if an input text t contains p, then there is indeed a sequence of choices causing t to be accepted. Essentially, M chooses to loop around the start state until p appears, when it chooses the transition to q. This choice inexorably leads to the final state and, therefore, acceptance. If p does not appear in t no such choice can ever be made; therefore, t is rejected. In other words, $L(M) = \{a,b\}^*p\{a,b\}^*$.

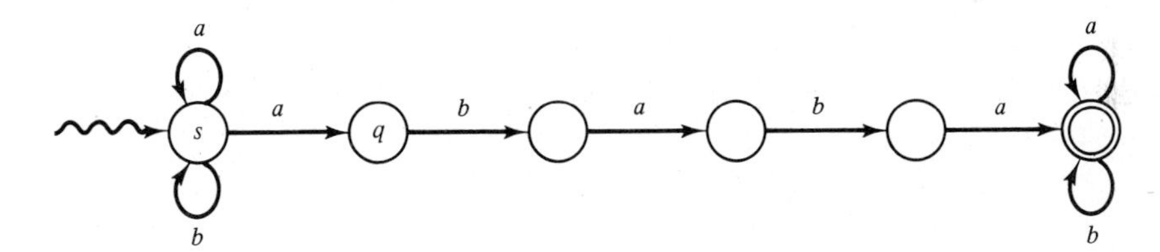

Figure 2.3.4

Try to define a *DFA* for $L(M)$ before looking at the continuation of Example 2.3.2 below.

Whatever choices are made by an *NFA* in a configuration sequence we know that at each move it reads one symbol, no more and no less. After k moves on an input word x the *NFA* will have read the first k symbols of x and, corresponding to the various possible choices it can be in one of a finite set of states. This leads us to the notion of a *super-configuration* and a *super-configuration sequence*, in which we keep track of all possible states that can be reached at each move.

Example 2.3.2C In M_2 we have

$$0aab \vdash 1ab \vdash 1b \vdash 2$$

$$0aab \vdash 3ab \vdash 3b \vdash 5$$

Combining these gives

$$\{0\}aab \vdash \{1,3\}ab \vdash \{1,3\}ab \vdash \{2,5\}$$

a super-configuration sequence.

More formally we have

Definition Let $M = (Q,\Sigma,\delta,s,F)$ be an NFA. A *super-configuration* of M has the form Kx where $K \subseteq Q$ and x is in Σ^*. It means that K is the current set of states and x is the unread portion of the input word. The starting super-configuration for a word x in Σ^* is $\{s\}x$.

We extend $\vdash$ to super-configurations as follows. We say $Kx \vdash Ny$, for $K,N \subseteq Q$ and x,y in Σ^*, if $x = ay$, for some a in Σ, y in Σ^*, and $N = \{q : (p,a,q)$ is in δ, for some p in $K\}$. We obtain $\vdash^+$ and $\vdash^*$ as before and if $Kx \vdash^* Ny$ we call it a *super-configuration sequence.*

Note that $K \vdash^* K$, for all $K \subseteq Q$ and $\varnothing x \vdash^* \varnothing$, for all x in Σ^*.

Let x in Σ^* be an arbitrary input word to an NFA $M = (Q,\Sigma,\delta,s,F)$. We show that $Kx \vdash^* N$ and $Kx \vdash^* P$ implies $N = P$. In other words, there is a unique super-configuration sequence, beginning with K, for each input word that is read completely.

Lemma 2.3.1 *Let $M = (Q,\Sigma,\delta,s,F)$ be an NFA. Then for all words x in Σ^* and for all $K \subseteq Q$*

$$Kx \vdash^* N \text{ and } Kx \vdash^* P \text{ implies } P = N.$$

Hence, M is deterministic with respect to super-configuration sequences.

Proof: (by induction on the length of x)

Basis: $|x| = 0$, that is, $x = \lambda$. Immediately, $K \vdash^* N$ implies $K = N$, and $K \vdash^* P$ implies $K = P$. Therefore, $P = N$ in this case.

Induction Hypothesis: Assume the lemma holds for all words x in Σ^* with $|x| \leq k$, for some $k \geq 0$.

Induction Step: Consider an arbitrary word x in Σ^*, with $|x| = k+1$, and an arbitrary $K \subseteq Q$. Assume $Kx \vdash^* N$ and $Kx \vdash^* P$, for some N, $P \subseteq Q$. Since $k+1 \geq 1$, x contains at least one symbol; hence, let $x = ay$, where a is in Σ and y is in Σ^*.

Now by construction $Ka \vdash T$, for some $T \subseteq Q$; therefore, $Kay \vdash Ty \vdash^* N$ and $Kay \vdash Ty \vdash^* P$.

But $|y| = k$ so the induction hypothesis applies, therefore $Ty \vdash^* N$ and $Ty \vdash^* P$ implies $P = N$.

This completes the induction step and, by the principle of induction, the lemma also. ■

We now relate super-configuration sequences to configuration sequences, and conversely.

Lemma 2.3.2 *Let $M = (Q,\Sigma,\delta,s,F)$ be an NFA. Then for all words x inΣ^*, and for all q in Q, $qx \vdash^* p$ if and only if $\{q\}x \vdash^* P$, for some P with p in P.*

Proof: (by induction on the length of x)

Basis: $|x| = 0$, that is, $x = \lambda$. $q \vdash^* p$ iff $q = p$, and $\{q\} \vdash^* P$ if and only if $P = \{q\}$, so q is in P as desired.

Induction Hypothesis: Assume the lemma holds for all x in Σ^* with $|x| \leq k$, for some $k \geq 0$.

Induction Step: Consider an arbitrary word x in Σ^* with $|x| = k+1$. Then $|x| \geq 1$, so x can be written as ya, with a in Σ and y in Σ^*. If $qx \vdash^* p$, then there is an r in Q, such that $qya \vdash^* ra \vdash p$. By the induction hypothesis $\{q\}y \vdash^* R$ with r in R. By the definition of super-configuration sequences $ra \vdash p$ implies for all R containing r $Ra \vdash P$, for some P containing p. But this means that we have the super-configuration sequence

$$\{q\}ya \vdash^* Ra \vdash P$$

for some P containing p, as desired.

On the other hand, if $\{q\}x \vdash^* P$ for some P containing p, then $\{q\}ya \vdash^* Ra \vdash P$, for some $R \subseteq Q$, $R \neq \varnothing$, by the definition of super-configuration sequences. But this implies, by the induction hypothesis, that $qy \vdash^* r$, for each r in R. Moreover, by definition, $Ra \vdash P$ implies there is some r in R such that $ra \vdash p$. Thus, putting these implications together we have $qya \vdash^* ra \vdash p$, as desired. This completes the induction step and the lemma. ■

We split x differently in the proofs of Lemmas 2.3.1 and 2.3.2. Try to prove each lemma by splitting x in the opposite way to see why we made the choice we did.

Since a word x is accepted by an NFA, $M = (Q,\Sigma,\delta,s,F)$, if $sx \vdash^* p$, for some p in F, Lemma 2.3.2 implies, in this case, that $\{s\}x \vdash^* P$, for some P containing p, that is, $P \cap F \neq \varnothing$. Conversely, if $\{s\}x \vdash^* P$, for some P containing a final state p, then Lemma 2.3.2 tells us $sx \vdash^* p$, that is, x is accepted. This motivates the following

Definition Let $M = (Q,\Sigma,\delta,s,F)$ be an *NFA* and x a word in Σ^*. We say x is *accepted* by M if $\{s\}x \vdash^* P$, for some P with $P \cap F \neq \varnothing$, otherwise x is *rejected.*

By the previous remarks this is equivalent to the original definition of acceptance and rejection for *NFAs*. Since we can view an *NFA* as a deterministic machine if we use super-configuration sequences, it should come as no surprise that this is the basis for constructing an equivalent *DFA*.

Algorithm *NFA to DFA — The Subset Construction.*
On entry: An *NFA*, $M = (Q,\Sigma,\delta,s,F)$.
On exit: A *DFA*, $M' = (Q',\Sigma,\delta',s',F')$ satisfying $L(M) = L(M')$.

begin Let $Q' = 2^Q$, $s' = \{s\}$, and $F' = \{K : K \subseteq Q \text{ and } K \cap F \neq \varnothing\}$. We define $\delta' : Q' \times \Sigma \to Q'$ by

For all $K \subseteq Q$ and for all a in Σ,
$\delta'(K,a) = N$ if $Ka \vdash N$ in M

end of Algorithm.

Observe that the *DFA*, M', has as many states as Q has subsets, and that each state in M' corresponds to a unique subset of Q, and vice versa. Consider our example *NFA*, M_1

Example 2.3.1B M_1 has 5 states $\{0,1,2,3,4\}$. The 32 subsets of Q, numbered $0-31$, for convenience, are

$s' = 0 = \{0\}$;	$1 = \{1\}$;	$2 = \{2\}$;
$3 = \{3\}$;	$4 = \{4\}$;	$5 = \{0,1\}$;
$6 = \{0,2\}$;	$7 = \{0,3\}$;	$8 = \{0,4\}$;
$9 = \{1,2\}$;	$10 = \{1,3\}$;	$11 = \{1,4\}$;
$12 = \{2,3\}$;	$13 = \{2,4\}$;	$14 = \{3,4\}$;
$15 = \{0,1,2\}$;	$16 = \{0,1,3\}$;	$17 = \{0,1,4\}$;
$18 = \{0,2,3\}$;	$19 = \{0,2,4\}$;	$20 = \{0,3,4\}$;
$21 = \{1,2,3\}$;	$22 = \{1,2,4\}$;	$23 = \{1,3,4\}$;
$24 = \{2,3,4\}$;	$25 = \{0,1,2,3\}$;	$26 = \{0,1,2,4\}$;
$27 = \{0,1,3,4\}$;	$28 = \{0,2,3,4\}$;	$29 = \{1,2,3,4\}$;
$30 = \{0,1,2,3,4\}$;	$31 = \varnothing$.	

These are the states of M_1'. There are 24 final states, namely, those which contain 2, 4, or both. δ' is defined by the transition table given in Table 2.3.1.

Example 2.3.1 illustrates one crucial problem with the subset construction, namely, the exponential increase in the number of states. This situation can often be ameliorated by only including states that can be *reached* from s'. We say that a state q is *reachable* from the start state s in a *DFA*, $M = (Q,\Sigma,\delta,s,F)$, if there

Table 2.3.1

	a	b	c		a	b	c
0	10	31	31	16	10	2	4
1	1	2	31	17	10	2	4
2	31	2	31	18	10	2	4
3	3	31	4	19	10	2	4
4	31	31	4	20	10	31	4
5	10	2	31	21	10	2	4
6	10	2	31	22	1	2	4
7	10	31	4	23	10	2	4
8	10	31	4	24	3	2	4
9	1	2	31	25	10	2	4
10	10	2	4	26	10	2	4
11	1	2	4	27	10	2	4
12	3	2	4	28	10	2	4
13	31	2	4	29	10	2	4
14	3	31	4	30	10	2	4
15	10	2	31	31	31	31	31

is a word x in Σ^* such that $sx \vdash^* q$ (see Section 0.2). An unreachable state is an irrelevant state; we can remove it without affecting the language accepted by the *DFA*; see Section 2.4.

Algorithm *NFA to DFA 2 — The Iterative Subset Construction.*
On entry: An *NFA*, $M = (Q, \Sigma, \delta, s, F)$.
On exit: A *DFA*, $M' = (Q', \Sigma, \delta', s', F')$ satisfying $L(M) = L(M')$.

begin Let $Q_0 = \{s\}$ be the zeroth subset of Q and 0 be the corresponding state in Q', let i be 0, and *last* be 0;
  while $i \leq last$ **do**
  begin
    for each symbol of a in Σ **do**
      if $\delta(Q_i, a) \neq Q_j$, for some j, $0 \leq j \leq last$ **then**
      begin $last := last + 1$;
        $\delta'(i, a) := last$; $Q_{last} := \delta(Q_i, a)$
      end;

$i := i+1$
end {while};
$Q' := \{i : 0 \leq i \leq last\}$;
$F' := \{i : Q_i \cap F \neq \varnothing$ and $0 \leq i \leq last\}$
end of Algorithm.

We introduce a tabular technique of executing this algorithm manually. The same technique can be applied also to the previous algorithm.

Example 2.3.1C The major difficulty with both subset algorithms is a bookkeeping one, namely, keeping track of the subsets considered and correctly computing δ' with respect to them. We overcome this difficulty to a large extent by using a tabular representation of δ. This is a "transition table" in which the entries are sets of states. From Figure 2.3.2 we obtain Table 2.3.2 for M_1 and δ.

Table 2.3.2

δ	a	b	c
0	$\{1,3\}$	$\varnothing$	$\varnothing$
1	$\{1\}$	$\{2\}$	$\varnothing$
2	$\{\varnothing\}$	$\{2\}$	$\varnothing$
3	$\{3\}$	$\{\varnothing\}$	$\{4\}$
4	$\{\varnothing\}$	$\{\varnothing\}$	$\{4\}$

Beginning with $\{0\}$ we begin to compute the relevant portion of δ' using Table 2.3.2. The first row, for $\{0\}$, can be filled in immediately. This introduces two new states in M_1', namely, $\{1,3\}$ and $\varnothing$. We add these states in the left-hand column, indicating that they must be considered; see Table 2.3.3. This corresponds to the statement $Q_{last} := \delta(Q_i, a)$ in the algorithm.

Table 2.3.3

δ'	a	b	c
$\{0\}$	$\{1,3\}$	$\varnothing$	$\varnothing$
$\varnothing$			
$\{1,3\}$			

From the definition of δ', $\delta'(\varnothing,d) = \varnothing$, for all d in $\{a,b,c\}$; therefore, the row for $\varnothing$ is easily filled in. However, although computing the entries for $\{1,3\}$ is not a difficult task, it requires some care. By definition, for d in $\{a,b,c\}$, $\delta'(\{1,3\},d) = \{q : 1d \vdash q$ or $3d \vdash q$ in $M_1\}$. But this means, for each d in $\{a,b,c\}$, $\delta'(\{1,3\},d) = \delta(1,d) \cup \delta(3,d)$. In other words, the row for $\{1,3\}$ is the union of the rows for 1 and 3 in Table 2.3.2. This gives Table 2.3.4, since $\{2\}$ and $\{4\}$ are new.

Table 2.3.4

δ'	a	b	c
$\{0\}$	$\{1,3\}$	$\varnothing$	$\varnothing$
$\varnothing$	$\{\varnothing\}$	$\{\varnothing\}$	$\{\varnothing\}$
$\{1,3\}$	$\{1,3\}$	$\{2\}$	$\{4\}$
$\{2\}$			
$\{4\}$			

We copy the entries for 2 and 4 from Table 2.3.2, which introduce no new sets. Therefore, the algorithm terminates with Table 2.3.5.

Table 2.3.5

δ'	a	b	c
$\{0\}$	$\{1,3\}$	$\varnothing$	$\varnothing$
$\varnothing$	$\{\varnothing\}$	$\{\varnothing\}$	$\{\varnothing\}$
$\{1,3\}$	$\{1,3\}$	$\{2\}$	$\{4\}$
$\{2\}$	$\varnothing$	$\{2\}$	$\varnothing$
$\{4\}$	$\varnothing$	$\varnothing$	$\{4\}$

We have introduced 5 states, rather than 32, and there are no other reachable states. The state diagram of the DFA, M_1', can now be drawn easily (see Figure 2.3.5), where we have once more used the abbreviation a, b, c on the transition from 2 to itself to denote three separate transitions. It is also easily seen that

$$L(M_1') = \{a^i b^j : i,j \geq 1\} \cup \{a^i c^j : i,j \geq 1\}.$$

that is, $L(M_1') = L(M_1)$ in this specific case, as claimed.

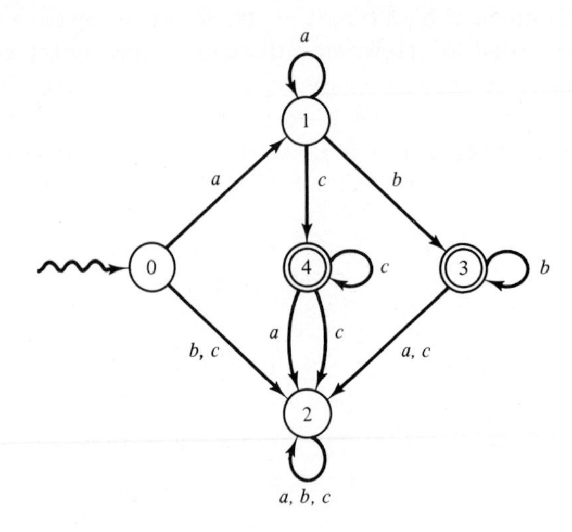

Figure 2.3.5

Although $L(M_1') = L(M_1)$ in Example 2.3.1 we need to prove that this equivalence holds in general. To this end we wish to prove:

Theorem 2.3.3 *Given an NFA,* $M = (Q, \Sigma, \delta, s, F)$, *then the DFA,* $M' = (Q', \Sigma, \delta', s', F')$, *obtained by either subset construction satisfies* $L(M') = L(M)$.

Proof: The theorem statement can be reformulated as

For all x in Σ^*:

$\{s\}x \vdash^* P$ in M, for some P with $P \cap F \neq \emptyset$ iff

$\{s\}x \vdash^* P$ in M', for some P in F'

Observing that $P \cap F \neq \emptyset$ if and only if P is in F' in either of the two constructions, we are left with proving that

For all x in Σ^*:

$\{s\}x \vdash^* P$ in M if and only if $\{s\}x \vdash^* P$ in M'

But this follows from the method of construction and Lemmas 2.3.1 and 2.3.2. ■

In summary we have proved

Theorem 2.3.4 $\boldsymbol{L}_{DFA} = \boldsymbol{L}_{NFA}$. *Thus, NFAs have the same expressive power as DFAs.*

Proof: By Theorem 2.3.1 we have proved that every language accepted by an

NFA is also accepted by some *DFA*, that is, $\mathbf{L}_{NFA} \subseteq \mathbf{L}_{DFA}$. The converse inclusion is obtained immediately from the definitions, since every *DFA* is an *NFA*, that is, $\mathbf{L}_{DFA} \subseteq \mathbf{L}_{NFA}$. ■

This theorem presents the solution to the MACHINE COMPARISON problem, discussed in Section 1.1, for *DFAs* and *NFAs*. These problems are discussed in more detail in Chapter 8.

We illustrate a possible application for *NFAs* with our pattern matching example.

Example 2.3.3B A typical recursive pattern search proceeds as follows

```
function Match(p,t,i);
    {Match is given the value 0 if p does not appear in t from position i onward,
    and otherwise the position of the first match of p in t from position i
    onward}
begin
    if |p|+i-1 > |t| then Match := 0 else
    if p matches t at position i, that is, p_1, . . . = t_i, t_{i+1}, . . .
                          then Match := i else
                               Match := Match(p,t,i+1)
end {of Match}.
```

At each call *Match* attempts to match p with t_i onward and if it fails it tries again with p at t_{i+1} onward. With the example pattern $p = abaa$, and text $t = aaababbaaa$ and $i = 3$ it only determines a match isn't possible at t_3 after comparing all four letters in p, where † indicates a match

```
    ↓
a a a b a b b a a a
    a b a a
    † † †
```

Match then begins again with $i = 4$

```
      ↓
a a a b a b b a a a
      a b a a
```

and with $i = 5$

```
        ↓
a a a b a b b a a a
        a b a a
        † †
```

but when $i = 3$, it has already been determined that $t_4 = b$ and $t_5 = a$ (and even $t_6 \neq a$, which in our two-letter case means $t_6 = b$!). Hence, it is already known, at this stage, that there is no reason to compare p with t_4 onward (it cannot match!). After the failure at t_3 *Match* can skip immediately to comparing t_6 with p_2 onward. This skipping is predetermined by *the pattern alone*, when there is a match of t_i, t_{i+1}, and t_{i+2} with p_1, p_2, and p_3, respectively, but $t_{i+3} \neq p_4$; then *Match* should continue by comparing t_{i+3} with p_2, t_{i+4} with p_3, etc., in an attempt to find a match at t_{i+2}. This yields a more efficient pattern matcher than the simple recursive one. The question that remains, however, is how do we construct it? We demonstrate how to produce a *DFA* which operates in this efficient manner.

Observe, as before, that a pattern p occurs in a text t if t is in $\Sigma^*\{p\}\Sigma^*$ and the pattern match terminates successfully when t_1p has been found, where $t = t_1pt_2$, for some t_1, t_2 in Σ^*, and p does not occur in any proper prefix of t_1p. The pattern match terminates unsuccessfully for all t in $\Sigma^* - \Sigma^*\{p\}\Sigma^*$.

Now we require that a pattern matching machine reading t a symbol at a time should terminate when p is found for the first time. For this purpose we relax our definition of acceptance in a finite automaton $M = (Q,\Sigma,\delta,s,F)$ to: x is accepted if $sx \vdash^* qy$, where q is in F and $x = wy$, for some w in Σ^*.

An *NFA*, M, for $p = abaa$, is displayed in Figure 2.3.6 and is a simple modification of the *NFA* in Figure 2.3.4. Note that $L(M) = \Sigma^*\{p\}$ as required, where our modified notion of acceptance may leave some of the input unread. Now we apply the *Iterative Subset Construction* to obtain the *DFA*, M', shown in Figure 2.3.7.

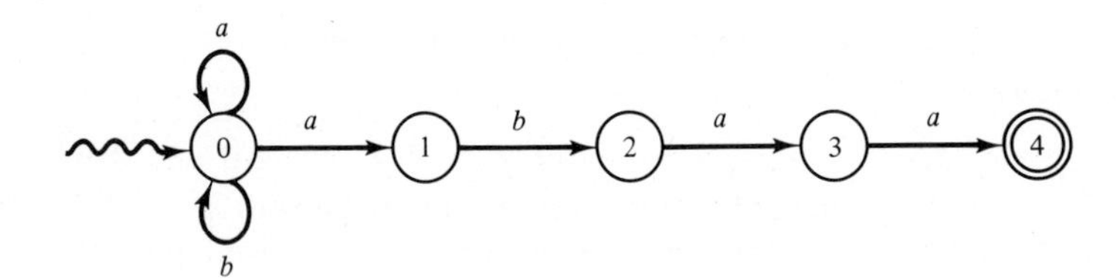

Figure 2.3.6

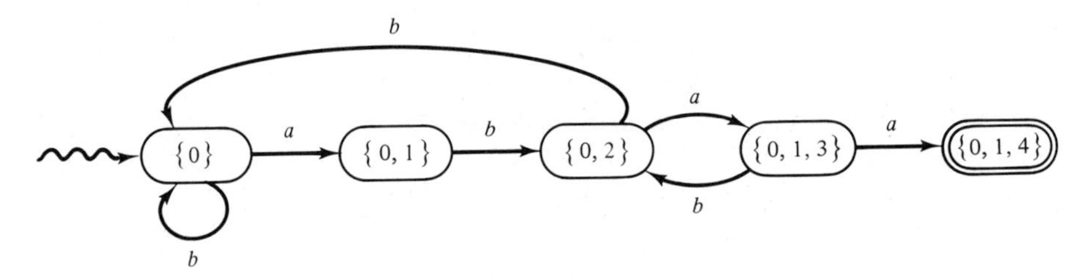

Figure 2.3.7

Consider the operation of M' on the particular text t given above

$$\{0\}aaababbaaa \vdash \{0,1\}aababbaaa \vdash \{0,1\}ababbaaa$$
$$\vdash \{0,1\}babbaaa \vdash \{0,2\}abbaaa \vdash \{0,1,3\}bbaaa$$
$$\vdash \{0,2\}baaa \vdash \{0\}aaa \vdash \{0,1\}aa \vdash \{0,1\}a \vdash \{0,1\}$$

In particular, the configuration $\{0,1,3\}bbaaa$ corresponds to $t_3 = a$, $t_4 = b$, $t_5 = a$. Also notice that the transition to $\{0,2\}baaa$ corresponds to recognizing that $t_5 = a$, $t_6 = b$; in other words, skipping from t_3 to t_5 has taken place implicitly here. At this stage if we have $t' = aaababaaaa$ rather than t as input we find that

$$\{0,1,3\}baaaa \vdash \{0,2\}aaaa \vdash \{0,1,3\}aaa \vdash \{0,1,4\}aa$$

and t' is accepted.

The *DFA* has the advantage that it never repeats a text comparison. This implies that the *DFA*-based pattern matcher requires at most $|t|$ comparisons, whereas the recursive one can be quite bad, as the following pathological example shows.

Let $t = a^{m+n-1}b$ and $p = a^{m-1}b$, where $m \geq 1$ and $m < n$. Then in the first n positions in t, p requires m comparisons to reject or accept each position and, hence requires mn comparisons overall rather than $m+n$ comparisons.

We, of course, need to construct the *DFA* corresponding to p, but its particularly simple structure leads to efficient algorithms for this task; see the Exercises.

2.4 MINIMIZATION AND SIMPLIFICATION

An *FA*, whether it is deterministic or nondeterministic, may have useless states, that is, states that can never be reached whatever the input word. We say such states are *inaccessible* or *unreachable.* They were already introduced in the previous section, but we now reconsider them. We discuss briefly how such states may be detected and eliminated from an *FA*, *simplifying* it and *reducing* it in size. We also discuss similar notions for input symbols and final states. However, the main purpose of this section is to demonstrate the ultimate simplification, namely, the construction of the *minimal-state DFA* equivalent to a given *DFA*.

First of all, however, we define the notions of reachability, usefulness, and symbol reachability in an *NFA*.

Definition Let $M = (Q,\Sigma,\delta,s,F)$ be an *NFA*. A state q in Q is *reachable* or *accessible* if there exists a word x in Σ^* such that $sx \vdash^* q$. If no such word exists q is said to be *unreachable* or *inaccessible.* A state q in Q is *useful* if there exists a word x in Σ^* such that $qx \vdash^* r$, for some state r in F. Similarly q is *useless* if no such word exists. A symbol a in Σ is (*symbol*) *reachable* if there exists a reachable state q in Q for which $\delta(q,a) \neq \varnothing$.

The first two of these notions can be viewed as variants of reachability in the underlying digraph of the *NFA*. Therefore, it is a straightforward matter to adapt the node connectedness algorithm of Section 0.2 to provide algorithms which detect unreachable and useless states and, hence, unreachable symbols. We leave the construction of these algorithms to the Exercises.

We turn to the construction of a minimal *DFA* from a given *DFA* by way of distinguishable and indistinguishable states.

Definition Let $M = (Q,\Sigma,\delta,s,F)$ be a *DFA*. For two distinct states p and q in Q we say p and q are *distinguishable* if there exists an x in Σ^* such that $px \vdash^* f$, $qx \vdash^* g$ and exactly one of f and g is in F. The word x *distinguishes p* from q. If no such word exists then p and q are *indistinguishable* and we write $p \equiv q$, since they are equivalent in their action as far as acceptance is concerned. Indeed, they can be merged without affecting the language accepted by M. This forms the basis of the minimization for which we need to refine the notion of indistinguishable. Given an integer $k \geq 0$, we say two distinct states p and q in Q are *k-distinguishable* if there is a word x in Σ^*, $|x| \leq k$, which distinguishes p from q. If there is no such word, then we say that p and q are *k-indistinguishable* and we write $p \overset{k}{\equiv} q$. These notations are appropriate since both $\equiv$ and $\overset{k}{\equiv}$ are equivalence relations over Q.

Clearly two states are distinguishable if they are k-distinguishable, for some $k \geq 0$. We first show that for a given $M = (Q,\Sigma,\delta,s,F)$ with $\#Q = m$, two distinct states are indistinguishable if and only if they are $(m-2)$-indistinguishable. This implies that only a finite number of words need be examined, that is, indistinguishability is decidable.

If we cannot distinguish p and q with words of length at most $k+1$, for some $k \geq 0$, then we cannot distinguish them with words of length at most k. In other words, $p \overset{k+1}{\equiv} q$ implies $p \overset{k}{\equiv} q$, for all $k \geq 0$, that is, $\overset{k+1}{\equiv} \subseteq \overset{k}{\equiv}$. We say $\overset{k+1}{\equiv}$ is a *refinement* of $\overset{k}{\equiv}$. Similarly, we obtain $\equiv \subseteq \overset{k}{\equiv}$, for all $k \geq 0$.

These remarks demonstrate that the equivalence relations form an ascending chain

$$\equiv \subseteq \cdots \subseteq \overset{k}{\equiv} \subseteq \cdots \subseteq \overset{1}{\equiv} \subseteq \overset{0}{\equiv}$$

There are at most two equivalence classes formed by $\overset{0}{\equiv}$; F and $Q-F$. We can construct $\equiv$ by refining $\overset{0}{\equiv}$ to give $\overset{1}{\equiv}$, $\overset{1}{\equiv}$ to give $\overset{2}{\equiv}$, and so on. In Lemma 2.4.1 we prove that this algorithm terminates within $m-2$ steps, where $m = \#Q$. Therefore, $\equiv \; = \; \overset{m-2}{\equiv}$. To refine $\overset{i}{\equiv}$ to give $\overset{i+1}{\equiv}$ we use the observation

$$\begin{array}{l} \textit{For two distinct states } p \textit{ and } q \textit{ with } p \overset{i}{\equiv} q, \\ \quad p \overset{i+1}{\equiv} q \textit{ if and only if, for all } a \textit{ in } \Sigma, \ \delta(p,a) \overset{i}{\equiv} \delta(q,a) \end{array} \tag{2.4.1}$$

Assume p and q are $(i+1)$-indistinguishable. Then for all a in Σ and x in Σ^* with $|x| \leq i$, $pax \vdash^* f$ and $qax \vdash^* g$, for some f and g in Q. Now either f is in F and g is in $Q-F$, or conversely. But this implies that $\delta(p,a)$ and $\delta(q,a)$ are i-indistinguishable.

Conversely, if we assume that for all a in Σ, $\delta(p,a) \stackrel{i}{\equiv} \delta(p,a)$, then immediately $p \stackrel{i+1}{\equiv} q$. So equation (2.4.1) holds.

Lemma 2.4.1 *Let $M = (Q,\Sigma,\delta,s,F)$ be a DFA with $\#Q = m$. Then two distinct states p and q are indistinguishable if and only if they are $(m-2)$-indistinguishable.*

Proof:

if: We prove that $(m-2)$-indistinguishability implies indistinguishability via the basic properties of $\stackrel{k}{\equiv}$ and $\equiv$, which we now establish. We have already observed that

$$\equiv \subseteq \cdots \subseteq \stackrel{k}{\equiv} \subseteq \cdots \subseteq \stackrel{1}{\equiv} \subseteq \stackrel{0}{\equiv}$$

We need to show that whenever $\stackrel{k+1}{\equiv} = \stackrel{k}{\equiv}$ we obtain equality from thereon in. If $\stackrel{k+1}{\equiv} = \stackrel{k}{\equiv}$, then $p \stackrel{k+1}{\equiv} q$ implies $\delta(p,a) \stackrel{k+1}{\equiv} \delta(q,a)$, for all a in Σ, using equation (2.4.1). But this ensures that $p \stackrel{k+2}{\equiv} q$, for all p and q such that $p \stackrel{k+1}{\equiv} q$, from equation (2.4.1); that is, $\stackrel{k+2}{\equiv} = \stackrel{k+1}{\equiv}$. By induction $\stackrel{k+i}{\equiv} = \stackrel{k}{\equiv}$, for all $i \geq 1$.

Now consider the ascending chain

$$\equiv \subseteq \cdots \subseteq \stackrel{k}{\equiv} \subseteq \cdots \subseteq \stackrel{1}{\equiv} \subseteq \stackrel{0}{\equiv}$$

If $F = \varnothing$ or $F = Q$, then the chain collapses with $\stackrel{0}{\equiv} = \equiv$ having one equivalence class Q or F, respectively. In this case all states are indistinguishable. Otherwise, $0 < \#F < m$ and strict inequality can hold for at most the first $m-1$ terms. This follows by considering the equivalence classes of $\stackrel{i}{\equiv}$. $\stackrel{0}{\equiv}$ has two nonempty equivalence classes, under the assumption on the size of F; therefore, $\stackrel{1}{\equiv}$ must have three, and, in general, $\stackrel{i}{\equiv}$ must have $i+2$. But for each of these to be nonempty there can be at most m equivalence classes with one state per class. Now $i+2 \leq m$ implies $i \leq m-2$, as desired.

only if: If p and q are indistinguishable, then they are, trivially, $(m-2)$-indistinguishable. ■

Before presenting the main theorem of this section we work through an example. In this example we introduce a tabular method for computing $\stackrel{k+1}{\equiv}$ from $\stackrel{k}{\equiv}$.

Example 2.4.1 Let M have the state diagram of Figure 2.4.1.

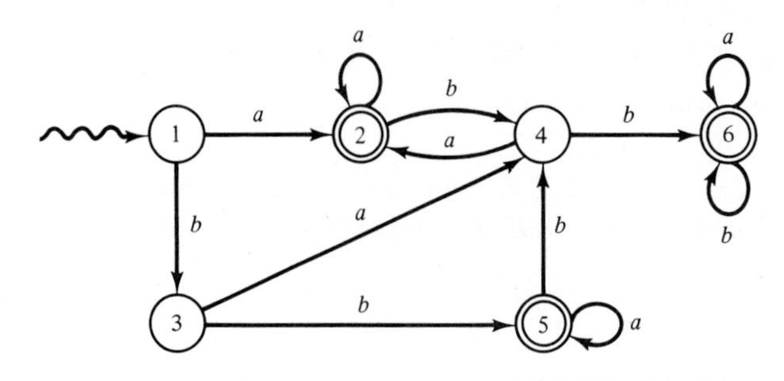

Figure 2.4.1

We prefer to work with the transition table of M, Table 2.4.1. Let $[p]_k$ denote the equivalence class of p with respect to $\overset{k}{\equiv}$. We also write $[p]$ if $\overset{k}{\equiv}$ is understood. Immediately, from the definition of $\overset{0}{\equiv}$, $[1]_0 = [3]_0 = [4]_0 = \{1,3,4\}$ and $[2]_0 = [5]_0 = [6]_0 = \{2,5,6\}$. We display this information in the modified transition table, Table 2.4.2, in which each entry is replaced by its equivalence class and we group entries with respect to their equivalence classes. The reason for doing this is that when computing $\overset{1}{\equiv}$, we make use of equation (2.4.1); that is, $p \overset{1}{\equiv} q$ if and only if $p \overset{0}{\equiv} q$, $\delta(p,a) \overset{0}{\equiv} \delta(q,a)$, and $\delta(p,b) \overset{0}{\equiv} \delta(q,b)$. But this holds if and only if $[p]_0 = [q]_0$, $[\delta(p,a)]_0 = [\delta(q,a)]_0$, and $[\delta(p,b)]_0 = [\delta(q,b)]_0$. In other words, if and only if rows p and q are in the same group in Table 2.4.2 and these rows are equal.

Table 2.4.1

δ	a	b
1	2	3
2	2	4
3	4	5
4	2	6
5	5	4
6	6	6

Table 2.4.2

	$\overset{0}{\equiv}$	a	b
$Q - F$	1	[2]	[3]
	3	[4]	[5]
	4	[2]	[6]
F	2	[2]	[4]
	5	[5]	[4]
	6	[6]	[6]

Consider $[1]_0 = \{1,3,4\}$ first. By inspection, every pair of rows from this class are distinct. More precisely $[\delta(1,a)]_0] \neq [\delta(3,a)]_0$, $[\delta(1,b)]_0 \neq [\delta(4,b)]_0$, and $[\delta(3,a)]_0 \neq [\delta(4,a)]_0$. This implies 1, 3, and 4 form singleton equivalence classes in $\overset{1}{\equiv}$.

On the other hand, rows 2 and 5 are equal and different from row 6. Thus, we obtain Table 2.4.3. To construct $\overset{2}{\equiv}$, we only need examine the rows 2 and 5, since all other equivalence classes are singletons. Once more these two rows are equal. (Note that [ ] means $[\ ]_1$ here rather than $[\ ]_0$.) Hence $\overset{2}{\equiv} = \overset{1}{\equiv} = \equiv$, and we are finished.

Table 2.4.3

$\overset{1}{\equiv}$	a	b
1	[2]	[3]
3	[4]	[5]
4	[2]	[6]
2	[2]	[4]
5	[5]	[4]
6	[6]	[6]

This means we can merge states 2 and 5 in M to give M'; see Figure 2.4.2.

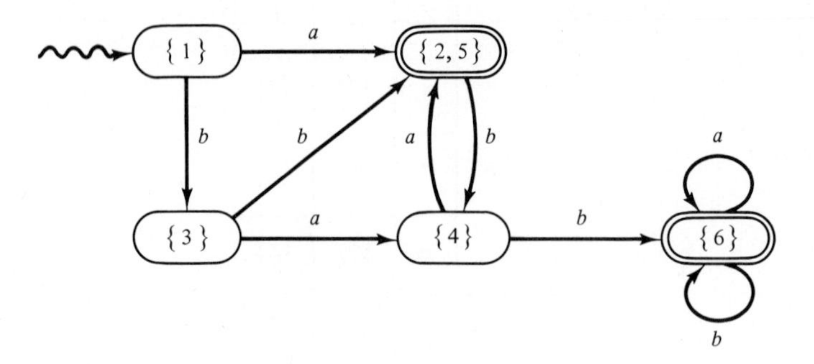

Figure 2.4.2

We claim M' is the minimal *DFA* equivalent to M. Although this can be proved for this specific example we prefer to give a general proof below.

The construction of $\equiv$ by starting from $\overset{0}{\equiv}$ and successively refining it, using equation (2.4.1), until $\overset{k+1}{\equiv} = \overset{k}{\equiv}$, should be clear from the above example. Let $M = (Q,\Sigma,\delta,s,F)$, $\equiv$ be $\overset{k}{\equiv}$ and construct a *DFA*, $M' = (Q',\Sigma,\delta',s',F')$, from $\overset{k}{\equiv}$ as follows.

Let $Q' = \{[p] : p$ is in Q and $[p]$ is the equivalence class of p in $\equiv\}$, $s' = [s]$, $F' = \{[p] : p$ is in $F\}$, and define δ' by

For all $[p]$ in Q', for all a in Σ, $\delta'([p],a) = [\delta(p,a)]$

This construction is seen in the above example. We now prove M' is minimal.

Theorem 2.4.2 *Let $M = (Q,\Sigma,\delta,s,F)$ be a complete DFA having only reachable states and $M' = (Q',\Sigma,\delta',s',F')$ be the corresponding DFA constructed from $\equiv$ for M. Then M' is a minimal complete DFA equivalent to M.*

Proof: (by contradiction)
Assume M' is not minimal. Then there exists a minimal complete *DFA* $M'' = (Q'',\Sigma,\delta',s'',F'')$ with $L(M'') = L(M') = L(M)$ and $m'' < m' \leq m$, where $m'' = \#Q''$, $m' = \#Q'$, and $m = \#Q$.

By associating each word x in Σ^* with the state it reaches from s'', we partition Σ^* into m'' classes with respect to M''. The same holds for M' except that we obtain m' classes. Because M' only consists of reachable states, by construction, each class with respect to M' is nonempty. Because $m' > m''$ there must be at least one class with respect to M'' which contains at least two words x and y with x and y in different classes with respect to M'. This is an argument which is similar to the pigeonhole principle; see Section 0.5. In other words, $s'x \vdash^* p$ and

$s'y \vdash^* q$ with $p \neq q$, but $s''x \vdash^* r$ and $s''y \vdash^* r$, for some r. The states p and q must be distinguishable in M'. (Otherwise $p = q$, a contradiction). Hence, there is a z in Σ^* such that $pz \vdash^* f$ and $qz \vdash^* q$ with exactly one of f and g in F, that is either xz or yz is in $L(M') = L(M)$. But in M'', $s''xz \vdash^* h$ and $s''yz \vdash^* h$, for some h which implies both xz and yz are in $L(M'')$ or both are not in $L(M'')$. In either case $L(M'') \neq L(M') = L(M)$, which provides the required contradiction, and establishes the theorem. ■

The above algorithm for constructing M' is not the most efficient. The question of efficiently constructing M' is left to the Exercises.

The proof of Theorem 2.4.2 also yields the following:

Corollary 2.4.3 *Let $M = (Q,\Sigma,\delta,s,F)$ be a complete DFA and M' and M'' be two minimal complete DFAs equivalent to M. Then M' and M'' are equal apart from a renaming states, that is, there is an isomorphism between M' and M''.*

Proof: This is left to the Exercises. ■

This corollary implies that

DFA EQUIVALENCE
INSTANCE: Two *DFAs* M_1 and M_2.
QUESTION: Is $L(M_1) = L(M_2)$?

is decidable; see the Exercises and Chapter 10.

2.5 *DFAs* AND TRIES

Although the definition of *DFA* cannot be further restricted without restricting the languages accepted by them, there are two restrictions that are of interest. We say a *DFA*, $M = (Q,\Sigma,\delta,s,F)$, has a *loop* if there is a state q in Q and a nonempty word x in Σ^+ such that $qx \vdash^* q$; q is said to be a *looping state.* Such a loop gives the star of x. M is said to be *loop-free* if it has no looping states, in other words, the state diagram has no cycles. Furthermore, we say M is a *tree* if it is loop-free and for all states q in Q, $q \neq s$, there is one and only one state p with $\delta(p,a) = q$, for some a in Σ. In this case the state diagram is a tree of arity $\#\Sigma$.

Example 2.5.1 Consider the PASCAL reserved words found in repetitive statements

for downto to do repeat until while

then we might specify them with the state diagram of Figure 2.5.1, where *blank* is used as a terminator. M is *loop-free*, but not a tree. However we might also construct the state diagram of Figure 2.5.2 and in this case M' is a tree *DFA*. A tree *DFA* is known as a *trie* in data structures —from re*trie*val. In such a structure

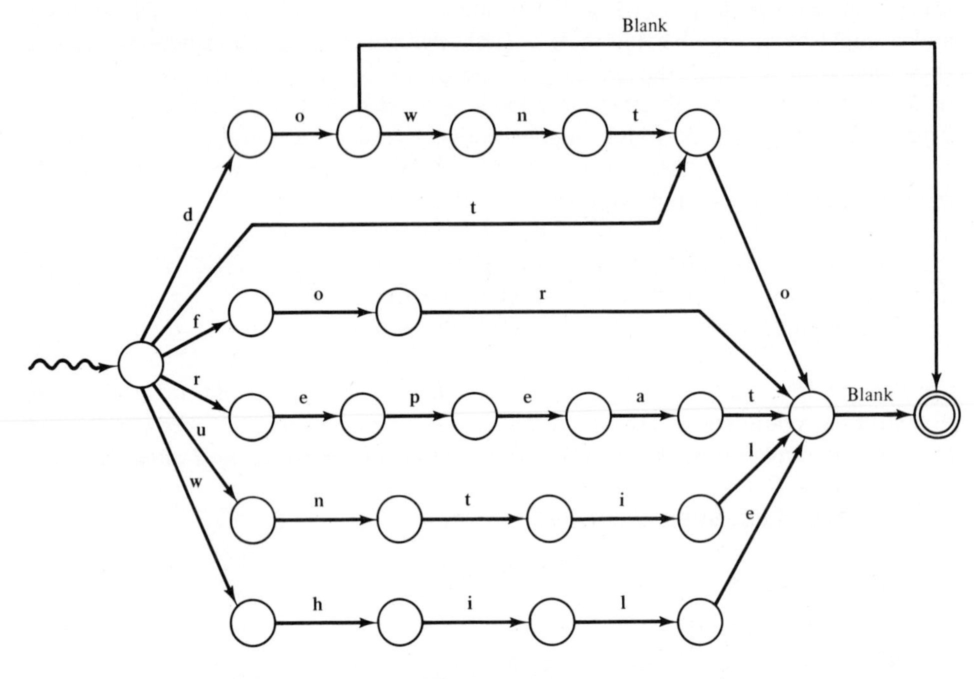

Figure 2.5.1

words with the same initial symbols share a common path, while a loop-free *DFA* may have shared paths for terminating symbols, thus saving even more space in a corresponding data structure.

These restricted *DFAs* are also restricted in their language acceptance capability, simply because the state diagram is a directed acyclic graph or *DAG*; see Section 0.2. This means there is a longest path from the start to a final state and therefore every input word longer than this cannot be accepted. In both M and M' of Example 2.5: no word of length greater than 7 can be accepted. This means that such machines can only accept finite languages. On the other hand, it is easy to construct *DFAs*, M, which accept infinite languages; see the Exercises and Section 8.1.

Since loop-free *DFAs* accept only finite languages, this raises the question: Is every finite language accepted by some loop-free *DFA*? This is indeed the case; its proof we leave to the Exercises. Now even if a *DFA* has loops it may still accept a finite language; therefore, we have the decision problem

DFA FINITENESS
INSTANCE: A *DFA*, $M = (Q,\Sigma,\delta,s,F)$.
QUESTION: Is $L(M)$ finite?

A positive response implies that there is a loop-free *DFA* accepting $L(M)$. This

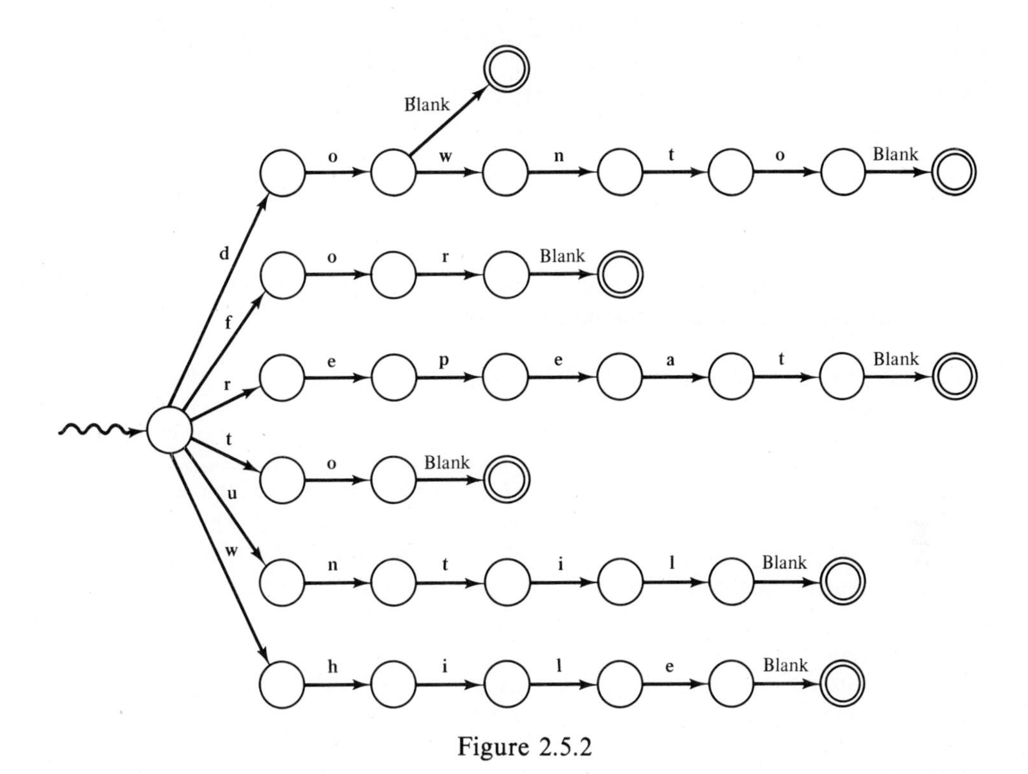

Figure 2.5.2

decision problem is shown to be decidable in Chapter 10. Essentially $L(M)$ is finite if and only if every state, which is reachable from s and can reach some final state, is not in any loop.

2.6 λ-*FAs* AND LAZY *FAs*

While in the previous section we considered how a *DFA* could be further restricted, here we consider how *FAs* can be further generalized. We allow words, including λ, rather than symbols in the transition function. Some other generalizations are discussed in the Exercises. Throughout this section we give sketches rather than details of the various models. The generalizations we consider are not introduced for their own sake but rather because they are either useful when constructing *FAs* or model natural file processing situations.

2.6.1 λ-*FAs* and λ-Transitions

In the *FAs* we have discussed so far we require that the reading head move at each step. We now relax this condition by allowing the reading head to remain at a cell during a transition. We call such a transition a *λ-transition*; the input symbol is ignored.

Definition $M = (Q,\Sigma,\delta,s,F)$ is a *λ-NFA*, or *λ-transition NFA*, if $\delta \subseteq Q \times (\Sigma \cup \{\lambda\}) \times Q$. We require, on a λ-transition, that the reading head doesn't move and that the input symbol is ignored.

Given a configuration px in $Q\Sigma^*$ we write

$$px \vdash qy$$

if $x = Ay$, for some A in $\Sigma \cup \{\lambda\}$, and (p,A,q) is in δ; $\vdash$ is extended to $\vdash^+$ and $\vdash^*$ as before.

Thus, at each configuration there is not only the choice of transition usual in *NFAs*, but also the additional possibility of taking λ-transitions.

There are two situations in which λ-transitions are useful as the following examples demonstrate.

Example 2.6.1 You are designing an *NFA* construction kit, which will provide some primitive *NFAs* together with methods of combining them. What methods of combining *NFAs* should be provided for, and how should the combinations be formed? It turns out that λ-transitions are very useful in this context.

The Plus of an *NFA*

Given the *NFA*, M [see Figure 2.6.1(a)], we can feed the final states back to the start state using λ-transitions to give M^+ [see Figure 2.6.1(b)].

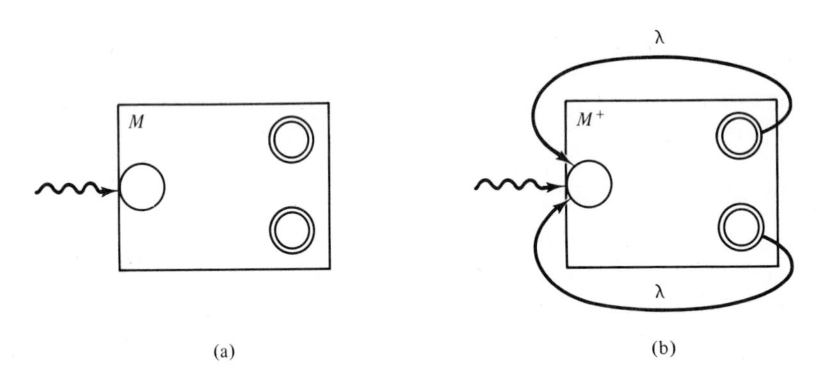

Figure 2.6.1

A little thought shows that if $x_1, \ldots, x_m$ are words accepted by M, then $x_1 \cdots x_m$ is accepted by M^+, and conversely.

The Union of Two *NFAs*

Given the *NFAs*, M_1 and M_2 [see Figures 2.6.2(a) and (b)], we can form their union M [see Figure 2.6.2(c)]. M consists of M_1 and M_2 together with a new start state which via λ-transitions leads to the start states of M_1 and M_2.

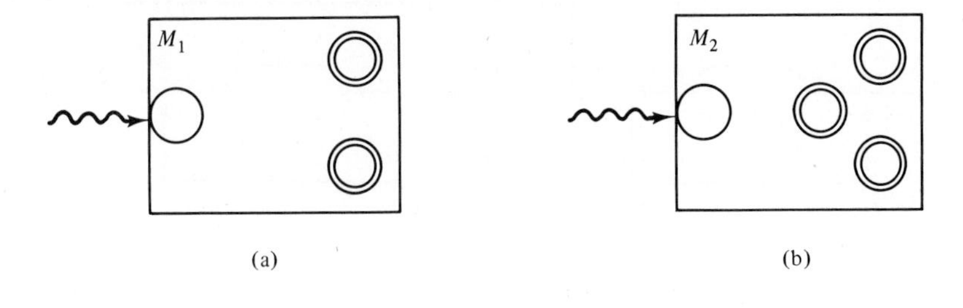

(a) (b)

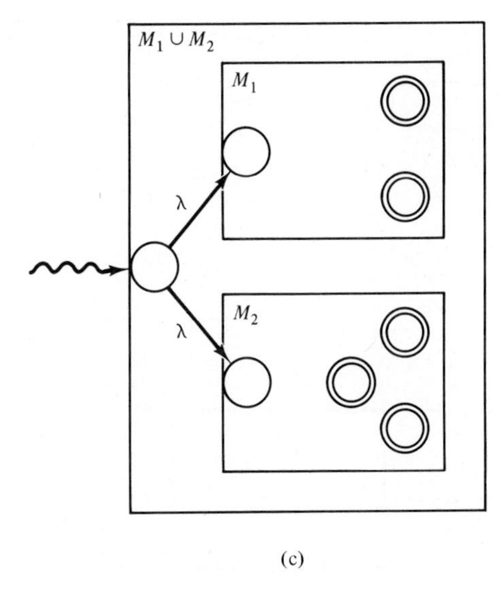

(c)

Figure 2.6.2

Again a little thought should show that if x is accepted by M_1, then x is accepted by M and if x is accepted by M_2, then x is accepted by M. The converse also holds.

These constructions are also useful theoretically as we shall demonstrate in Chapters 3 and 8.

Example 2.6.2 In the introduction to Chapter 4 we discuss syntax diagrams, which look somewhat like state diagrams. We might define an unsigned decimal number using the syntax diagram of Figure 2.6.3.

Now transforming this syntax diagram into a state diagram by taking its dual, we obtain Figure 2.6.4 and λ-transitions occur naturally. Indeed, they are present implicitly in the original syntax diagram, that is, we should really have drawn Figure 2.6.5.

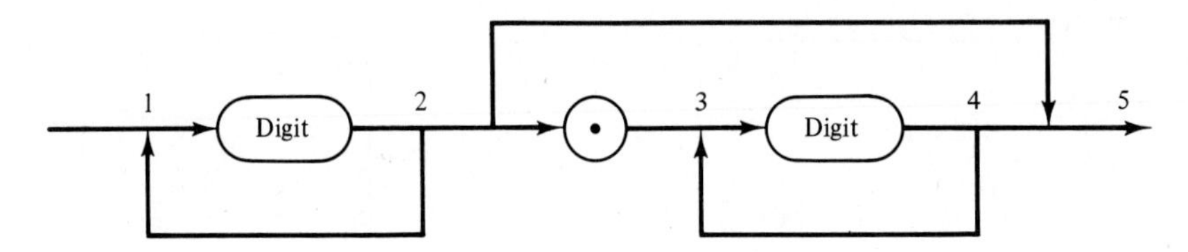

Figure 2.6.3

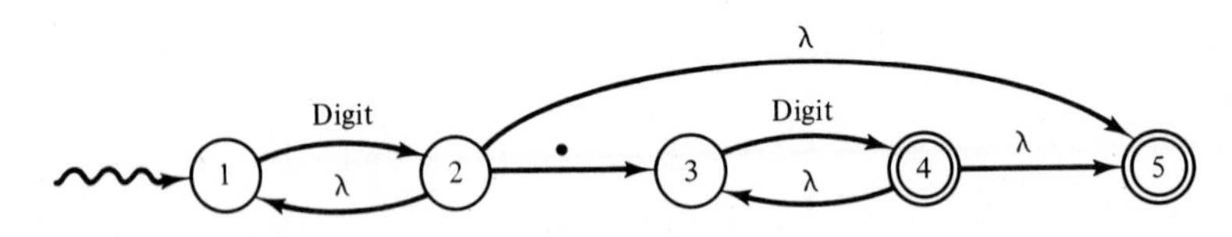

Figure 2.6.4

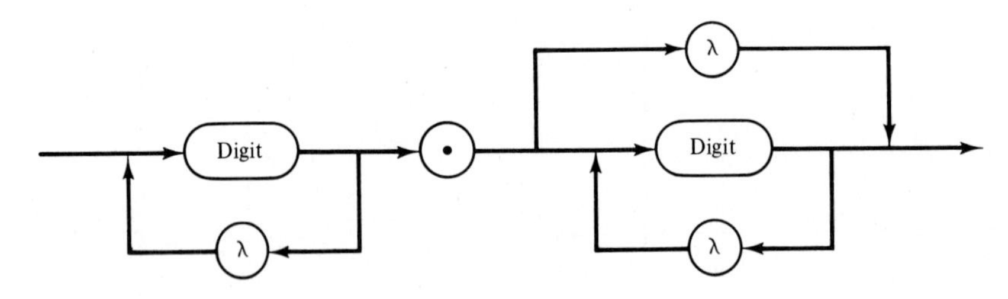

Figure 2.6.5

Example 2.6.3 Let M be as displayed in Figure 2.6.6. Then observe that a can be accepted by either

$$0aa \vdash 1aa \vdash 2a \vdash 2 \vdash 3$$

or

$$0aa \vdash 7a \vdash 7$$

Clearly, every NFA is a λ-NFA, but surprisingly the expressive power of λ-$NFAs$ is no more than that of $NFAs$.

Theorem 2.6.1 *Let $M = (Q, \Sigma, \delta, s, F)$ be a λ-NFA; then an equivalent NFA, M', can be constructed from M.*

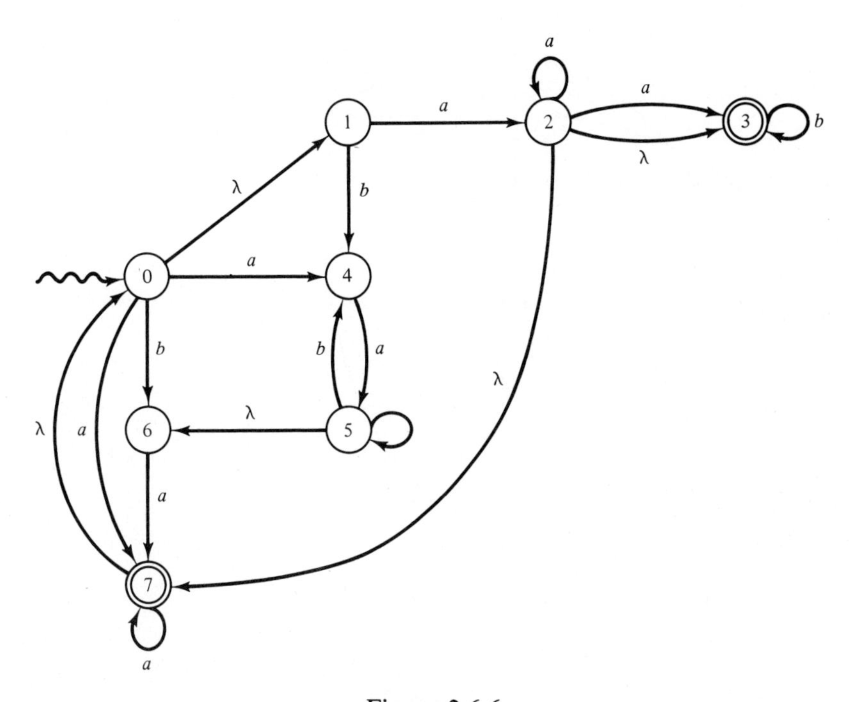

Figure 2.6.6

Proof: We give a two-stage construction of an equivalent *NFA*, M', from M. The detailed proof of correctness is left to the Exercises.

(1) We modify M so that whenever there is a configuration sequence $p \vdash^i q$, for some $i > 1$, M has a transition (p,λ,q) and whenever a state can reach a final state by way of a λ-transition it too is made final. More formally, construct $M'' = (Q,\Sigma,\delta',s,F'')$, where

$$\delta' = \delta \cup \{(p,\lambda,q) : p \vdash^i q \text{ in } M, \text{ for some } i > 1\}$$

and

$$F'' = F \cup \{p : (p,\lambda,f) \text{ is in } \delta, \text{ for some } f \text{ in } F\}$$

Clearly, $L(M) \subseteq L(M'')$. $L(M'') \subseteq L(M)$ is also easy to prove, so we leave it to the Exercises. The λ-*NFA* of Figure 2.6.6 gives Figure 2.6.7 under this construction.

(2) We now remove λ-transitions from M'' to give the required *NFA*, M'. Whenever we have a configuration sequence

$$pax \vdash qax \vdash rx$$

in M'', we add the transition (p,a,r) to M''. This gives

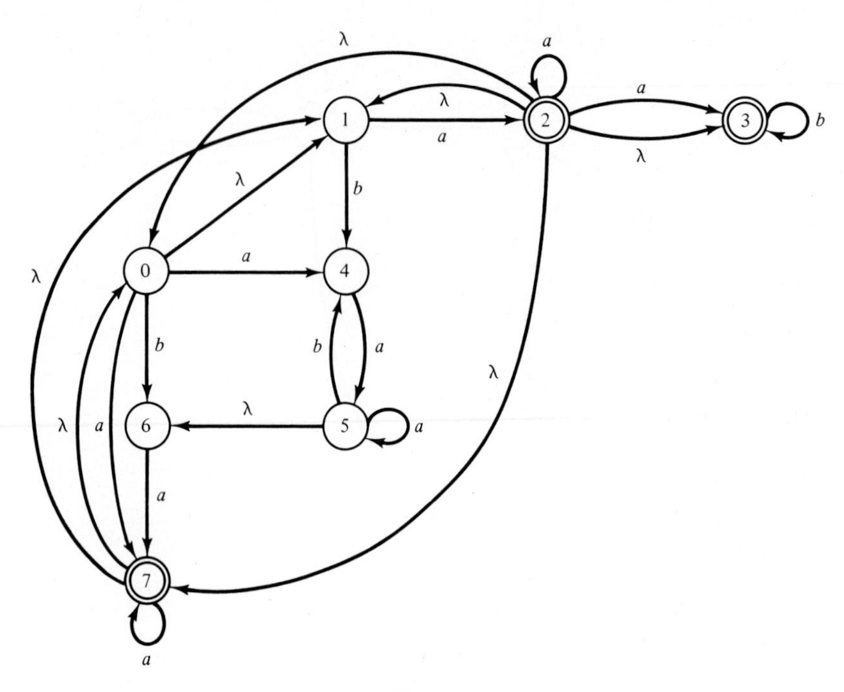

Figure 2.6.7

$$pax \vdash rx$$

directly, by this short cut. The construction of M'' implies that we do not need to examine configuration sequences of the form

$$pax \vdash^{i} qax \vdash rx$$

for $i > 1$. Finally, we remove all λ-transitions.

Formally, define M' as (Q,Σ,δ',s,F''), where δ' is defined as follows

$$\delta' = \delta' - \{(p,\lambda,q) : p,q \text{ in } Q\}$$
$$\cup \{(p,a,r) : (p,\lambda,q) \text{ and } (q,a,r) \text{ are in } \delta,$$
$$\text{for some } p,q, \text{ and } r \text{ in } Q \text{ and } a \text{ in } \Sigma\}$$

The proof of equivalence of M' and M'' is left to the Exercises. Figure 2.6.8 displays the *NFA* obtained from the λ-*NFA* of Figure 2.6.6. ■

2.6.2 Lazy Finite Automata

Definition A *lazy finite automaton*, M, is a quintuple (Q,Σ,δ,s,F) where Q, Σ, s, and F are as for *FAs*, but $\delta \subseteq Q \times \Sigma^* \times Q$, a finite transition relation. In other words, transitions are allowed with words rather than just symbols and the

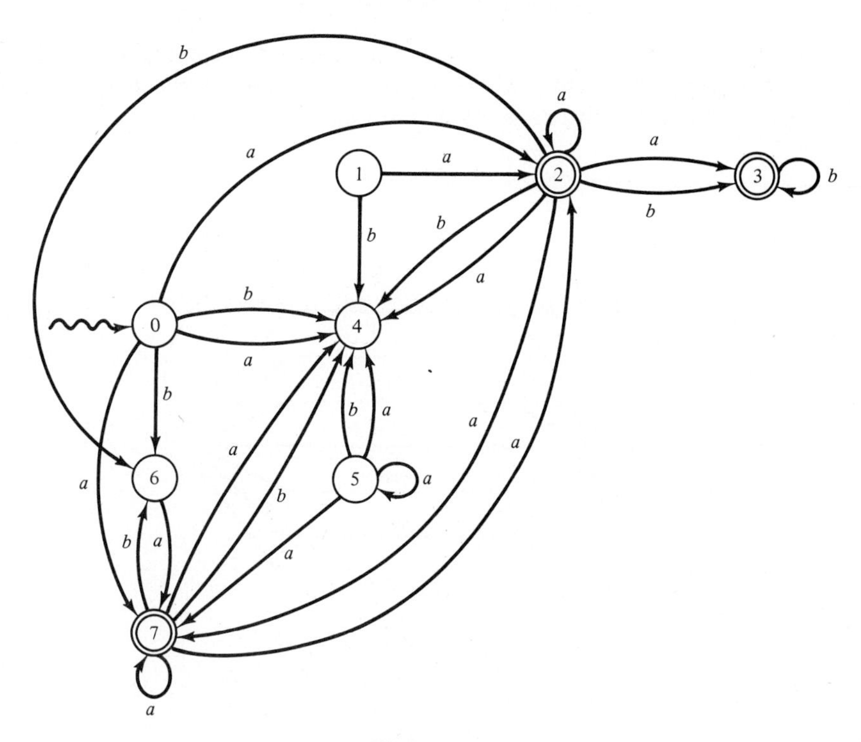

Figure 2.6.8

empty word. A single move is defined by

$$px \vdash qy$$

if $x = wy$, for some w in Σ^*, and (p, w, q) is in δ.

When w is in Σ, M operates as before, when $w = \lambda$, M ignores the input as in a λ-*NFA*, and when $|w| > 1$, M reads w and the reading head moves $|w|$ cells to the right.

The lazy *FA* is useful when demonstrating that a language is accepted by some *FA*, since we have:

Theorem 2.6.2 *Let M be a lazy FA. Then there is a λ-NFA, M', with $L(M') = L(M)$.*

In other words, lazy *FAs* are no more powerful than *FAs*. The simple proof of this theorem is left to the Exercises. The lazy *FA* is also a useful starting point for constructing an *FA*, deterministic or otherwise.

Example 2.6.4 Consider the reserved words for the transfer of control statements of PASCAL, namely,

case of end if then else goto

and PASCAL identifiers. We can construct an *FA* to accept sequences of such words and identifiers by starting with the lazy automaton shown in Figure 2.6.9. This can then be, successively, transformed until an equivalent *DFA* is obtained.

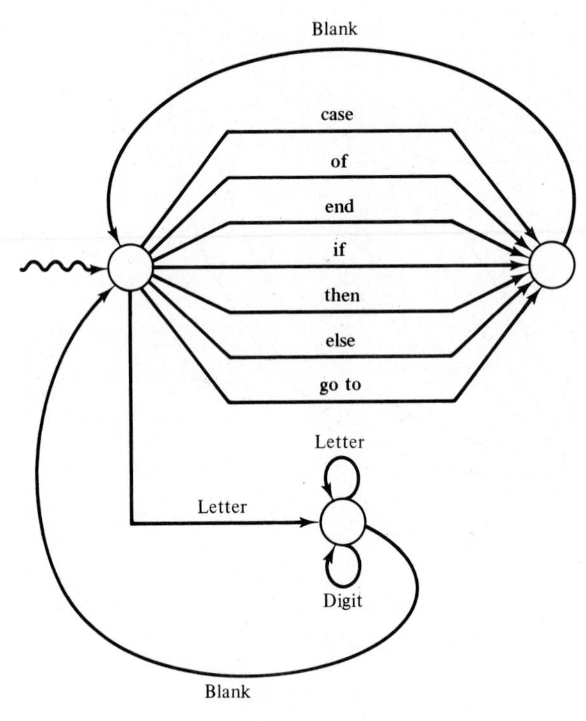

Figure 2.6.9

2.7 ADDITIONAL REMARKS

2.7.1 Summary

We have introduced DETERMINISTIC FINITE AUTOMATA and *DFA* LANGUAGES as well as NONDETERMINISTIC FINITE AUTOMATA and *NFA* LANGUAGES. We have proved constructively that $\boldsymbol{L}_{DFA} = \boldsymbol{L}_{NFA}$ and have shown, by example, how *NFAs* can be used for PATTERN MATCHING. We have demonstrated that for every *DFA* there is an EQUIVALENT MINIMAL STATE *DFA* which is unique up to a renaming of states. Finally, we have introduced a number of variants of the basic models: LOOP-FREE *DFAs*, TREE *DFAs*, λ-TRANSITION *NFAs*, and LAZY *FAs*.

2.7.2 History

The article of Hayes (1983) is a splendid introduction to finite automata, while Barnes (1972) is an older, but still useful, introduction. Chomsky and Miller (1958) is a pioneering study of *FA* languages. Finite automata first emerged as a model of neural nets in the work of McCulloch and Pitts (1943). See also the collection of papers in Shannon and McCarthy (1956). Finite automata are a natural extension of switching circuits — see Harrison (1965) and Brzozowski and Yoeli (1976) — and therefore are a basic model for computers, albeit a limited one. We return to this issue in Chapter 8.

The pattern-matching example is based on the challenging paper of Knuth et al. (1977). Consult Wirth (1984) for an excellent introduction to this approach to pattern matching. Aho (1980) is a good source for recent results. Cohen (1979) and Floyd (1967b) discuss nondeterministic programming.

The construction of a minimal *DFA* is based on the work of Moore (1956), while Hopcroft (1971) presents a more efficient algorithm.

In Example 2.5.1 the notion of the *trie* was introduced as a restricted class of finite automata. Tries are a useful data structure for representing a set of words which contains many common prefixes. See Knuth (1973) and Gonnet (1984) for a general introduction and Comer (1979, 1981) for more specialized results. Rabin and Scott (1959), and Shepherdson (1959) proved the basic result for two-way finite automata; see Exercise 6.5.

2.8 SPRINGBOARD

2.8.1 The Syntactic Monoid

In Section 2.4 an equivalence relation, *indistinguishability*, was introduced. A machine-independent definition of an equivalence relation for a language $L \subseteq \Sigma^*$ is

For all x, y in Σ^*:
$x \equiv y$ iff for all w and z in Σ^*, wxz is in L iff wyz is in L.

Letting $[x]$ denote the equivalence class of x under $\equiv$, we may define a binary operation $\circ$ on equivalence classes by

$$[x] \circ [y] = [xy]$$

It is easy to show that $[x] \circ [\lambda] = [\lambda] \circ [x] = [x]$, for all x in Σ^*; hence, the set of equivalence classes under $\circ$ forms a monoid, *the syntactic monoid*. Moreover, L is a *DFA* language if and only if its syntactic monoid is finite.

Explore the connection between the structure of the syntactic monoid of a *DFA* language and its minimal *DFA* representation. Can subclasses of the *DFA* languages be characterized by properties of their syntactic monoid? What are the syntactic monoids of finite languages like? As a starting point for your investigation, see McNaughton and Papert (1971). Eilenberg (1974, 1976) has become the fundamental reference work for the algebraic study of *FAs*.

2.8.2 Life and Cellular Automata

The game of life is one example of cellular automata, that is it consists of a infinite array (or cells) of identical *DFAs* (or *NFAs*). Each *FA* takes as its input, at each step, the states of its four neighbors; see Figure 2.8.1.

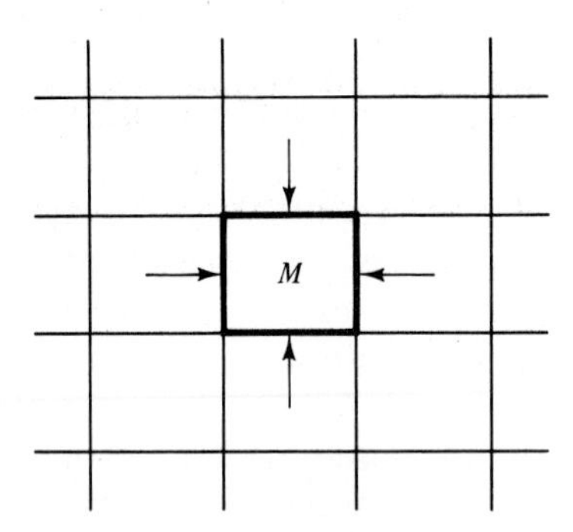

Figure 2.8.1

The transitions at all cells are assumed to be simultaneous, which causes the beauty and intricacy of life. In graphics one can envisage an array of processors, one per pixel, carrying out the basic graphics tasks. In image processing such an array could be used to implement a number of image smoothing and pattern recognition processes, for example. More recently such automata have been used to probe the origins of life. Consult Berlekamp et al. (1982), and Gardner (1983) for an introduction to life, and Burks (1970) and Hayes (1984) for cellular automata. See also Wolfram et al. (1984).

2.8.3 Communication Protocols

Computer networking has gained in prominence in the last ten years, a natural outgrowth of the decentralization of computing. In its wake it has become necessary to design protocols for communication between two or more machines, that is, how the communication of a message is established. In social gatherings we shake hands, say hi, etc. Since computers can be viewed as finite-state automata, protocols can be studied theoretically by considering protocols between finite state automata. Consult Gouda and Rosier (1985).

2.8.4 Security

Every operating system must provide some basic level of security to each user so that someone's file cannot be read or modified either accidentally or intentionally by someone else. In the movie *War Games* (MGM, 1981) a student modifies a file containing his grades, by gaining unauthorized access to it. Real-life examples of this occur much too frequently. Usually files have various permissions associated with them which detail who can do what. It has been observed (see Harrison et al., 1976), that such protection systems can be modeled by finite-state automata.

Consult Harrison et al. (1976) and DeMillo et al. (1978).

2.8.5 Tree Automata

All of the automaton models discussed in detail in this text take as their input a word or words. However, one interesting line of development has been to consider nonlinear input, for example, trees, dags, and graphs. A tree automaton is a finite automaton which takes a node-labeled tree as its input. In the top-down model the automaton descends down the tree a level at a time, keeping with each node the current state of the node. Since the transition function is the same as in a *DFA* the automaton acts in parallel at every node in a level, but deals with each node separately. For example, given the tree of Figure 2.8.2(a) and the automaton of Figure 2.8.2(b) we obtain Figure 2.8.3(a) after one step, since $\delta(0,a) = 1$. After two more steps we obtain Figure 2.8.3(b) if we extend the tree in the standard way.

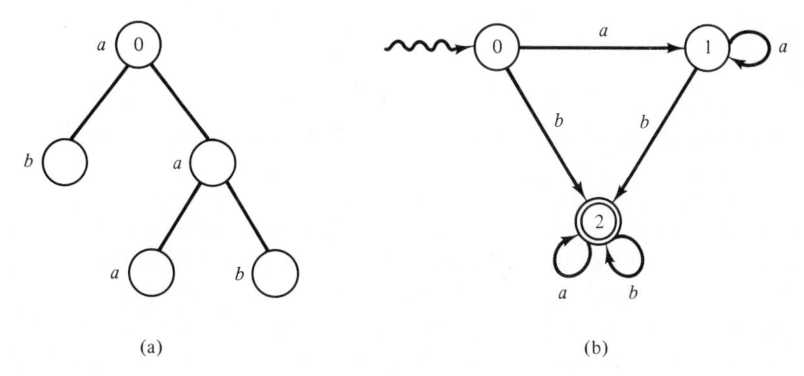

Figure 2.8.2

We say an input tree is accepted if at least one extended node ends up in a final state. Such tree automata can be used to model various tree walking algorithms and various notions of translation (see Chapter 7). They have served, theoretically, to unify apparently disparate entities; for example, the sets of syntax trees of *CFGs* (see Section 4.1) can be characterized as the sets of trees accepted by tree automata. See Thatcher (1967, 1973), the survey of Engelfriet (1980), and the book by Gecseg and Steinby (1984).

2.8.6 Avoiding Spaghetti Programming

In text editors the effect of a key on the terminal keyboard depends on the previous keys and in some editors a single key may have many different effects. For example in Vi, a text editor available under UNIX, the letter *d* can mean: delete a character; delete the current word; delete the previous word; or delete a line. These are indicated by typing *d*, *dw*, *db*, or *dd*, respectively, at command level.

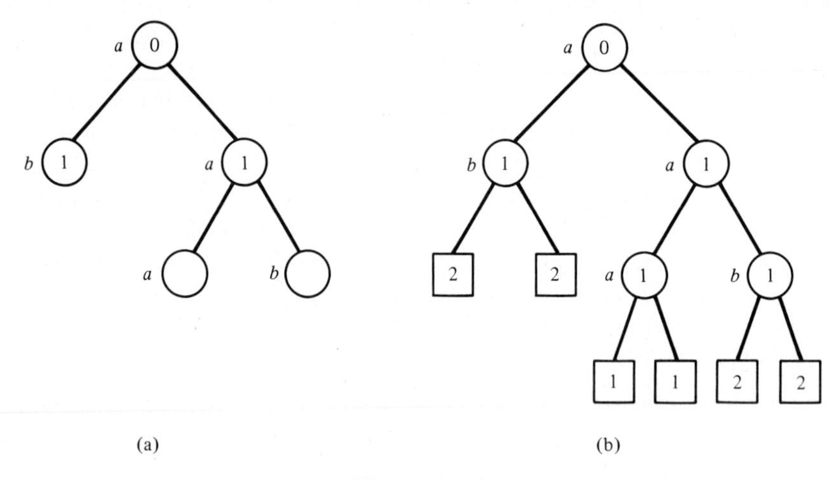

Figure 2.8.3

A telephone has a somewhat similar dependency in that a digit may be part of a country code, an area code, an exchange, or a local number depending on the previous digits.

The rules for deciphering command sequences in Vi or digit sequences in a telephone call are deceptively complex. For this reason, writing a command interpreter for Vi can become an exercise in what is sometimes called "sphaghetti programming." An unstructured approach to writing these kinds of programs usually results in highly intertwined and convoluted code, reminiscent of a bowl of sphaghetti. However we can use state diagrams to provide structure, and, perhaps, to automate the production of code. The state diagram approach gives rise to highly structured programs that are straightforward to code, maintain, and modify.

For example, assuming we are in command mode we have the state diagram of Figure 2.8.4.

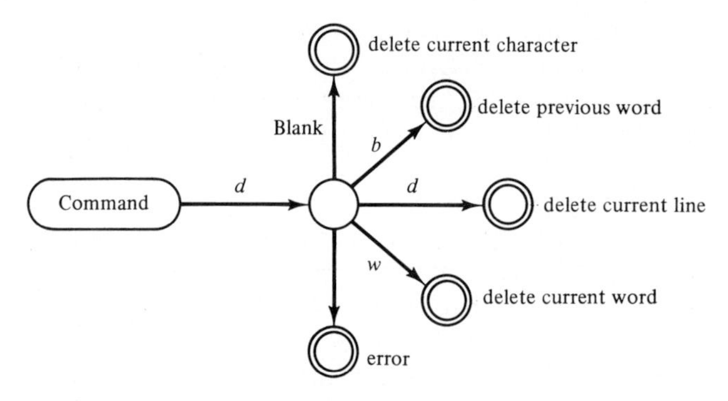

Figure 2.8.4

Of course, Vi is more complex than this indicates, because we can write *d*), *d*(, *d*}, *d*{, *d*]], *d*[[, *dL*, *d*3*L*, *d/pattern/*, and many other sequences, all of which have different meanings. However this only serves to confirm the soundness of the state diagram approach. Without it the code of the text editor could indeed end up as a bowl of spaghetti.

EXERCISES

1.1 Construct a state diagram for a simplified remote controller for TV that allows for ten channels, on and off, and four volume levels.

1.2 Observe your local elevator and construct a state diagram which models its actions.

1.3 Find ten different mechanisms or situations in your local community which can be modeled by a state diagram. For example the 2×2×2 Rubik's Cube.

2.1 Given the following *DFA*, M,

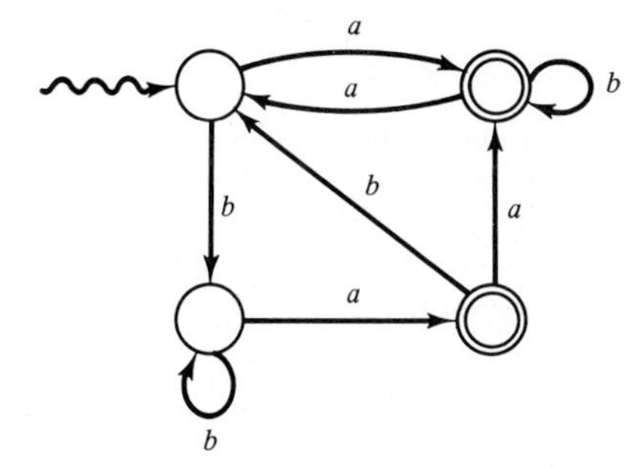

(i) give four words accepted by M and their configuration sequences;
(ii) give four words rejected by M.

2.2 Given the following *DFA*, M,

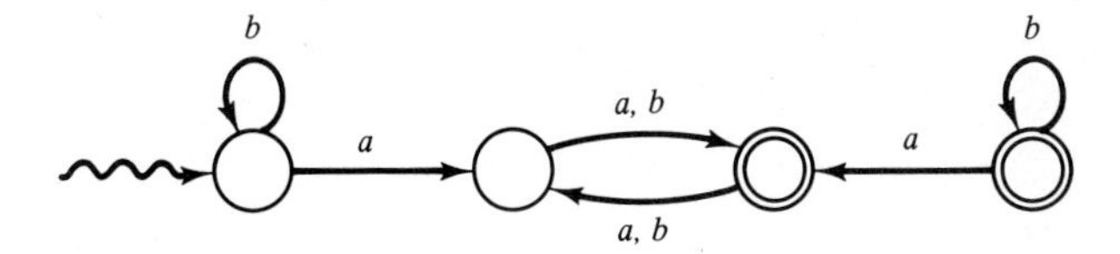

what is $L(M)$? Prove it.

2.3 Given the *DFA*, M,

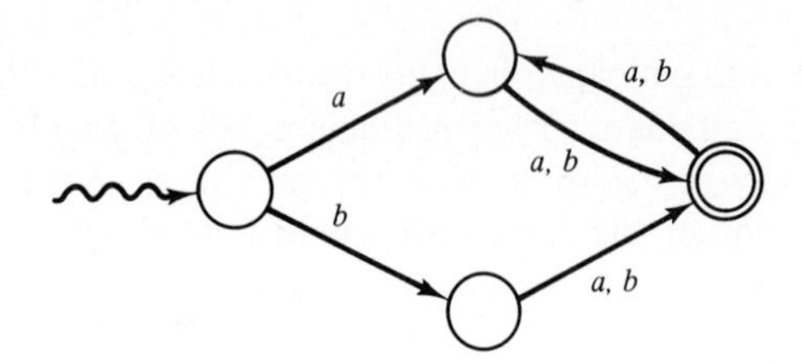

what is $L(M)$? Prove it.

2.4 Given the following *DFA*, M,

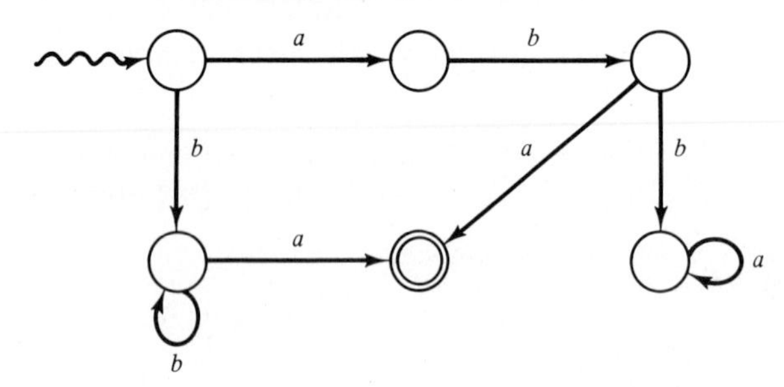

is M complete or incomplete? If it incomplete, then complete it.

2.5 Let $M = (Q, \Sigma, \delta, s, F)$ be a *DFA* and $E \subseteq L(M)$ be the set of all words in $L(M)$ of even length. Then is there a *DFA*, M', accepting E? If so, then demonstrate how to construct it, and if not, prove it.

2.6 The language $L \subseteq \{a,b\}^*$ consists of all words having an odd number of a's and even number of b's. Thus *aaa*, *abbabba*, and *abb* belong to L. Construct a *DFA* accepting L and prove that it is correct.

2.7 Construct a *DFA* accepting the language $L \subseteq \{a,b\}^*$ consisting of all words x that satisfy, simultaneously,

(i) the length of x is divisible by 3;
(ii) x begins with a and ends with b;
(iii) *aaa* is not a subword of x.

3.1 Given the *NFA*, M,

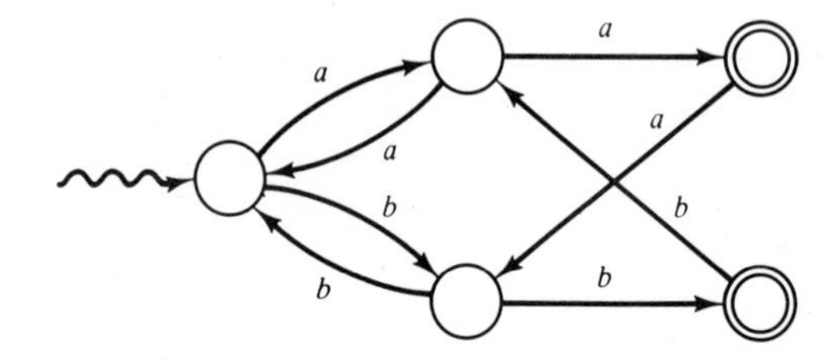

give examples of

(i) four words accepted by M and accepting configuration sequences for them; and
(ii) four words rejected by M.

3.2 Prove that *NFA*, M_2, of Example 2.3.2 accepts the language $L = \{a^i b^j, a^i c^j : i,j \geq 1\}$.

3.3 Given the *NFA*, M,

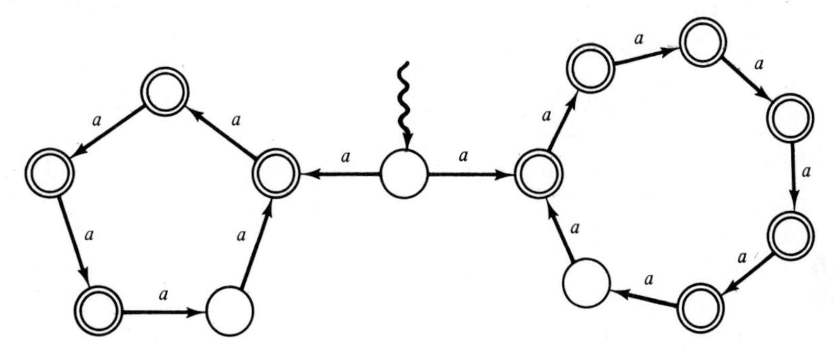

construct an equivalent *DFA*, M', which has only reachable states.

3.4 Construct an *NFA* to recognize the pattern *ababa* in a text from $\{a,b\}^*$ and convert it into an equivalent *DFA*.

3.5 Given the *NFAs* of Exercises 3.1 and 3.3 demonstrate how they can be viewed as deterministic machines using control words.

3.6 Is the following statement true?

> Whenever the subset construction (either one) is applied to an *NFA* with 5 states, the resulting *DFA* has at most 32 states.

4.1 Given the *NFA*, M,

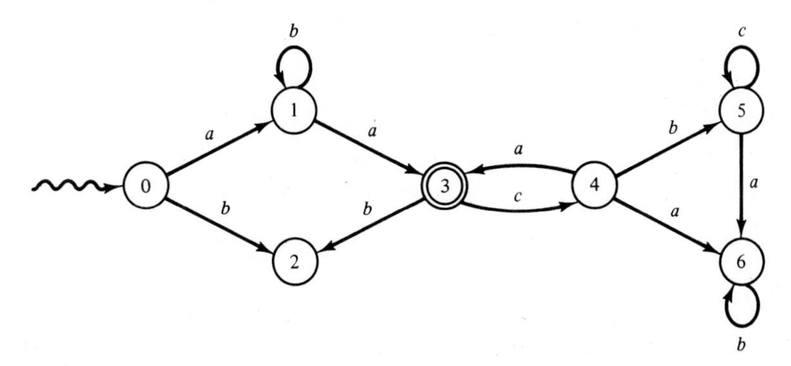

what are

(i) the reachable states of M;
(ii) the reachable symbols of M; and
(iii) the useful states of M.

4.2 Design algorithms to detect reachable and useful states and reachable symbols in an arbitrary *NFA*. Prove the correctness of your algorithms.

4.3 Given the following *DFAs*, construct minimal-state *DFAs* equivalent to them.

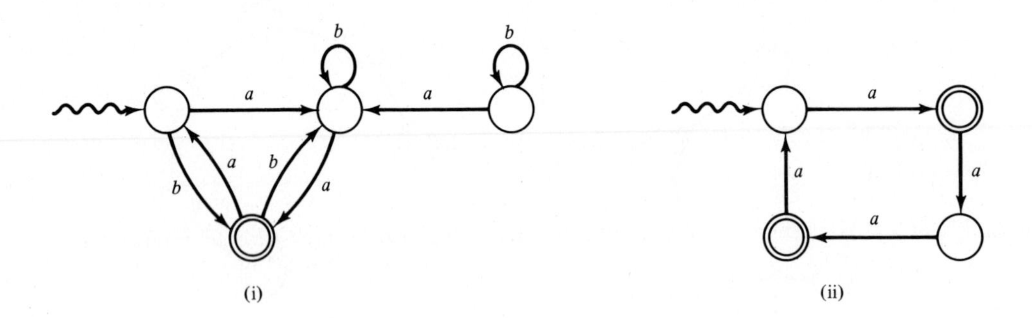

(i) (ii)

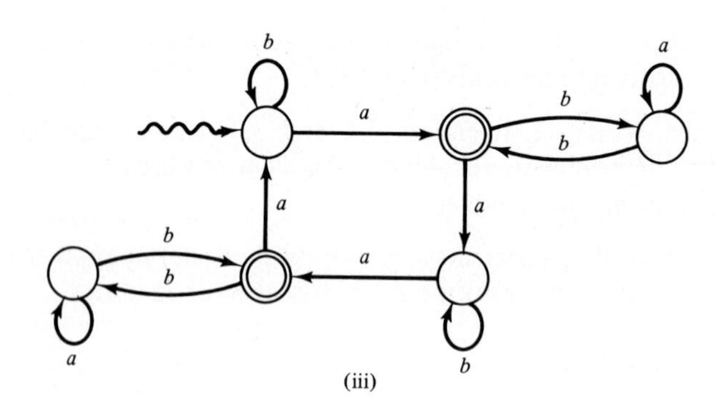

(iii)

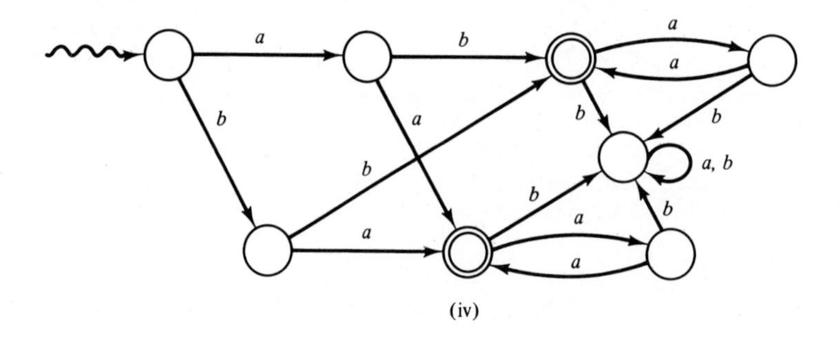

(iv)

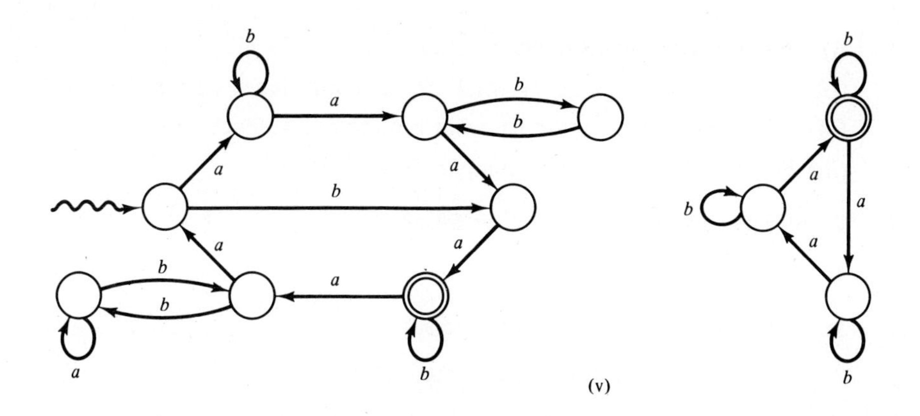

(v)

4.4 The algorithm discussed in the text to compute an equivalent minimal-state *DFA* from a given *DFA* can be implemented to require $O(mn^2)$ time, where m is the number of input symbols and n is the number of states in the original *DFA*. However, a more efficient construction can be obtained which requires $O(mn \log n)$ time; see Hopcroft (1972). Design such a construction.

4.5 Given the two *DFAs*, M_1 and M_2,

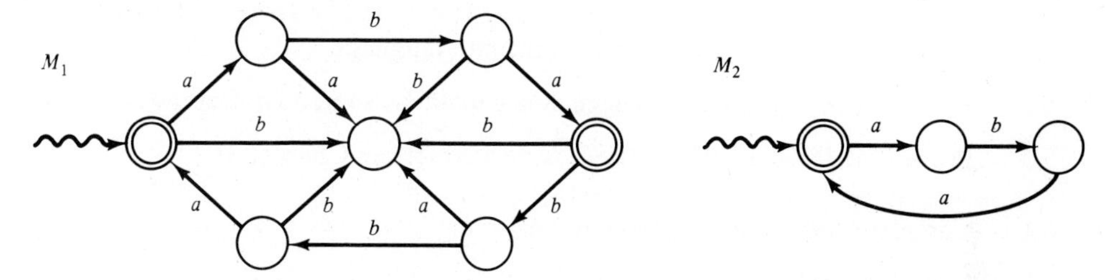

prove they are either equivalent or inequivalent by way of Corollary 2.4.3.

4.6 As in Section 2.7.1 define a relation $\equiv$ for words. For $L \subseteq \Sigma^*$, define a relation $\equiv$ by

For all x, y in Σ^*, $x \equiv y$ if and only if
for all w, z in Σ^*, wxz is in L if and only if wyz is in L

(i) Prove that $\equiv$ is an equivalence relation.
(ii) Find the equivalence classes of $\equiv$ for $L = \{a^3b^{2i}a : i \geq 0\}$ and for $L = \{abaa, aab\}$.

4.7 For $L \subseteq \Sigma^*$, define a relation $\underset{L}{\equiv}$ as follows:

For all x, y in Σ^*, $x \underset{L}{\equiv} y$ if and only if
for all z in Σ^*, xz is in L if and only if yz is in L

(i) Prove $\equiv_L$ is an equivalence relation.
(ii) For $L = \{a^i b^i : i \geq 0\}$ what are the equivalence classes of $\equiv_L$.
(iii) For $L = \{a^i b^j : i,j \geq 0\}$ what are the equivalence classes of $\equiv_L$.
(iii) For L a *DFA* language compare $\equiv$ and $\equiv_L$.

4.8 Is the following statement true or false?

$L = \{ba^{3i} : i \geq 0\}$ and $\equiv_L$ has exactly five equivalence classes.

4.9 Is the following statement true or false?

$L = \{a\}^*$ and $\equiv_L$ has exactly one equivalence class.

4.10 Are the following statements about the equivalence classes of $\equiv_L$ true or false?

The number of equivalence classes of $\equiv_L$

(i) for $L = \{ba^{3i} : i \geq 0\}$ is five, where $L \subseteq \{a,b\}^*$;
(ii) for $L = \{a^i : i \geq 0\}$ is one, where $L \subseteq \{a\}^*$;
(iii) for $L = \{ba^i : i \geq 0\}$ is three, where $L \subseteq \{a,b\}^*$;
(iv)* for $L = \{a^p : p \text{ is prime}\}$ is infinite, where $L \subseteq \{a\}^*$.

5.1 Construct a *DFA* that accepts an infinite language.

5.2 Prove that every finite language is accepted by some loop-free *DFA*.

5.3 Prove that *DFA* FINITENESS is decidable; see Chapter 10.

6.1 Complete the proof of Theorem 2.6.1.

6.2 Prove Theorem 2.6.2.

6.3 Convert the lazy automaton of Example 2.6.4 to an equivalent λ-*NFA* and then convert the λ-*NFA* to an equivalent *NFA*.

6.4 Given the λ-*NFA*, M,

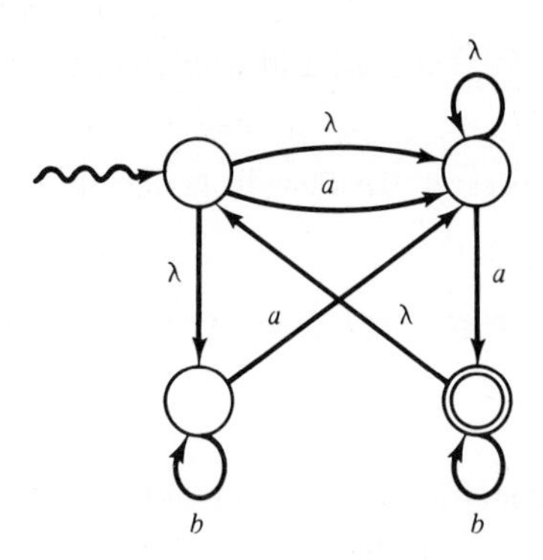

either construct an equivalent *NFA*, M', using the construction in the text, or prove your resulting *FA*, M', accepts $L(M)$.

6.5 We can generalize λ-*NFAs* even further by allowing the read head to move to the left, move to the right, or stay where it is. To prevent the head from falling of the input tape we add left and right endmarkers, $\boldsymbol{l}$ and $\boldsymbol{r}$, respectively. A move left over $\boldsymbol{l}$ and a move right over $\boldsymbol{r}$ are forbidden. We accept by final state when the read head is over $\boldsymbol{r}$. The transition relation $\delta \subseteq Q \times (\Sigma \cup \{\boldsymbol{l},\boldsymbol{r}\} \times Q \times \{L,R,\lambda\}$, where L means move left one position, R means move right by one position, and λ means stay in the same position. This is called a *two-way finite automaton.* Prove that its power is the same as that of λ-*NFAs*.

PROGRAMMING PROJECTS

P1.1 Write a program to simulate the actions of one of your local coffee vending machines (or other type of vending machine). The visual display is a crucial aspect of this programming project.

P1.2 Write a program to simulate the action of a bank of elevators in one of your local buildings.

P2.1 Write a program for a specific *DFA* which provides a visual display of the action of the *DFA* on a user-specified input word. For example this could be done, in its simplest form, by providing the configuration sequence.

P2.2 Write a program which given the specification of a *DFA* produces as output a program of the form specified in Exercise P2.1, a *DFA* compiler.

P2.3 Provide the software tools for displaying the state diagram of an arbitrary *DFA* at your terminal and on various output devices. Bear in mind that the whole state diagram may not fit on the screen in any readable form, so you need to be able to drive around it.

P2.4 Based on the tools provided in Exercise P2.3 design and implement an editor for state diagrams.

P2.5 Write a program which given an arbitrary *DFA* will generate example words accepted by the *DFA*.

P3.1 Generalize Exercise P2.1 to deal with *NFAs*. Think carefully about how to deal with nondeterminism. One approach is to work through all possible configuration sequences exhaustively.

P3.2 Write a program which given the specification of an *NFA* yields an equivalent *DFA*.

P3.3 Implement a pattern-matching program based on the ideas in Example 2.3.3. Consider how to extend this to finding the first appearance of one of a finite number of patterns.

P3.4 Design and implement a *lexical checker* for PASCAL, ADA, etc. That is, it checks that the given sequence of input characters forms a sequence of valid lexical units, for example, identifiers, constants, word delimiters, and punctuation.

P4.1 Write a program to carry out the elimination of unreachable states and symbols and useless states from an *NFA*.

P4.2 Write a program to perform the minimization of a *DFA*. Use either the approach taken in the text or that mentioned in Exercise 4.4.

P4.3 Write a program to perform equivalence testing for two given *DFAs* based on minimization.

P5.1 Write a program which given a *DFA* decides whether or not it is loop-free and if it is whether or not it is a tree.

P6.1 Write a set of programs to convert λ-*NFAs* to *NFAs* and lazy *FAs* to λ-*NFAs*.

Chapter 3

Regular Expressions

3.1 MOTIVATION AND DEFINITION

Searching for an appearance of *baa* in a text file t consisting only of the letters a and b can be formulated as

Is t in $\{a,b\}^*\{baa\}\{a,b\}^*$?

In other words, can t be decomposed into three subwords, the second being *baa*, the first and third being arbitrary words over the alphabet of t, possibly empty. If we wish to obtain the first appearance of *baa*, if any, then the search can be formulated as

Is t in $(\{a,b\}^* - \{a,b\}^*\{baa\}\{a,b\}^*)\{baa\}\{a,b\}^*$?

The language expressions, appearing here, are more usually denoted by regular expressions

Is t in the language denoted by $[a \cup b]^*baa[a \cup b]^*$?

Is t in the language denoted by $[a^*[bb^*a]^*b^*]baa[a \cup b]^*$?

where the second regular expression is more complicated than its corresponding language expression, because the difference operator is unavailable.

Similarly, in searching a bibliography on trees one may wish to find all papers on *AVL* trees, in which case a single word is insufficient. We might use the regular expression

$$A^*[avl \cup hb \cup [height\ balanced]]A^*.$$

where A is the underlying alphabet.

Again to find all papers by some particular authors in a bibliography we have a similar situation, for example,

$$A^*[gonnet \cup mehlhorn \cup ottmann]A^*$$

but if we only wish to list joint papers, then

$$A^*[gonnet\ A^*\ mehlhorn \cup gonnet\ A^*\ ottmann$$
$$\cup\ mehlhorn\ A^*\ ottmann]A^*$$

is, perhaps, more appropriate, assuming the authors are listed in alphabetical order.

In UNIX, a widely available programming environment, three pattern matchers are available: *grep*, *egrep*, and *fgrep*. Each of them search for the appearance of a given pattern in some line of a given text file. The major difference between them is that *fgrep* is given a single word, *egrep* is given a regular expression, and *grep* is given a restricted regular expression. Although *egrep* can deal with everything that *grep* and *fgrep* can, they can be converted into more efficient pattern matchers. We have shown how *fgrep* can be efficiently implemented in Section 2.3. In this chapter we demonstrate one technique for converting a regular expression to a *DFA*, which is a spin off from one of the main theorems. Apart from introducing regular expressions as a more convenient notation than the language expressions of Section 1.2 we prove that they define the same family of languages as finite automata.

We define legal regular expressions recursively.

Definition Let Σ be an alphabet. Then a *regular expression* E over Σ is defined recursively as one of the following types.

(1) $\varnothing$;
(2) λ;
(3) a, where a is in Σ;
(4) $[E_1 \cup E_2]$, where E_1 and E_2 are regular expressions;
(5) $[E_1 \cdot E_2]$, where E_1 and E_2 are regular expressions; or
(6) E_1^*, where E_1 is a regular expression.

The set of all regular expressions over Σ is denoted by $\boldsymbol{R}_\Sigma$. We usually omit the symbol $\cdot$ in regular expressions.

Example 3.1.1A Let $\Sigma = \{a,b\}$ be an alphabet. Then

$$\varnothing,\ \lambda,\ a,\ b$$

are four regular expressions over Σ by (1), (2), and (3) in the above definition.

By (4)

$$[a \cup b]$$

is a regular expression over Σ, since a and b are. We often describe the construction of $[a \cup b]$ from its component parts with an expression tree; see Figure 3.1.1.

Since square brackets are used indicate which regular expressions are combined, they are unnecessary in this tree representation and are, therefore, omitted. Thus, the regular expression

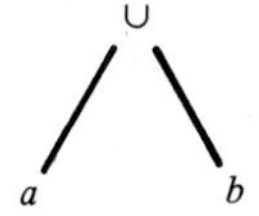

Figure 3.1.1

$$[[a\,a]b]$$

is represented by the tree of Figure 3.1.2. Observe how the order of application of $\cdot$ is indicated here. We also have the regular expression

$$[a[ab]]$$

which yields the tree of Figure 3.1.3.

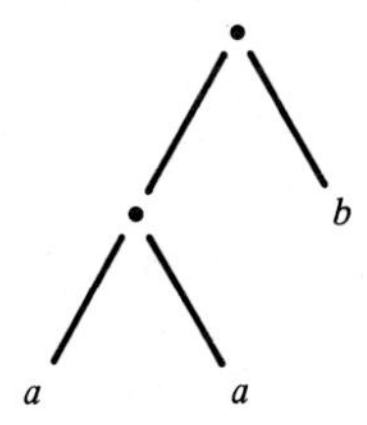

Figure 3.1.2

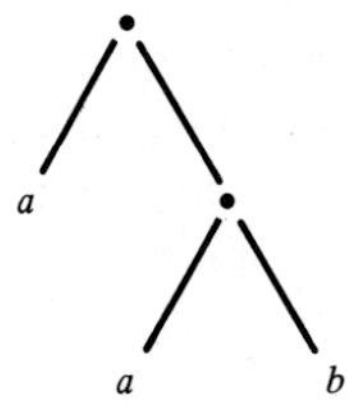

Figure 3.1.3

As we shall see, $[[aa]b]$ and $[a[a\,b]]$ denote the same language, namely, $\{aab\}$. For this reason we will usually write these regular expressions as

$$[aab]$$

or, even,

$$aab$$

when it is clear what is intended. We discuss this issue in more detail below.

Each branching of the tree corresponds to an application of one construction rule. So

$$[[a \cup b][ab]]^*$$

gives the tree of Figure 3.1.4, since it is broken down into its components using the rules indicated in Figure 3.1.4.

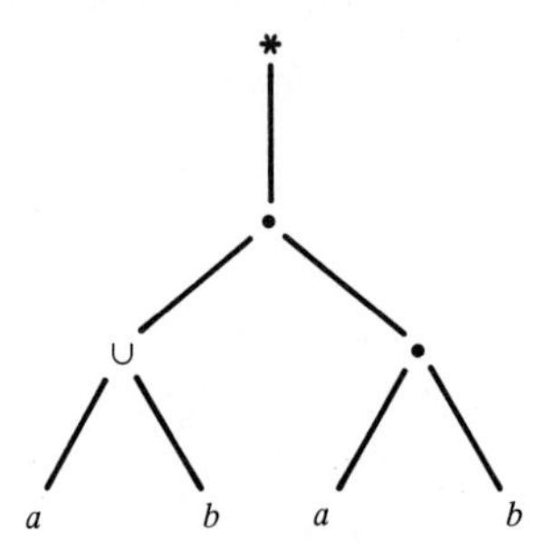

Figure 3.1.4

The bracketing required by the definition of regular expressions often obscures their meaning. For example, we must write $[[[[ab]c]d]e]$ or something similar to denote the language $\{abcde\}$. For this reason we prefer to omit brackets when the meaning is clear, simply writing *abcde*, for example. As with arithmetic expressions in PASCAL and other programming languages we assign priorities to the operators, so that brackets can be omitted. We assign * the highest priority, $\cdot$ the next highest, and $\cup$ the lowest priority. Brackets are used to clarify and override these priorities.

Example 3.1.2 The term ab^* stands for $[ab^*]$ and denotes the language $\{ab^i : i \geq 0\}$, whereas $[ab]^*$ denotes the language $\{(ab)^i : i \geq 0\}$.

The expression $a \cup bc$ stands for $[a \cup [bc]]$ and denotes the language $\{a, bc\}$.

The expression $a \cup b \cup c \cup d$ stands for $[a \cup [b \cup [c \cup d]]]$ or $[[a \cup b] \cup [c \cup d]]$ and denotes the language $\{a, b, c, d\}$.

Example 3.1.3A Let E be

$$[+ \cup - \cup \lambda][1[0 \cup 1]^*] \cup 0$$

Then E is a regular expression over $\{0, 1, +, -\}$ since

$$E = E_1 \cup E_2$$

where $E_1 = [+ \cup - \cup \lambda][1[0 \cup 1]^*]$ and $E_2 = 0$. Now

$$E_1 = E_3 E_4$$

where $E_3 = [+ \cup - \cup \lambda]$ and $E_4 = [1[0 \cup 1]^*]$. Breaking down E_3 we see that

$$E_3 = [E_5]$$

where $E_5 = + \cup - \cup \lambda$, and, hence,

$$E_5 = E_6 \cup E_7$$

where $E_6 = +$ and $E_7 = - \cup \lambda$. Finally, we obtain

$$E_7 = E_8 \cup E_9$$

where $E_8 = -$ and $E_9 = \lambda$. E_4 can also be broken down into its constituent parts; see Figure 3.1.5. In breaking E into its constituent parts we chose to break it at the $\cup$ symbol.

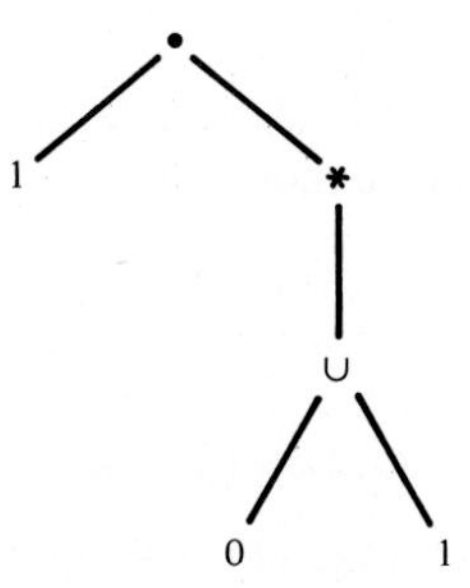

Figure 3.1.5

Apparently, we could also have broken it down as

$$E = E'E''$$

where $E' = [+ \cup - \cup \lambda]$ and $E'' = [1[0 \cup 1]^*] \cup 0$. However, this is not legal, because it would go against the relative priorities of $\cdot$ and $\cup$. If we chose to break E down in this fashion it would be equivalent to being given E as

$$[+ \cup - \cup \lambda][[1[0 \cup 1]^*] \cup 0]$$

E is intended to denote signed and unsigned binary integers with leading zeros suppressed. It could also be written as

$$[+ \cup - \cup \lambda][1[0 \cup 1]^{***}] \cup \lambda^*0$$

since, as we shall see, E^* and E^{**} denote the same language, and λ^* denotes the same language as λ, that is, $\{\lambda\}$.

Each regular expression denotes a language that is also defined recursively.

Definition Let E be a regular expression over Σ. Then the *language* $L(E)$ *denoted by* E is defined as follows.

(1) If $E = \varnothing$ then $L(E) = \varnothing$;
(2) If $E = \lambda$ then $L(E) = \{\lambda\}$;
(3) If $E = a$, for some a in Σ, then $L(E) = \{a\}$;
(4) If $E = [E_1 \cup E_2]$, then $L(E) = L(E_1) \cup L(E_2)$;
(5) If $E = [E_1 E_2]$, then $L(E) = L(E_1)L(E_2)$; and
(6) If $E = E_1^*$, then $L(E) = L(E_1)^*$.

A language $L \subseteq \Sigma^*$ is said to be a *regular language* if there is a regular expression E over Σ with $L = L(E)$. We say L is a *REGL* in this case. The *family of regular languages* is denoted by $\boldsymbol{L}_{REG}$ and defined by

$$\boldsymbol{L}_{REG} = \{L : L = L(E) \text{ for some regular expression } E\}.$$

Example 3.1.1B The regular expressions

$$\varnothing, \lambda, a, b$$

denote the languages

$$\varnothing, \{\lambda\}, \{a\}, \{b\}$$

respectively. These follow from rules (1), (2), and (3), respectively. In a similar manner, $[a \cup b]$ denotes $\{a,b\}$ from rule (4), while $[[aa]b]$ denotes $L([aa])L(b) = \{aa\}\{b\} = \{aab\}$. Again, $L([a[ab]])$ also denotes $\{aab\}$, since $L(a)L([ab]) = \{a\}\{ab\}$. Finally, $L([[a \cup b][ab]]^*)$ denotes

$$[L([a \cup b])L([ab])]^* = [\{a,b\}\{ab\}]^* = \{aab, bab\}^*$$

Example 3.1.3B Given $E = [+ \cup - \cup \lambda]\,[1[0 \cup 1]^*] \cup 0$, we find that $L(E) = L(E_1) \cup L(0)$, where $E_1 = [+ \cup - \cup \lambda][1\,[0 \cup 1)^*]$, by rule (4). $L(E_1) = L(E_2)\,L(E_3)$, where $E_2 = [+ \cup - \cup \lambda]$ and $E_3 = [1\,[0 \cup 1)^*]$, by (5). Now

$$\begin{aligned} L(E_2) &= L(+) \cup L(- \cup \lambda), && \text{by (4)} \\ &= L(+) \cup L(-) \cup L(\lambda), && \text{by (4) and associativity of } \cup, \\ &= \{+\} \cup \{-\} \cup \{\lambda\}, && \text{by (2) and (3),} \\ &= \{+,-,\lambda\} \end{aligned}$$

Similarly,

$$\begin{aligned} L(E_3) &= L(1)L([0 \cup 1]^*), && \text{by (5),} \\ &= L(1)L(0 \cup 1)^*, && \text{by (6)} \\ &= \{1\}\{0,1\}^*, && \text{by (4)} \end{aligned}$$

Hence,

$$\begin{aligned} L(E) &= \{+,-,\lambda\}(\{1\}\{0,1\}^*) \cup \{0\} \\ &= \{+1,-1,1\}\{0,1\}^* \cup \{0\} \end{aligned}$$

By (6) if $E = E_1^*$, then $L(E) = L(E_1)^*$ and, therefore, if $E = E_1^{**}$, then $L(E) = L(E_1^*)^* = (L(E_1)^*)^* = L(E_1)^*$, that is, E^{**} and E^* denote the same language. Similarly, $E \cup \varnothing$ denotes the same language as E, and $E \cup \varnothing^*$ the same as $E \cup \lambda$, since $\varnothing^* = \{\lambda\}$, by definition. This implies we could omit λ in the definition of regular expressions, since it can always be replaced by $\varnothing^*$. We include λ because it is convenient rather than necessary. These simplifications of regular expressions are purely formal, they are independent of the underlying alphabet and of the specific expressions. The above discussion leads to the following:

Definition Let E_1 and E_2 be two regular expressions over Σ. Then E_1 and E_2 are *equivalent* if $L(E_1) = L(E_2)$.

As we now show, a language is denoted by a regular expression if and only if it is accepted by a *DFA*. Thus, these two models are equivalent with respect to their expressive power. Regular expressions are an alternative method of specifying a language when a machine to accept it is not needed. When such a machine is needed the construction of the next section provides one means of obtaining it.

3.2 REGULAR EXPRESSIONS INTO FINITE AUTOMATA

We demonstrate that $L(E)$, for E a regular expression, is accepted by a λ-*NFA* M_E, which is constructed recursively from E. To this end we define three operations on λ-*NFAs*.

Definition Let $M_1 = (Q_1,\Sigma_1,\delta_1,s_1,F_1)$ and $M_2 = (Q_2,\Sigma_2,\delta_2,s_2,F_2)$ be two λ-*NFAs* with disjoint state sets as displayed in Figure 3.2.1.

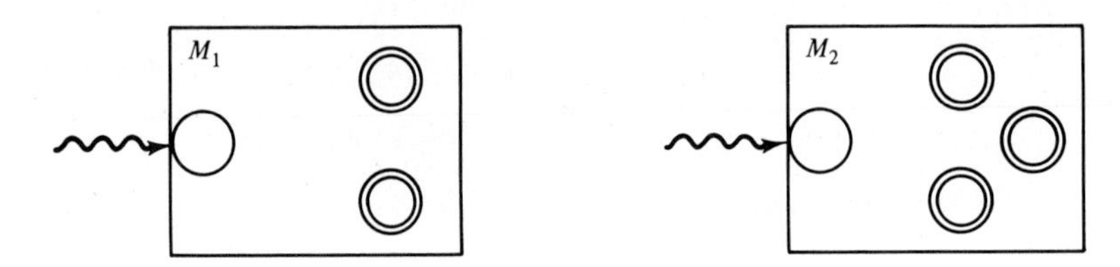

Figure 3.2.1

(a) $M_1 \cup M_2$, the *union of* M_1 *and* M_2, is defined as the λ-*NFA*, $M = (Q, \Sigma_1 \cup \Sigma_2, \delta, s, F)$, where $Q = Q_1 \cup Q_2 \cup \{s\}$, $\delta = \delta_1 \cup \delta_2 \cup \{(s, \lambda, s_1), (s, \lambda, s_2)\}$, s is a new start state, and $F = F_1 \cup F_2$; see Figure 3.2.2.

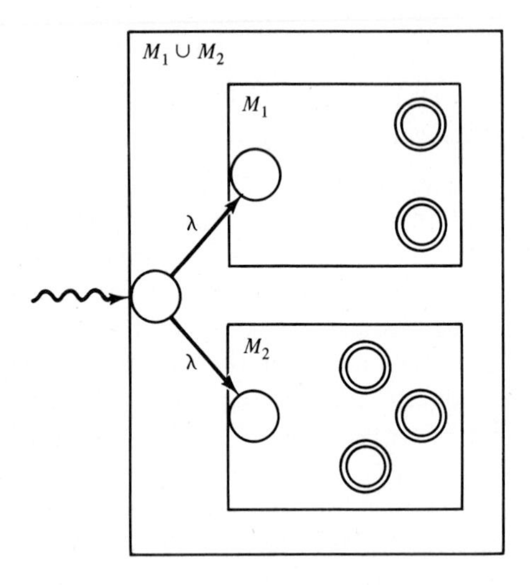

Figure 3.2.2

(b) $M_1 M_2$, *the product of* M_1 and M_2, is defined as the λ-*NFA*, $M = (Q, \Sigma_1 \cup \Sigma_2, \delta, s_1, F_2)$, where $Q = Q_1 \cup Q_2$ and $\delta = \delta_1 \cup \delta_2 \cup \{(f, \lambda, s_2) :$ f is in $F_1\}$; see Figure 3.2.3.

(c) M_1^*, the *star of* M_1, is defined as the λ-*NFA*, $M = (Q_1, \Sigma_1, \delta, s, F_1 \cup \{s\})$, where $\delta = \delta_1 \cup \{(s, \lambda, s_1)\} \cup \{(f, \lambda, s_1) : f$ is in $F_1\}$ and s is a new start and final state; see Figure 3.2.4.

The importance of these constructions comes from the following:

Theorem 3.2.1 *Let* M_1 *and* M_2 *be two* λ-*NFAs with disjoint state sets. Then*

(a) $L(M_1 \cup M_2) = L(M_1) \cup L(M_2)$;

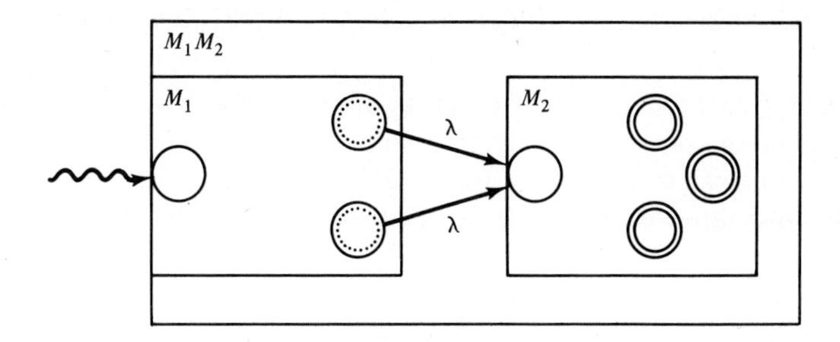

Figure 3.2.3

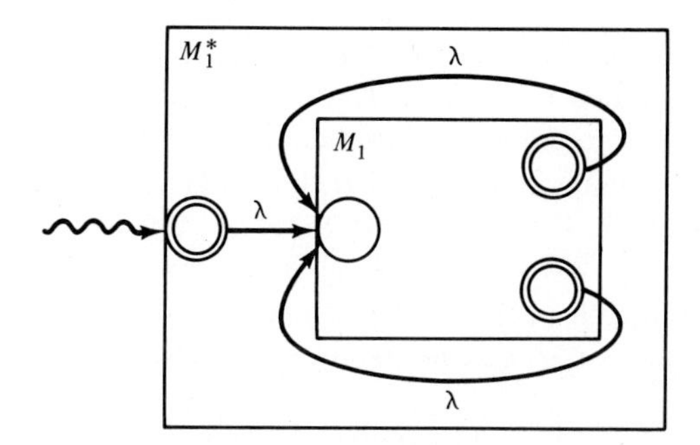

Figure 3.2.4

(b) $L(M_1M_2) = L(M_1)L(M_2)$; *and*
(c) $L(M_1^*) = L(M_1)^*$.

Proof: Each statement is proved in a similar manner, by showing containment each way using the generic approach. Let $M_1 = (Q_1,\Sigma_1,\delta_1,s_1,F_1)$ and let $M_2 = (Q_2,\Sigma_2,\delta_2,s_2,F_2)$, where $Q_1 \cap Q_2 = \varnothing$.

(a) Let x be any word in $L(M_1 \cup M_2)$. By construction this implies $sx \vdash^* f$, for some f in $F_1 \cup F_2$. Now either f is in F_1 or f is in F_2. We only consider f in F_1, since f in F_2 gives rise to a similar argument. Now either $sx \vdash s_1x$ or $sx \vdash s_2x$ in the configuration sequence $sx \vdash^* f$. Because $Q_1 \cap Q_2 = \varnothing$, f is not reachable from s_2; therefore, we must have $sx \vdash s_1x \vdash^* f$. But this implies $s_1x \vdash^* f$ in M_1, since the only transitions added to form $M_1 \cup M_2$ are from s to s_1 and s_2. In other words, we have shown that x is in $L(M_1)$ if f is in F_1. By a similar argument, x is in $L(M_2)$ if f is in F_2. Therefore, $L(M_1 \cup M_1) \subseteq L(M_1) \cup L(M_2)$.

Conversely, if x is in $L(M_1) \cup L(M_2)$, then $s_1x \vdash^* f_1$ in M_1, for some f_1 in F_1, or $s_2x \vdash^* f_2$ in M_2, for some f_2 in F_2. (Note that both may hold.) This implies, by construction, that $sx \vdash s_ix \vdash^* f_i$ in $M_1 \cup M_2$, for

$i = 1$ or $i = 2$. That is, $L(M_1) \cup L(M_2) \subseteq L(M_1 \cup M_2)$, and equality has been proved.

(b) Any word x in $L(M_1M_2)$ has a configuration sequence $s_1x \vdash^* f$, where f is in F_2. Now the only transitions from states in M_1 to states in M_2 are λ-transitions from final states in M_1 to s_2. Hence, this configuration sequence can be split into three parts $s_1x \vdash^* f_1y \vdash s_2y \vdash^* f$, where f_1 is in F_1 and $x = wy$, for some w in Σ_1^* and y in Σ_2^*. Immediately, w is in $L(M_1)$ and y is in $L(M_2)$; therefore, x is in $L(M_1)L(M_2)$. This is the central part of the proof; we leave its completion to the Exercises.

(c) Similarly, to the proof of (b) a configuration sequence $sx \vdash^* f$ in M_1^* can be split into parts — added λ-transitions and configuration sequences in M_1. The details are left to the Exercises. ■

If M_1 and M_2 do not have disjoint state sets, then the states of M_2, for example, can always be renamed to achieve this.

Theorem 3.2.1 has an important implication. It demonstrates that given two *DFA* languages, their union is also a *DFA* language, as is their catenation. Moreover, the third part of the theorem shows that the star of a *DFA* language is once again a *DFA* language. We say that $\boldsymbol{L}_{DFA}$ is *closed* under union, catenation, and star. A detailed study of such closure properties is to be found in Chapter 9, where their importance is also discussed. In summary:

Theorem 3.2.2 $\boldsymbol{L}_{DFA}$ *is closed under union, catenation, and star.*

We now have

Theorem 3.2.3 *Let E be a regular expression over Σ. Then a DFA, M, with $L(M) = L(E)$ can be constructed from E.*

Proof: We demonstrate how to construct a λ-*NFA*, M, from E with $L(M) = L(E)$. This is sufficient to prove the theorem, since we have already shown in Chapter 2 how to obtain an equivalent *DFA* from M.

The construction of a λ-*NFA*, M, from E is defined recursively following the definition of regular expressions.

(1) If $E = \varnothing$, then $M = (\{s\},\Sigma,\delta,s,\varnothing)$, where $\delta(s,a) =$ ***undef***, for all a in Σ; see Figure 3.2.5(a).
(2) If $E = \lambda$, then $M = (\{s\},\Sigma,\delta,s,\{s\})$, where $\delta(s,a) =$ ***undef***, for all a in Σ; see Figure 3.2.5(b).
(3) If $E = a$, then $M = (\{s,f\},\Sigma,\delta,s,\{f\})$, where $\delta(s,a) = f$, and δ is undefined otherwise; see Figure 3.2.6.
(4) If $E = [E_1 \cup E_2]$, then $M = M_1 \cup M_2$, where M_i is constructed from E_i, for $i = 1,2$.

Figure 3.2.5

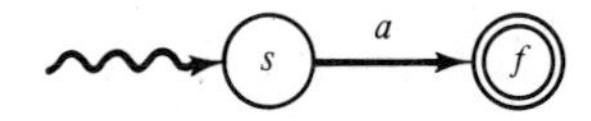

Figure 3.2.6

(5) If $E = [E_1 E_2]$, then $M = M_1 M_2$, where M_i is constructed from E_i, for $i = 1,2$.
(6) If $E = E_1^*$, then $M = M_1^*$, where M_1 is constructed from E_1.

Claim: For E an arbitrary regular expression over Σ, the λ-*NFA*, M, constructed as above satisfies $L(M) = L(E)$.

Proof of Claim: (by structural induction)
For a regular expression E, let $Op(E)$ be the total number of $\cup$, $\cdot$, and $*$ operations in E. Clearly, $Op(E) \geq 0$, equality being obtained when $E = \varnothing$, λ, or a, for some a in Σ. We base the structural induction on $Op(E)$.

Basis: $Op(E) = 0$. Then $E = \varnothing$, λ, or a in Σ. Clearly, the corresponding M given by the above construction satisfies $L(E) = L(M)$.

Induction Hypothesis: Assume the claim holds for all E with $Op(E) \leq k$, for some $k \geq 0$.

Induction Step: Consider an arbitrary regular expression E with $Op(E) = k+1$. Since $k \geq 0$, we have $k+1 \geq 1$, and E contains at least one operator $\cup$, $\cdot$, or $*$. So E must have one of the following forms, for some expressions E_1 and E_2

$$E = [E_1 \cup E_2]$$

$$E = [E_1 \cdot E_2]$$

$$E = E_1^*$$

In each case $Op(E_1) \leq k$ and, in the first two cases, $Op(E_2) \leq k$. By the induction hypothesis this implies that there are λ-*NFAs* M_1 and M_2 with $L(M_1) = L(E_1)$ and $L(M_2) = L(E_2)$.

To complete the induction step we need to construct a λ-*NFA*, M, such that $L(M) = L(E)$. If $E = [E_1 \cup E_2]$, we define M to be $M_1 \cup M_2$, if $E = [E_1 \cdot E_2]$, we define M to be $M_1 M_2$, and if $E = E_1^*$, we define M to be M_1^*. Theorem 3.2.1 implies, in all three cases, that $L(M) = L(E)$ as desired.

This completes the induction step and the proof of the claim. ■

The proof of the theorem now follows. ■

Example 3.2.1 Let $E = [+ \cup - \cup \lambda][1[0 \cup 1]^*] \cup 0$. Define M_+, M_-, M_λ, M_0, and M_1 as shown in Figures 3.2.7(a)-(e), respectively. From $+ \cup - \cup \lambda$ we obtain Figure 3.2.8 and from $1[0 \cup 1]^*$ we have Figure 3.2.9. These give the M of Figure 3.2.10.

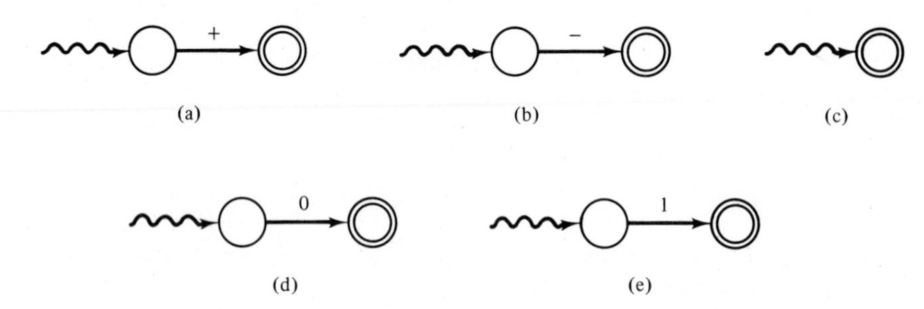

Figure 3.2.7

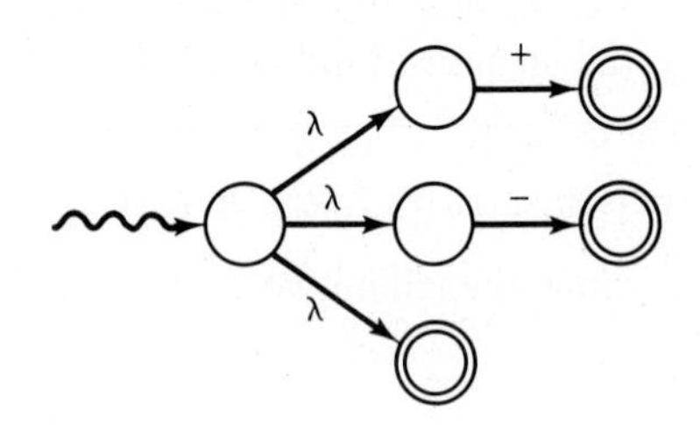

Figure 3.2.8

3.3 EXTENDED FINITE AUTOMATA INTO REGULAR EXPRESSIONS

Not only can we obtain a *DFA* accepting the same language as a given regular expression, but the converse also holds. This demonstrates that every *DFA* language is regular, and vice versa; that is, $\boldsymbol{L}_{DFA} = \boldsymbol{L}_{REG}$. In other words, regular expressions and *DFAs* have exactly the same expressive power.

To prove that every *DFA* language is regular we introduce an extension of finite automata which allow regular expressions in transitions.

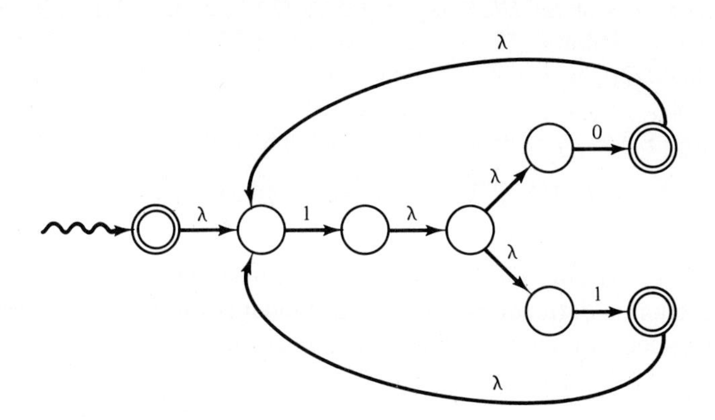

Figure 3.2.9

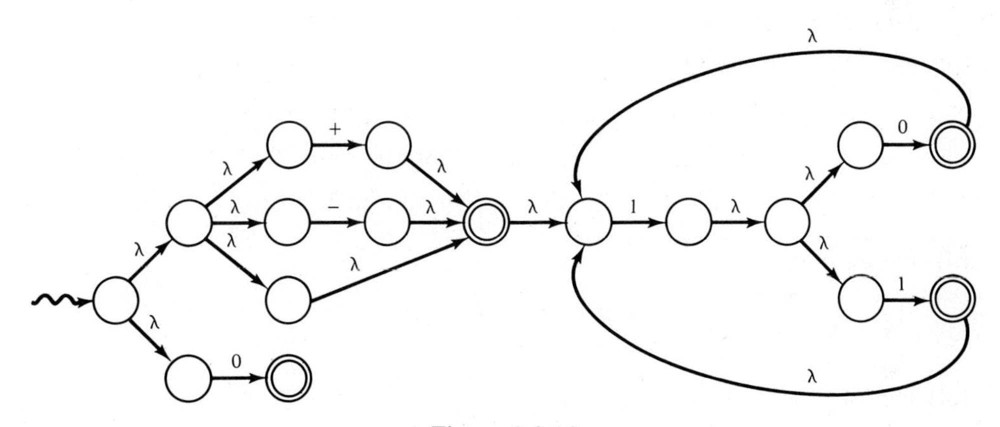

Figure 3.2.10

Definition An *extended finite automaton*, or *EFA*, M, is a quintuple $(Q, \Sigma, \delta, s, f)$, where

Q is a finite, nonempty set of *states*;
Σ is an *input alphabet;*
$\delta : Q \times Q \to \boldsymbol{R}_\Sigma$ is a *total extended transition function*;
s in Q is a *start* state; and
f in Q is a *final* state.

Note that in a transition $\delta(p,q) = E$, E is a regular expression over Σ. In particular, if $\delta(p,q) = \varnothing$, then M cannot move from p to q in one step.

A *configuration* of an *EFA* is a word in $Q\Sigma^*$, as it is for *FAs*, and a single move is defined as follows. We write $px \vdash qy$ in M if $\delta(p,q) = E$ and $x = wy$, for some w in $L(E)$.

We obtain $\vdash^i$, $\vdash^+$, and $\vdash^*$ as before. Thus, the *language accepted by an EFA* $M = (Q,\Sigma,\delta,s,f)$ is denoted by $L(M)$ and is defined by

$$L(M) = \{x : x \text{ is in } \Sigma^* \text{ and } sx \vdash^* f\}$$

It is not immediately clear that *EFAs are* indeed a generalization of *FAs*, since each *EFA* has only one final state, the transition function is total, and the transition function is defined differently. But such is the case as we now prove for *DFAs*. The proof can be extended easily to accommodate *NFAs*, λ-*NFAs*, and lazy *FAs*.

Lemma 3.3.1 *Every DFA language is an EFA language. Therefore, every regular language is an EFA language.*

Proof: Let $M = (Q,\Sigma,\delta,s,F)$ be a given *DFA*. Define an *EFA*, $M' = (Q \cup \{f\},\Sigma,\delta',s,f)$, as follows, where f is a new state not in Q.

(i) For all p and q in Q,

$$\delta'(p,q) = \begin{cases} \emptyset, \text{ if } \delta(p,a) \neq q, \text{ for all } a \text{ in } \Sigma. \\ a_1 \cup \cdots \cup a_r, \quad \text{if } \delta(p,a_i) = q,\ 1 \leq i \leq r, \text{ and} \\ \qquad \delta(p,a) \neq q, \text{ for all } a \text{ in } \Sigma - \{a_1,\ldots,a_r\}. \end{cases}$$

(ii) For all p in Q,

$$\delta'(p,f) = \begin{cases} \emptyset, & \text{if } p \text{ is not in } F \\ \lambda, & \text{if } p \text{ is in } F \end{cases}$$

(iii) For all p in $Q \cup \{f\}$,

$$\delta'(f,p) = \emptyset.$$

This construction is illustrated in Example 3.3.1 below.

We leave to the Exercises the formal proof that $L(M') = L(M)$. Accepting configuration sequences in M are carried over unchanged into M' except for the addition of a terminating λ-transition to ensure acceptance. ■

Example 3.3.1 Given the *DFA* of Figure 3.3.1 we obtain the *EFA* of Figure 3.3.2 using the construction given in the proof of Lemma 3.3.1. We have omitted all $\emptyset$-transitions.

We now prove directly that every *EFA* language is a regular language, that is, *EFAs* and *DFAs* have the same expressive power. The proof is constructive and simple, illustrating yet again that it is often easier to prove a more general result

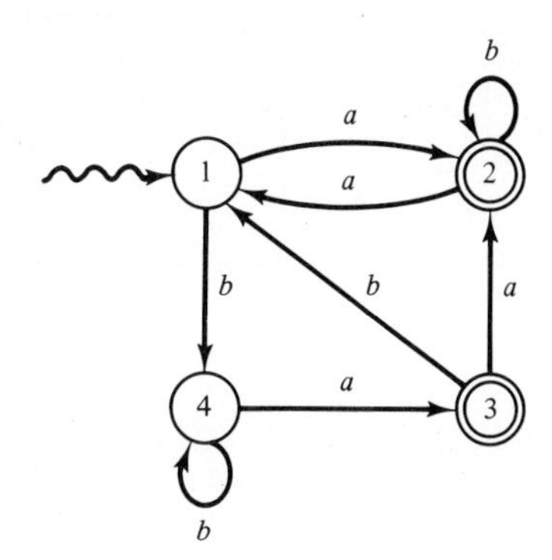

Figure 3.3.1

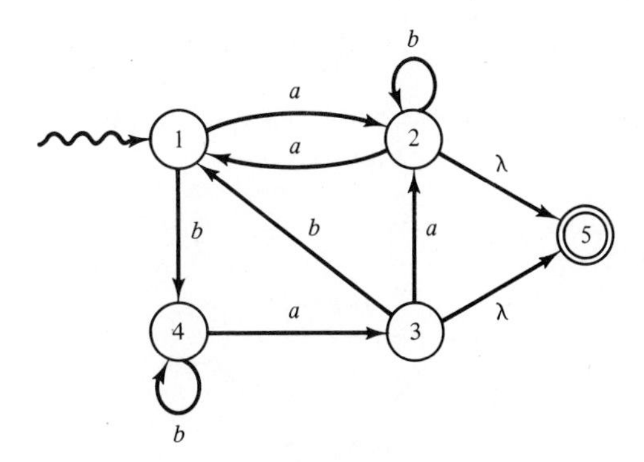

Figure 3.3.2

than a more restricted one. For every *EFA*, M, we demonstrate that there is a regular expression E with $L(E) = L(M)$.

Given an *EFA*, $M = (Q,\Sigma,\delta,s,f)$, we find it convenient to require that $\delta(p,s) = \varnothing$, for all p in Q. If M does not satisfy this requirement simply introduce a new start state s' which has only a λ-transition to s, the original start state. All other transitions involving s' are $\varnothing$-transitions. The *EFA*, M', obtained in this way is clearly equivalent to M; see Figure 3.3.3.

We now demonstrate how to construct a regular expression E for an *EFA*, M with a such start state such that $L(E) = L(M)$. Let $M = (Q,\Sigma,\delta,s,f)$ be an *EFA* such that $\delta(p,s) = \varnothing$, for all p in Q and let q be in $Q - \{s,f\}$. (If $Q = \{s,f\}$ the construction is complete.) We will eliminate q from M by providing short cuts or bypasses around it. Let p and r be states distinct from q. Given the situation shown in Figure 3.3.4 we add a short cut to M, directly from p to r to give Figure 3.3.5. Intuitively, if we originally had a configuration sequence

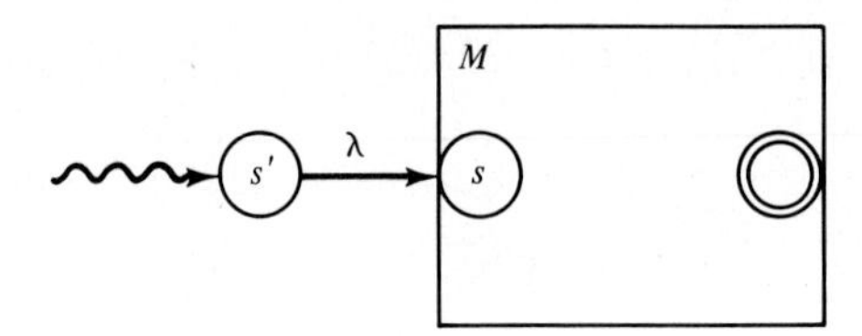

Figure 3.3.3

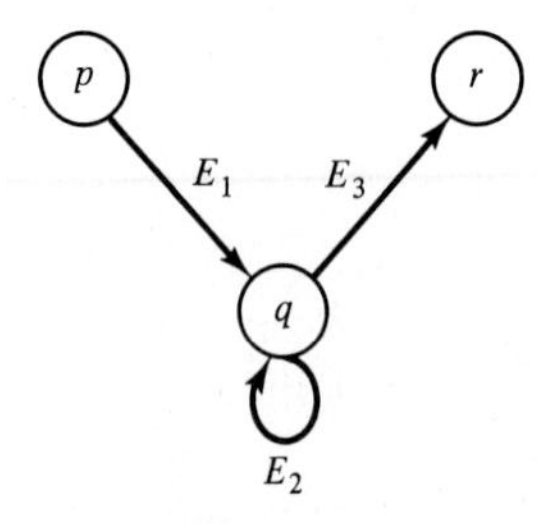

Figure 3.3.4

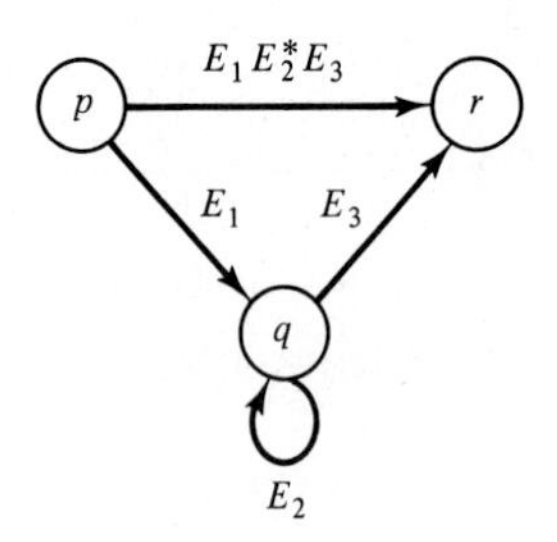

Figure 3.3.5

$$pw \vdash qx \vdash^* qy \vdash r$$

where $qx \vdash^* qy$ using the transition $\delta(q,q) = E_2$, then we now also have the transition

$$pw \vdash r$$

Conversely, if $pw \vdash r$, where w is in $E_1E_2^*E_3$, then there is a configuration sequence

$$pw \vdash qx \vdash^* qy \vdash r$$

using the transition $\delta(q,q) = E_2$.

Because q is neither the start state nor the final state this implies any word accepted by M before adding the short cuts is still accepted, and conversely.

After adding all the short cuts for all pairs p and r, we remove q and the transitions to and from q. By the following lemma the resulting EFA, M', has $L(M') = L(M)$ and M' has one less state than M.

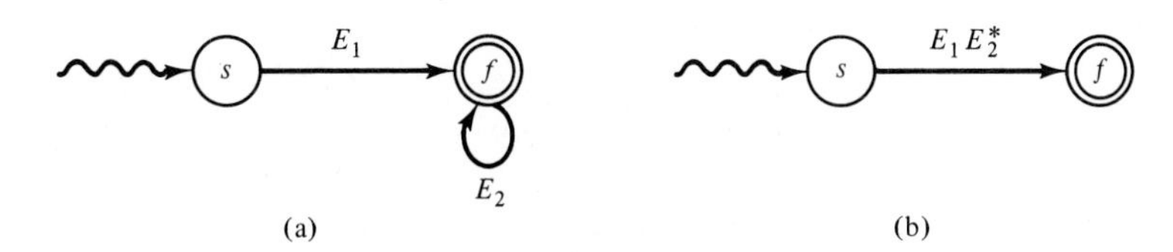

Figure 3.3.6

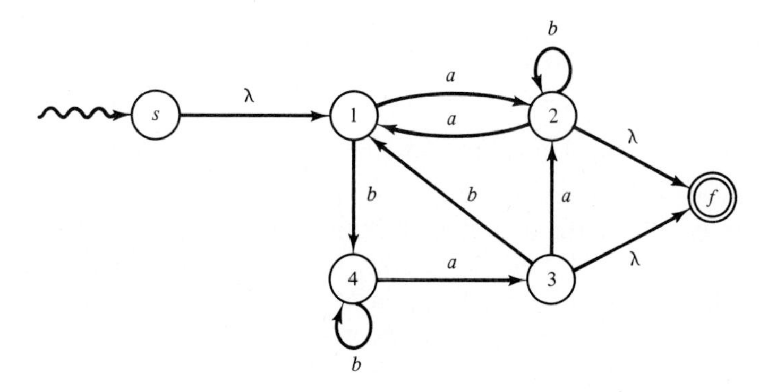

Figure 3.3.7

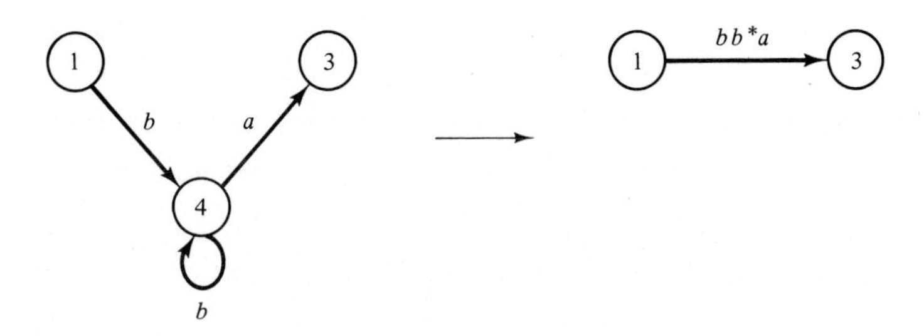

Figure 3.3.8

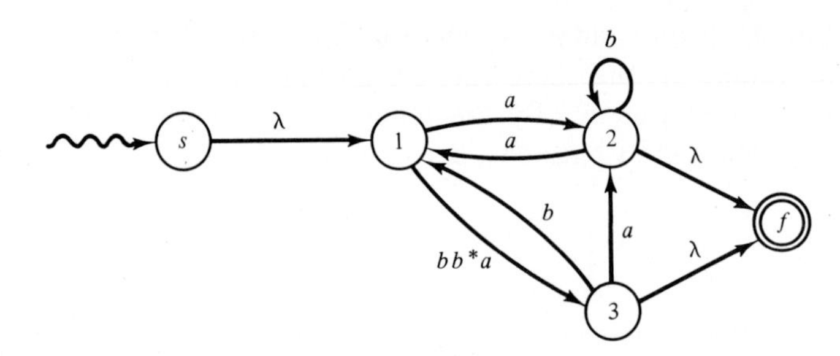

Figure 3.3.9

Lemma 3.3.2 *Let $M = (Q,\Sigma,\delta,s,f)$ be an EFA and q be a state in $Q - \{s,f\}$. Define an EFA, $M' = (Q - \{q\},\Sigma,\delta',s,f)$, where for all p and r in $Q - \{q\}$, $\delta'(p,r) = \delta(p,r) \cup \delta(p,q)\delta(q,q)^*\delta(q,r)$. Then $L(M') = L(M)$.*

Proof: The proof is in two parts. First, we show that $L(M') \subseteq L(M)$ and, second, that $L(M) \subseteq L(M')$. In both cases we use the generic approach.

(i) $L(M') \subseteq L(M)$. Consider any word x in $L(M')$. It has a configuration sequence

$$p_0x_1 \cdots x_m \vdash p_1x_2 \cdots x_m \vdash \cdots \vdash p_{m-1}x_m \vdash p_m$$

in M', where $p_0 = s$, $p_m = f$, and $x = x_1 \cdots x_m$, for some x_i in Σ^*, $1 \leq i \leq m$. From the definition of δ', each step $p_{i-1}x_i \vdash p_i$ in M' either is a transition in M or it corresponds to a configuration sequence $p_{i-1}u_iv_iw_i \vdash qv_iw_i \vdash^* qw_i \vdash p_i$ in M, where $x_i = u_iv_iw_i$, for some u_i, v_i, and w_i in Σ^*, $1 \leq i \leq m$. In other words, $sx \vdash^* f$ in M, x is in $L(M)$, and, hence, $L(M') \subseteq L(M)$.

(ii) $L(M) \subseteq L(M')$. Again consider any word x in $L(M)$. We must show that x is in $L(M')$. Since x is in $L(M)$ it has a configuration sequence

$$p_0x_1 \cdots x_m \vdash p_1x_2 \cdots x_m \vdash \cdots \vdash p_{m-1}x_m \vdash p_m$$

in M, where $p_0 = s$, $p_m = f$, and $x = x_1 \cdots x_m$, for some x_i in Σ^*, $1 \leq i \leq m$.

On the one hand, if q does not appear in this sequence, then it is a valid configuration sequence in M'. Therefore, in this case x is also in $L(M')$. On the other hand, if q does appear, then we need to replace transitions to and from q with transitions in M' to obtain a configuration sequence in M'. For this purpose, let $(p_i, \ldots, p_j)$ be a *q-block*, if $0 < i \leq j < m$, $p_i = p_{i+1} = \cdots = p_j = q$, $p_{i-1} \neq q$, and $p_{j+1} \neq q$. By definition, $p_0 \neq q$ and $p_m \neq q$, so every q-block has a preceding and succeeding state in the configuration sequence. If $(p_i, \ldots, p_j)$ is the first q-block in the sequence, then we have

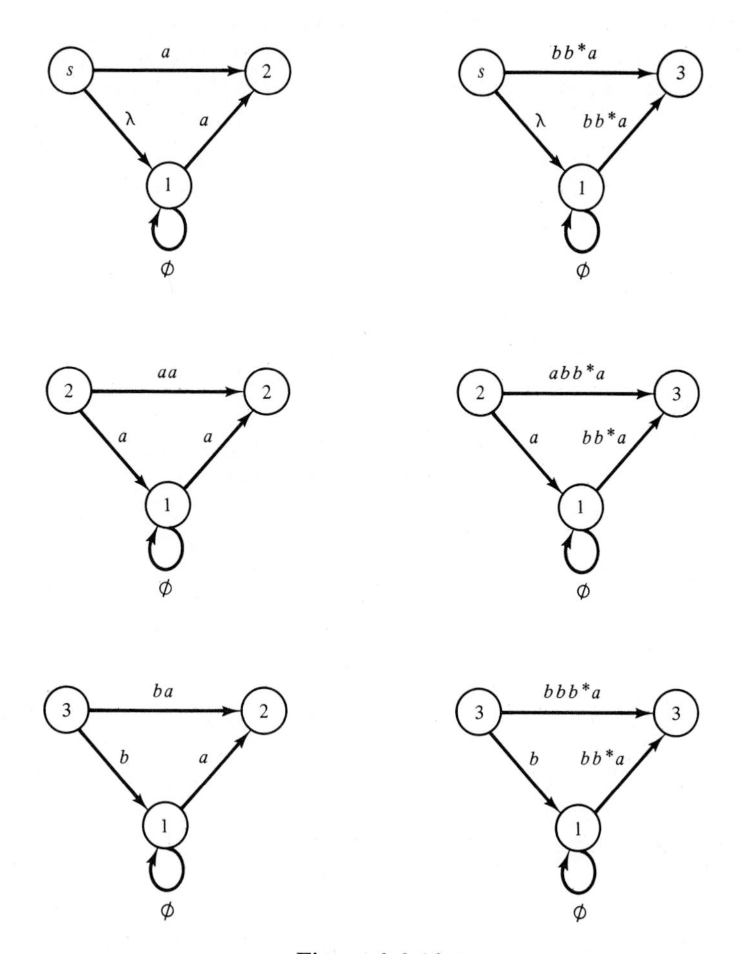

Figure 3.3.10

$$p_0x_1 \cdots x_m \vdash^* p_{i-1}x_i \cdots x_m$$

in M and M', and

$$p_{i-1}x_i \cdots x_m \vdash p_{j+1}x_{j+2} \cdots x_m$$

in M', by construction. The remaining portion

$$p_{j+1}x_{j+2} \cdots x_m \vdash^* p_m$$

of the original configuration sequence in M has one fewer q-block than the original. This process can be repeated until no q-blocks remain, when we have obtained a configuration sequence for x in M'. A proof by induction on the number of q-blocks can be formalized easily. We leave this to the Exercises. Therefore, $L(M) \subseteq L(M')$ as desired. ∎

To complete the construction we repeat the elimination process if $Q - \{q,f,s\}$ is nonempty. Eventually we are left with an *EFA*, $\overline{M}$, which has exactly two states, as shown in Figure 3.3.6(a). Note that, because of our assumptions regarding s and f, $\bar{\delta}(s,s) = \bar{\delta}(f,s) = \varnothing$. Immediately, we can construct the *EFA* $\overline{\overline{M}}$ of Figure 3.3.6(b) which is equivalent to $\overline{M}$. Moreover, x is in $L(\overline{\overline{M}})$ if and only if x is in $L(E_1E_2^*)$; hence, $E_1E_2^*$ is the desired regular expression. To summarize, we have the following:

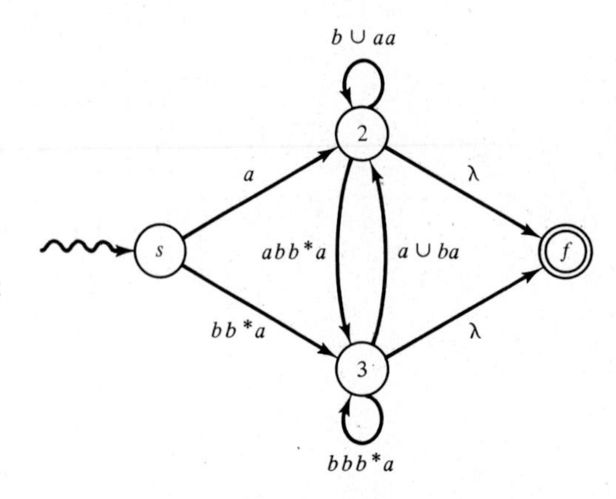

Figure 3.3.11

Theorem 3.3.3 *Let L be an EFA language. Then L is a regular language. This implies* $\boldsymbol{L}_{DFA} = \boldsymbol{L}_{REG}$.

Proof: Since L is accepted by some *EFA*, M, we can carry out the above construction to obtain a regular expression E. By Lemma 3.3.2 $L(E) = L(M) = L$; therefore, L is regular. The second statement follows from Lemma 3.3.1 and Theorem 3.2.3. ■

Example 3.3.2 Given the *DFA*, M, of Figure 3.3.1, we obtain the *EFA* of Figure 3.3.7 when adding new start and final states to it. Eliminating 4 is particularly simple, since 1 and 3 are the only pair of states involving 4; see Figure 3.3.8. This gives the *EFA* of Figure 3.3.9. Eliminating 1 is more difficult since we have six pairs of states to deal with; see Figure 3.3.10. After removing 1 we obtain the *EFA* of Figure 3.3.11. Eliminating 2 we have four pairs of states; see Figure 3.3.12. After removing 2 we obtain Figure 3.3.13. Although the regular expressions involved are, by now, quite complex, the elimination of 3 is straightforward; see Figures 3.3.14 and 3.3.15. So the resulting regular expression E is

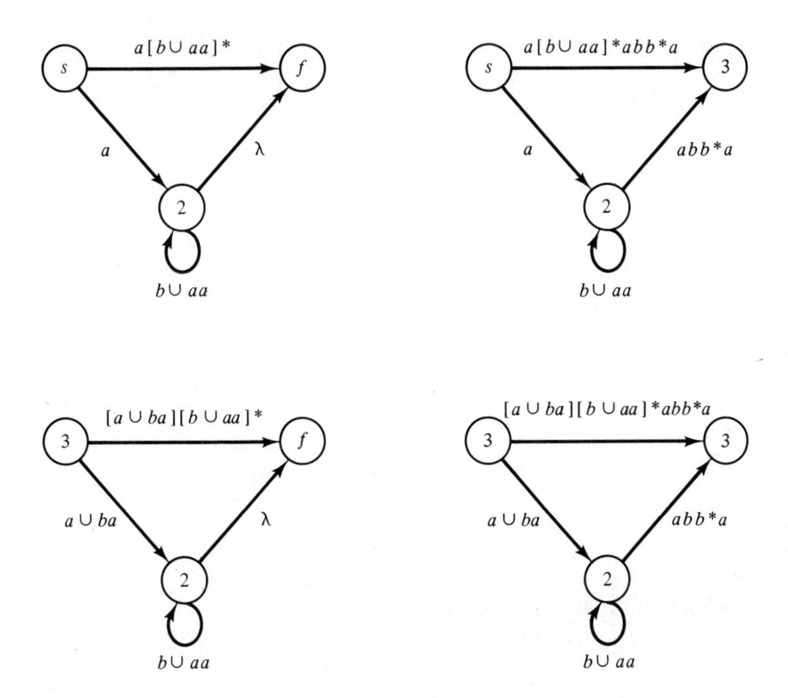

Figure 3.3.12

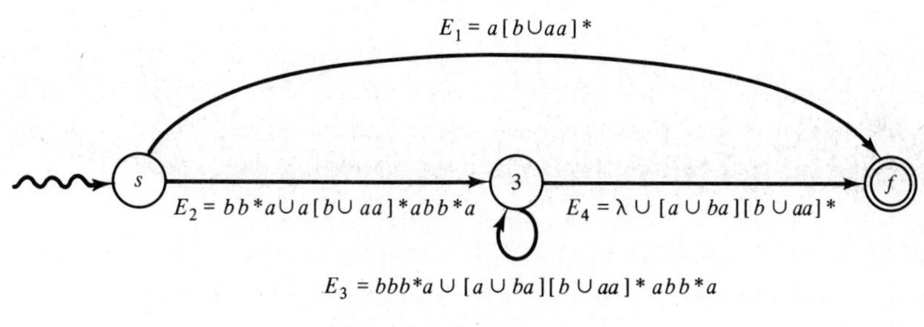

Figure 3.3.13

$$a[b \cup aa]^* \cup [bb^*a \cup a[b \cup aa]^*abb^*a]$$
$$[bbb^*a \cup ba[b \cup aa]^*abb^*a]^*[\lambda \cup ba[b \cup aa]^*]$$

3.4 EXTENDED REGULAR EXPRESSIONS

Regular expressions, as defined, only allow the operations of union, catenation, and iteration. However, it is often convenient to use *extended regular expressions*

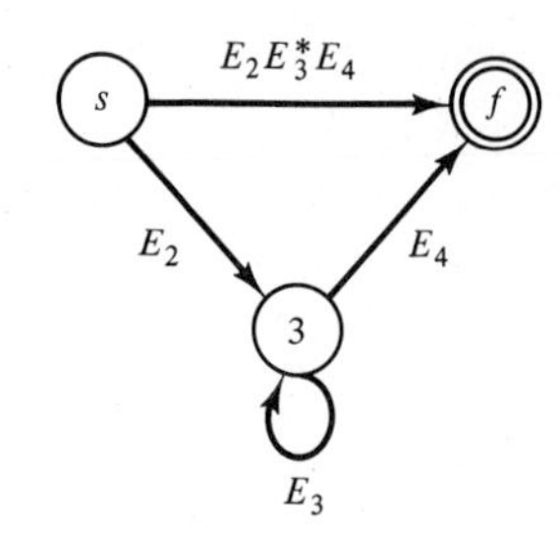

Figure 3.3.14

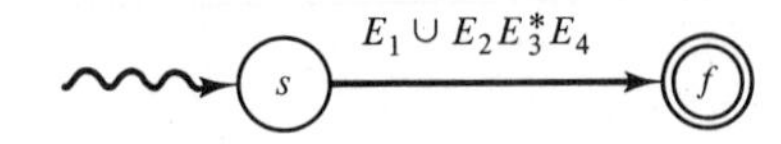

Figure 3.3.15

which also allow complement, intersection, and difference. In the introduction we could have used such an expression to denote the appearance of a pattern p in a text t by writing

$$t \text{ belongs to } L(\neg\, \varnothing\; p\; \neg\, \varnothing)$$

where $\neg$ denotes complementation. This expression is much simpler than

$$[a \cup b \cup c \cup \cdots]^*\, p[a \cup b \cup \cdots]^*$$

where $\Sigma = \{a,b,c, \ldots\}$. Introducing complementation, intersection, and difference as operators does not affect the generative power of regular expressions, since $\boldsymbol{L}_{REG}$ is closed under these operations, see Section 9.1.

So in the recursive definition of a regular expression add

(7) $\neg [E_1]$, where E_1 is an extended regular expression; or
(8) $[E_1 \cap E_2]$, where E_1 and E_2 are extended regular expressions; or
(9) $[E_1 - E_2]$, where E_1 and E_2 are extended regular expressions;

and replace "regular expression" by "extended regular expression" throughout.

Also, in the definition of $L(E)$, the language denoted by an extended regular expression, add

(7) if $E = \neg [E_1]$, then $L(E) = \Sigma^* - L(E_1)$;
(8) if $E = [E_1 \cap E_2]$, then $L(E) = L(E_1) \cap L(E_2)$; or
(9) if $E = [E_1 - E_2]$, then $L(E) = L(E_1) - L(E_2)$.

To illustrate the descriptive power of extended regular expressions we consider a final example.

Example 3.4.1 Boggle *Boggle* (Parker Bros., 1976) is an English word game in which a 4×4 or 5×5 matrix of random letters is given and the aim is to find as many words as possible of three or more letters within a given time limit. A word appears in the matrix if it corresponds to an acyclic path through adjacent cells of the matrix. Two cells are adjacent if they have an edge or corner in common. Given the 4×4 matrix in Figure 3.4.1, then *Open* is a valid word as is *Slay* (see Figure 3.4.2), but *Sons* is not since the path has a cycle; see Figure 3.4.3.

Z	E	O	E
U	S	N	P
L	A	B	T
M	Y	A	P

Figure 3.4.1

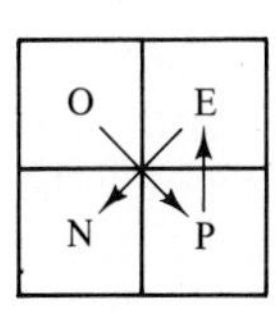

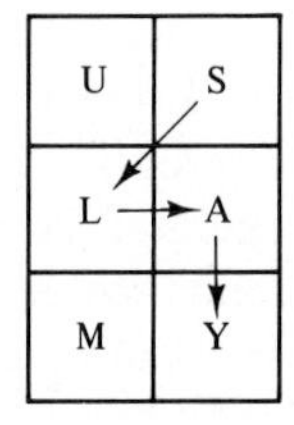

Figure 3.4.2

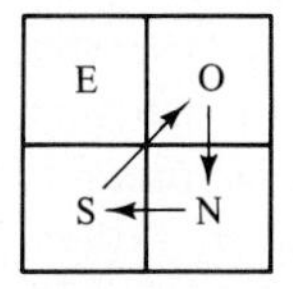

Figure 3.4.3

We consider how we might use a computer to help us solve a given instance of Boggle. To simplify our discussion we only consider a 3×3 matrix, whose cells we number in row-major order as in Figure 3.4.4.

Now, when a word appears in the matrix it traces out a sequence of adjacent cells, none of which are repeated. In other words, it corresponds to a sequence without repetitions over $\{1, \ldots, 9\}$. Of course, not all sequences correspond to

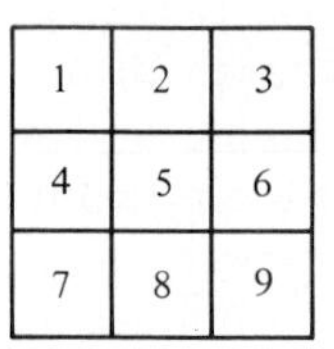

Figure 3.4.4

acyclic paths. If we can generate all such sequences of size at least 3 and at most 9, then to solve an instance of Boggle we simply substitute each digit with its corresponding letter and check the resulting list for valid words. Once the size of the Boggle matrix is known, then the cell sequences can be generated once and for all and held in a file *Cellsequences*, say. Under UNIX we might then execute

tr Cellsequences trlist || *sort* || *common*

where *trlist* is a file specifying the correspondence between cells and letters and *tr* performs the corresponding transliteration or morphism of letters for digits. *Sort* sorts the word sequences into alphabetical order and *common* checks them against an on-line dictionary reporting all legal words.

But how do we generate all valid cell sequences? Here is one method. First, generate a file of all $9! = 362880$ permutations of $1, \ldots, 9$. There are standard programs to do this. Second, form all prefixes of these sequences of length at least three. This is a simple programming task. Third, carry out a pattern match against this file with a regular expression that specifies the adjacency relationships. We consider its construction below. The resulting file contains exactly the cell sequences we require. For example, consider the permutation 659874123, which satisfies the adjacency condition. Clearly, 659, 6598, 65987, 659874, 6598741, and 65987412 also satisfy the adjacency condition. In contrast consider the permutation 412587396 which does not satisfy the adjacency condition. The prefixes 412, 4125, 41258, and412587 all satisfy the adjacency condition and are never prefixes of a permutation which satisfies the condition. That the resulting file contains every acyclic cell sequence follows by observing that each such sequence is either a permutation or a proper prefix of some permutation.

We wish to specify a regular expression that filters out valid Boggle cell sequences from invalid ones. However, rather than giving a regular expression we provide an extended regular expression E. The simplicity of E demonstrates the descriptive power of intersection and complement.

E is an extended regular expression over $\{1, \ldots, 9\}$. We specify it by restricting the set $\{1, \ldots, 9\}^*$ of all sequences in a stepwise manner. We do this by specifying the sequences we don't want. Let A be the regular expression

$$[1 \cup 2 \cup 3 \cup 4 \cup 5 \cup 6 \cup 7 \cup 8 \cup 9]$$

First, we don't want those sequences that have repeated cells, that is,

$$E_1 = [A^*1A^*1A^* \cup A^*2A^*2A^* \cup \cdots \cup A^*9A^*9A^*]$$

Second, we don't want those sequences that have invalid adjacencies that is,

$$E_2 = [A^*13A^* \cup A^*16A^* \cup A^*17A^* \cup A^*18A^* \cup A^*19A^* \cup \cdots \cup A^*91A^* \cup A^*92A^* \cup A^*93A^* \cup A^*94A^* \cup A^*97A^*]$$

Third, we want only those sequences that have neither repeated cells nor invalid adjacencies, that is,

$$E_3 = [\neg E_1 \cap \neg E_2]$$

Fourth, and, finally, we only want sequences of length at least 3, that is,

$$E = E_3 \cap AAAA^*$$

By back substitution E can be expressed in terms only of A and if necessary in terms only of $\{1, \ldots, 9\}$. It is, of course, horrendously large, but it can be composed quite easily with the help of a text editor.

The approach we have taken is not the only one, another approach is to be found in the Exercises.

3.5 ADDITIONAL REMARKS

3.5.1 Summary

We have introduced REGULAR EXPRESSIONS, a notation for denoting REGULAR LANGUAGES. We have proved that $\boldsymbol{L}_{REG} = \boldsymbol{L}_{DFA}$ demonstrating that regular expressions have the same power as finite automata, by using EXTENDED FINITE AUTOMATA. Finally, we defined EXTENDED REGULAR EXPRESSIONS which have the same power as regular expressions but also allow complementation, intersection, and difference. They enable regular languages to be defined more simply.

3.5.2 History

Regular expressions were introduced by Kleene (1956) who also proved that they are equivalent in expressive power to finite automata. The proof given here is similar to that of McNaughton and Yamada (1960). Brzozowski (1962, 1964) developed the theory of regular expressions. He introduced the notion of derivative (see the Springboard section below), and extended finite automata. The proof of Theorem 3.3.3 is based on Brzozowski and McCluskey (1963). Bourne (1983) is a good source for details of the UNIX system, while Kernighan and Plauger (1976) discuss various text processing programs.

3.6 SPRINGBOARD

3.6.1 A Differential Calculus

Let L be a language of Σ and a be a letter of Σ. Then the *left quotient* of L with respect to a, denoted by $a\backslash L$, is defined as

$$a\backslash L = \{x : ax \text{ is in } L\}$$

Therefore, $\{a\}(a\backslash L)$ is the set of words in L that begin with the letter a.

In analogy with this we recursively define *the derivative of a regular expression E over Σ with respect to a letter a in Σ denoted by $\partial_a E$, as follows.*

(1) If $E = \varnothing$, then $\partial_a E = \varnothing$;

(2) If $E = \lambda$, then $\partial_a E = \varnothing$;

(3) If $E = b$ in Σ, then $\partial_a E = \lambda$ if $b = a$, otherwise $\partial_a E = \varnothing$;

(4) If $E = [E_1 \cup E_2]$, then $\partial_a E = [\partial_a E_1 \cup \partial_a E_2]$;

(5) If $E = [E_1 E_2]$, then $\partial_a E = [[(\partial_a E_1)E_2] \cup [\delta(E_1)\partial_a E_2]]$; and

(6) If $E = E_1^*$, then $\partial_a E = [(\partial_a E_1)E_1^*]$.

The function δ appearing in part (5) has the value λ if λ is in $L(E_1)$ and the value $\varnothing$ otherwise; hence,

$$\delta(E_1)\partial_a E_2 = \begin{cases} \partial_a E_2, & \text{if } \lambda \text{ is in } L(E_1) \\ \varnothing, & \text{otherwise} \end{cases}$$

We can now recursively extend the notion of derivative to a derivation with respect to a word as follows.

Let x be a word over Σ and E a regular expression over Σ. Then $\partial_x E$ denotes the derivative of E with respect to the word x and is defined recursively by

$$\partial_x E = \begin{cases} E, & \text{if } x = \lambda \\ \partial_y(\partial_a E), & \text{if } x = ay, \text{ for some } a \text{ in } \Sigma \text{ and } y \text{ in } \Sigma^*. \end{cases}$$

It can be proved that $\partial_x E = x\backslash L(E)$, where the notion of left quotient is extended in a similar recursive manner. Moreover, it can also be shown that each regular expression E, has only a finite number of dissimilar derivatives and that these correspond to the states of a minimal *DFA* accepting $L(E)$.

Investigate the relationship of regular expressions, derivatives, and finite automata; see Brzozowski (1962, 1964). Write a program to find all dissimilar derivatives of a regular expression.

3.6.2 Descriptional Complexity

One measure of the descriptive complexity of a regular expression is the maximum number of nested stars. This is called the *star height* of a regular expression and is a reasonable measure since stars correspond to cycles in a corresponding *FA*. For example, the three regular expressions

$$a \cup bc, \quad a \cup b^*c, \quad \text{and} \quad (a \cup b^*)^*c \cup ca^*b$$

have star height 0, 1, and 2, respectively. A recursive definition of star height can easily be given. Moreover, for $i \geq 0$, let $\boldsymbol{L}_i$ be the family of languages denoted by regular expressions of star height at most i. It can be proved that $\boldsymbol{L}_0 \subset \boldsymbol{L}_1 \subset \boldsymbol{L}_2 \subset \cdots$, but if extended regular expressions are allowed, then it is a longstanding open problem of whether or not an infinite proper hierarchy exists. Indeed, it is open whether or not there is a regular language of extended star height two but not of extended star height one. In a similar manner the *dot depth* of a regular expression can be defined.

Investigate star height, see Salomaa (1969), and dot depth; see Brzozowski (1980) and the references therein. What other reasonable measures of descriptive complexity are there?

3.6.3 VLSI

Recent progress in VLSI design has lead to the investigation of how regular expressions can be converted or compiled into VLSI designs. Consult Floyd and Ullman (1984), Ullman (1984), and the references therein.

3.6.4 Path Expressions

Consider the following program fragment.

```
i := 0; notfound := true                          — 1
while (i < n) and notfound do                     — 2
begin i := i+1;                                   — 3
    if A[i] = requiredvalue then                  — 4
    begin {print message}
        notfound := false                         — 5
    end;
end;
if notfound then {print message};                 — 6
```

We view it, abstractly, as a digraph which shows the flow of control; a *flow graph*. This is shown in Figure 3.6.1, where the numbered nodes correspond to the numbered code segments in the above program fragment.

One approach to testing such a program segment is based on examining the paths in the flow graph, since paths correspond to transfers of control. For this reason it is necessary to be able to finitely specify these paths. Regular expressions turn out to be one approach for this task. For example, given the flowgraph of Figure 3.6.1 with edges labeled with $a, b, c, \ldots$, as shown in Figure 3.6.2, then each path through the flowgraph is given by a sequence of these letters. The set of all paths is given by a regular expression or *path expression*, as it is called in this context. Hence, we obtain the path expression

$$E = a[[bd[eg \cup f]]^* \cup c]$$

Beizer (1983) is a good introduction to this approach to program testing; see

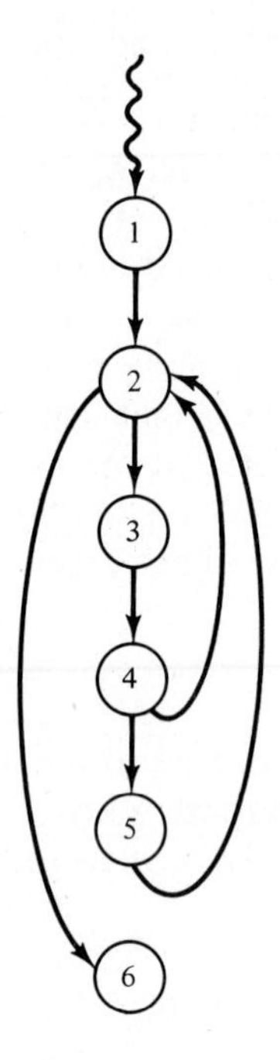

Figure 3.6.1

Fosdick and Osterweil (1976), also.

Path expressions are also used as the basis for specifying concurrent tasks. In this setting we have some basic operations a, b, c, . . . , for example, read a block, eat a course, and deposit cash. These can then be combined using union, catenation, and star. Here $a \cup b$ means perform exactly one of a or b, ab means perform a before b, and a^* means repeatedly perform a. The COSY notation is one example of such an approach; see Lauer et al. (1979). Finally, path expressions have been used to solve various path problems on digraphs; see Tarjan (1981).

EXERCISES

1.1 Given the regular expression

$$E = c^*[\lambda \cup a[a \cup b \cup c]^* \cup [a \cup b \cup c]^*b]c^*$$

(i) Is $L(E) = \{a,b,c\}^*$?
(ii) Is $L(EE) = \{a,b,c\}^*$?
In each case justify your answer.

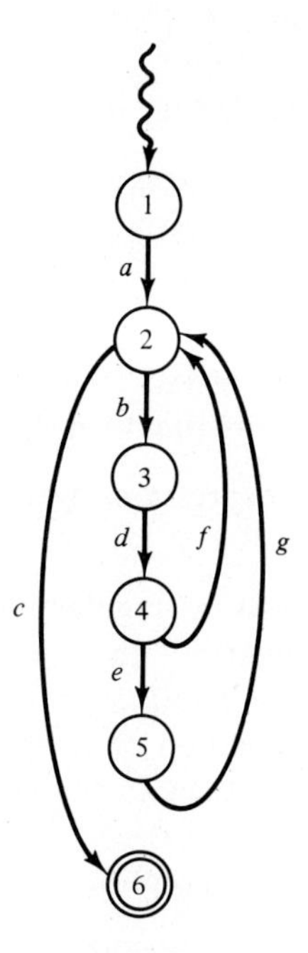

Figure 3.6.2

1.2 Let $E = [[\varnothing^* \cup a]^* \cup \varnothing a[b^*]][a \cup b]^*$. Is λ in $L(E)$? Justify your answer.

1.3 Given the regular expression over $\{a,b,c\}$

$$[ca[ab^* \cup c \cup \lambda]^*[c^*b \cup a]]^*$$

(i) give four words in $L(E)$;
(ii) give four words not in $L(E)$.

1.4 Construct a regular expression denoting the language $L \subseteq \{a,b,c\}^*$, where L consists of all words x in $\{a,b,c\}^*$ that satisfy, simultaneously, that

(i) there are an even number of a's in x;
(ii) there are $4i+1$ b's in x for some $i \geq 0$;
(iii) $|x| = 3i$, for some $i \geq 0$;
(iv) the subword abc does not appear in x.

1.5 A regular expression is said to be *ambiguous* if there is a word which can be obtained from the given regular expression in at least two distinct ways. For example, given $a^*bb^* \cup b^*bc^*$ or b^*bb^*.

For each of the following regular expressions indicate whether or not they are ambiguous and justify your claim.

(i) $a[[ab]^*cb^*]^* \cup a[ababcb^*]^*a^*$.
(ii) $aab^*[ab]^* \cup ab^* \cup a^*bba^*$.
(iii) $aaba^* \cup aaaba \cup aabba^* \cup a$.

1.6 Let E_1 and E_2 be the following regular expressions over $\{a,b,c\}$. Prove or disprove that E_1 and E_2 are equivalent.

(i) $E_1 = [a \cup b]^*a^*$, $E_2 = [[a \cup b]a]^*$.
(ii) $E_1 = \varnothing^{**}$, $E_2 = \varnothing^*$.
(iii) $E_1 = [[a \cup b]c]^*$, $E_2 = [ac \cup bc]^*$.
(iv) $E_1 = b[ab \cup ac]$, $E_2 = [ba \cup ba][b \cup c]$.

1.7 Let E_1 and E_2 be regular expressions over the same alphabet Σ, such that λ is not in $L(E_1)$.

(i) Prove that there exists a regular expression E such that E and $E_1E \cup E_2$ are equivalent, that is, $L(E) = L(E_1)L(E) \cup L(E_2)$. Such an E is said to be a *solution* of the *equation* $E = E_1E \cup E_2$.
(ii) Prove that if E is a solution of $E = E_1E \cup E_2$ and $L \subseteq \Sigma$ satisfies $L = L(E_1)L \cup L(E_2)$, then $L = L(E)$.

2.1 Construct a finite automaton accepting the language denoted by

(i) $a^*ba^*ab^*$.
(ii) $a^*bb^*[a \cup b]ab^*$.
(iii) $b[[aab^* \cup a^4]b]^*a$.

2.2 Is the following statement true or false? Every language accepted by a λ-*NFA* is also denoted by a regular expression. Justify your answer.

2.3 Complete the proof of Theorem 3.2.1.

3.1 Given the following *DFAs* construct equivalent regular expressions.

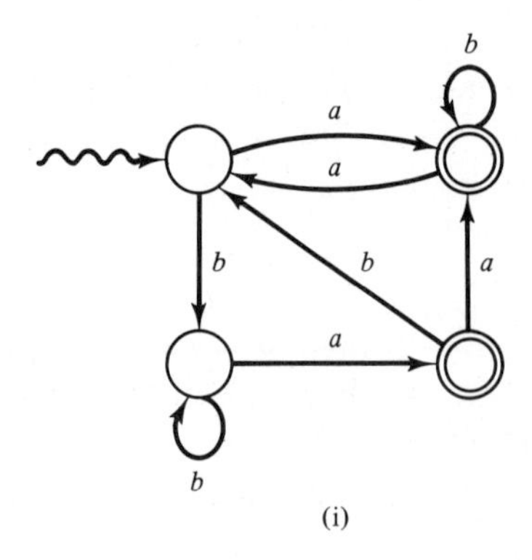

(i)

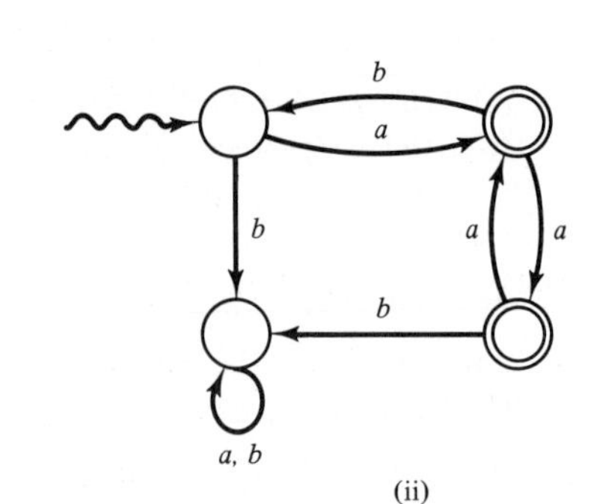

(ii)

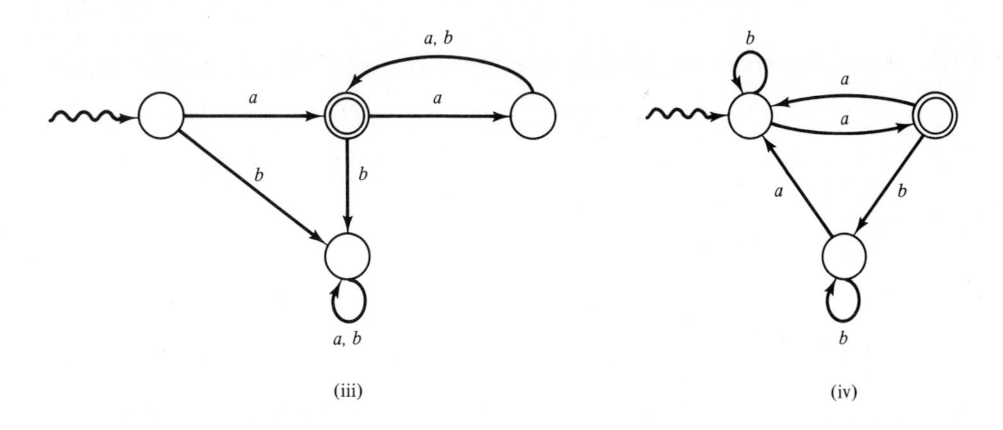

3.2 Complete the proof of Lemma 3.3.1.

3.3 Formalize the proof by induction in the second part of the proof of Lemma 3.3.2.

4.1 $L = L(E)$ for the regular expression $E = [ababa]^*$. Show that L can be obtained by a finite number of applications of the operations union, catenation, and complement from the languages $\varnothing$, $\{a\}$, and $\{b\}$.

4.2 Given the following extended regular expressions over $\{a,b,c\}$, derive equivalent regular expressions.

(i) $\neg[\varnothing^*]$.
(ii) $\neg\varnothing$.
(iii) $[\neg\varnothing]^*$.
(iv) $\neg[a^*[a \cup b]cc[\neg[\varnothing^*]]]$.

4.3 Extend the construction of a λ-*NFA* from a regular expression given in Section 3.2 to include extended regular expressions.

4.4 Extended regular expressions can be further extended by allowing E^i for any integer $i \geq 0$, and adding the symmetric difference operator Δ. The symmetric difference of two sets A and B, denoted by $A \,\Delta\, B$, is defined by $A \,\Delta\, B = (A - B) \cup (B - A)$. Does this extension add any power to extended regular expressions? Justify your answer.

PROGRAMMING PROJECTS

P1.1 Implement a package to symbolically manipulate regular expressions using transformation discussed here and in Salomaa (1969).

P1.2 Write a program to check that given regular expressions have the correct structure; see Chapter 4.

P1.3 Write a program to generate example words specified by a regular expression.

P2.1 Write a program to convert a regular expression into an equivalent λ-*NFA* and then into a *DFA*.

P2.2 Write a program to check whether or not two regular expressions are similar; see Brzozowski (1964).

P2.3 Write a program to check whether or not two regular expressions are equivalent.

P3.1 Write a program to convert a *DFA* into an equivalent regular expression.

P4.1 Write a program which attempts to convert extended regular expressions into equivalent regular expressions.

Chapter 4

Context-Free Grammars

4.1 BASICS OF CONTEXT-FREE GRAMMARS

An examination of programming manuals and texts demonstrates that context-free grammars and their variants are now universally accepted as the most widely used specification tool for the syntactic structure of programming languages. Most PASCAL texts, for example, include syntax diagrams, which are similar to state diagrams, to define PASCAL's syntactic structure. Although they are more readable than the traditional notation for context-free grammars, they are easily obtained from them. For example, the syntax diagram of Figure 4.1.1 is a specification of the structure of simple expressions similar to those in PASCAL. *Simple expression*, *term*, and *factor* are syntactic classes, or *nonterminals* as they are usually called in context-free grammars. Each syntax diagram has an entering edge on the left and an exiting edge on the right. It can be viewed as the dual of a digraph (see Section 0.2), that is, edges correspond to nodes and nodes to edges. Strictly speaking, this implies that edges cannot lead to or from edges as we have in the syntax diagrams above. Rather we should draw *simple expression*, for example, as shown in Figure 4.1.2. However, the conventional representation of syntax diagrams is well entrenched and more pleasing to the eye, so we will stick with them. Each path through a state diagram which goes from the entering to exiting edges gives rise to a sequences of boxes. An oval box contains a terminal symbol or word, that is, part of the language being specified, whereas a rectangular box contains a nonterminal, denoting a call of the corresponding syntax diagram. For example, representing boxes by their contents we obtain

factor, ∗ , *factor*, **div**, *factor*, **mod**, *factor*

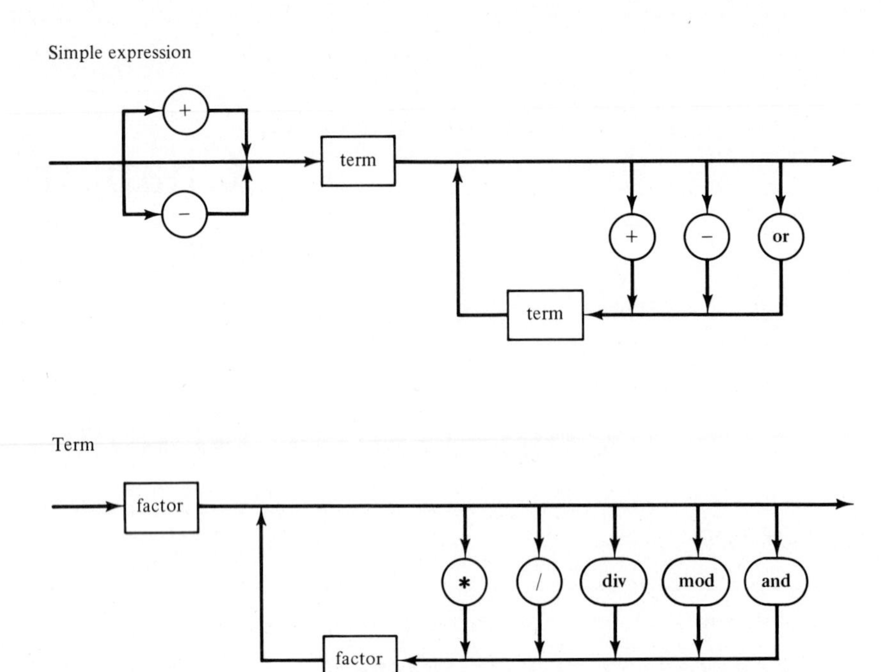

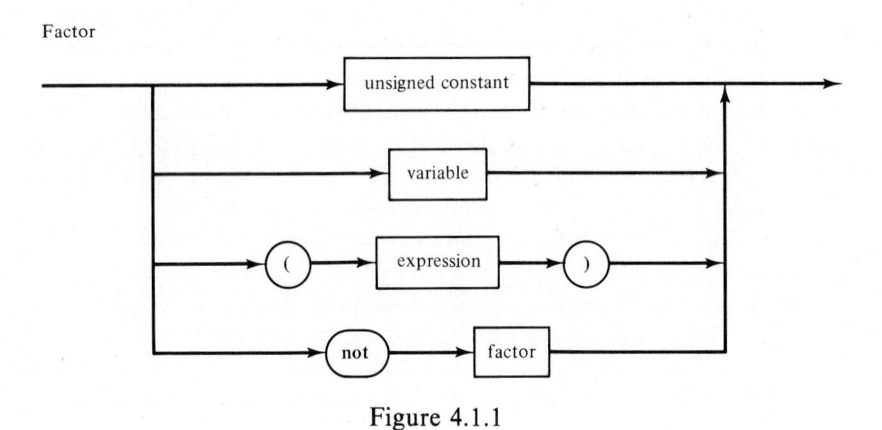

Figure 4.1.1

as the boxes in one path through *term*. Each nonterminal in this sequence gives rise to a call of the corresponding syntax diagram; the resulting sequence is substituted for it. We have left *unsigned constant* and *variable* unspecified for simplicity.

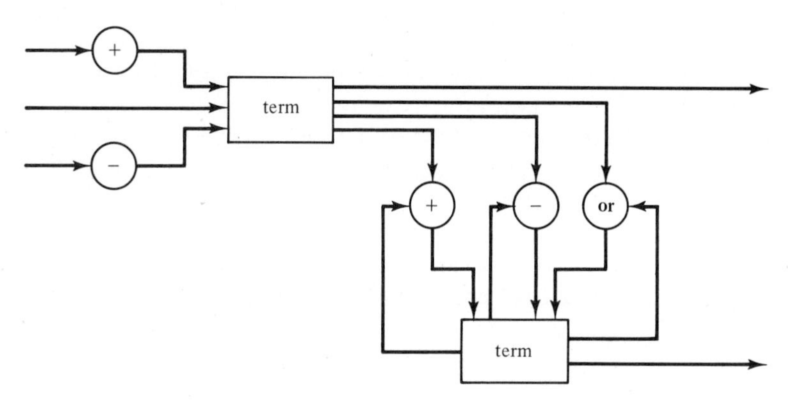

Figure 4.1.2

Another popular specification tool is the Backus-Naur form, or *BNF*, as it is usually known. The same simple expressions can be specified in *BNF* as

<simple expression> ::= <term> | + <term> | − <term> |
<simple expression> + <term> |
<simple expression> − <term> |
<simple expression> **or** <term>

<term> ::= <factor> | <term> * <factor> |
<term> / <factor> | <term> **div** <factor> |
<term> **mod** <factor> | <term> **and** <factor>

<factor> ::= <unsigned constant> | <variable> |
(<simple expression>) | **not** <factor>

The nonterminals in *BNF* are indicated with angle brackets < >; the symbol ::= should be read as "is defined as"; and | indicates alternative definitions. *BNF* is said to be a *metalanguage*, since it is a language that can be used to specify languages. Since *BNF* is itself a language one can ask whether or not *BNF* can be used to specify *BNF*. This exercise is not a completely theoretical one, since compiler-compilers or syntax-directed compilers are often given the specification of a programming language using a *BNF*-like notation. Therefore, they need to be able to check the syntax of the specification itself. From this viewpoint *BNF* leaves much to be desired; see the Exercises. A result of this deficiency has been the proliferation of "practical" syntactic metalanguages. In order to combat this proliferation a standard metalanguage has been introduced in the United Kingdom. We specify simple expressions in the British Standard syntactic metalanguage as

simple expression = ['+' | '−'], term, {('+' | '−' | '*or*'), term};

term = factor, {('*' | '/' | 'div' | 'mod' | 'and'), factor};

factor = unsigned constant | variable | '(', simple expression, ')'
| 'not', factor;

This metalanguage is much richer than that of *BNF*, while bearing much similarity to it. Terminals are placed in either apostrophes or string quotes rather than nonterminals in angle brackets. Catenation is explicitly identified by use of a comma and there are three kinds of brackets. Parentheses group items together as in arithmetic expressions; square brackets indicate a group that can be ignored; and braces indicate a group that can be repeated any number of times, including zero, that is, braces denote the star operation.

Later on in this chapter we sketch a proof of the equivalence of these three notations, with respect to their expressive power, to the notation of context-free grammars which forms our final example. With context-free grammars we obtain

$S \rightarrow T;\quad S \rightarrow +T;\quad S \rightarrow -T;\quad S \rightarrow S+T;$

$S \rightarrow S-T;\quad S \rightarrow S \text{ **or** } T$

$T \rightarrow F;\quad T \rightarrow T*F;\quad T \rightarrow T/F;\quad T \rightarrow T \text{ **div** } F$

$T \rightarrow T \text{ **mod** } F;\quad T \rightarrow T \text{ **and** } F$

$F \rightarrow U;\quad F \rightarrow I;\quad F \rightarrow (S);\quad F \rightarrow \text{ **not** } F$

which is an almost direct transliteration of the *BNF* specification. S, T, and F denote *simple expression*, *term*, and *factor*, while U and I denote *unsigned constant* and *variable* (or *identifier*). We do not use V since it has a reserved meaning throughout this chapter. Alternatives have been separated into distinct productions rather than being grouped together. Apart from this syntactic "sugar" *BNF* and context-free grammars are identical.

We have already indicated the usefulness of these equivalent notations when describing the syntactic structure of a programming language, but this is not their only use. Programs written in a high-level programming language have, usually, to be compiled into some lower-level language before they can be executed. This can only be done if the original program and its statements are structurally, that is, syntactically, correct. Checking the syntactic structure, that is, parsing, depends upon having a rigorous description of the structure. Moreover, such syntax checkers can be generated automatically from a syntactic specification in any one of above notations.

We begin a more detailed investigation of context-free grammars with their formal definition.

Definition A *context-free grammar*, or *CFG*, G is specified by a quadruple (N,Σ,P,S), where

- N is the *nonterminal* or *variable* alphabet;
- Σ is the *terminal* alphabet and N and Σ are disjoint;
- $P \subseteq N \times (N \cup \Sigma)^*$ is a *finite* set of *productions* or *rules*; and
- S in N is the *sentence* symbol.

A production is usually written as $A \rightarrow \alpha$ where A is in N and α is in $(N \cup \Sigma)^*$. It is said to be an *A-production* or production for A, and α is the *right-hand side* of the production. It is convenient to speak of the *vocabulary* of G which is denoted by V and defined as $N \cup \Sigma$.

Example 4.1.1A Let $G = (\{S,T,F\},\{a,(,),+,*\},P,S)$ be the *CFG* whose productions are

$$P = \{S \rightarrow T;\ \ S \rightarrow S+T;\ \ T \rightarrow F;\ \ T \rightarrow F*T;\ \ F \rightarrow a;\ \ F \rightarrow (S)\}$$

This is a simplified version of an expression grammar for PASCAL or any other typical computer language such as C or ALGOL.

To avoid verbosity we introduce some notational conventions which are used throughout this chapter, except when explicitly stated otherwise.

Notational Conventions

Nonterminals are represented by uppercase letters $A,B,C,\ldots,S$, except when stated otherwise. In particular, S denotes the sentence symbol. Terminals are represented by early lowercase letters $a,b,c,\ldots$. Late uppercase letters, $\ldots,X,Y,Z$ are used to mean either a nonterminal or a terminal. Terminal words are represented by late lowercase letters, $\ldots,x,y,z$, while lowercase Greek letters are used to represent arbitrary words of terminals and nonterminals. In all cases these symbols can appear with subscripts and/or superscripts.

In examples we typically use the *BNF* symbol | to group together all productions for a single nonterminal.

Example 4.1.1B We may, using these conventions, simply specify G by

$$S \rightarrow T \mid S+T;\ \ T \rightarrow F \mid F*T;\ \ F \rightarrow a \mid (S)$$

since, by convention, S is the sentence symbol, S,T and F are the nonterminals, and, hence, $+, *, (,)$, and a are the terminals.

A *CFG* is used to *generate* a language by *rewriting*, which we now define.

Definition Let $G = (N,\Sigma,P,S)$ be a *CFG* and β a word in V^*. If β can be decomposed into $\beta_1 A \beta_2$ for some A in N and for some β_1 and β_2 in V^*, and there is an A-production $A \rightarrow \alpha$ in P, then β can be *rewritten* as $\beta_1\alpha\beta_2$ or $\beta_1\alpha\beta_2$ is said to be *derived* from β. We have *replaced* A by α or *substituted* α for A. In this case we write

$$\beta \Rightarrow \beta_1\alpha\beta_2 \text{ in } G$$

or simply

$$\beta \Rightarrow \beta_1\alpha\beta_2$$

if G is understood.

Observe that for β to be rewritten it must contain at least one nonterminal, because only nonterminals have productions associated with them. However, there may be many nonterminals in β and/or many appearances of one nonterminal. In this case β can be rewritten in many different ways. Similarly, when a specific appearance of a nonterminal in β is chosen to be rewritten there may be more than one production available for it and, hence, once more, more than one way of rewriting it. In other words, rewriting is a nondeterministic action.

Example 4.1.1C Let $\beta = T+F*T$; then β has three positions at which it can be rewritten. Letting $\beta_1 = \lambda$ and $\beta_2 = +F*T$, then $\beta = \beta_1 T \beta_2$ and we can replace T by either F or $F*T$ giving either

$$T+F*T \Rightarrow F+F*T$$

or

$$T+F*T \Rightarrow F*T+F*T$$

Letting $\beta_1 = T+$ and $\beta_2 = *T$, $\beta = \beta_1 F \beta_2$ and we can rewrite β as either

$$T+F*T \Rightarrow T+a*T$$

or

$$T+F*T \Rightarrow T+(S)*T$$

When necessary we use an underscore indicate where rewriting has occurred. So with the above derivations we obtain

$$\underline{T}+F*T \Rightarrow F+F*T$$

$$\underline{T}+F*T \Rightarrow F*T+F*T$$

$$T+\underline{F}*T \Rightarrow T+a*T$$

$$T+\underline{F}*T \Rightarrow T+(S)*T$$

Let us look at another example which is a grammar for a fragment of English.

Example 4.1.2A The grammar G is given by

$S \rightarrow MQ$

$M \rightarrow ADN$

$Q \rightarrow sat\ O$

$O \rightarrow on\ M$

$A \rightarrow a \mid the$

$D \rightarrow W \mid WD$

$W \rightarrow small \mid large \mid heavy \mid very \mid \lambda$

$N \to cat \mid mat \mid elephant$

The lowercase words *sat*, *on*, etc., are assumed to be indivisible terminal symbols. Now

$$\underline{S} \Rightarrow MQ$$
$$\underline{M}Q \Rightarrow ADNQ$$
$$the\ cat\ sat\ \underline{Q} \Rightarrow the\ cat\ sat\ on\ M$$

are examples of rewriting with this grammar.

The notion of rewriting is comparable to that of a move in an *FA*, and in the same way that we obtained a configuration or move sequence for *FAs* we obtain a rewriting or derivation sequence here.

Definition Let $G = (N, \Sigma, P, S)$ be a *CFG* and α be a word in V^*.
If there are words $\alpha_0, \ldots, \alpha_{n-1}$, $\beta_0, \ldots, \beta_{n-1}, \beta$ in V^* and nonterminals $A_0, \ldots, A_{n-1}$, for some $n \geq 1$, such that

$$\alpha_0\underline{A}_0\beta_0 \Rightarrow \alpha_1\underline{A}_1\beta_1 \Rightarrow \cdots \Rightarrow \alpha_{n-1}\underline{A}_{n-1}\beta_{n-1} \Rightarrow \beta$$

where $\alpha = \alpha_0 A_0 \beta_0$, then we write

$$\alpha \Rightarrow^n \beta$$

We say that β can be derived from α in n rewriting steps, that is, $\alpha \Rightarrow^+ \beta$ is an *n-step derivation* or, more simply, a *derivation*. Also, the sequence $\alpha_0\underline{A}_0\beta_0, \ldots, \alpha_{n-1}\underline{A}_{n-1}\beta_{n-1}, \beta$ is a *derivation sequence*, it details how the derivation occurs, the productions that are used, and to which appearances of symbols they are applied. We write $\alpha \Rightarrow^+ \beta$ if there is an $n \geq 1$ and an n-step derivation starting from α and leading to β. Also, we write $\alpha \Rightarrow^* \beta$ if either $\alpha = \beta$ or $\alpha \Rightarrow^+ \beta$, and in the former case we may also write $\alpha \Rightarrow^0 \beta$, the *0-step derivation.*

Example 4.1.1D We have already shown that $T + F * T \Rightarrow^+ F * T + F * T$ by a 1-step derivation. Now $T + F * T \Rightarrow^+ a * a + a * a$ by a 7-step derivation since

$$\underline{T} + F * T \Rightarrow \underline{F} * T + F * T \Rightarrow a * T + \underline{F} * T \Rightarrow a * \underline{T} + a * T$$
$$\Rightarrow a * \underline{F} + a * T \Rightarrow a * a + a * \underline{T} \Rightarrow a * a + a * \underline{F} \Rightarrow a * a + a * a$$

Moreover, $T + F * T \Rightarrow^* T + F * T$ since $T + F * T \Rightarrow^0 T + F * T$.

Example 4.1.2B In this *CFG* we may obtain

$$\underline{S} \Rightarrow \underline{M}Q \Rightarrow \underline{A}DNQ \Rightarrow the\ \underline{D}NQ \Rightarrow the\ \underline{W}NQ$$

$$\Rightarrow the\ \underline{N}Q \Rightarrow the\ cat\ \underline{Q} \Rightarrow the\ cat\ sat\ \underline{O}$$

$$\Rightarrow the\ cat\ sat\ out\ \underline{M} \Rightarrow the\ cat\ sat\ on\ \underline{A}DN$$

$$\Rightarrow the\ cat\ sat\ on\ the\ \underline{D}N \Rightarrow the\ cat\ sat\ on\ the\ \underline{W}N$$

$$\Rightarrow the\ cat\ sat\ on\ the\ \underline{N} \Rightarrow the\ cat\ sat\ on\ the\ mat$$

We call "the cat sat on the mat" a sentence since it can be derived from or generated by the sentence symbol S.

Having described rewriting we can now define how a language is associated to a *CFG*.

Definition Let $G = (N,\Sigma,P,S)$ be a *CFG*. A word α in V^* is a *sentential form* if there exists a derivation $S \Rightarrow^* \alpha$. A derivation $S \Rightarrow^* \alpha$ is called a *sentential derivation.* A sentential form α is *terminal* if α is in Σ^*. A terminal sentential form is also called a *sentence. The language of G*, denoted by $L(G)$, is the set of all sentences, namely,

$$L(G) = \{x : S \Rightarrow^+ x \text{ and } x \text{ is in } \Sigma^*\}.$$

A language $L \subseteq \Sigma^*$ is said to be a *context-free language*, or *CFL*, if there is a *CFG*, G, with $L = L(G)$. The *family of CFLs*, denoted by $\mathbf{L}_{CF}$, is defined by

$$\mathbf{L}_{CF} = \{L : L \text{ is a } CFL\}.$$

Finally, we say that two *CFGs*, G_1 and G_2, are *equivalent* if $L(G_1) = L(G_2)$, that is, they specify the same language.

Example 4.1.1E $T+F*T$ is a sentential form since $\underline{S} \Rightarrow T+\underline{S} \Rightarrow T+\underline{T} \Rightarrow T+F*T$ and $a*a+a*a$ is a sentence. Other sentences are $a,(a)$, and $(a+a)*a$ since

$$\underline{S} \Rightarrow \underline{T} \Rightarrow \underline{F} \Rightarrow a$$

$$\underline{S} \Rightarrow \underline{T} \Rightarrow \underline{F} \Rightarrow (\underline{S}) \Rightarrow (\underline{T}) \Rightarrow (\underline{F}) \Rightarrow (a)$$

$$\underline{S} \Rightarrow \underline{T} \Rightarrow F*\underline{T} \Rightarrow F*\underline{F} \Rightarrow \underline{F}*a \Rightarrow (\underline{S})*a$$

$$\Rightarrow (\underline{T}+S)*a \Rightarrow (\underline{F}+S)*a \Rightarrow (a+\underline{S})*a$$

$$\Rightarrow (a+\underline{T})*a \Rightarrow (a+\underline{F})*a \Rightarrow (a+a)*a$$

Example 4.1.2C Other sentences in $L(G)$ are

the cat sat on the elephant
the very large elephant sat on the very small cat

Construct derivations for them.

Before continuing our discussion of derivations we consider two further examples.

Example 4.1.3 A classical *CFL* is the language $L = \{a^i b^i : i \geq 0\}$. This language can be generated by a very simple grammar indeed. Since the a's and b's must be matched in number this implies that we must arrange for a b to be deposited when and only when an a has been deposited. The simplest *CFG* is

$$S \rightarrow aSb \mid \lambda$$

Observe that $S \Rightarrow^i a^i S b^i \Rightarrow a^i b^i$, for all $i \geq 0$.

On the other hand, if we wish to generate almost all of L, but miss a finite number of words, for example, $L - \{a^2b^2, a^7b^7, a^{13}b^{13}\}$, then we ensure that the matching loop does not take place on the sentence symbol. One example *CFG* is

$$S \rightarrow A \mid \lambda \mid ab \mid a^3b^3 \mid a^4b^4 \mid a^5b^5 \mid a^6b^6$$
$$a^8b^8 \mid a^9b^9 \mid a^{10}b^{10} \mid a^{11}b^{11} \mid a^{12}b^{12}$$
$$A \rightarrow aAb \mid a^{14}b^{14}$$

Observe that the terminating production for A is large enough to ensure none of the omitted words are generated and small enough to ensure no included words are omitted. As a final example, consider $L \cup L^R$. We must not match the symbols with the sentence symbol, since, for example, the *CFG*

$$S \rightarrow aSb \mid bSa \mid \lambda$$

not only generates all words in $L \cup L^R$ but also words outside it, for example, $abab$. One *CFG* for $L \cup L^R$ is

$$S \rightarrow A \mid B;\ A \rightarrow aAb \mid \lambda;\ B \rightarrow bBa \mid \lambda$$

Example 4.1.4 Proofs about *CFGs* are usually inductive. We illustrate this by proving that the language of the following *CFG*, G,

$$S \rightarrow \lambda \mid aSb \mid bSa \mid SS$$

is the language L defined by

$$L = \{x : x \text{ is in } \{a,b\}^*, |x|_a = |x|_b\}$$

L contains all words that have an equal number of a's and b's.

We split the proof into two parts. First, we prove that $L(G) \subseteq L$ and, second, that $L \subseteq L(G)$.

I. $L(G) \subseteq L$.

We claim a stronger result, namely, that for all α, where $S \Rightarrow^* \alpha$, $|\alpha|_a = |\alpha|_b$. This is proved by induction on the length n of sentential derivations.

Basis: $n = 0$. The only α is S since $S \Rightarrow^0 S$. Clearly $|S|_a = |S|_b$ in this case.

Induction Hypothesis: Assume the claim holds for all sentential derivations of length at most n, for some $n \geq 0$.

Induction Step: Consider a sentential derivation $S \Rightarrow^{n+1} \alpha$.

Since $n \geq 0$ and $n+1 \geq 1$ this implies there is some β such that

$$S \Rightarrow^n \beta \Rightarrow \alpha$$

By the induction hypothesis $|\beta|_a = |\beta|_b$ and, by inspection of the productions, either no a's and no b's are added to β or one a and one b are added in the derivation $\beta \Rightarrow \alpha$. Hence, $|\alpha|_a = |\alpha|_b$.

By the principle of induction, we have established the claim and, hence, $|x|_a = |x|_b$, for all x in Σ^*, where $S \Rightarrow^* x$. Thus, $L(G) \subseteq L$.

II. $L \subseteq L(G)$.

This is also established by induction, but in this case on the number of a's in the words of L. We have to demonstrate that there is a sentential derivation in G for every word x in L. This is our claim.

Basis: $n = |x|_a = 0$. Also, $x = \lambda$ and $S \rightarrow \lambda$ in G; therefore, λ is in $L(G)$.

Induction Hypothesis: Assume the claim holds for all words x in L with $|x|_a \leq n$, for some $n \geq 0$.

Induction Step: Consider a word x in L, $|x|_a = n+1$.

From the productions in G we see that we should aim to reduce x to yz, ayb, or bya, where y and z are in L with $|y|_a < n$ and $|z|_a < n$.

If $x = ayb$, then $|y|_a = |y|_b$, y is in L, and $|y|_a = n$. Immediately, $S \Rightarrow^* y$ in G, by the induction hypothesis, and in this case we obtain $S \Rightarrow aSb \Rightarrow^* ayb$ and, therefore, x is in $L(G)$.

If $x = bya$, then we find $S \Rightarrow bSa \Rightarrow^* bya$ and x is in $L(G)$.

Finally, if $x \neq ayb$ and $x \neq bya$, we wish to decompose x into yz, where $y \neq \lambda \neq z$ and y and z are in L. We prove that such a decomposition exists. Since the first and last letters of x are not different they must be the same, so either $x = aya$ or $x = byb$. We assume $x = aya$; the other subcase is symmetric.

Since x contains at least two a's, we have $n \geq 2$. Consider the $2n+1$ prefixes of x, $x_0, \ldots, x_{2n}$, where $x = x_i y_i$, for some y_i, $0 \leq i \leq 2n$, and $|x_i| = i, 0 \leq i \leq 2n$. Now the sequence

$$|x_0|_a - |x_0|_b, \ |x_1|_a - |x_1|_b, \ldots, |x_{2n}|_a - |x_{2n}|_b$$

of integers begins with 0 and ends with 0. We argue that there is some i, $0 < i < n$, such that $|x_{2i}|_a - |x_{2i}|_b = 0$. Assume no such i exists, that is, $|x_{2i}|_a - |x_{2i}|_b > 0$, for all i, $0 < i < n$. In particular, for $z = x_{2n-2}$, $|z|_a - |z|_b > 0$. Now either $x = zaa$, in which case

$$|x|_a - |x|_b = |z|_a - |z|_b + 2$$

or $x = zba$, in which case

$$|x|_a - |x|_b = |z|_a - |z|_b$$

In both cases we conclude that $|x|_a - |x|_b > 0$, a contradiction.

Therefore, there is an i, $0 < i < n$, with $|x_{2i}|_a = |x_{2i}|_b$. Let $x_{2i} = y$ and $x = yz$. Clearly $|z|_a = |z|_b$, so both y and z are in L. Now $y \neq \lambda$, $z \neq \lambda$, $S \Rightarrow^* y$, and $S \Rightarrow^* z$ by the induction hypothesis, hence, $S \Rightarrow SS \Rightarrow^* yS \Rightarrow^* yz$ in G and x is in $L(G)$.

This completes the induction step and establishes the claim. Therefore, $L \subseteq L(G)$.

Returning to our discussion of derivations we observe that most sentential forms have many different derivation sequences.

Example 4.1.1F

$$\underline{S} \Rightarrow \underline{S} + T \Rightarrow T + T$$

Consider the possible derivation sequences for $a + a$ from $T + T$. There are six of them

(i) $\underline{T} + T \Rightarrow \underline{F} + T \Rightarrow a + \underline{T} \Rightarrow a + \underline{F} \Rightarrow a + a.$
(ii) $\underline{T} + T \Rightarrow F + \underline{T} \Rightarrow \underline{F} + F \Rightarrow a + \underline{F} \Rightarrow a + a.$
(iii) $\underline{T} + T \Rightarrow F + \underline{T} \Rightarrow F + \underline{F} \Rightarrow \underline{F} + a \Rightarrow a + a.$
(iv) $T + \underline{T} \Rightarrow \underline{T} + F \Rightarrow \underline{F} + F \Rightarrow a + \underline{F} \Rightarrow a + a.$
(v) $T + \underline{T} \Rightarrow \underline{T} + F \Rightarrow F + \underline{F} \Rightarrow \underline{F} + a \Rightarrow a + a.$
(vi) $T + \underline{T} \Rightarrow T + \underline{F} \Rightarrow \underline{T} + a \Rightarrow \underline{F} + a \Rightarrow a + a.$

This multiplicity of derivation sequences corresponding to a given derivation, even when the same productions are applied to the same appearances of the same symbols, is unfortunate. Can we really be sure we have explored all possible derivation sequences, when attempting to generate a proposed sentence. Fortunately, we can avoid this problem by, first, using the *syntax* or *derivation tree* and, second, using *canonical* derivations. We deal with these approaches one at a time.

Example 4.1.1G Consider the derivation sequence

$$\underline{S} \Rightarrow \underline{S} + T \Rightarrow \underline{T} + T \Rightarrow F + \underline{T} \Rightarrow F + \underline{F} \Rightarrow \underline{F} + a \Rightarrow a + a$$

We can capture the derivation and its sequence with an ordered, directed tree whose frontier is (or external nodes are) labeled with a, $+$, and a from left to right, whose root is labeled with S, and whose internal nodes are labeled with appropriate nonterminals; see Figure 4.1.3. The syntax tree displays the derivational structure of the sentence, that is, which productions are applied to which appearances of which symbols. At the same time it suppresses the order of application of the productions.

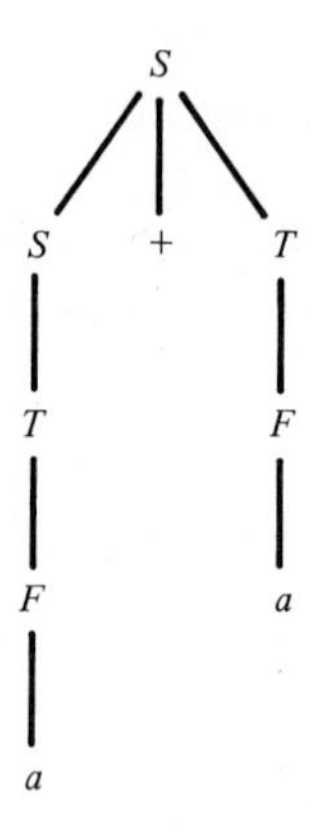

Figure 4.1.3

Given a derivation sequence a simple iterative algorithm serves to construct its corresponding syntax tree. We illustrate this using the above derivation sequence. Initially, we have a tree with a single external node labeled with S; see Figure 4.1.4(a). Now $\underline{S} \Rightarrow S + T$ appends (external) nodes S, $+$, and T to the partial tree, giving Figure 4.1.4(b). $\underline{S} + T \Rightarrow T + T$ appends T, since this is the result of rewriting S, to give Figure 4.1.5 and $\underline{T} + T \Rightarrow F + T$ appends F to the leftmost T to give Figure 4.1.6, and so on.

Note that at each stage the frontier is an intermediate sentential form. Moreover, it is easy to show that the other five derivation sequences for $a + a$ also produce the same syntax tree.

Before going any further we define syntax trees for a given grammar in their full generality.

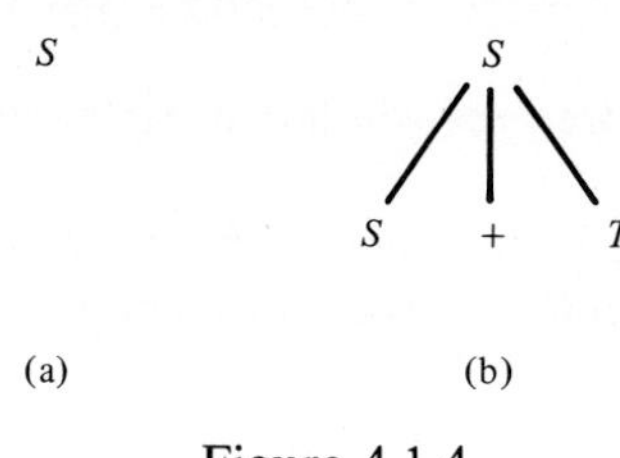

Figure 4.1.4

Figure 4.1.5

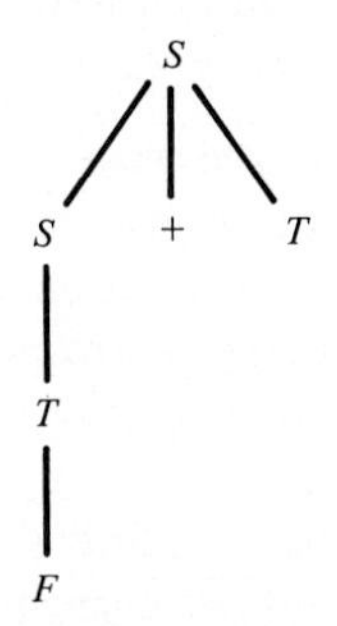

Figure 4.1.6

Definition Let $G = (N,\Sigma,P,S)$ be a *CFG* and let T be an ordered, directed tree whose nodes are labeled with either symbols from V or the empty word. Then T is a *syntax tree* (*with respect to* G) if every node u in T satisfies the following condition:

If u is an external node, then it is labeled with a nonterminal symbol, a terminal symbol, or with λ, denoting the empty word, and in this latter case it is the only child of its parent,

otherwise u is an internal node labeled with a nonterminal symbol A, and if it has k children, $k \geq 1$, labeled with $X_1, \ldots, X_k$, from left to right, then the production $A \rightarrow X_1 \cdots X_k$ is in P.

In Example 4.1.1, above, the reader should check that the syntax tree does indeed satisfy this definition.

The frontier of a syntax tree read in left to right order gives a word, the *yield* of the tree, denoted by $y(T)$.

In the above example we have indicated how a syntax tree can always be constructed from a derivation sequence. We now demonstrate the converse, thus proving that:

> *If the root of a syntax tree is labeled with the sentence symbol, then its yield is a sentential form and if its yield is a terminal word, then it is a sentence.*

Lemma 4.1.1 *Let $G = (N,\Sigma,P,S)$ be a CFG and T be a syntax tree, with respect to G, having A in N as its root label and $\alpha = y(T)$. Then $A \Rightarrow^* \alpha$ in G with a derivation sequence that uses the same productions applied to the same appearances of the same symbols that appear in T.*

Proof: (by induction on the number m of internal nodes in T)

Basis: $m = 0$. T consists solely of one external node, so $A \Rightarrow^* \alpha$ and the remaining conditions are satisfied vacuously.

Induction Hypothesis: Assume the lemma statement holds for all syntax trees T with at most n internal nodes, for some $n \geq 0$.

Induction Step: Let T be a syntax tree with $m = n+1$ internal nodes. Now T must have at least one internal node v, with all of its children belonging to the frontier (a basic property of trees; see Section 0.2). Let B be the nonterminal labeling v and $B \rightarrow \beta$ be the production corresponding to v's children. Let T' be T with v's children removed. T' is a syntax tree with one fewer internal node than T. By the induction hypothesis there is a derivation sequence corresponding to $A \Rightarrow^* y(T')$ which satisfies the conditions of the lemma. Extend this by rewriting in $y(T')$ the appearance of B corresponding to v using $B \rightarrow \beta$. This gives a derivation $A \Rightarrow^* y(T)$ which satisfies the conditions of the lemma, concluding the induction step and the lemma. ∎

Lemma 4.1.1 demonstrates that to every syntax tree there corresponds a derivation sequence. In other words, derivation sequences give rise to *all* syntax trees; none are omitted.

Given a syntax tree any traversal of it produces a derivation sequence. However, there are two specific traversals that are of particular interest: preorder and reverse preorder traversals. These correspond, as we shall see, to the following *canonical* derivation sequences.

Definition Let $G = (N,\Sigma,P,S)$ be a CFG. Let β and γ be in V^* such that $\beta = \beta_1 A \beta_2$, $\gamma = \beta_1 \alpha \beta_2$ and $A \rightarrow \alpha$ is in P, that is, $\beta_1 \underline{A} \beta_2 \Rightarrow \beta_1 \alpha \beta_2$ in G.

If β_1 is in Σ^* we write $\beta \overset{L}{\Rightarrow} \gamma$; β is *rewritten in the leftmost position* to give γ. Similarly, if β_2 is in Σ^* we write $\beta \overset{R}{\Rightarrow} \gamma$; β is *rewritten in the rightmost position* to give γ.

We extend $\overset{L}{\Rightarrow}$ and $\overset{R}{\Rightarrow}$ to $\overset{L}{\Rightarrow}^n$ and $\overset{R}{\Rightarrow}^n$, $\overset{L}{\Rightarrow}^+$ and $\overset{R}{\Rightarrow}^+$, and $\overset{L}{\Rightarrow}^*$ and $\overset{R}{\Rightarrow}^*$, respectively. We speak of *leftmost* and *rightmost derivations*, respectively.

Example 4.1.1H Compare

$$S \overset{L}{\Rightarrow} S+T \overset{L}{\Rightarrow} T+T \overset{L}{\Rightarrow} F*T+T \overset{L}{\Rightarrow} a*T+T$$

$$\overset{L}{\Rightarrow} a*F+T \overset{L}{\Rightarrow} a*a+T \overset{L}{\Rightarrow} a*a+F \overset{L}{\Rightarrow} a*a+a$$

and

$$S \overset{R}{\Rightarrow} S+T \overset{R}{\Rightarrow} S+F \overset{R}{\Rightarrow} S+a \overset{R}{\Rightarrow} T+a$$

$$\overset{R}{\Rightarrow} F*T+a \overset{R}{\Rightarrow} F*F+a \overset{R}{\Rightarrow} F*a+a \overset{R}{\Rightarrow} a*a+a$$

Examining the two derivations in detail it can be seen that the same productions are applied to the same appearances of nonterminals, but their order of application is different.

We wish to prove that whenever there is a sentential derivation of some terminal word in a *CFG*, then there is always a leftmost and rightmost sentential derivation of the same terminal word. To achieve this result we need to strengthen Lemma 4.1.1 and to this end we require the following notions.

Definition Let $G = (N,\Sigma,P,S)$ be a *CFG* and let β and γ be in V^* such that

$$D : \beta = \beta_0 \Rightarrow \cdots \Rightarrow \beta_n = \gamma$$

is a derivation in G for some $n \geq 0$. A nonterminal in γ is clearly not rewritten in D — we say it is *passive*. A nonterminal which is rewritten in D is said to be *active*. To generalize the notion of a leftmost derivation we require that the derivation is leftmost with respect to the active nonterminals.

We say D is a *pre-leftmost derivation* if at each derivation step the leftmost active nonterminal is rewritten. In this case we write $\beta \overset{l}{\Rightarrow}^n \gamma$, $\beta \overset{l}{\Rightarrow}^+ \gamma$, or $\beta \overset{l}{\Rightarrow}^* \gamma$. *Pre-rightmost derivations* are defined similarly, giving rise to the notation $\beta \overset{r}{\Rightarrow}^n \gamma$, $\beta \overset{r}{\Rightarrow}^+ \gamma$, and $\beta \overset{r}{\Rightarrow}^* \gamma$.

A pre-leftmost derivation is a leftmost derivation if the derived word is terminal, since in this case all nonterminals are rewritten, that is, they are all active.

Example 4.1.11 The derivation

$$F * \underline{T} + T \Rightarrow F * F + \underline{T} \Rightarrow F * F + \underline{F} \Rightarrow F * F + a$$

is pre-leftmost since the first two appearances of F are both passive.

We now have

Lemma 4.1.2 *Let $G = (N,\Sigma,P,S)$ be a CFG and T be a syntax tree, with respect to G, having A in N as its root label and $\alpha = y(T)$. Then $A \overset{l}{\Rightarrow}^* \alpha$ in G ($A \overset{r}{\Rightarrow}^* \alpha$ in G) with a pre-leftmost (pre-rightmost) derivation that uses the same productions applied to the same appearances of the same symbols that appear in T.*

Proof: (by induction on the number m of internal nodes of T)
The proof is similar to that of Lemma 4.1.1 so we only give the induction step for pre-leftmost derivations. Since $m \geq 1$ there is at least one production displayed in T and, hence, there is a rightmost internal node v labeled with B, all of whose children belong to the frontier. Remove v's children, giving a new syntax tree T' with $m-1$ internal nodes. There is a pre-leftmost derivation corresponding to T' in which this appearance of B is passive. Now applying the production determined by v in T to this appearance of B also gives a pre-leftmost derivation. This is because all symbols to B's right are either terminal or passive, by the choice of v.■

We now obtain the following important theorem.

Theorem 4.1.3 *Let $G = (N,\Sigma,P,S)$ be a CFG. Then for all x in Σ^*, the following three conditions are equivalent*

(i) $S \Rightarrow^+ x$ *in G;*

(ii) $S \overset{L}{\Rightarrow}^+ x$ *in G; and*

(iii) $S \overset{R}{\Rightarrow}^+ x$ *in G.*

Proof: Since leftmost and rightmost derivations *are* derivations, this implies that if $S \overset{L}{\Rightarrow}^+ x$ or $S \overset{R}{\Rightarrow}^+ x$ in G, then $S \Rightarrow^+ x$ in G. Therefore, it only remains to prove that (i) implies (ii) and (i) implies (iii). Since these proofs are similar we only prove (i) implies (ii).

Assume $S \Rightarrow^+ x$ in G. Then there is a derivation $D : S \Rightarrow \cdots \Rightarrow x$ in G, and therefore there is a syntax tree T corresponding to D. But Lemma 4.1.2 implies there is a pre-leftmost derivation corresponding to T, which in this case is a leftmost derivation since x is terminal. That is, $S \overset{L}{\Rightarrow}^+ x$ in G. ■

This theorem means that we need consider either only leftmost derivations or only rightmost derivations when generating sentences.

Syntax trees, leftmost derivations, and rightmost derivations enable us to spot immediately when two derivational structures for the same sentence are really different.

Example 4.1.5 Let $G = (\{S,T,F\},\{a,(,),+,*\},P,S)$ where

$$P = \{S \to T \mid S+T \mid S*F;\ T \to F \mid F*T;\ F \to a \mid (S)\}.$$

G is a slightly modified version of the grammar from the previous example. Consider the sentence $a+a*a$. It has two distinct syntax trees as shown in Figures 4.1.7 and 4.1.8. The word $a+a*a$ also has two distinct leftmost derivations and two distinct rightmost derivations.

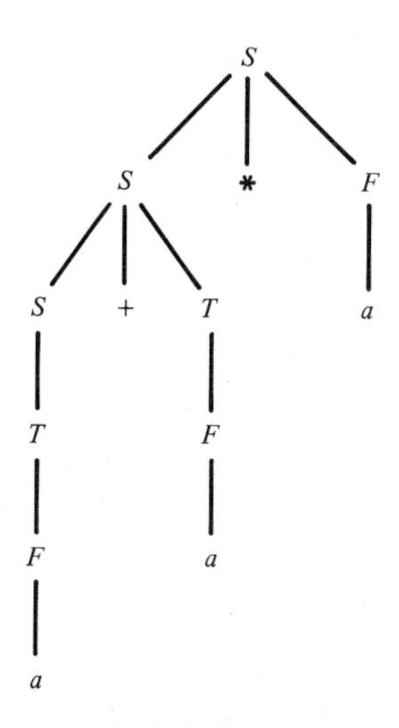

Figure 4.1.7

The multiplicity of different derivations leads to the following

Definition Let $G = (N,\Sigma,P,S)$ be a *CFG*. Let x be a sentence with two distinct syntax trees (leftmost derivations or rightmost derivations). Then x is said to be *ambiguous* with respect to G. G is *ambiguous* if there is at least one ambiguous sentence derived from S. We say a *CFL*, L, is (inherently) *ambiguous* if for all *CFGs*, G, with $L(G) = L$, G is ambiguous.

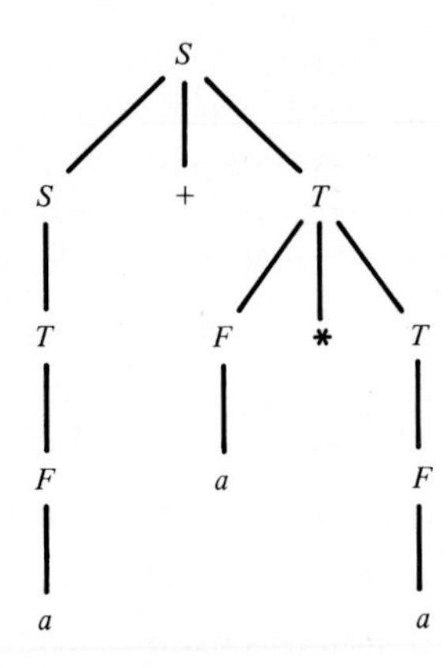

Figure 4.1.8

Ambiguity is important, in practice, because it indicates that a sentence may have more than one meaning. In natural languages this causes difficulty in communication, and in the legal arena causes lawsuits. In programming languages ambiguity indicates the possibility that a program which is compiled with different compilers can then be executed to give different results with the same data. We illustrate each of these possibilities with classical examples.

Example 4.1.6 Consider the English sentence

Time flies like an arrow

We usually interpret this as a simile to mean time passes as rapidly as an arrow flies. Time and arrow are both nouns, and flies is the verb. Its structure is the same as in the sentence

Jonas walks like a crab

or

Nancy walks like a queen

However, we may consider flies to be a noun, time an adjective, and like to be the verb, as in the sentence

Monkeys like a banana

In this case time flies are a specific kind of fly, as are blow flies. For some reason time flies really like an arrow.

There are, at least, two other meanings attached to this five-word sentence. Try to discover them and the different syntactic structures corresponding to them.

Example 4.1.7 The language ALGOL 60, perhaps best known as a precursor of PASCAL, was the first programming language whose syntax was defined using *BNF*. It should have been expected, but in general wasn't, that there would be errors and omissions in this definition, particularly in the meaning of the language constructs. For example, there was a major syntactic error in the definition of the **if-then-else** statement, which was corrected in the revised version, released three years later.

Essentially, the ALGOL 60 grammar contained the following productions

<conditional statement> ::= <if part> | <if part> **else** <statement>
<if part> ::= **if** <Boolean expression> **then** <statement>
<statement> ::= <conditional statement> |
<assignment statement> |
<for statement> | . . .

Now the conditional statement:

if $a = b$ **then if** $b = c$ **then** $x := 1$ **else** $x := 2$

is ambiguous, since the dangling **else** may belong to either the first or the second **if**. It can be viewed as:

if $a = b$ **then**
    if $b = c$ **then** $x := 1$
else $x := 2$

where we have used indentation to indicate that the **else** is considered as the alternative to the first **if**. In this case when $a \neq b$, x is given the value 2. Alternatively, the **else** can be considered as the alternative to the second **if**, that is, as:

if $a = b$ **then**
    if $b = c$ **then**
        $x := 1$
    else $x := 2$

In this case when $a \neq b$, the value of x is unchanged. The problem of the dangling **else** was resolved grammatically in the revised version of ALGOL 60 by redefining <if part> as

<if part> ::= **if** <Boolean expression> **then** <nonconditional statement>

The same ambiguity occurs in PASCAL but it is resolved semantically by always grouping the dangling **else** with the last unmatched **if**.

This example gives rise to some natural problems. First, we have

CFG AMBIGUITY
INSTANCE: A *CFG*, $G = (N, \Sigma, P, S)$.
QUESTION: Is G ambiguous?

since it is obviously important to know whether a given grammar for a programming language is ambiguous. Unfortunately, this is an undecidable decision problem; see Section 10.4. Second, when we know that a given *CFG* is ambiguous we have

UNAMBIGUOUS *CFG* CONSTRUCTION
INSTANCE: An ambiguous *CFG*, $G = (N, \Sigma, P, S)$.
QUESTION: Can an equivalent unambiguous *CFG* be constructed?

This problem is closely related to

CFL AMBIGUITY
INSTANCE: An ambiguous *CFG*, $G = (N, \Sigma, P, S)$.
QUESTION: Does there exist an equivalent unambiguous *CFG*?

These issues are discussed further in Chapter 10.

Having demonstrated the existence of ambiguous grammars, we discuss the existence of ambiguous languages using a classical example.

Example 4.1.8 The language, L, defined as follows:

$$L = \{a^i b^j c^k : i,j,k \geq 1 \text{ and } (i = j \text{ or } j = k)\}$$

is an ambiguous *CFL*.

The proof of this result is beyond the scope of this introductory text. However, it is discussed further in Section 4.5.2. For now we try to understand why L should be ambiguous.

We can split L into L_1 and L_2 according to the second condition in the specification of L. We obtain

$$L_1 = \{a^i b^j c^k : i,j,k \geq 1 \text{ and } i = j\}$$

and

$$L_2 = \{a^i b^j c^k : i,j,k \geq 1 \text{ and } j = k\}$$

from which $L = L_1 \cup L_2$. Note that the condition $i = j$ does not prohibit $j = k$ from also being satisfied, that is, $i = j = k$. So not only does L_1 (and L) contain words such as

$$a^5 b^5 c^{26}, \quad a^2 b^2 c$$

but also words

$$a^5 b^5 c^5, \quad a^{27} b^{27} c^{27}$$

Similarly, L_2 contains words such as

$$a b^6 c^6, \quad a^{10} b^4 c^4$$

and words

$$a^5b^5c^5, \quad abc$$

So $L_1 \cap L_2 \neq \emptyset$.

Giving *CFGs* for L_1 and L_2 is straightforward and similar to the grammars in Example 4.1.3.

Let G_1 have productions

$$S \to A_1C_1$$
$$A_1 \to aA_1b \mid ab$$
$$C_1 \to cC_1 \mid c$$

satisfying $L(G_1) = L_1$. Similarly, let G_2 have productions

$$S \to A_2B_2$$
$$A_2 \to aA_2 \mid a$$
$$B_2 \to bB_2c \mid bc$$

satisfying $L(G_2) = L_2$.

A *CFG* for L is now easily obtained by taking all these productions, call it G. Now $S \Rightarrow A_1C_1$ and $S \Rightarrow A_2B_2$ are the only possible one-step derivations from S. Taking the first one yields words in L_1 and taking the second yields words in L_2. Since G_1 and G_2 do not have nonterminals in common apart from S and S only appears at the first step in a derivation, no other words are generated. Thus, $L = L(G)$.

But observe that for all $i \geq 1$

$$S \overset{L}{\Rightarrow} A_1C_1 \overset{L}{\Rightarrow}{}^{+} a^ib^ic^i$$

and

$$S \overset{L}{\Rightarrow} A_2B_2 \overset{L}{\Rightarrow}{}^{+} a^ib^ic^i$$

are distinct leftmost derivations for the same sentence. So G is ambiguous.

This is a demonstration that *one* grammar for L is ambiguous, not all grammars. However, the basic reason for the ambiguity of G, that L must be split into an L_1 and L_2 which have words in common, points the way to a proof. A *CFG* for L must match as and bs and, at the same time, exclude words which also match them with cs.

This section would be incomplete without mentioning the fundamental decidability problem for *CFGs*, namely,

CFG MEMBERSHIP
INSTANCE: A *CFG*, $G = (N,\Sigma,P,S)$, and a word x in Σ^*.
QUESTION: Is x in $L(G)$?

This is the problem any compiler (or interpreter) for any programming language must solve so that only syntactically correct programs are allowed to be compiled. Usually more information is required than a simple yes or no answer. On the one hand, if a program is syntactically correct, then a compiler needs the syntactic structure of the program, since this reflects, for example, the order of execution of the operators in an expression. Some representation of the syntax tree is sufficient for this purpose. On the other hand, if a program is not syntactically correct, then a meaningful error message needs to be produced and this, once more, requires structural information about the program up to the position at which an error was diagnosed. We only consider the yes case in the following problem

CFG PARSING
INSTANCE: A *CFG*, $G = (N,\Sigma,P,S)$, and a word x in Σ^*.
QUESTION: What are the syntax trees of x with respect to G?

We attack the membership and parsing problems in more detail in Chapters 5, 8, 10, and 11. We close this section with one approach to *CFG* PARSING for a restricted class of *CFGs*.

A *CFG*, $G = (N,\Sigma,P,S)$, is in *Greibach normal form* if for all productions $A \rightarrow \alpha$ in P, α is in ΣN^*. In the next section we demonstrate that, essentially, every *CFL* has a *CFG* in Greibach normal form.

Given such a *CFG*, G, note that if a word x in Σ^* is in $L(G)$, then $S \Rightarrow^n x$, where $n = |x|$. This is simply because each production that is applied deposits exactly one terminal symbol. Thus, we may explore, exhaustively, all sentential derivations of length $|x|$. To make this exploration simpler we only consider leftmost derivations, and we number the productions of each nonterminal arbitrarily, but uniquely from 1 upward. Given x we generate a *search tree* (see Figure 4.1.9), where $S \rightarrow a_i\alpha_i$ is the ith production of S, and $a_ia_{ij}\alpha_{ij}$ is the result of applying the jth production of the first symbol of α_i to $a_i\alpha_i \ldots$. If x is in $L(G)$, then $x = a_{i_1}a_{i_1i_2} \cdots a_{i_1i_2 \cdots i_n}$ for some $i_1, \ldots, i_n$, where $n = |x|$. Moreover, a path from S to x in level n is completely determined by the sequence $i_1, i_2, \ldots, i_n$ and gives a leftmost derivation for x. An example should clarify this idea.

Example 4.1.9 Let G be given by

$S \rightarrow aBA \mid (CBA \mid aA \mid (CA \mid aB \mid (CB \mid a \mid (C$

$T \rightarrow aB \mid (CB \mid a \mid (C$

$A \rightarrow +S$

$B \rightarrow *T$

$C \rightarrow aBAD \mid (CBAD \mid aAD \mid (CAD \mid aBD \mid (CBD \mid aD \mid (CD$

$D \rightarrow)$

where the productions of each nonterminal are numbered from 1 upward in left-to-right order. Clearly G is in Greibach normal form. Consider the word (a). We obtain the search tree, up to level 2, shown in Figure 4.1.10, where subtree (2) is

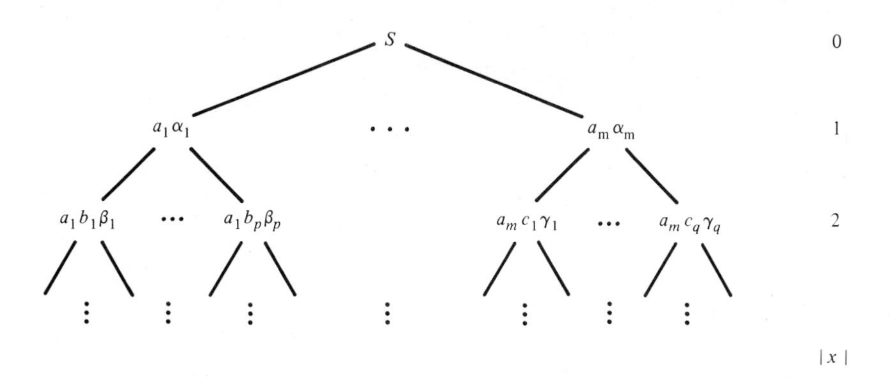

Figure 4.1.9

the same as (1) except the penultimate B is removed everywhere, (3) is the same as (1) with the final A removed everywhere, and (4) is the same as (3) with the final B removed everywhere. Already, at level 2, we have a frontier with 35 words! At level 3 it has size 160. As in most search problems we are faced with a combinatorial explosion in the size of the frontier, and, therefore, we need to prune the search tree whenever possible. An immediate pruning rule is

> *Whenever the terminal prefix of a leftmost sentential form does not match the input word, discontinue searching from its corresponding node in the search tree.*

At level 1 in the above tree we discontinue searching nodes 1, 3, 5, and 7, yielding the pruned tree of Figure 4.1.11. Similarly, four out of each of the eight children of each node can be discarded at level 2, yielding the subtrees of Figure 4.1.12. At level 3 the first three of the four nodes of each subtree are discontinued; thus, we obtain exactly the four nodes at level 3 shown in Figure 4.1.13 of which the final one matches the input word exactly. In other words, (a) is generated by the leftmost derivation

$$S \overset{L}{\Rightarrow} (C \overset{L}{\Rightarrow} (aD \overset{L}{\Rightarrow} (a)$$

Rather than discontinuing the search from a node once it has been generated it is more convenient to discard it before it can be generated, that is *predictively.* We obtain a modified pruning rule:

> *When a node is about to be generated, in the search tree, whose terminal prefix does not match the input word, discard it.*

With Greibach normal form grammars we can guarantee that each nonterminal, which generates terminal words, cannot generate a terminal word of length less than one. This implies that we can prune any node whose sentential form has length greater than the input word. Again we can do this predictively so that using this additional rule the portion of the search tree explored when testing (a) for

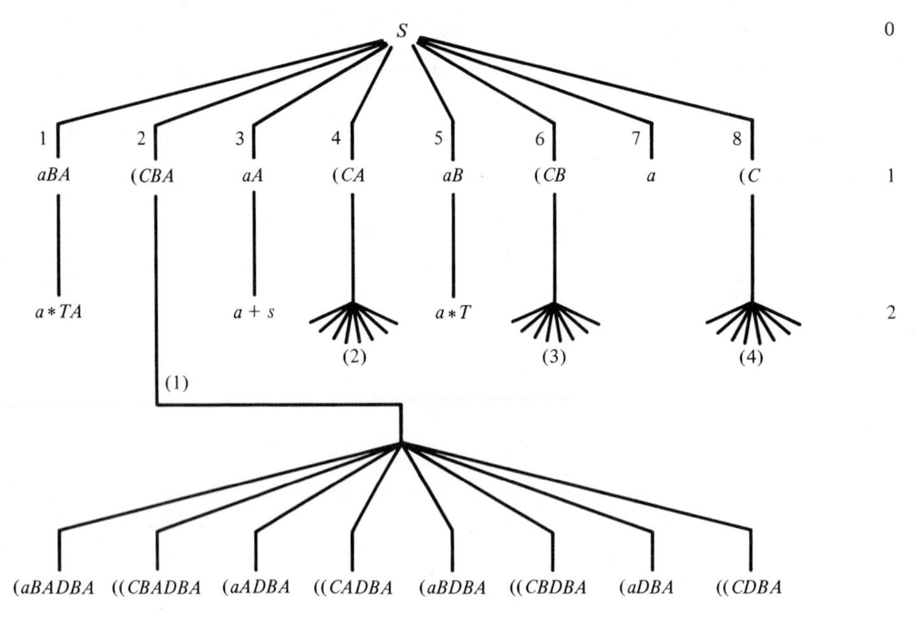

Figure 4.1.10

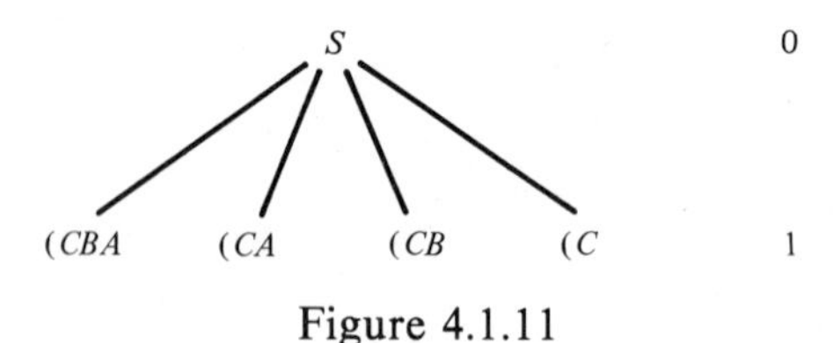

Figure 4.1.11

membership in $L(G)$ is given in Figure 4.1.14. This second rule can be stated as

> *When the sentential form of a node about to be generated in the search tree is longer than the input word, discard it.*

The worked example above provides the basis for a general technique to solve *CFG* MEMBERSHIP and *CFG* PARSING. However, it is extremely inefficient; in the worst cast it can explore p^n derivations, where n is the length of the given word and $p \geq 2$ is a constant. Fortunately, more efficient techniques exist. Methods of this kind are called top-down parsing methods since they attempt to build a syntax tree for an input word from the root to the frontier. The other fundamental approach attempts the converse, that is, to build the syntax tree from the frontier to the root, and is called bottom-up parsing. We explore parsing again in the next chapter as well as in Chapters 8 and 11.

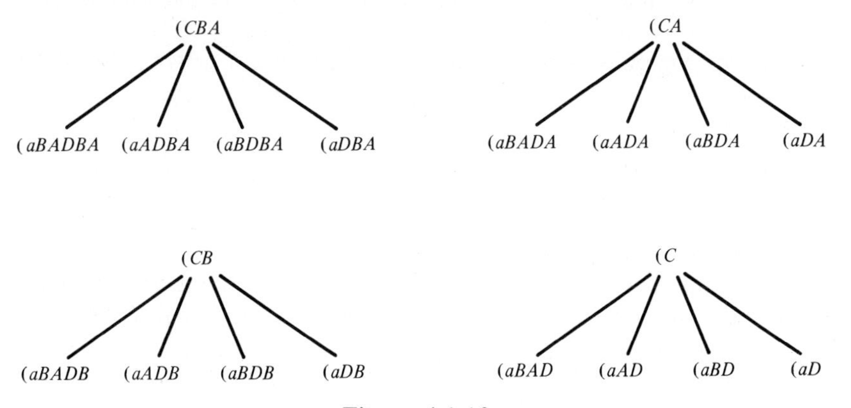

Figure 4.1.12

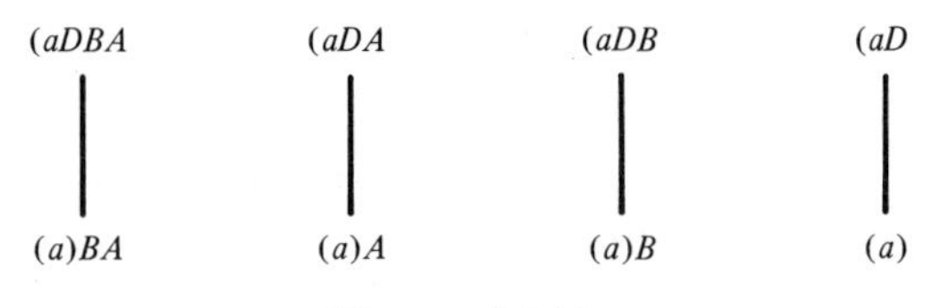

Figure 4.1.13

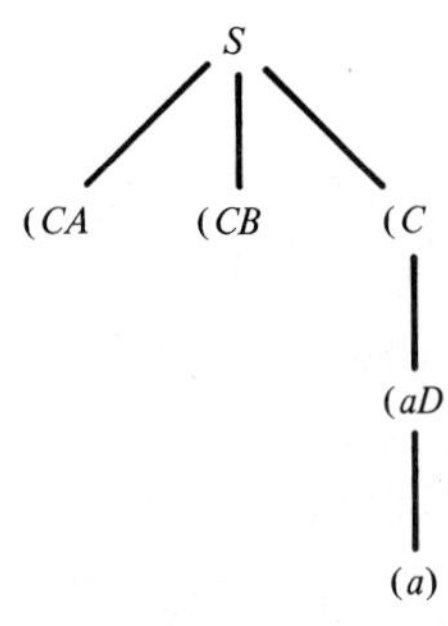

Figure 4.1.14

4.2 SIMPLIFICATIONS

Beginning with the most basic simplification, that is, removing redundant or useless symbols, we demonstrate how the classical simplifications or *normal forms*, as they are commonly called, are obtained. The major normal forms are the (Noam) Chomsky normal form and the (Sheila) Greibach normal form. These normal forms are useful from both a theoretical and practical viewpoint, indeed Greibach normal form, for example, arose from practical considerations in a natural-language parsing project, while Chomsky normal form is the basis of the Cocke-Kasami-Younger parsing algorithm (see Section 11.2). Sections 4.2.2-4.2.6 are devoted to proving these normal form results by way of other simplifications, which are of interest in their own right.

4.2.1 Redundant Symbols

Symbols in a *CFG* can be redundant because either they do not appear in any sentential form or, even if they do appear in a sentential form, they are unable to derive any terminal words. In the former case such symbols are said to be *unreachable* and in the latter they are said to be *nonterminating.* Correspondingly, a symbol that appears in a sentential form is said to be *reachable*, and one that derives a terminal word is said to be *terminating.* Clearly, the sentence symbol is always reachable and the terminal symbols are always terminating. A *CFG* with neither nonterminating nor unreachable symbols is said to be *reduced.*

We now give two algorithms, one to detect terminating symbols and the second to detect reachable symbols. These form the basis for eliminating redundant symbols from a *CFG*. The algorithm *Terminating Symbols* given below is based on the following idea

> *If a nonterminal A has a production consisting solely of terminating symbols, then A is itself terminating.*

In particular, if G generates a nonempty language, then there must be at least *one* nonterminal having a production with a terminal right-hand side. *Terminating Symbols* identifies these nonterminals, treats them as new terminals, and it then repeats the process until no new terminating symbols are found. Marking is used to identify the new terminating symbols.

Algorithm *Terminating Symbols.*
On entry: An arbitrary *CFG*, $G = (N,\Sigma,P,S)$.
On exit: The set T of terminating symbols.

begin {A *marking* algorithm is used}
  $T := \Sigma$;
  Mark, in some way, all appearances of terminal symbols in P;
  repeat
    Let Q be the set of unmarked nonterminals which have at least one production whose right-hand side consists solely of marked symbols. (Note that the right-hand side of a λ-production consists solely of marked symbols);
    Mark all appearances in P of the symbols in Q;
    $T := T \cup Q$
  until $Q = \varnothing$;
end of Algorithm.

Example 4.2.1A Let G be described by

$$S \rightarrow ASB \mid BSA \mid SS \mid aS \mid \lambda$$
$$A \rightarrow AB \mid B$$
$$B \rightarrow BA \mid A$$

Initially, $T = \{a\}$ and we mark the production

$$S \rightarrow \hat{a}S$$

using the hat symbol " ^ ".

But this means that S should be marked, since the right-hand side of $S \rightarrow \lambda$ is completely marked. S is the only new symbol to be marked, that is, $Q := \{S\}$.

We mark all appearances of S, namely,

$$\hat{S} \rightarrow A\hat{S}B \mid B\hat{S}A \mid \hat{S}\hat{S} \mid \hat{a}\hat{S}$$

let $T = \{a,S\}$ and repeat the process. This time, however, $Q := \varnothing$, and hence the loop and algorithm terminate with $T = \{a,S\}$, the only terminating symbols.

The algorithm *Terminating Symbols* always terminates since the loop is only repeated when at least one unmarked symbol is added to T. Therefore, after at most $\#N+1$ repetitions of the loop the algorithm must terminate.

That *Terminating Symbols* exits with T containing exactly the terminating symbols of the given grammar is not difficult to prove. Let T_G be the terminating symbols of G. Then $\Sigma \subseteq T_G$ and $\Sigma \subseteq T$; therefore, only consider the nonterminals in T_G.

It follows immediately that $T \subseteq T_G$, since whenever a nonterminal is marked in *Terminating Symbols* we have traced a terminating derivation for it, via the marked symbols and completely marked right-hand sides.

To prove $T_G \subseteq T$ it is sufficient to prove that if A is not in T on termination of *Terminating Symbols*, then A is not in T_G.

The details of the proof are left to the Exercises. Thus we have:

Lemma 4.2.1 *Let $G = (N,\Sigma,P,S)$ be an arbitrary CFG. Then* Terminating Symbols *halts when given G and correctly reports the terminating symbols of G.*

Clearly, the nonterminating symbols of a *CFG*, G, are exactly $N - T_G$. To simplify G we remove these symbols, apart from S, from N and all productions from P which contain them.

Example 4.2.1B We obtain from G the simplified *CFG*, G',

$$S \rightarrow SS \mid aS \mid \lambda$$

The productions retained in the simplified grammar are exactly those marked in *Terminating Symbols*. Since we have removed nonterminating symbols, which do not affect the language of the *CFG*, we have:

Theorem 4.2.2 *Let $G = (N,\Sigma,P,S)$ be an arbitrary CFG. Then there is an algorithm which constructs an equivalent CFG, $G' = (N',\Sigma,P',S)$, from G, where $N' \subseteq N$, $P' \subseteq P$, and $N' \cup \Sigma$ consists solely of terminating symbols, with the possible exception of S.*

We now turn to reachable symbols. The algorithm *Reachable Symbols* can be viewed as the node-connectedness problem for digraphs; see the Exercises. We give, however, an explicit algorithm which traces the symbols reachable from S, then those reachable from these, and so on. Note that if A is reachable then all symbols in α are reachable, where $A \rightarrow \alpha$ is a production in G.

Algorithm *Reachable Symbols.*
On entry: An arbitrary *CFG*, $G = (N,\Sigma,P,S)$.
On exit: The set R of reachable symbols.

begin
  $R := \{S\}$;
  $Q := \{S\}$; {The set of newly marked symbols}
  repeat
    Mark all symbols on the right-hand sides of all productions which have left-hand sides in Q;
    Let Q be the set of newly marked symbols, which are not already in R;
    $R := R \cup Q$
  until $Q = \varnothing$
end of Algorithm.

Example 4.2.2 Let G be

$$S \rightarrow aS \mid SB \mid SS \mid \lambda$$
$$A \rightarrow ASB \mid c$$
$$B \rightarrow b$$

Initially, $R = \{S\}$ and $Q = \{S\}$. Thus, we obtain

$$S \rightarrow \hat{a}\hat{S} \mid \hat{S}\hat{B} \mid \hat{S}\hat{S} \mid \hat{\lambda}$$

and a and B are the only newly marked symbols that are not in R, since S is already in R. Hence, $Q := \{a,B\}$ and $R := \{S,a,B\}$. We now obtain

$$B \rightarrow \hat{b}$$

and, hence, $Q := \{b\}$ and $R := \{S,a,B,b\}$. Since Q contains no nonterminals, executing the loop leaves $Q = \varnothing$ and R unchanged.

Hence, S,a,B, and b are reachable, but A and c are unreachable.

The marking used in *Reachable Symbols* is different from that used in *Terminating Symbols* in that it marks *all* symbols on the right-hand sides of productions and proceeds from left-hand sides to corresponding right-hand sides, rather than conversely. However, it again mirrors possible derivations in the given grammar, namely, derivations from the sentence symbol.

In spite of this observation the proofs of termination and correctness are similar, if not simpler than previously, so we leave these completely to the Exercises, yielding:

Lemma 4.2.3 *Let $G = (N,\Sigma,P,S)$ be an arbitrary CFG. Then* Reachable Symbols *halts when given G and correctly reports the reachable symbols of G.*

Again the unreachable symbols, by definition, cannot affect the language of the *CFG*, G. Thus, letting R_G be the reachable symbols, the unreachable nonterminals are $N - R_G$ and the unreachable terminals are $\Sigma - R_G$. We remove these symbols from N and Σ, respectively, and also remove all productions from P involving them. Thus, we obtain

Theorem 4.2.4 *Let $G = (N,\Sigma,P,S)$ be an arbitrary CFG. Then there is an algorithm which constructs an equivalent CFG, $G' = (N',\Sigma',P',S)$, where $N' \subseteq N$, $\Sigma' \subseteq \Sigma$, $P' \subseteq P$, and $N' \cup \Sigma'$ consists solely of reachable symbols.*

Finally, we combine both constructions to obtain

Theorem 4.2.5 *Let $G = (N,\Sigma,P,S)$ be an arbitrary CFG. Then there is an algorithm which constructs an equivalent, reduced CFG, $G' = (N',\Sigma',P',S)$, where $N' \subseteq N$, $\Sigma' \subseteq \Sigma$, and $P' \subseteq P$.*

Proof: First apply *Terminating Symbols* to find the nonterminating symbols to be eliminated giving, via Theorem 4.2.2, G'' equivalent to G. Second, apply *Reachable Symbols* to G'' to find the unreachable symbols and eliminate them to obtain G'. Since G'' is equivalent to G and G' is equivalent to G'', then G' is equivalent to G. ■

The order of elimination chosen in the above proof is crucial; see the Exercises.

4.2.2 Empty Productions

An *empty production* (λ*-production* or *null production*) is a production whose right-hand side is the empty word, that is, a production of the form

$$A \rightarrow \lambda$$

Given a *CFG*, $G = (N, \Sigma, P, S)$, if λ is in $L(G)$, then empty productions are necessary; otherwise λ cannot be generated at all. However, when λ is not in $L(G)$ any empty productions in G can be eliminated in such a way that $L(G)$ is unaffected. The elimination procedure only affects productions which contain, on their right-hand sides, nonterminals that can derive the empty word. We call such nonterminals *λ-nonterminals*. A nonterminal A is a λ-nonterminal if there is a derivation $A \Rightarrow^+ \lambda$ in G. Note that λ is in $L(G)$ if and only if S is a λ-nonterminal. If G has no λ-nonterminals it is said to be *λ-free*.

If G has λ-productions, then in a sentential derivation

$$S \Rightarrow \alpha_1 \Rightarrow \alpha_2 \Rightarrow \cdots \Rightarrow \alpha_n \text{ in } \Sigma^*$$

The lengths of the α_i may vary tremendously with respect to one another. This is because many nonterminals may be introduced which only contribute the empty word to α_n. However, if G has no λ-productions, then we know that

$$|S| = 1 \leq |\alpha_1| \leq |\alpha_2| \leq \cdots \leq |\alpha_n|$$

the lengths are monotonically increasing. This property is useful when testing whether or not a given word is generated by a *CFG*.

We first give an algorithm to find the λ-nonterminals of a *CFG*. It is almost identical to *Terminating Symbols*, since it based on the same idea, namely, if $A \Rightarrow^+ \lambda$, then there is an $A \rightarrow \alpha$ in G with $\alpha \Rightarrow^+ \lambda$.

Algorithm *λ-Nonterminals*.
On entry: An arbitrary *CFG*, $G = (N, \Sigma, P, S)$.
On exit: The set Λ of λ-nonterminals.

begin {A marking algorithm is used}
    $\Lambda := \varnothing$;
    Mark the right-hand sides of all empty productions in P;
    repeat
        Let Q be the set of unmarked nonterminals which have at least one production whose right-hand side consists solely of marked symbols;
        Mark all appearances in P of the symbols in Q;
        $\Lambda := \Lambda \cup Q$
    until $Q = \varnothing$
end of Algorithm.

Example 4.2.3A Let G be

$S \rightarrow aS \mid SS \mid bA$

$A \rightarrow BB$

$B \rightarrow CC \mid ab \mid aAbC$

$C \rightarrow \lambda$.

Initially, we have

$$C \to \hat{\lambda} \text{ and } \Lambda = \varnothing.$$

When executing the loop the first time, $Q = \{C\}$, and we obtain

$$B \to \hat{C}\hat{C} \mid aAb\hat{C},\ \hat{C} \to \hat{\lambda} \text{ and } \Lambda = \{C\}.$$

Since Q is nonempty a second iteration yields $Q = \{B\}$,

$$A \to \hat{B}\hat{B},\ \hat{B} \to \hat{C}\hat{C} \mid ab \mid aAb\hat{C}, \text{ and } \Lambda = \{B,C\}.$$

Again Q is nonempty; we find $Q = \{A\}$,

$$S \to b\hat{A},\ \hat{A} \to \hat{B}\hat{B},\ \hat{B} \to a\hat{A}b\hat{C}, \text{ and } \Lambda = \{A,B,C\}.$$

Once more Q is nonempty, but a further iteration does not yield any new symbols; therefore, the algorithm terminates. A,B, and C are the only λ-nonterminals in G, and, since S is not a λ-nonterminal, λ is not in $L(G)$.

Because of the strong similarity of *λ-Nonterminals* to *Terminating Symbols* we obtain immediately:

Lemma 4.2.6 *Let $G = (N,\Sigma,P,S)$ be an arbitrary CFG. Then* λ-Nonterminals *halts when applied to G and correctly reports G's λ-nonterminals.*

To eliminate λ-nonterminals from a *CFG* we use the following algorithm, which eliminates λ-nonterminals without introducing new ones. The strategy is based on the following idea

> Let A be a λ-nonterminal in G. Then conceptually we split it into two nonterminals A' and A'' such that A' generates all the nonempty terminal words that A does, and A'' only generates λ. Now in the right-hand side of a production in which A appears once, $B \to \alpha A\beta$ say, we replace it with two productions $B \to \alpha A'\beta$ and $B \to \alpha A''\beta$. Since A'' only generates λ we replace it by λ giving $B \to \alpha\beta$ and use A in place of A'. If $B \to \alpha A\beta A\gamma$, then we obtain all four combinations of replacements, and so on.

Before giving the algorithm we introduce a useful variant of equivalence. Let G_1 and G_2 be two *CFGs*; we say they are *λ-equivalent* if they generate the same language when the empty word is ignored. In other words, G_1 and G_2 are λ-equivalent if $L(G_1) - \{\lambda\} = L(G_2) - \{\lambda\}$. Clearly, equivalence always implies λ-equivalence, but the converse implication is not valid.

Algorithm *Eliminate λ-Nonterminals.*
On entry: $G = (N,\Sigma,P,S)$ is an arbitrary reduced *CFG* with at least one λ-nonterminal.
On exit: A reduced *CFG*, $G' = (N',\Sigma,P',S)$, λ-equivalent to G having no λ-nonterminals.

begin

For all λ-nonterminals A in N',
For all productions $B \to \beta$ in P with $|\beta|_A = m \geq 1$

Let $\beta = \beta_0 A \beta_1 \cdots A \beta_m$ where $|\beta_0 \cdots \beta_m|_A = 0$;
Replace $B \to \beta$ in P with the 2^m productions

$B \to \beta_0 A \beta_1 \cdots A \beta_m$ } all A's present

$B \to \beta_0 \beta_1 A \cdots A \beta_m$
$B \to \beta_0 A \beta_1 \beta_2 A \cdots A \beta_m$
.
.
.
$B \to \beta_0 A \beta_1 \cdots A \beta_{m-1} \beta_m$ } one A omitted

$B \to \beta_0 \beta_1 \beta_2 A \cdots A \beta_m$
.
.
.
$B \to \beta_0 A \cdots A \beta_{m-2} \beta_{m-1} \beta_m$ } two A's omitted

.
.
.

$B \to \beta_0 \beta_1 \cdots \beta_{m-1} \beta_m$ } all A's omitted.

We can describe this set of productions more succinctly as
$\{B \to \beta_0 X_m \beta_1 \cdots X_m \beta_m : X_i \text{ is in } \{\lambda, A\}, 1 \leq i \leq m\}$

Finally, remove all λ-productions from P giving P'. It is possible that a λ-nonterminal A only generated the empty word in G, in which case A and possibly other nonterminals are now nonterminating. So apply Theorem 4.2.5 to reduce this intermediate grammar to give the final G'.

end of Algorithm.

Example 4.2.3B We have

$S \to aS \mid SS \mid bA$
$A \to BB$

$B \rightarrow CC \mid ab \mid aAbC$

$C \rightarrow \lambda$

C is the only λ-nonterminal with a λ-production, while $B \rightarrow CC$ and $B \rightarrow aAbC$ are the only productions containing C. Thus, we obtain

$B \rightarrow CC$

$B \rightarrow C$

since omitting the first or second C leads to the same result.

$B \rightarrow \lambda$

$B \rightarrow aAbC$

$B \rightarrow aAb$

Since $C \rightarrow \lambda$ is the only C-production only $B \rightarrow \lambda$ and $B \rightarrow aAb$ are retained, yielding the new grammar

$S \rightarrow aS \mid SS \mid bA$

$A \rightarrow BB$

$B \rightarrow \lambda \mid ab \mid aAb$

We can repeat the elimination with $B \rightarrow \lambda$. Since only the production $A \rightarrow BB$ contains B we obtain

$A \rightarrow BB \mid B \mid \lambda$

and in this case since there are other B-productions these are all retained, giving

$S \rightarrow aS \mid SS \mid bA$

$A \rightarrow BB \mid B \mid \lambda$

$B \rightarrow ab \mid aAb$

Finally, eliminate $A \rightarrow \lambda$, that is,

$B \rightarrow aAb$

is replaced by

$B \rightarrow aAb \mid ab$

and

$S \rightarrow bA$

is replaced by

$S \rightarrow bA \mid b$

The final λ-free CFG is

$S \rightarrow aS \mid SS \mid bA \mid b$

$A \to BB \mid B$

$B \to ab \mid aAb$

Lemma 4.2.7 *Let $G = (N,\Sigma,P,S)$ be an arbitrary reduced CFG with at least one λ-nonterminal. Then* Eliminate λ-Nonterminals *halts with a reduced, λ-equivalent CFG, G', with no λ-nonterminals.*

Proof: This is left to the Exercises. ■

This leads, as we have seen in the above example, to

Theorem 4.2.8 *Let $G = (N,\Sigma,P,S)$ be an arbitrary reduced CFG. Then there is an algorithm which constructs a reduced, λ-free CFG, G', λ-equivalent to G. In other words, for every CFL, L, there is a λ-free CFG, G, with $L(G) = L - \{\lambda\}$.*

4.2.3 Unit Productions

A production $A \to \alpha$ in a *CFG*, $G = (N,\Sigma,P,S)$, is said to be a *unit production* if α is in N. A special case of a unit production is the production $A \to A$, a *circular production.* Such a production is similar to the transition (q,λ,q) in a λ-*NFA*, in that it can be removed immediately without affecting the language of the grammar. A *CFG* with no unit productions is said to be *unit-free.* The algorithm to eliminate one unit production is similar to the removal of λ-transitions in a λ-*NFA*.

Algorithm *Eliminate One Unit Production.*
On entry: $G = (N,\Sigma,P,S)$ is an arbitrary reduced *CFG* with no circular productions and at least one unit production.
On exit: G' is a reduced equivalent *CFG* with one fewer unit production than G.

begin
    Choose an arbitrary unit production $B \to C$ in G; there must be at least one;
    Replace $B \to C$ in P with $B \to \alpha$, for all α in $V^* - N$, satisfying $C \Rightarrow C_1 \Rightarrow \cdots \Rightarrow C_r \Rightarrow \alpha$, for some nonterminals $C_1, \ldots, C_r$.
end of Algorithm.

Example 4.2.4 Let G be

$$S \to T \mid S+T;\ \ T \to F \mid F*T;\ \ F \to a \mid (S) \mid T$$

Let $T \to F$ be the chosen unit production. Now

$$F \Rightarrow a;\ \ F \Rightarrow (S);\ \ F \Rightarrow T \Rightarrow F*T;\ \ F \Rightarrow T \Rightarrow F \Rightarrow \cdots$$

so replace $T \to F$ with

$T \to a \mid (S)$

giving

$$S \to T \mid S+T; \quad T \to a \mid (S) \mid F * T; \quad F \to a \mid (S) \mid T$$

Repeating the algorithm choose $S \to T$ and replace it with

$$S \to a \mid (S) \mid F * T$$

giving

$$S \to a \mid (S) \mid F * T \mid S+T; \quad T \to a \mid (S) \mid F * T; \quad F \to a \mid (S) \mid T$$

and, finally, replace $F \to T$ by $F \to a \mid (S) \mid F * T$, giving the unit-free *CFG*

$$S \to a \mid (S) \mid F * T \mid S + T$$
$$T \to a \mid (S) \mid F * T$$
$$F \to a \mid (S) \mid F * T$$

We leave the proofs of the following results to the Exercises.

Lemma 4.2.9 *Let $G = (N, \Sigma, P, S)$ be an arbitrary reduced CFG with no circular productions and having at least one unit production. Then* Eliminate One Unit Production *halts with a reduced, equivalent CFG, G', with one fewer unit production than G.*

We eliminate all unit productions by iterating *Eliminate One Unit Production* as in Example 4.2.4.

Theorem 4.2.10 *Let $G = (N, \Sigma, P, S)$ be an arbitrary reduced CFG. Then there is an algorithm which constructs an equivalent, unit-free, reduced CFG, G', from G.*

4.2.4 Binary Form and Chomsky Normal Form

A *CFG*, $G = (N, \Sigma, P, S)$, is said to be in *binary form* if for all $A \to \alpha$ in P, $|\alpha| \leq 2$. It is in *Chomsky normal form*, or *CNF*, if it is in binary form and there are only two kinds of productions

$$A \to a; \ a \text{ in } \Sigma, \quad \text{and} \quad A \to BC; \ B, C \text{ in } N$$

We prove that for every *CFG*, G, a λ-equivalent *CFG*, G', in *CNF* can be effectively constructed. We do this by first obtaining a binary form equivalent *CFG*. Letting $G = (N, \Sigma, P, S)$ be a *CFG* we define $Maxrhs(G) = \max\{ |\alpha| : A \to \alpha \text{ is in } P \}$. Clearly, G is in binary form if and only if $Maxrhs(G) \leq 2$. If $Maxrhs(G) > 2$, each right-hand side α of length $Maxrhs(G)$ is split into two halves. The second half is replaced by a new nonterminal giving a shorter right-hand side.

Algorithm *Reduce Largest Productions.*
On entry: An arbitrary reduced *CFG*, $G = (N, \Sigma, P, S)$, with $Maxrhs(G) \geq 3$.

On exit: An equivalent reduced CFG, G', with $Maxrhs(G') < Maxrhs(G)$.

begin

For all productions $A \to \alpha$ in P with $|\alpha| = Maxrhs(G)$

Letting $\alpha = \alpha_1\alpha_2$, where $|\alpha_1| = \lfloor |\alpha|/2 \rfloor \geq 1$ and $|\alpha_2| = \lceil |\alpha|/2 \rceil \geq 2$, replace $A \to \alpha$ in P by

$A \to \alpha_1[A\alpha]$ and $[A\alpha] \to \alpha_2$,

where $[A\alpha]$ is a new unique nonterminal.

$\{$Then $2 \leq |\alpha_1[A\alpha]| \leq \lceil(|\alpha|+1)/2\rceil < |\alpha|\}$

end of Algorithm.

Example 4.2.5 Let G be

$S \to T \mid S+T \mid (S)+T$

$T \to F \mid F*T$

$F \to a \mid (S)$

$Maxrhs(G) = 5$ from $S \to (S)+T$. Applying *Reduce Largest Productions*, it is replaced by

$S \to (SA$

$A \to)+T$

where A is a new nonterminal, giving G_1 with $Maxrhs(G_1) = 3$, from $S \to S+T$; $S \to (SA$; $A \to)+T$; $T \to F*T$; $F \to (S)$. These can be replaced, in turn, by

$S \to SB$;	$B \to +T$
$S \to (C$;	$C \to SA$
$A \to)D$;	$D \to +T$
$T \to FE$;	$E \to *T$
$F \to (H$;	$H \to S)$

where B, C, D, E, and H are new nonterminals, giving G_2 with $Maxrhs(G_2) = 2$. In other words, G_2 is in binary form.

Lemma 4.2.11 *Let $G = (N, \Sigma, P, S)$ be an arbitrary reduced CFG with $Maxrhs(G) \geq 3$. Then* Reduce Largest Productions *halts with a reduced equivalent CFG, G', satisfying $Maxrhs(G') < Maxrhs(G)$.*

Proof: This is left to the Exercises. ■

This algorithm yields the following by its iterative application, as in Example 4.2.5:

Theorem 4.2.12 *Let $G = (N, \Sigma, P, S)$ be an arbitrary reduced CFG. Then there is an algorithm that constructs an equivalent reduced CFG, G', in binary form from G.*

We now are able to prove:

Theorem 4.2.13 Chomsky Normal Form Theorem
Let $G = (N, \Sigma, P, S)$ be an arbitrary reduced CFG. Then there is an algorithm that constructs a λ-equivalent reduced CFG, G', in CNF from G. Thus, for every CFL, L, there exists a CFG, G, in CNF with $L(G) = L - \{\lambda\}$.

Proof: First, eliminate λ-productions and then unit productions from G to give a reduced λ-free, unit-free *CFG* G'', λ-equivalent to G.

Now apply Theorem 4.2.12 to obtain a λ-equivalent reduced *CFG*, G''', in binary form from G''. Observe that *Reduce Largest Productions* introduces neither λ-productions nor unit productions; hence, G''' only has productions of the forms

(1) $A \rightarrow BC$;
(2) $A \rightarrow a$;
(3) $A \rightarrow ab$;
(4) $A \rightarrow bC$;
(5) $A \rightarrow Bc$;

of which only (1) and (2) are desired. However, to eliminate the remaining three we simply introduce new productions and nonterminals

$$\bar{a} \rightarrow a$$

for all a in Σ''' appearing on the right-hand side of productions of types (3), (4), and (5), and replace these productions by

(3′) $A \rightarrow \bar{a}\bar{b}$;
(4′) $A \rightarrow \bar{b}C$;
(5′) $A \rightarrow B\bar{c}$.

This gives a G' in *CNF*, which is also reduced and λ-equivalent to G''' and hence λ-equivalent to G. ■

Example 4.2.6 Let G be

$$S \rightarrow T \mid S + T$$
$$T \rightarrow F \mid F * T$$
$$F \rightarrow a \mid (S)$$

Eliminating unit productions we obtain

$$S \rightarrow a \mid (S) \mid F * T \mid S + T$$
$$T \rightarrow a \mid (S) \mid F * T$$
$$F \rightarrow a \mid (S)$$

giving G_1 with $Maxrhs(G_1) = 3$. This is transformed into a binary form CFG, G_2

$$S \rightarrow a \mid (B \mid FA \mid SC; \qquad A \rightarrow *T$$
$$T \rightarrow a \mid (B \mid FA; \qquad B \rightarrow S)$$
$$F \rightarrow a \mid (B; \qquad C \rightarrow +T$$

Finally, we add

$$\bar{*} \rightarrow *;\ \bar{(} \rightarrow (;\ \bar{)} \rightarrow);\ \bar{+} \rightarrow +;$$

to obtain

$$S \rightarrow a \mid \bar{(}B \mid FA \mid SC; \qquad A \rightarrow \bar{*}T$$
$$T \rightarrow a \mid \bar{(}B \mid FA; \qquad B \rightarrow S\bar{)}$$
$$F \rightarrow a \mid \bar{(}B; \qquad C \rightarrow \bar{+}T$$

which is in CNF.

4.2.5 Greibach Normal Form and Two-Standard Normal Form

Our final normal form results are due to Sheila Greibach.

Let $G = (N, \Sigma, P, S)$ be a CFG. Then G is said to be in *Greibach normal form*, or GNF, if each production $A \rightarrow \alpha$ in P satisfies

$$\alpha \text{ is in } \Sigma N^*,$$

that is, its right-hand side begins with a terminal and is followed by a, possibly empty, sequence of nonterminals. For $m \geq 1$, G is said to be in *m-standard normal form*, or m-SNF, if each production $A \rightarrow \alpha$ in P satisfies

$$\alpha \text{ is in } \bigcup_{i=0}^{m} \Sigma N^i,$$

that is, it is in GNF, but there are never more than m nonterminals on the right-hand sides of productions. We are particularly interested in 2-standard normal form.

To obtain a Greibach normal form grammar from a given CFG we first remove the left-recursion.

Let $G = (N, \Sigma, P, S)$ be a CFG. We say that G is *left-recursive* if there is a nonterminal A in G and a derivation $A \overset{L}{\Rightarrow}{}^+ A\alpha$, for some α in V^*. We also say that such a nonterminal is *left-recursive*. If there is a derivation $A \overset{R}{\Rightarrow}{}^+ \alpha A$ we say A and G are *right-recursive*. If $A \rightarrow A\alpha$ is in P we say A is *directly left-recursive* and if $A \rightarrow \alpha A$ is in P, A is *directly right-recursive*. Finally, G is *non-*

left- (*non-right-*) *recursive* if it is not left- (right-) recursive.

Example 4.2.7 Let G be

$$S \to S+T \mid T*F \mid a \mid (S)$$
$$T \to T*F \mid S+T \mid a \mid (S)$$
$$F \to a \mid (S)$$

S and T are directly left-recursive, and therefore G is left-recursive. Moreover T is also directly right-recursive; thus G is also right-recursive.

We leave to the Exercises a proof of the following:

Theorem 4.2.14 *Let $G = (N,\Sigma,P,S)$ be a reduced, λ-free, unit-free CFG. Then there is an algorithm which constructs an equivalent non-left-recursive CFG, G', from G. In other words, for every CFL, L, there exists a non-left-recursive CFG, G, with $L(G) = L - \{\lambda\}$.*

Clearly, a similar result holds for non-right-recursive *CFGs* too. Returning to the construction of *CFGs* in *GNF* there is a crucial observation, as follows:

Observation 4.2.1 Let $G = (N,\Sigma,P,S)$ be a reduced, non-left-recursive *CFG* in *CNF*. Then there is a nonterminal A in N whose productions all begin with a terminal, that is, A only has *terminal left appearances*. For if we assume otherwise, then every nonterminal A_i has at least one production $A_i \to A_j A_k$, where $N = \{A_1, \ldots, A_n\}$ and $A_1 = S$. But this implies that we can obtain a leftmost derivation

$$S = A_1 \overset{L}{\Rightarrow} A_{i_1}A_{j_1} \overset{L}{\Rightarrow} A_{i_2}A_{j_2}A_{j_1} \overset{L}{\Rightarrow} \cdots \overset{L}{\Rightarrow} A_{i_m}A_{j_m}A_{j_{m-1}} \cdots A_{j_1}$$

for any $m \geq 1$. In particular, choose $m = n$ in which case $A_1, A_{i_1}, \ldots, A_{i_m}$ contain at least one repeated nonterminal $A = A_{i_j} = A_{i_k}$ say, by the pigeonhole principle. This implies A is left-recursive. But we assumed G to be non-left-recursive, so there must be a nonterminal with all of its productions starting with a terminal.

Example 4.2.8A Let G be the non-left-recursive *CFG*

$S \to TA \mid T;$ $\qquad B \to *T$

$T \to FB \mid F;$ $\qquad C \to SD$

$F \to a \mid (C;$ $\qquad D \to)$

$A \to +S;$

Then A, B, D and F have only terminal left appearances.

Given a λ-free, non-left-recursive *CFG* we begin to obtain an equivalent *GNF* grammar by carrying out the following substitution rule

> *For each nonterminal A having only terminal left appearances, replace each occurrence of A in the leftmost position of a production with its right-hand sides.*

So if $A \rightarrow a_1 \mid a_2 \mid a_3D$ and $B \rightarrow AC$ we obtain $B \rightarrow a_1C \mid a_2C \mid a_3DC$. After this has been carried out either all nonterminals have only terminal left appearances or there is some nonterminal B that has a production starting with a nonterminal.

Observation 4.2.2 Let $G = (N, \Sigma, P, S)$ be a reduced, λ-free, non-left-recursive *CFG*. Then there is a nonterminal B with B-productions having left appearances of either terminals or nonterminals which have only terminal left appearances.

This can be proved by contradiction along similar lines to the argument used for Observation 4.2.1. However, they are both special cases of a more general result, which we prefer to prove. Observation 4.2.1 shows the existence of nonterminals which in one leftmost derivation step produce a terminal on the left, while Observation 4.2.2 captures those that require two leftmost derivation steps.

We need the following definition before stating and proving the general result.

Definition Let $G = (N, \Sigma, P, S)$ be a non-left-recursive *CFG*. For all A in N, define $int(A)$ by

$$int(A) \text{ is the largest integer } k \text{ such that } A \overset{L}{\Rightarrow}{}^{k-1} B\alpha \overset{L}{\Rightarrow} a\beta,$$

$$\text{for some } a \text{ in } \Sigma,\ B \text{ in } N, \text{ and } \alpha, \beta \text{ in } (N \cup \Sigma)^*.$$

Let $intmax(G) = \max(\{int(A) : A \text{ is in } N\})$.

Lemma 4.2.15 *Let $G = (N, \Sigma, P, S)$ be a reduced, λ-free, non-left-recursive CFG. Then*

(i) *int is well-defined with $1 \leq intmax \leq \#N$;*
(ii) *for all i, $1 \leq i \leq intmax$, there is an A with $int(A) = i$; and*
(iii) *if $A \rightarrow B\alpha$ is in P then $int(A) > int(B)$, that is, int provides a partial ordering of N.*

Proof: Part (i) follows from observing that the longest leftmost derivation, of the kind considered, can have at most $\#N$ steps. Otherwise left-recursion is present by the pigeonhole principle.

Parts (ii) and (iii) follow from observing that

$$int(A) = \begin{cases} 1, & \text{if } A \text{ has only terminal left appearances,} \\ 1 + \max\{int(B) : A \to B\alpha \text{ in } P\}, & \text{otherwise} \end{cases}$$

and proceeding inductively on *intmax*. ∎

Example 4.2.8B Simple calculation yields

$$\begin{array}{ll} int(S) = 3 & int(B) = 1 \\ int(T) = 2 & int(C) = 4 \\ int(F) = 1 & int(D) = 1 \\ int(A) = 1 & \end{array}$$

so $intmax = 4$. Now since $T \to FB$ we expect from Lemma 4.2.15 that $int(T) > int(F)$ and this is indeed the case since $2 > 1$.

The algorithm to produce *GNF* is now straightforward.

Algorithm *GNF Conversion.*
On entry: A reduced, λ-free, non-left-recursive *CFG* $G = (N,\Sigma,P,S)$.
On exit: A reduced, equivalent *CFG* $G' = (N',\Sigma,P',S)$ in Greibach normal form.

begin
  $T := \{A : A \text{ is in } N \text{ and } int(A) = 1\}$;
  $P' := \{A \to \alpha : A \to \alpha \text{ is in } P \text{ and } int(A) = 1\}$;
  $P'' := P - P'$;
  for $i := 2$ **to** *intmax* **do**
  begin
    for all A **in** T **do**
      replace all left appearances of A in P'' by the
      right-hand sides of all A-productions in P';
    $T := \{A : A \text{ is in } N \text{ and } int(A) = i\}$;
    $P' := P' \cup \{A \to \alpha : A \to \alpha \text{ is in } P'' \text{ and } int(A) = i\}$;
    $P'' := P'' - P'$
  end;
  It is possible that some nonterminals in N are now unreachable from S using the productions in P'. In this case remove the unreachable nonterminals from N, giving N', and their associated productions from P', to give the new P'. Otherwise $N' = N$ and P' is as computed. Finally, it is possible that there are right-hand sides having the form $A \to a\alpha b\beta$, where b is terminal. Modify these productions to give $A \to a\alpha\bar{b}\beta$, and $\bar{b} \to b$, where $\bar{b}$ is a new nonterminal to be added to N'.

end of Algorithm.

Example 4.2.8C

$T := \{A,B,D,F\}$

$P' := \{A \to +S;\ B \to *T;\ D \to);\ F \to a \mid (C\}$

$P'' := \{C \to SD;\ S \to TA \mid T;\ T \to FB \mid F\}$

$i := 2;$ {Replace all left appearances of $A,B,D,$ and F}

$P'' := \{C \to SD;\ S \to TA \mid T;\ T \to aB \mid (CB \mid a \mid (C\}$

$T := \{T\}$

$P' := \{A \to +S;\ B \to *T;\ D \to);\ F \to a \mid (C;\ T \to aB \mid (CB \mid a \mid (C\}$

$P'' := \{C \to SD;\ S \to TA \mid T\}$

$i := 3;$ {Replace all left appearances of T }

$P'' := \{C \to SD;\ S \to aBA \mid (CBA \mid aA \mid (CA \mid aB \mid (CB \mid a \mid (C\}$

$T := \{S\}$

$P' := (P' \cup P'') - \{C \to SD\}$

$P'' := \{C \to SD\}$

$i := 4;$ {Replace all left appearances of S}

giving the final GNF grammar G'

$S \to aBA \mid (CBA \mid aA \mid (CA \mid aB \mid (CB \mid a \mid (C$
$T \to aB \mid (CB \mid a \mid (C$
$F \to a \mid (C$
$A \to +S;\ \ B \to *T;\ \ D \to)$
$C \to aBAD \mid (CBAD \mid aAD \mid (CAD \mid aBD \mid (CBD \mid aD \mid (CD$

which is also in 4-standard normal form.

We now have:

Theorem 4.2.16 Greibach Normal Form Theorem
Given an arbitrary CFG, $G = (N,\Sigma,P,S)$, *there is an algorithm that constructs a* λ*-equivalent CFG,* G', *in GNF from it. In other words, for every CFL,* L, *there exists a CFG,* G, *in GNF with* $L(G) = L - \{\lambda\}$.

Proof: The first statement follows from the previous discussion, observing that *GNF Conversion* only uses nonterminal substitution, which by Lemma 4.2.17, below, preserves the language generated by the grammar. The second statement follows from Theorem 4.2.14 and the first statement. ■

Lemma 4.2.17 The Nonterminal Substitution Lemma
Let $G = (N,\Sigma,P,S)$ be an arbitrary CFG, $A \rightarrow \alpha_1 B \beta_2$ be a production in P, and

$$B \rightarrow \beta_1 \mid \cdots \mid \beta_r$$

be the B-productions in P. Let $G' = (N,\Sigma,P',S)$ be a new CFG in which $P' = P$ except that $A \rightarrow \alpha_1 B \alpha_2$ is replaced by

$$A \rightarrow \alpha_1\beta_1\alpha_2 \mid \cdots \mid \alpha_1\beta_r\alpha_2$$

Then $L(G') = L(G)$.

Proof: The proof is straightforward; see the Exercises. ■

It is possible to prove a stronger version of the *GNF* theorem, namely, it is sufficient to allow only productions of the following types

(1) $A \rightarrow a$;
(2) $A \rightarrow aB$;
(3) $A \rightarrow aBC$;

where B and C are not necessarily distinct from A or each other. In other words, 2-*SNF* is sufficient to capture all *CFLs*. We leave to the Exercises the proof of this remarkable theorem and others of a similar nature.

Theorem 4.2.18 Greibach 2-Standard Normal Form Theorem
Let G be an arbitrary reduced CFG. Then there is an algorithm that constructs a reduced λ-equivalent CFG, G', in 2-SNF from G. In other words, for every CFL, L, there exists a 2-SNF CFG, G, with $L(G) = L - \{\lambda\}$.

If only productions of types (1) and (2) are allowed we do not obtain all *CFLs* but only the *REGLs*, as we shall see in the next section.

4.3 LINEAR AND REGULAR GRAMMARS

Theorem 4.2.18 tells us that every *CFG* has a λ-equivalent *CFG* in 2-standard normal form, which raises the natural question: why not 1-standard normal form? We show that 1-standard normal form grammars generate $\boldsymbol{L}_{DFA}$ and, hence, $\boldsymbol{L}_{REG}$. Because $\{a^i b^i : i \geq 1\}$ is a *CFL* (see Example 4.1.3), and is not in $\boldsymbol{L}_{DFA}$ (see Example 2.2.4), this implies $\boldsymbol{L}_{DFA} \subset \boldsymbol{L}_{CFL}$. So 1-standard normal form grammars are restrictive in their expressive power.

Theorem 4.3.1 *For every reduced CFG, G, in 1-SNF there exists an NFA, M, with $L(M) - \{\lambda\} = L(G)$, and conversely.*

Proof:

(1) Given a *CFG*, $G = (N,\Sigma,P,S)$, in 1-*SNF* we prove that there is an *NFA*, M, with $L(M) = L(G)$.

Consider a word $a_1 \cdots a_n$ in $L(G)$, where a_i is in Σ, $1 \leq i \leq n$, and $n \geq 1$. Let its derivation sequence be

$$\begin{aligned} A_0 &\Rightarrow a_0A_1 \\ &\Rightarrow a_1a_2A_2 \\ &\quad \vdots \\ &\Rightarrow a_1a_2 \cdots a_{n-1}A_{n-1} \\ &\Rightarrow a_1 \cdots a_n \end{aligned}$$

where $A_0 = S$ and A_i is in N, $0 \leq i \leq n-1$. Observe that this is similar to an accepting configuration sequence of $a_1 \cdots a_n$ for some *NFA*, M, in which only the part of the input read so far is displayed. This is the basic idea behind the construction of an *NFA*, M, with $L(M) = L(G)$.

More formally the *NFA*, $M = (N \cup \{f\},\Sigma,\delta,S,\{f\})$, where f is a new state and δ is defined as follows.

(i) For all A and B in N and for all a in Σ, if $A \rightarrow aB$ is in P, then add (A,a,B) to δ, and

(ii) For all A in N and for all a in Σ, if $A \rightarrow a$ is in P, then add (A,a,f) to δ.

Given the derivation above we obtain the corresponding accepting configuration sequence

$$\begin{aligned} A_0 &\Rightarrow A_1a_2 \cdots a_n \\ &\Rightarrow A_2a_3 \cdots a_n \\ &\quad \vdots \\ &\Rightarrow A_{n-1}a_n \\ &\Rightarrow f \end{aligned}$$

Conversely, given this accepting configuration sequence in M, we obtain the derivation sequence given above.

The formal proof that $L(M) = L(G)$ is based on this correspondence and requires induction. We leave this to the Exercises.

(2) Given an NFA, $M = (Q,\Sigma,\delta,s,F)$, we construct a CFG, G, in 1-SNF such that $L(G) = L(M)$. Essentially, the inverse of the above construction is used. Let $G = (Q,\Sigma,P,s)$, where P contains the following productions.

(i) For all p and q in Q and for all a in Σ, if (p,a,q) is in δ, then add $p \to aq$ to P, and
(ii) For all p in Q, for all f in F, and for all a in Σ, if (p,a,f) is in δ, then add $p \to a$ to P.

Again we leave to the Exercises the proof that $L(G) = L(M)$. ■

1-standard normal form grammars are sometimes called *regular grammars*, or *REGG*.

Let $G = (N,\Sigma,P,S)$ be a reduced CFG; then it is said to be *linear*, or a *LING*, if all productions in P are either of the form $A \to x$ or $A \to xBy$, for some terminal words x and y. G is said to be *right linear*, or a *RLING*, if its productions are either of the form $A \to x$ or $A \to xB$. *Left linear CFGs*, or *LLINGs*, are defined similarly. Clearly, a CFG in 1-SNF is a $RLING$ and a $LING$. A CFL, L, is said to be a *linear* language or $LINL$ if $L = L(G)$, for some $LING$, G.

Again, it is not difficult to prove:

Theorem 4.3.2 *Let G be a RLING or a LLING. Then there exists a lazy NFA, M, with $L(M) = L(G)$, and conversely.*

Regular grammars model tail-end recursion in computer programs and their transformation into lazy *NFAs* corresponds to the transformation of tail-end recursion into iteration.

Example 4.3.1 Consider the PASCAL function:

```
function F : boolean;
begin
    if match('a') then F := true else
    if not match('b') then F := false else F := F
end;
```

where *match* compares its character argument with the current character of the standard input file. If these match, then the value **true** is returned and the input file is advanced one character, otherwise the value **false** is returned. Thus, F tests for the occurrence of $b^i a$, $i \geq 0$, from the initial position in the input file.

Now F is tail-end recursive, since it is recursive at the last position in the body of F. Therefore, we can model F by the simple regular grammar

$$F \to a \mid bF$$

We can transform the function F into iterative form as

```
function IF : boolean;
begin
    while match('b') do;
    if match('a') then IF := true else IF := false
end;
```

which corresponds to transforming the regular grammar into the state diagram of Figure 4.3.1.

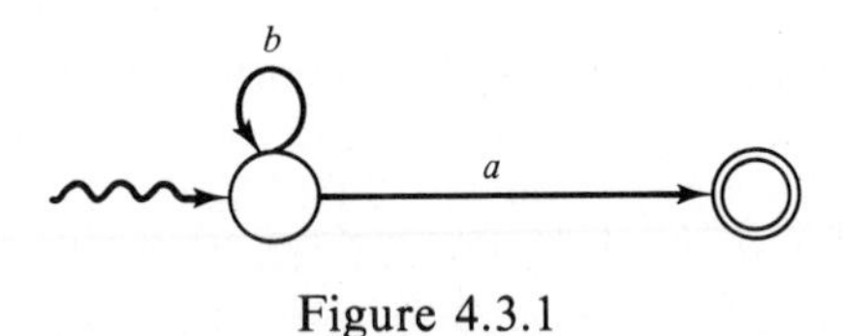

Figure 4.3.1

Linear grammars, on the other hand, are more powerful than *FAs* and less powerful than unrestricted *CFGs*. The *family of linear languages* is denoted by $\boldsymbol{L}_{LIN}$, and defined by $\boldsymbol{L}_{LIN} = \{L : L = L(G) \text{ for some } LING, G\}$, has attained a central place in the theory of *CFLs*, as we shall see in Chapters 8, 9, and 10. Note that the ambiguous language of Example 4.1.8 is linear.

4.4 EXTENDED CONTEXT-FREE GRAMMARS

In a *CFG* not only are the terminal and nonterminal alphabets finite, but the set of productions is also finite, and there is only a single sentence symbol. We could consider extensions in which infinite nonterminal and terminal alphabets are allowed and also arbitrary sets of words rather than a single sentence symbol. Although these may be of interest theoretically, we prefer to examine practically motivated extensions. To this end we extend the notion of a *CFG* to allow it to contain infinitely many productions, since this enables us to easily mimic *BNF* notation and syntax diagrams as we shall see.

In a *CFG*, $G = (N,\Sigma,P,S)$, each nonterminal A has a finite set of productions

$$A \rightarrow \alpha_1 \mid \cdots \mid \alpha_r$$

The extension we consider depends upon viewing these productions as a single super production

$$A \rightarrow \{\alpha_1, \ldots, \alpha_r\}$$

since the right-hand side of this super A-production is now a finite subset of V^*, that is,

$$A \rightarrow F_A$$

where $F_A \subseteq V^*$ is a finite language. We extend this by allowing the right-hand side to be a regular language over V, that is,

$$A \to R_A$$

where $R_A \subseteq V^*$ and R_A is a *REGL*. In practice, R_A can be denoted by a regular expression, while theoretically we deal directly with the regular language. Hence, we define an *extended CFG*, or *ECFG*, $G = (N,\Sigma,P,S)$ to be similar to a *CFG* except that P is a set of super productions for all nonterminals, namely, for all A in N, $A \to R_A$, for some $R_A \subseteq V^*$, where R_A a regular language. Note that $R_A = \varnothing$ is legal, corresponding to A having no productions at all, and that every *CFG* is an *ECFG*.

The notions of $\Rightarrow, \Rightarrow^+, \Rightarrow^*, \overset{L}{\Rightarrow}, \overset{L}{\Rightarrow}^+, \overset{L}{\Rightarrow}^*, \overset{R}{\Rightarrow}, \overset{R}{\Rightarrow}^+$, and $\overset{R}{\Rightarrow}^*$ are the same as for a *CFG*, except that there may be an infinite set of choices for the rewriting of any nonterminal. Therefore, the language of an *ECFG*, G, denoted by $L(G)$, is defined by

$$L(G) = \{x : S \Rightarrow^* x \text{ in } G\}$$

as before for a *CFG*.

Surprisingly *ECFGs* only generate *CFLs* as we now show.

Theorem 4.4.1 *Let $G = (N,\Sigma,P,S)$ be an ECFG. Then there is a CFG $G' = (N',\Sigma,P',S)$ with $L(G') = L(G)$.*

Proof: For all super productions $A \to R_A$ in P, if R_A is finite then carry these productions unchanged into P'; otherwise we know that there exists a *RLING* $G_A = (N_A, N \cup \Sigma, P_A, S_A)$ with $L(G_A) = R_A$, where N_A is a new set of nonterminals. Now take P_A into P' together with the production $A \to S_A$ and take N_A into N'.

Finally, take N into N', that is,

$$N' = N \cup \bigcup_{A \text{ in } N} N_A$$

where $N_A \cap N_B = \varnothing$, for $A \neq B$, and

$$P' = \{A \to S_A : A \text{ is in } N \text{ and } R_A \text{ is infinite}\}$$
$$\cup \{A \to R_A : A \text{ is in } N \text{ and } R_A \text{ is finite}\}$$
$$\cup \bigcup_{A} P_A$$

where the final union is taken over all A in N with R_A infinite. Clearly G' is a *CFG* since it has a *finite number* of productions. Now observe that

$$S \overset{R}{\Rightarrow}^+ x \text{ in } G \text{ implies } S \overset{R}{\Rightarrow}^+ x \text{ in } G'$$

since each derivation step $\beta A w \overset{R}{\Rightarrow} \beta\alpha w$ in G can be simulated by the two-step derivation $\beta A w \overset{R}{\Rightarrow} \beta S_A w \overset{R}{\Rightarrow}^{+} \beta \alpha w$ in G'. Conversely, every derivation $S \overset{R}{\Rightarrow}^{+} x$ in G' can be shown to be the simulation of a derivation $S \overset{R}{\Rightarrow}^{+} x$ in G. Thus, $L(G) = L(G')$. Finally, the construction of G' is effective if each R_A is denoted by a regular expression E_A. This follows by constructing an *NFA*, M_A, from E_A with $L(M_A) = L(E_A)$, and then constructing the *RLING*, R_A, from M_A with $L(G_A) = L(M_A) = R_A$; see Theorems 4.3.1 and 3.2.1. ■

Example 4.4.1A The *CFG* of Example 4.1.1 with multiplication denoted by $\times$ to avoid confusion is

$$S \rightarrow T \mid S+T; \quad T \rightarrow F \mid F \times T; \quad F \rightarrow a \mid (S)$$

It can be rewritten as the *ECFG*

$$S \rightarrow T[+T]^*; \quad T \rightarrow [F\times]^*F; \quad F \rightarrow a \mid (S)$$

from which we obtain the syntax diagrams of Figure 4.4.1 after noticing that $[F\times]^*F \equiv F[\times F]^*$. These should be compared with the syntax diagrams given in the Section 4.1. This transliteration can be automated.

Equivalently, this *CFG* can be transformed into the *extended BNF grammar*

$$<S> ::= <T>\{+<T>\}$$

$$<T> ::= \{<F>\times\}<F>$$

$$<F> ::= a \mid (<S>)$$

where $\{\cdots\}$ denotes "zero or more of." This is discussed further in the Exercises.

Since a syntax diagram is, essentially, the dual of a state diagram, constructing a syntax diagram for a production $A \rightarrow R_A$ corresponds to constructing an *NFA* from E_A (see Section 3.3) and then taking the dual (see Section 0.2). Conversely, given a syntax diagram for a nonterminal A, constructing its regular expression E_A can be done by first taking the dual of the syntax diagram and then using the construction of Section 3.2 to obtain the associated regular expression.

Example 4.4.1B From the syntax diagrams of Figure 4.4.1 we obtain as their duals the state diagrams of Figure 4.4.2. We have labeled entering edges "entry" and exiting edges "exit"; the other edges have been labeled uniquely with positive integers. There is an edge in the dual if the source and target nodes correspond to edges entering and leaving a box or oval in the original syntax diagram. Furthermore, the edge label is then the same as the symbol in the box or oval. Clearly the resulting diagrams are state diagrams, in the sense of Chapter 2, corresponding to *NFAs*. The state "entry" is the start state and the state "exit" the only final state.

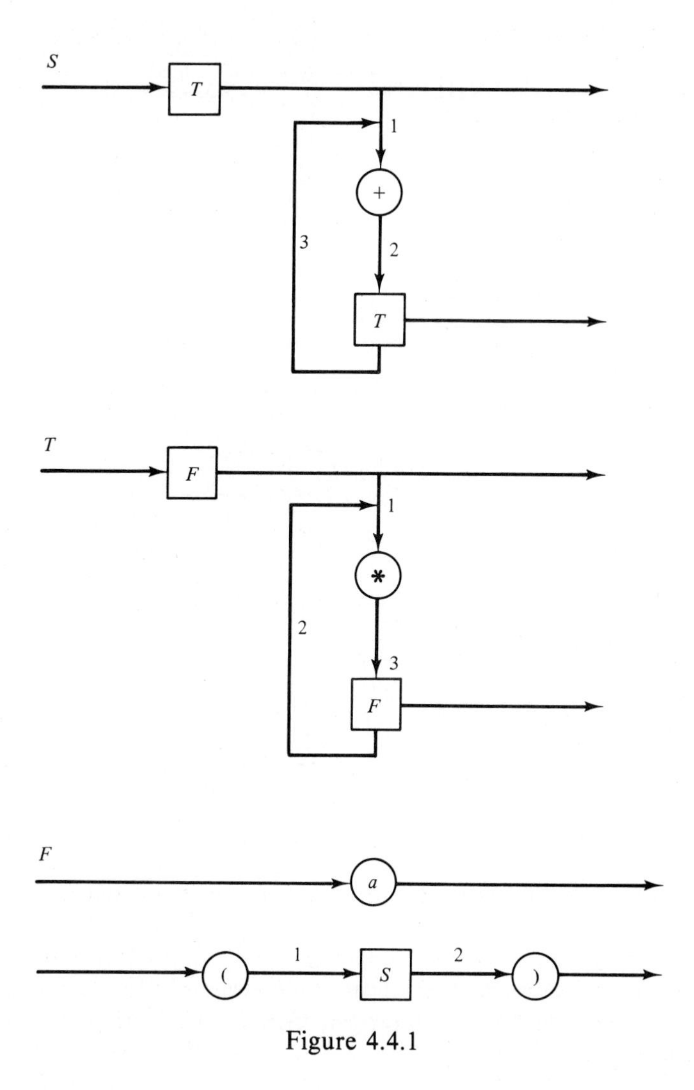

Figure 4.4.1

This construction can easily be reversed, but we leave the details to the Exercises.

Thus, we have the following.

Theorem 4.4.2 *Let D be a set of syntax diagrams specifying a language $L(D)$. Then there is an algorithm to construct an ECFG, G, from D with $L(D) = L(G)$, and conversely. Thus every CFL can be specified by a set of syntax diagrams and every set of syntax diagrams specifies a CFL.*

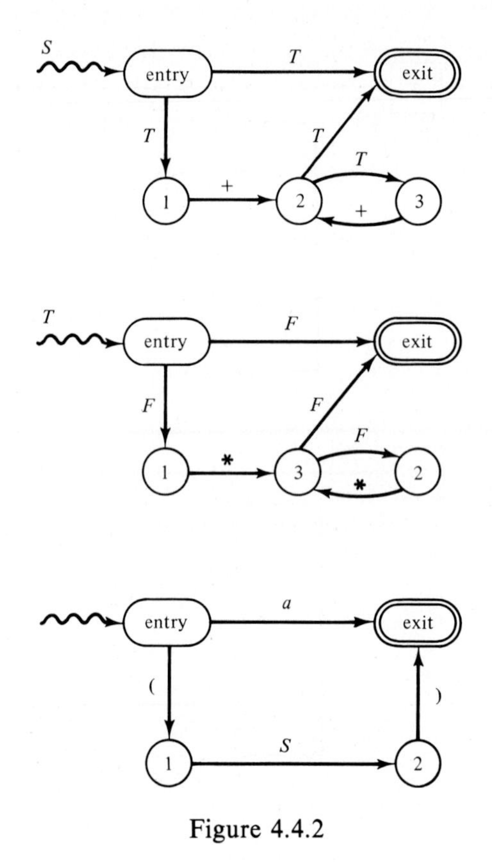

Figure 4.4.2

Proof: The details of the proof by construction and induction are left as Exercises. ■

Furthermore, it follows that extended *BNF* grammars are also equivalent to *ECFGs* and, therefore, to *CFGs*; see the Exercises.

4.5 ADDITIONAL REMARKS

4.5.1 Summary

We have introduced a generating model for specifying languages, namely, CONTEXT-FREE GRAMMARS. Paralleling configuration sequences in automata we have DERIVATION SEQUENCES or DERIVATIONS of which the restricted LEFTMOST and RIGHTMOST DERIVATIONS are of fundamental importance because of their close relationship with SYNTAX TREES. Syntax trees enable us to capture in a formal way the twin notions of AMBIGUITY and UNAMBIGUITY of grammars and languages.

Various simplifications or normal forms for context-free grammars have been discussed, in particular, CHOMSKY and GREIBACH NORMAL FORMS play an important rôle in the theory and practice of context-free grammars.

Finally, we have considered the restricted classes of grammars, namely, LINEAR and REGULAR GRAMMARS, both being less powerful than context-free grammars, and, introduced one extension of *CFGs*, EXTENDED CONTEXT-FREE GRAMMARS which are closely related to SYNTAX DIAGRAMS and the EXTENDED BACKUS-NAUR FORM.

4.5.2 History

Noam Chomsky originated the formalism for context-free grammars (see Chomsky, 1956, 1963), during his studies in linguistics (see Chomsky, 1957). *BNF* was first used in describing ALGOL in Backus (1959) and Naur (1960). Syntax diagrams first appear as state diagrams in Conway (1963), and have been used in their present form to specify PASCAL; see Wirth (1973). The British Standard metalanguage BS 6154 is defined in Scowen (1983). However, the underlying notion of a rewriting system first appears in the work of the Norwegian mathematician Thue (see Thue, 1914), and subsequently in Post (1947).

Chomsky normal form is a weakened version of Chomsky's original definition and proof; see Chomsky (1959) and Exercise 4.2.17. Greibach normal form and Greibach 2-standard normal form were first proved in Greibach (1965). The partial proof given here is based on Wood (1969a). Rosenkrantz (1967) and Wood (1969a) give matrix methods for the removal of left-recursion and a recursive elimination technique due to M.C. Paull is to be found in Hopcroft and Ullman (1979). In Kuich and Salomaa (1985) a formal power series approach, along the lines of Rosenkrantz (1967), is to be found.

Extended context-free grammars were first introduced in van Leeuwen (1974a), whereas extended *BNF* appears much earlier.

Cantor (1962), Chomsky and Schutzenberger (1963), Floyd (1962a), and Greibach (1963) were the first to formally investigate ambiguity.

Cohen and Gotlieb (1970) suggested a data structure for representing *CFGs*, see Programming Project P1.1.

4.6 SPRINGBOARD

4.6.1 Recognition and Parsing

Membership testing is a fundamental problem for *CFGs* since this is the problem faced by any compiler designer. Although this is discussed in Chapters 5, 8, and 11 there is an overabundance of material which must be excluded. Consider one approach to membership testing based on restricted classes of *CFGs* given by various notions of precedence grammars. These give rise to the so-called *bottom-up parsers* and *recognizers* in which a syntax tree is constructed from the frontier to the root. Alternatively, they can be viewed as building up a rightmost derivation backward from the input word. In precedence-based parsers each pair (X,Y) of symbols from the given *CFG* is related by at most one of $X \lessdot Y$, $X \doteq Y$, and $X \gtrdot Y$. These relations are assigned in such a way that in any word $\alpha = X_1 \cdots X_n$ if $\cdots X_i \lessdot X_{i+1} \doteq \cdots \doteq X_j \gtrdot X_{j+1} \cdots$ for some i and j, then $X_{i+1} \cdots X_j$ is either the right-hand side of a production in the given grammar or a syntax error has been found.

Investigate precedence grammars and parsing. A good place to begin is Aho and Ullman (1972).

4.6.2 Ambiguity

We have introduced the basic notion of ambiguity for *CFGs* and *CFLs*; however, these notions may be refined as follows. Let $k \geq 1$ be an integer. A *CFG*, $G = (N, \Sigma, P, S)$, is *ambiguous of degree k* or is *k-ambiguous* if: (i) for all x in $L(G)$, x has at most k distinct syntax trees; and (ii) there is at least one x in $L(G)$ which has exactly k distinct syntax trees. Similarly, a *CFL*, $L \subseteq \Sigma^*$, is *k-ambiguous* if every *CFG* for L is k_1-ambiguous for some $k_1 \geq k$. Clearly a *CFG* and a *CFL* are unambiguous if they are 1-ambiguous. We say a *CFG* is *ambiguous of infinite degree* or *∞-ambiguous* if it is not k-ambiguous for any $k \geq 1$. Again a *CFL* is ∞-ambiguous if every *CFG* for L is ∞-ambiguous.

Investigate k- and ∞-ambiguity, consulting Harrison (1978).

4.6.3 *W*-Grammars

CFGs are useful for specifying the syntactic structure of programming languages such as ALGOL 60, PASCAL, etc. However, they are unable to capture the descriptions of these programming languages exactly; they usually are intended to generate all valid programs together with many invalid ones. They are said to specify an *envelope* of the corresponding programming language. They are unable to capture *context-dependent features*, for example, matching the usage of a variable with its type, matching the usage of a variable with its declaration, preventing multiple declarations, etc. For example, using the syntax diagram definition of PASCAL we may obtain:

```
program p;
const a = a;
type a = record a : integer; a : a end;
var a : integer;
begin
    read(a,a);  write(a[a])
end
```

This problem was recognized even with ALGOL 60's definition and since *CFLs* were sometimes known as ALGOL-like languages, this led to papers with apparently self-contradictory titles such as "ALGOL is not an ALGOL-like language"; see Floyd (1962b), for example.

During the 1960s a search began for some better way to specify programming languages which would allow identifier matching of various kinds to be specified. Of these attempts *W-grammars*, or *van Wijngaarden grammars*, were most successful in that they were used in the formal definition of ALGOL 68. (Some people would maintain that they helped to obscure a beautiful programming language and ensure its early extinction.) W-grammars have two levels of generation. The first is the generation of nonterminals and their associated productions based on generic patterns, and the second is the generation of words using these nonterminals and productions. This enables a W-grammar to specify an infinite number of nonterminals and productions although only a finite subset are required at any time.

Investigate the theory and application of W-grammars. The text of Cleaveland and Uzgalis (1977) on the topic is unique and, happily, readable. The definition of ALGOL 68 in van Wijngaarden (1969) and the revised definition van Wijngaarden et al (1974) form the major application, but recently in Gonnet and Tompa (1983) and Gonnet (1984) W-grammars have been used to classify and enumerate data structures.

4.6.4 Simplifications and Efficiency

In Section 4.2 we introduced a number of simplifications of *CFGs*. Since each simplification can be carried out algorithmically we have two issues of efficiency to address. The first is the efficiency of the construction or algorithm. Is it the best possible? The second is the efficiency of the transformation itself on the input *CFG*. Does it provide the smallest output *CFG* which satisfies the requirements? Is there a transformation that performs better? For this to be measured we need to decide on a reasonable measure of the size of a *CFG*. One such measure is simply the number of symbols in its productions.

Investigate simplification efficiency in both senses. Consult Harrison (1978).

4.6.5 Context-Free Expressions

Just as there are regular expressions corresponding to regular grammars there are context-free expressions corresponding to *CFGs*. Investigate context-free expressions by way of Gruska (1971), McWhirter (1971), and Yntema (1971).

4.6.6 E0L Grammars and Languages

In a *CFG* rewriting occurs at exactly one position in a word at a time — it is *sequential rewriting*. However, we may force rewriting in a *CFG* to take place at every position in a word — *parallel rewriting*. Furthermore, if there is no production available for some symbol in the word to be rewritten, then rewriting is undefined, for example, terminal symbols have no productions. We call such a *CFG* an $E0L$ grammar ($E0L$ stands for Extended zero-sided Lindenmayer). It can be shown that every *CFL* can be generated by an $E0LG$; see the Exercises. Moreover, there exist simple $E0L$ languages which cannot be generated by any *CFG*, for example, $\{a^{2^i} : i \geq 0\}$. In this setting it is reasonable to allow terminal symbols also to have productions, but it can be shown that this does not increase the power of these grammars. Investigate $E0L$ grammars and their restricted forms $0L$ and $D0L$ grammars. Consult Rozenberg and Salomaa (1980) and read the early papers of Aristid Lindenmayer, the theoretical biologist who introduced these grammars to model biological development; see Lindenmayer (1968, 1971). More recently these grammars have been used, in computer graphics, to generate realistic pictures of trees and plants; see Smith (1985).

4.6.7 ALGOL-Like Languages

Since *BNF* specifications were first used for the specification of ALGOL 60, the associated languages became known as ALGOL-like languages. The interpretation of *BNF* differs from that of *CFGs* in that a *BNF* specification is considered to be a set of simultaneous word equations which are to be solved, rather than as a rewriting system. This point of view is taken in Ginsburg and Rice (1962) where it is proved that the solution of such a system of equations is exactly the language generated by the corresponding *CFG*. An extension of this approach was taken later in Herman (1973) where $E0L$ languages are characterized in a similar manner.

EXERCISES

1.1 Given the *CFG*, G,

$$S \rightarrow aSbSa \mid \lambda.$$

(i) give four words in $L(G)$, and some derivation for each of them;
(ii) give four words not in $L(G)$, justifying your answers;
(iii) give all syntax trees of the word $aababaababaa$.

1.2 Given the *CFG*, G,

$$S \rightarrow SS \mid aSa' \mid aa' \mid bSb' \mid bb' \mid cSc' \mid cc'$$

give all syntax trees of the word

$$cc'bcaa'bb'cc'c'b'aa'.$$

1.3 Given the *CFG*, G,

$$S \rightarrow aAaS \mid bSa \mid b$$
$$A \rightarrow Ab \mid aaa$$

and the labeled trees below, which of them are syntax trees with respect to G? Justify your answers.

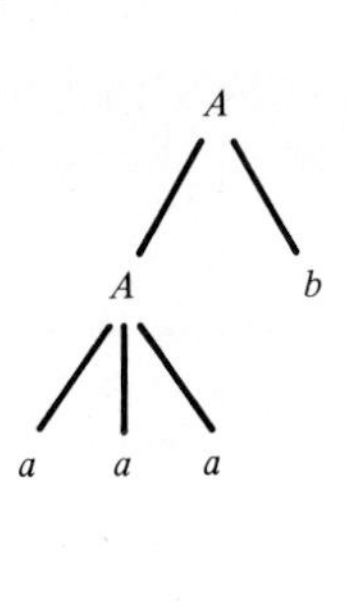

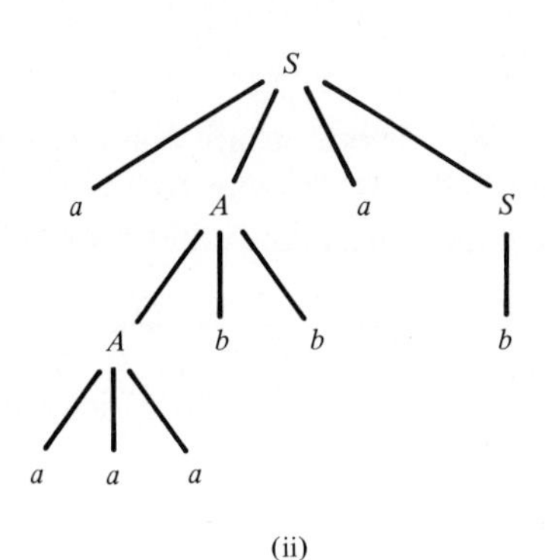

(i) (ii)

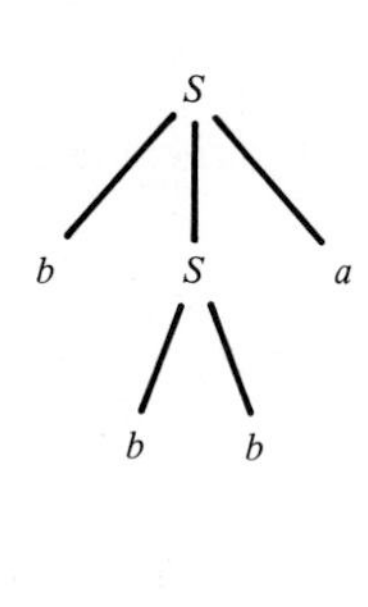

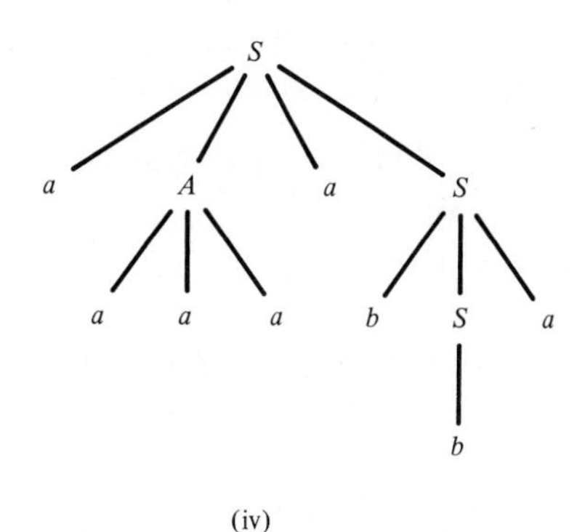

(iii) (iv)

1.4 Give a *CFG* for each of the following languages, justifying your answers informally.

(i) $\{a^{i+3}b^{2i+1} : i \geq 0\} \cup \{a^{2i+1}b^{3i} : i \geq 0\}$.
(ii) $\{a^i b^j c^j d^i e^3 : i,j \geq 0\}$.
(iii) $\{a^i b^j : i,j \geq 0$ and either $i = 2j$ or $2i = j\}$.
(iv) $\{a^i b^j c^k d^l : i,j,k,l \geq 0, i < l,$ and $j \neq k\}$.
(v) $\{a^i b^j b^i a^j : i,j \geq 0\}$.

1.5 A *pure CFG*, G, has only terminal symbols; thus, it is specified by a triple (Σ, P, σ), where $P \subseteq \Sigma \times \Sigma^*$ is a finite set of productions, σ in Σ^+ is the *initial word*, *starting word*, or *axiom*, and $L(G) = \{x : \sigma \Rightarrow^* x\}$. Prove that for every pure *CFG*, G, there is a *CFG*, H, with $L(G) = L(H)$.

1.6 Prove that the language $\{a, aa\}$ cannot be generated by any pure *CFG*.

1.7 Let $G = (N, \Sigma, P, S)$ be a *CFG* and let $SF(G) = \{\alpha : S \Rightarrow^* \alpha\}$, the language of sentential forms. Prove that $SF(G)$ is a *CFL*.

1.8 We demonstrated how to decide if a given word is in the language of a *CFG*, for a *CFG* in Greibach normal form; see Example 4.1.9. To this end, we constructed a search tree for the given word and the given *CFG*. What difficulties arise if we allow arbitrary *CFGs*? We also introduced two pruning rules. These ensure for Greibach normal form *CFGs* that the generated portion of the search tree is finite. Are these rules applicable in general? Do they guarantee finiteness in general?

1.9 Let $L = L(G)$ for the *CFG* of Exercise 1.2. A word x in L is said to be *prime* if there is no word y in L such that $|y| < |x|$ and $x = yz$, for some z. For example, $caa'bb'c'$ is prime but $cc'aa'baa'b'$ is not.

(i) Prove that every word in L has a unique decomposition as a catenation of prime words.

(ii) Let L_a, L_b, and L_c be the sets of prime words beginning with a, b, and c, respectively. By associating nonterminals A, B, and C with L_a, L_b, and L_c, respectively, construct an unambiguous *CFG*, with nonterminals S, A, B, and C, generating L.

1.10 Given the following regular expressions construct *CFGs* generating the languages they denote.

(i) $a^*c^*b^* + (a^*d)^*cb^*$.
(ii) $a^*ba^*ab^*$.
(iii) $a((ab)^*cb^*)^* + a(ababcb^*)^*a^*$.

1.11 Let G be given by,

$$S \to SS \mid aaSb \mid bSaa \mid \lambda$$

Is $L(G) = \{x : x \text{ in } \{a,b\}^*, |x|_a = 2|x|_b\}$? Justify your answer.

1.12 Consider the *CFG* of Example 4.1.7. Construct the portion of the leftmost derivation tree corresponding to the following input words

(i) a.
(ii) $(a)a$.
(iii) $(a)(a)$.
(iv) $(a + a)$.

2.1 Given the *CFG*, G,

$$S \to ABB \mid CAC$$
$$A \to a$$
$$B \to Bc \mid ABB$$
$$C \to bB \mid a$$

(i) What are the terminating symbols of G?
(ii) What are the reachable symbols of G?
(iii) Construct a reduced *CFG*, G', equivalent to G?

2.2 Prove Lemmas 4.2.1 and 4.2.3.

2.3 Rephrase the problem of finding reachable symbols in a *CFG* as a node-connectedness problem for digraphs. Can you rephrase the problem of finding terminating symbols in terms of digraphs?

2.4 Given the *CFG*

$$S \to aSASb \mid Saa \mid AA$$
$$A \to caA \mid Ac \mid bca \mid \lambda$$

(i) What are the λ-nonterminals of G?
(ii) Is λ in $L(G)$?
(iii) Construct a λ-free *CFG* G' equivalent to G.

2.5 Given two arbitrary *CFGs* G_1 and G_2,

(i) If G_1 and G_2 are equivalent, then are they always, never, or sometimes λ-equivalent?
(ii) If G_1 and G_2 are λ-equivalent are they always, never, or sometimes equivalent?

2.6 Prove Lemmas 4.2.6 and 4.2.7.

2.7 Given an integer $k \geq 1$ consider the *CFG* G_k

$$S \to A_2 \mid aA_2$$
$$A_i \to A_{i+1} \mid aA_{i+1},\ 1 < i < k$$
$$A_k \to a_k$$

(i) What are the unit productions of G_k?
(ii) Construct an equivalent unit-free *CFG*.

2.8 Prove Lemma 4.2.9 and Theorem 4.2.10.

2.9 Let G be a *CFG* with $L(G)$ infinite. Provide an algorithm to construct an equivalent *CFG*, G', in which every nonterminal generates infinitely many terminal words. Prove that your algorithm is correct.

2.10 Let G be the *CFG*

$$S \to aSASb \mid Saa \mid b$$
$$A \to caA \mid Ac \mid bca$$

(i) Construct an equivalent *CFG* in binary form.
(ii) Construct an equivalent *CFG* in Chomsky normal form.

2.11 Prove Lemma 4.2.11.

2.12 Prove Theorem 4.2.14.

[*Hint*: Carry out an induction construction based on the number of left-recursive symbols. Given a directly left-recursive nonterminal A, observe that $A \to A\alpha \mid \beta$ can be replaced by $A \to \beta \mid \beta X$; $X \to \lambda \mid \alpha X$.]

2.13 Given the CFG, G,

$$S \to AAS \mid CBCb \mid CSS$$
$$A \to CACB \mid BBC$$
$$B \to aB \mid CC$$
$$C \to abCba \mid aba$$

(i) What is $int(A)$ for all nonterminals A?
(ii) What is $intmax$?
(iii) Construct an equivalent CFG in Greibach normal form.

2.14 Given the CFG, G, in GNF

$$S \to aCBCB \mid aAAS \mid bSSSS$$
$$A \to bACCB \mid aBBCB$$
$$B \to aBB \mid bACA \mid bCC$$
$$C \to aBCBA \mid bSBSBBS \mid a$$

construct an equivalent CFG in 2-standard normal form.

2.15 Prove Lemma 4.2.17.

2.16 Prove Theorem 4.2.18.

[*Hint*: Assume without loss of generality that the given G is in GNF. Let $A \to aA_1 \cdots A_m$ be a production in G which has a maximum length right-hand side. If $m \leq 2$ then there is nothing to prove, so assume $m \geq 3$. First, for all production $B \to bB_1 \cdots B_n$ in G with $n \geq 1$ replace $B \to bB_1 \cdots B_n$ with $B \to b[B_1 \cdots B_n]$, where $[B_1 \cdots B_n]$ is a new nonterminal. Second, for all $B_1 \cdots B_r$ in N^* with $1 \leq r \leq m$ add productions $[B_1 \cdots B_r] \to c[\gamma]\,[B_2 \cdots B_r]$ to G', for all $c[\gamma]$, where $B_1 \to c\gamma$ is in the original G. In all cases $[\ ]$ denotes the empty word.]

2.17 Prove that every CFL, L, without λ can be generated by a Chomsky normal form CFG, G, in which the following three restrictions hold.

(i) If $A \to BC$ is in G for some A, B, C, then $B \neq C$.
(ii) If $A \to BC$ is in G, then $A \to DB$ is not in G for any D.
(iii) If $A \to BC$ is in G, then $A \to CD$ is not in G for any D.

3.1 Complete the proof of Theorem 4.3.1.

3.2 Given the NFA

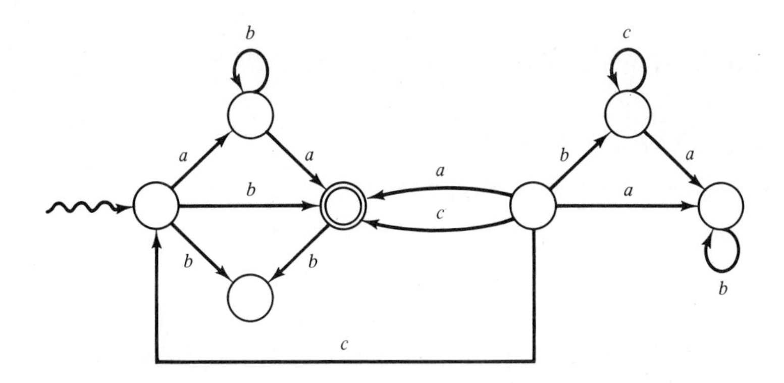

construct an equivalent regular grammar.

3.3 Given the right linear grammar

$$S \rightarrow aaA \mid bB \mid cbba$$
$$A \rightarrow ab \mid B \mid abbA$$
$$B \rightarrow cbc \mid aB \mid bA$$

construct an equivalent *NFA*.

3.4 Prove Theorem 4.3.2.

3.5 For a given $n \geq 1$ and a given alphabet $\{a,b\}$, let $x_1, \ldots, x_n$ and $y_1, \ldots, y_n$ be nonempty words over $\{a,b\}$. Let L be the set of all words

(i) $x_{i_1} \cdots x_{i_r} y_{i_r}^R \cdots y_{i_1}^R$, for $r \geq 1$ and for all j, $1 \leq i_j \leq n$;

(ii) $i_1 \cdots i_r x_{i_r}^R \cdots x_{i_1}^R$ and $i_1 \cdots i_r y_{i_r}^R \cdots y_{i_1}^R$, for $r \geq 1$ and for all j, $1 \leq j \leq r$, $1 \leq i_j \leq n$, where the alphabet is $\{a,b,1,\ldots,n\}$;

(iii) over $\{a,b\}$ not of form (i).

In each case prove that L is a linear language.

3.6 Let L be a *LINL*. Prove that L^R is also a linear language (see Chapter 9).

3.7 Let $G = (N,\Sigma,P,S)$ be a *LING*. Let *Left*(G) denote the set of words

$$Left(G) = \{x : S \Rightarrow^* xAy \text{ and } A \rightarrow z \text{ is in } P \text{ for some } z \text{ in } \Sigma^*\}.$$

Prove that *Left*(G) is a *REGL*. Define *Right*(G) in a similar manner and prove *Right*(G) is also a *REGL*.

3.8 Provide a generic construction that converts an arbitrary *RLING* into an equivalent *LLING*.

4.1 Prove Theorem 4.4.2.

4.2 Prove that if a language is generated by Extended *BNF*, then it is generated by a context-free grammar, and conversely.

4.3 Let G be the *ECFG*

$$S \to [T[+ \cup -]]^* T \cup T[\mathbf{or}\, T]^*$$
$$T \to [a \cup b \cup c]^*[a \cup b \cup c] \cup (S)$$

(i) Construct an equivalent *CFG*.
(ii) Construct the equivalent syntax diagrams.

4.4 Which of the simplifications of Section 4.2 carry over to *ECFGs*? For those that do carry over provide algorithms paralleling those for *CFGs*. For those that don't carry over try to suggest similar simplifications that do carry over.

4.5 We may extend *ECFGs* even further by allowing the right-hand side of a production to be a *LINL* or *CFL* to give *LECFGs* and *CECFGs*, respectively. Prove that they also generate exactly the *CFLs*.

4.6 In a natural way we can speak of *EREGGs*, *ELINGs*, as well *ELLINGs* and *ERLINGs*. Do they generate exactly the *REGLs*, *LINLs*, *REGLs*, and *REGLs*, respectively?

PROGRAMMING PROJECTS

P1.1 Design a data structure for representing *CFGs*.

P1.2 Write a program to display at a terminal or other output device the steps of a sentential derivation. Extend it to display the corresponding syntax tree.

P1.3 Write a program to generate all leftmost or rightmost derivations of a specified length.

P2.1 Write a suite of programs to check if a given *CFG* satisfies the various simplifications.

P2.2 Write a suite of programs to carry out the various simplifications introduced in Section 4.2.

P3.1 Write a program to test if a *CFG* is left linear, right linear, linear, or regular.

P3.2 Write programs to convert

(i) a left linear grammar into an equivalent right linear grammar;
(ii) a right linear grammar into an equivalent regular grammar; and
(iii) a regular grammar into an equivalent *NFA*.

P4.1 Write programs to convert (i) an *ECFG* into an equivalent *CFG*, and (ii) an *ECFG* into an equivalent set of syntax diagrams.

P5.1 Programming exercises will be discovered during your investigations. For example, write a program to check that a *CFG* is simple precedence.

Chapter 5

Pushdown Automata

5.1 DETERMINISTIC PUSHDOWN AUTOMATA

Pushdown stores or stacks are found in cafeterias as a method of storing and supplying plates and soup bowls, on warships as a method of storing battle helmets, and, doubtless, are to be found in many other areas of daily life. In computing pushdown stores, LIFO (last-in, first-out) stores, stacks, or nesting stores, as they have been known, have been in use since the 1950s, sometime appearing as hardware devices, for example, in the English Electric Leo Marconi KDF9 computer, but, more usually, as data structures or data types implemented at the software level. They are used to evaluate arithmetic expressions in postfix form and to generate code for their evaluation; to keep track of the return addresses and environments invoked by subprogram calls, particularly recursive ones; to keep track of locally defined variables when executing block-structured languages. Stacks are, indeed, one of the most frequently chosen data structures.

It should not be too surprising, therefore, that they form the basis of a fundamental and important automata-theoretic model. Their importance derives from two properties. The first property is that the nondeterministic pushdown automata accept exactly the context-free languages. The second property, which is somewhat related to the first, is that they lead to automatic methods of obtaining membership testers for *CFGs*. (A membership tester decides for a given *CFG* and a given input word whether or not the word is generated by the given *CFG*.)

We begin by introducing deterministic pushdown automata, before examining the nondeterministic variant. Finally, we examine one extended and one restricted version of the basic model.

A deterministic pushdown automaton is essentially a finite automaton with some additional machinery. First, we add an unbounded file or tape of symbols, *the pushdown*, and a *read-write head* for the pushdown; see Figure 5.1.1. In order to simulate the stack data structure, only the topmost symbol can be read, that is, the one before the end-of-file or *eof*, and a write is only allowed if the pushdown is at *eof*. For simplicity we assume that these conditions are always met automatically by the read-write head. The read-write head executes a POP operation by reading the topmost symbol and erasing it. A PUSH operation is executed by simply writing symbols. This implementation corresponds to a typical array-based implementation of a stack, except that the array is assumed to be extendible to the right. To allow the automaton to check for an empty stack the first location is reserved for an *initial pushdown symbol* $\perp$, pronounced as "bottom". This symbol is never changed during a computation; when the pushdown only contains $\perp$ it is said to be *empty*.

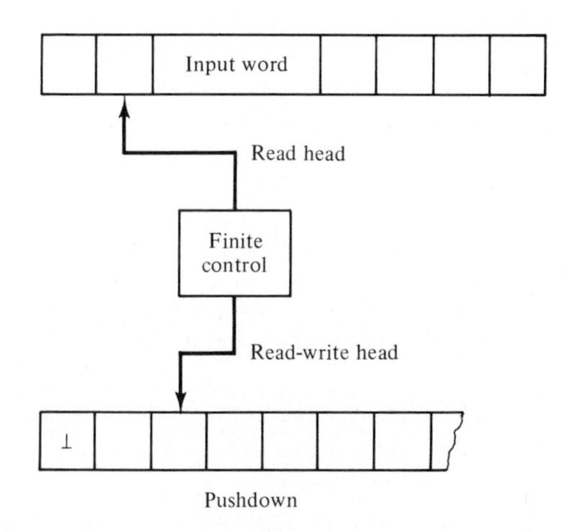

Figure 5.1.1

Second, the transition functions for *DFAs* are extended to allow them to depend on the topmost pushdown symbol.

Before giving an example of a pushdown automaton we provide a formal definition of the model.

Definition A *deterministic pushdown automaton*, or *DPDA*, M is specified by a sextuple $M = (Q, \Sigma, \Gamma, \delta, s, F)$, where

Q is a finite set of *states*;
Σ is an *input alphabet*;
Γ is a *pushdown alphabet*, $\perp$ is in Γ;

$\delta: Q \times \Sigma \times \Gamma \to Q \times \Gamma^*$ is a (*partial*) *transition function*;
s in Q is a *start state*; and
$F \subseteq Q$ is a set of *final states*.

Because $\perp$ is used to mark the first cell of the pushdown, the transition function must neither destroy it nor replicate it. For all p in Q and a in Σ, if $\delta(p,a,\perp)$ is defined, then $\delta(p,a,\perp) = (q,\perp,g)$, for some q in Q and g in $(\Gamma - \{\perp\})^*$. Similarly, for all p in Q, a in Σ, and c in $\Gamma - \{\perp\}$, if $\delta(p,a,c)$ is defined, then $\delta(p,a,c) = (q,p)$, for some q in Q and g in $(\Gamma - \{\perp\})^*$. This forces each *DPDA* to maintain the position of $\perp$.

The more common definition of a *DPDA* allows a transition to occur without advancing the input. However, the definition we have chosen is not only simpler, but also it suffices to introduce the concept of a *DPDA*. We return to this issue at the end of the section.

A *DPDA* is initialized by

(i) giving it an input word x;
(ii) setting the read head over the first symbol of the input;
(iii) placing $\perp$ as the only symbol in the pushdown;
(iv) setting the read-write head over $\perp$;
(v) setting the current state of the *DPDA* to s; and
(vi) starting the *DPDA*.

The basic execute cycle of the *DPDA* is

(i) read the symbol under the read head, that is, the *current symbol*;
(ii) read the symbol under the read-write head, that is, the *topmost pushdown symbol*;
(iii) compute the next state and replacement pushdown word from the current state, current symbol, and current pushdown symbol;
(iv) move the read head one cell to the right;
(v) replace the topmost pushdown symbol with the computed pushdown word, resetting the read-write head to be over the new topmost pushdown symbol; and
(vi) let the current state be the next state.

To be able to discuss computations formally, that is, configuration sequences, we require the notion of a configuration of a *DPDA*.

Definition Let $M = (Q,\Sigma,\Gamma,\delta,s,F)$ be a *DPDA*. We say that a word gqx in $\perp(\Gamma - \{\perp\})^* Q \Sigma^*$ is a *configuration*. Intuitively, g represents the current contents of the pushdown, the rightmost symbol being the topmost element, q the current state, and x the remaining, that is, unread, input.

This notion of configuration should be compared with the configuration of a *DFA*; see Section 2.2. Initially a *DPDA* has configuration $\perp sx$, where x is the input word. Note that if Γ and Q, and Σ and Q have elements in common, then a configuration can contain two or more states. Such a configuration does not determine a unique situation in the given *DPDA*, so we forbid it by assuming $\Gamma \cap Q = \varnothing$ throughout.

Example 5.1.1A Let $\Sigma_n = \{1, \ldots, n\} \cup \{\bar{1}, \ldots, \bar{n}\}$, $n \geq 1$, and consider the *CFG*, G_n,

$$S \rightarrow SS \mid \lambda \mid iS\bar{i}, \quad 1 \leq i \leq n$$

Assume we have n procedures (or functions) $p_1, \ldots, p_n$ which are totally, mutually recursive, that is, p_i contains a call of p_j, for all i and j. Now to each procedure call there is a corresponding return or exit. Let the symbols $1, \ldots, n$ correspond to calls of procedures $p_1, \ldots, p_n$ and the symbols $\bar{1}, \ldots, \bar{n}$ correspond to their exits. Now each word in $L(G_n)$ corresponds to a valid sequence of procedure calls and exits.

Recall that if such a sequence occurs during the execution of a PASCAL program, for example, then the run-time manager keeps track of the return addresses, corresponding to each call, in a run-time stack, before transferring control to the called procedure. On meeting an exit the manager transfers control to the return address given on the top of the stack and removes it. We construct a *DPDA*, M_n, which carries out the stacking and unstacking required for a sequence of calls of and exits from $p_1, \ldots, p_n$. Therefore, M_n ends up with an empty pushdown, that is, its pushdown only contains $\perp$, when the input word is in $L(G_n)$.

We construct M_n as follows. Whenever a call is read it is transferred to the pushdown, and whenever an exit is read it cancels the call on the top of the pushdown — if they correspond in name.

Let $M_n = (\{s\}, \Sigma_n, \Gamma_n, \delta_n, s, \{s\})$, where $\Gamma_n = \{1, \ldots, n, \perp\}$, and δ_n is defined by

(i) $\delta_n(s, i, \perp) = (s, \perp i)$, $1 \leq i \leq n$.
(ii) $\delta_n(s, i, j) = (s, ji)$, $1 \leq i, j \leq n$.
(iii) $\delta_n(s, \bar{i}, j) = (s, \lambda)$, if $i = j$, $1 \leq i, j \leq n$.

Because it is easier to understand a state diagram rather than the corresponding transition function we introduce the *pushdown state diagram* of a *DPDA* (and, later, of an *NPDA*.) The edges are labeled with an input symbol, a pushdown symbol, and a pushdown word corresponding to the change in the pushdown. The state change is indicated by the edge.

For example, with $n = 2$, M_2 and δ_2 may be specified as in Figure 5.1.2. Observe how the transition $\delta(s, \bar{1}, 1) = (s, \lambda)$ becomes the labeled edge of Figure 5.1.3. A slash (/) is used to separate the input data, that is, the input and pushdown symbols, from the output data, that is, the pushdown word.

Example 5.1.1 demonstrates that the input and pushdown alphabets in a *DPDA* need not, necessarily, be disjoint. The languages $L(G_n)$ of Example 5.1.1 turn out to be fundamental in the study of context-free languages. They are called

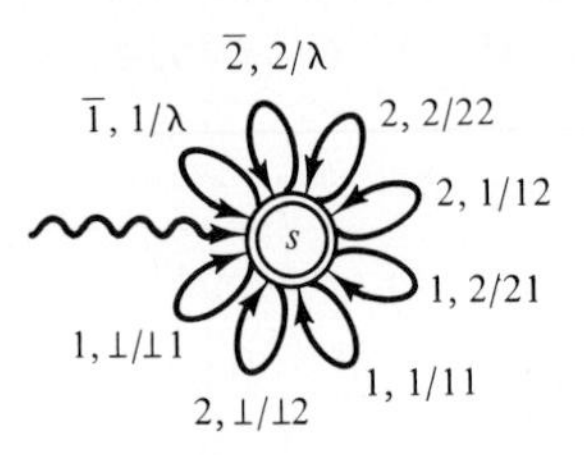

Figure 5.1.2

Figure 5.1.3

Dyck languages. Each language $L(G_n)$ corresponds to a set of well-formed parenthesized expressions over n pairs of parentheses. Hence, $L(G_1)$ corresponds to the parenthesization found in arithmetic expressions, for example, in $((a+b)*(c+d)+e)\div f$. And, $L(G_2)$ corresponds to the parenthesization found in subscript expressions in ALGOL or PASCAL, for example, in $a[b[c*(d+e)]+(f-g)*h[i]]$.

We now define configuration or transition sequences and acceptance for *DPDAs*.

Definition Let $M = (Q,\Sigma,\Gamma,\delta,s,F)$ be a *DPDA*. Given a configuration $gcpax$, where g is in Γ^*, c is in Γ, p is in Q, a is in Σ, x is in Σ^*, and we have the transition $\delta(p,a,c) = (q,\alpha)$, then $g\alpha qx$ is obtained from $qcpax$ in M. We say $gcpax$ *yields* $g\alpha qx$ *in one step* and we denote it by $gcpax \vdash g\alpha qx$.

Apart from the additional dependence on the topmost pushdown symbol this one step computation is similar to that for *DFAs*; see Section 2.2.

Definition Let $M = (Q,\Sigma,\Gamma,\delta,s,F)$ be a *DPDA*, and C_1 and C_2 be two configurations; then we write

$$C_1 \vdash^i C_2, \text{ for some } i \geq 1$$

if either $i = 1$ and $C_1 \vdash C_2$ or there is a configuration C_3 such that $C_1 \vdash C_3$ and $C_3 \vdash^{i-1} C_2$. This is an *(i-step) configuration sequence.*

As for *DFAs* we extend $\vdash$ to $\vdash^+$ and $\vdash^*$. We write $C_1 \vdash^+ C_2$ if there is an i such that $C_1 \vdash^i C_2$ and we write $C_1 \vdash^* C_2$ if either $C_1 = C_2$ or $C_1 \vdash^+ C_2$.

Finally, we say a word x in Σ^* is *accepted* by M if

$$\bot sx \vdash^* \bot f$$

for some f in F. Thus, acceptance occurs when all the input has been read, the pushdown only contains $\bot$ once more, and M is in some final state. A configuration sequence which begins with $\bot sx$, for some x in Σ^*, and ends with $\bot f$, for some f in F, is said to be an *accepting configuration sequence.*

Now the *language accepted by M* is denoted by $L(M)$ and is defined by

$$L(M) = \{x : x \text{ is in } \Sigma^* \text{ and } M \text{ accepts } x\}$$

Example 5.1.1B With input $12\bar{2}\bar{1}$ we have the configuration sequence of Figure 5.1.4. Here $\bot s12\bar{2}\bar{1} \vdash^* \bot s$ and $12\bar{2}\bar{1}$ is accepted. However, $121\bar{2}$ is rejected, since the *DPDA* hangs as shown in Figure 5.1.5.

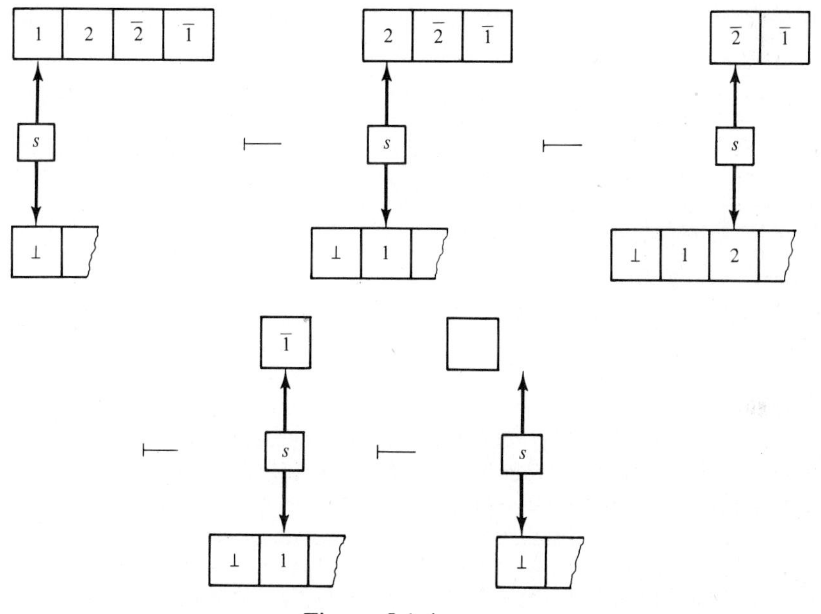

Figure 5.1.4

Indeed it can be proved that $L(M_n) = L(G_n)$; see the Exercises.

The principal importance of *DPDAs* stems from their ability to model efficient membership testing for a subclass of *CFGs* and *CFLs*. We first introduce a restricted class of *CFGs*.

Definition Let $G = (N,\Sigma,P,S)$ be a *CFG* in Greibach normal form (see

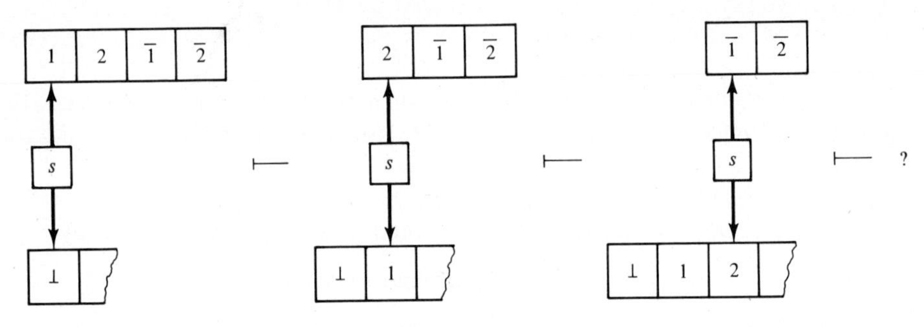

Figure 5.1.5

Section 4.2.5). We say G is a *simple deterministic CFG*, or an *s-CFG*, if it satisfies

For all A in N, for all a in Σ,
if $A \rightarrow a\alpha$ and $A \rightarrow a\beta$ are both in P, for some α and β in N^*, then $\alpha = \beta$.

In other words, every pair (A,a) in $N \times \Sigma$ corresponds to at most one production $A \rightarrow a\alpha$, for some α in G.

Example 5.1.2A Let G be

$S \rightarrow [AC \mid a$
$A \rightarrow [AA \mid]B$
$B \rightarrow [A \mid a$
$C \rightarrow a \mid [AC$

then G is an *s-CFG* and, moreover, if the appearances of a in every word generated by G are ignored, then the words in $L(G)$ correspond to fully parenthesized expressions.

The importance of the above definition is demonstrated when determining whether or not a given word is generated by such a *CFG*, (see Sections 4.1 and 10.1), that is, when solving

s-CFG MEMBERSHIP
INSTANCE: An arbitrary *s-CFG* $G = (N,\Sigma,P,S)$ and an arbitrary word x in Σ^*.
QUESTION: Is x in $L(G)$?

Example 5.1.2B Given the word x = [ ][ ]aa, is it generated by G?

Observe that

$$S \overset{L}{\Rightarrow}^{+} [\ ][\ ]aa \text{ if and only if } AC \overset{L}{\Rightarrow}^{+}][\ ]aa$$

since $S \to [AC$ is the *only* production corresponding to the pair (S,[) which deposits a [in the leftmost position. Now

$$AC \overset{L}{\Rightarrow}^{+}][\ ]aa \text{ if and only if } BC \overset{L}{\Rightarrow}^{+} [\ ]aa$$

since $A \to]B$ is the *only* production corresponding to the pair (A,]), and, hence, $AC \overset{L}{\Rightarrow}]BC$. Again

$$BC \overset{L}{\Rightarrow}^{+} [\ ]aa \text{ if and only if } AC \overset{L}{\Rightarrow}^{+}]aa$$

since $B \to [A$ is the only production for (B,[) and $BC \overset{L}{\Rightarrow} [AC$. Once more

$$AC \overset{L}{\Rightarrow}^{+}]aa \text{ if and only if } BC \overset{L}{\Rightarrow}^{+} aa$$

since $A \to]B$ is the only production for (A,]) and $AC \overset{L}{\Rightarrow}]BC$. Now

$$BC \overset{L}{\Rightarrow}^{+} aa \text{ if and only if } C \overset{L}{\Rightarrow}^{+} a$$

since $B \to a$ is the only production for (B,a) and $BC \overset{L}{\Rightarrow} aC$. Finally,

$$C \overset{L}{\Rightarrow}^{+} a$$

since $C \to a$ is a production for C.

We have proved that $S \overset{L}{\Rightarrow}^{+}$ [ ][ ]aa, that is [ ][ ]aa is in $L(G)$ using a technique which depends on there being at most one alternative at each step. Similarly we can prove that [ ] is not in $L(G)$

$$\begin{aligned}[\ ] \text{ is in } L(G) &\text{ if and only if } S \overset{L}{\Rightarrow}^{+} [\ ] \text{ in } G \\ &\text{ if and only if } AC \overset{L}{\Rightarrow}^{+}] \\ &\text{ if and only if } BC \overset{L}{\Rightarrow}^{+} \lambda\end{aligned}$$

and this is clearly impossible.

We discuss the *CFG* MEMBERSHIP problem in more detail in Section 5.2 and we show it is decidable in Section 10.1. The above example demonstrates that it is decidable for *s-CFGs*. Furthermore the proof technique used in the example can be fully automated for *s-CFGs*. We demonstrate this by defining a generic construction of a *DPDA*, M, from an *s-CFG*, G, such that $L(M) = L(G)$ and the computations of M correspond to proofs of membership or nonmembership in $L(G)$. This should be compared to the parser that can be constructed using Theorem 7.4.1.

Construction 5.1.1 *s-CFG to DPDA.*
Given an *s-CFG*, $G = (N,\Sigma,P,S)$, we construct a *DPDA*, $M = (\{s,f\},\Sigma,N \cup \{\bot\},\delta,s,\{f\})$, from G by defining δ as follows

(i) For all a in Σ, $\delta(s,a,\bot) = (f,\bot\alpha^R)$ if $S \to a\alpha$ is in P, for some α, where for a word x in Σ^*, the *reverse* of x, denoted by x^R, is defined by $x^R = x$ if $x = \lambda$ otherwise $x^R = a_n \cdots a_1$ if $x = a_1 \cdots a_n$ for some a_i in Σ, $1 \le i \le n$.

(ii) For all A in N, $\delta(f,a,A) = (f,\alpha^R)$ if $A \to a\alpha$ is in P, for some α. ■

Theorem 5.1.1 *Let $G = (N,\Sigma,P,S)$ be an s-CFG. Then the DPDA, M, obtained from G with Construction 5.1.1 satisfies $L(M) = L(G)$.*

Proof: By induction on the length of derivations and configuration sequences, see the Exercises. ■

Example 5.1.2C Applying Construction 5.1.1 we obtain $M = (\{s,f\},\{[,],a\},\{S,A,B,C,\bot\},\delta,s,\{f\})$, where δ is displayed in Figure 5.1.6. Consider the configuration sequence of M for input [ ][ ]*aa* shown in Figure 5.1.7 and compare this with the earlier proof that x is in $L(G)$. In particular, observe how each computation step of M corresponds to one step in the proof.

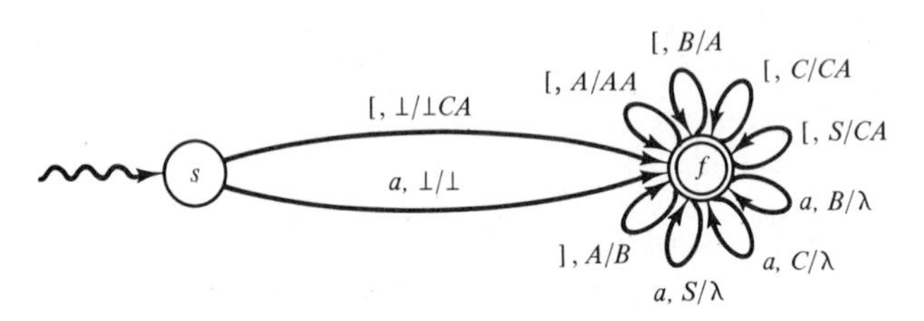

Figure 5.1.6

Similarly, with [ ] we have the configuration sequence of Figure 5.1.8 and the computation halts with a nonempty pushdown; hence [ ] is not in $L(G)$.

Finally, before closing this section we modify the definition of *DPDAs* to allow λ-transitions, that is, transitions which do not read an input symbol, and therefore do not advance the read head.

Definition A *λ-transition DPDA*, or *λ-DPDA*, M, is specified by a sextuple

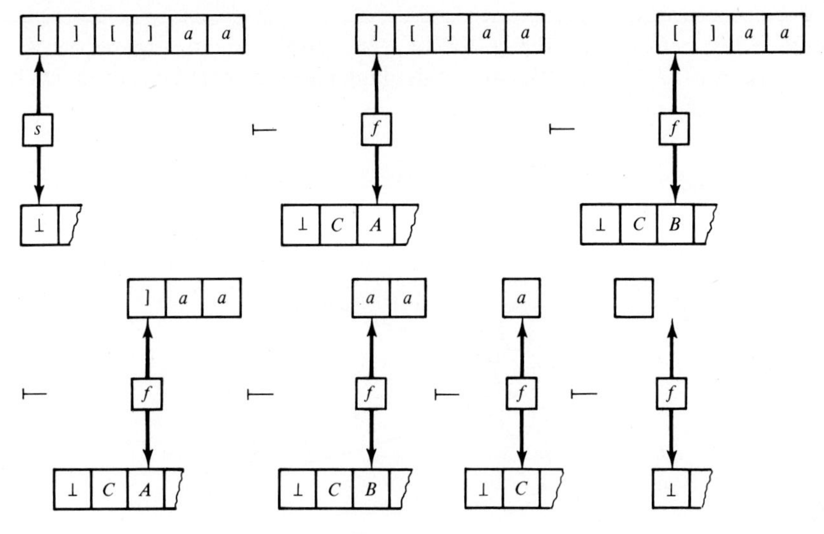

Figure 5.1.7

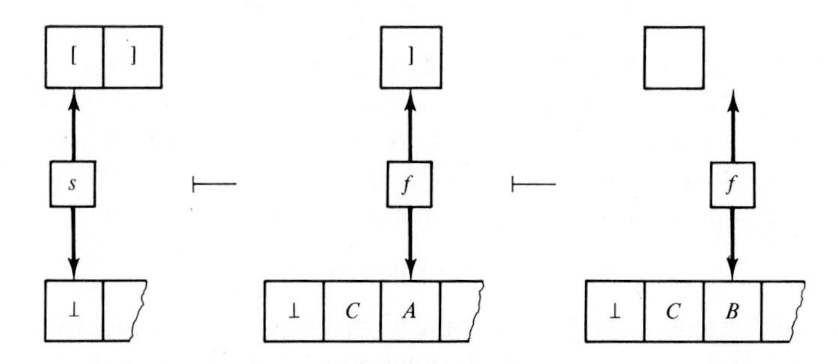

Figure 5.1.8

$M = (Q, \Sigma, \Gamma, \delta, s, F)$ as for a *DPDA* except that the transition function δ has domain $Q \times (\Sigma \cup \{\lambda\}) \times \Gamma$ rather than $Q \times \Sigma \times \Gamma$. To ensure that the λ-*DPDA* operates deterministically we also require

For all q in Q and for all c in Γ,

$$\delta(q, \lambda, c) = \textbf{\textit{undef}} \text{ if, for some } a \text{ in } \Sigma, \ \delta(q, a, c) \neq \textbf{\textit{undef}}.$$

This restriction on δ means that when a λ-*DPDA*, M, is in some state q and the topmost symbol in the pushdown is c a transition can take place either by reading an input symbol or by not reading an input symbol. Both possibilities cannot occur since this would introduce choice, that is, nondeterminism. We define acceptance as before, in which case every *DPDA* language is a λ-*DPDA* language. That the converse does not hold is shown with the following example.

Example 5.1.3 Consider the λ-*DPDA*, M, displayed in Figure 5.1.9. Then $L(M) = \{a^i b^j c : i \geq j \geq 0\}$, and this language cannot be accepted by any *DPDA*, as we now prove.

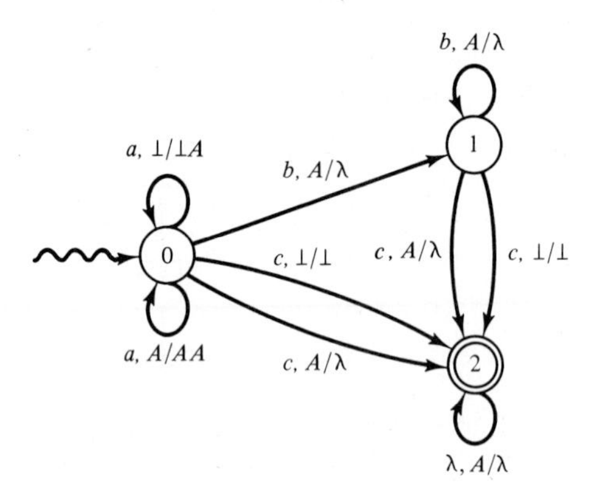

Figure 5.1.9

Claim: $L = \{a^i b^j c : i \geq j \geq 0\}$ *is not accepted by any DPDA.*

Proof of Claim: Assume L is accepted by some *DPDA*, that is, $L = L(M)$ for some *DPDA* $M = (Q, \{a,b,c\}, \Gamma, \delta, s, F)$. Since $a^i c$ is in L, for all $i \geq 1$, we must have the accepting configuration sequence

$$\bot s a^i c \vdash^+ \bot \alpha p c \vdash \bot f$$

for some α in Γ^*, some p in Q, and some f in F. But this implies $\alpha = \lambda$ or α is in Γ from the definition of a transition function for a *DPDA*. Therefore, there are only a finite number of intermediate configurations of the form $\bot \alpha p c$, which in one move lead to acceptance. This implies that there are at least two words $a^i c$ and $a^j c$, $1 \leq i < j$, that give rise to the same intermediate configuration $\bot \alpha p c$.

In other words, we have

$$\bot s a^i c \vdash^+ \bot \alpha p c \vdash \bot f$$

and

$$\bot s a^j c \vdash^+ \bot \alpha p c \vdash \bot f$$

Finally, consider the word $a^j b^j c$ in L. Since M is deterministic we obtain

$$\bot s a^j b^j c \vdash^+ \bot \alpha p b^j c \vdash^+ \bot f'$$

for some f' in F.

But this implies that we also obtain

$$\bot s a^i b^j c \vdash^+ \bot \alpha p b^j c \vdash^+ \bot f'$$

that is, $a^i b^j c$ is in $L(M)$ and, hence, in L. This is a contradiction since $j > i$, thus $L \neq L(M)$, for any *DPDA*, M. ■

Up until now we have defined acceptance by final state and empty pushdown. However, we may define acceptance by only final state or by only empty pushdown. For a given λ-*DPDA* more words are usually accepted by these less restrictive modes of acceptance. Fortunately, however, they do not increase the expressive power of λ-*DPDAs*. We consider this issue further in the Exercises. For this reason we base the definition of *deterministic CFLs* on λ-*DPDAs* with acceptance by both final state and empty pushdown.

Definition A *CFL*, L, is a *deterministic CFL*, or a *DCFL*, if it is accepted by some λ-*DPDA*, M, that is, $L = L(M)$. The *family of DCFLs* is denoted by $\mathbf{L}_{DCF}$ and forms an important subclass of $\mathbf{L}_{CF}$. We return to this topic in Sections 8.2 and 11.3.

5.2 NONDETERMINISTIC PUSHDOWN AUTOMATA

We extend *DPDAs* by allowing nondeterministic transitions, which results in nondeterministic *PDAs*. This extension is fundamental since these automata characterize $\mathbf{L}_{CF}$ exactly, while *DPDAs* do not; see Section 8.2. The proof of this result is the main purpose of this section, but along the way we consider how nondeterministic *PDAs* provide a general model for membership testing for *CFGs* as we did for *DPDAs* and membership testing for *s-CFGs*.

The hardware and specification of a nondeterministic *PDA* is the same as for a *DPDA*, except that the transition function is replaced by a finite transition relation. Thus we have the following.

Definition A *nondeterministic pushdown automaton*, or *NPDA*, M is specified by a sextuple $M = (Q, \Sigma, \Gamma, \delta, s, F)$, where

- Q is a finite set of *states*;
- Σ is an *input alphabet*;
- Γ is a *pushdown alphabet*, $\bot$ is in Γ;
- $\delta \subseteq Q \times \Sigma \times \Gamma \times Q \times \Gamma^*$ is a *finite transition relation*;
- s in Q is a *start* state; and
- $F \subseteq Q$ is a set of *final states*.

Again, many definitions of an *NPDA* allow λ-transitions, but in this case no gain in accepting power is achieved; see the Exercises. The transition relation is restricted with respect to $\bot$ in a similar manner to the restriction on transition functions in *DPDAs*.

A *configuration* of an *NPDA*, M, is a word gqx in $\perp(\Gamma - \{\perp\})^*Q\Sigma^*$ as for *DPDAs*. Configuration sequences are defined in a similar manner except that at each step more than one successor configuration may exist. The notations $\vdash$, $\vdash^i$, $\vdash^+$, and $\vdash^*$ are also used as before. Acceptance in *NPDAs* is defined in a similar manner to acceptance in *NFAs*.

Definition We say that a word x in Σ^* is *accepted* by an *NPDA*, $M = (Q,\Sigma,\Gamma,\delta,s,F)$ if there is *some* accepting configuration sequence $\perp sx \vdash^* \perp f$, for some f in F. This existential notion of acceptance is required since there may be many different computations of M on x. But x is accepted as long as there is one configuration sequence which leads to a final state and empty pushdown when all of x has been read; see the discussion of acceptance for *NFAs* in Section 2.3.

The *language accepted by* M, denoted by $L(M)$, is defined as

$$L(M) = \{x : x \text{ is in } \Sigma^* \text{ and } M \text{ accepts } x\}$$

The *family of NPDA languages*, denoted by $\boldsymbol{L}_{NPDA}$, is then

$$\boldsymbol{L}_{NPDA} = \{L : \text{there is an } NPDA\ M \text{ with } L = L(M)\}$$

As mentioned above we wish to demonstrate that $\boldsymbol{L}_{CF} = \boldsymbol{L}_{NPDA}$, but before doing this we have an example.

Example 5.2.1 We construct an *NPDA*, M, as follows: Let $\Sigma = \{a,b\}$, $\Gamma = \{A,\perp\}$, $Q = \{0,1,2,3\}$, $s = 0$, $F = \{0,3\}$, and the transition relation δ consist of

$(0,a,\perp,1,\perp A)$	$(0,a,\perp,2,\perp AA)$
$(1,a,A,1,AA)$	$(2,a,A,2,AAA)$
$(1,b,A,3,\lambda)$	$(2,b,A,3,\lambda)$
$(3,b,A,3,\lambda)$	

It should be clear that we can also specify δ by using a pushdown state diagram. However, in this case we may have more than one edge with the same input symbol joining the same two states; see Figure 5.2.1. Now with input λ we obtain $\perp 0 \vdash^* \perp 0$, and hence λ is accepted by M. Otherwise M reads a sequence of n a's, $n \geq 1$, and depending upon the initial choice or guess of moving into either state 1 or 2 M deposits n A's or $2n$ A's in the pushdown. Finally, on meeting the first b it moves into state 3, a final state, only emptying the pushdown if there are as many b's as A's in the pushdown. Thus, M accepts only words of the form a^nb^n and a^nb^{2n}, that is,

$$L(M) = \{a^ib^i : i \geq 0\} \cup \{a^ib^{2i} : i \geq 0\}.$$

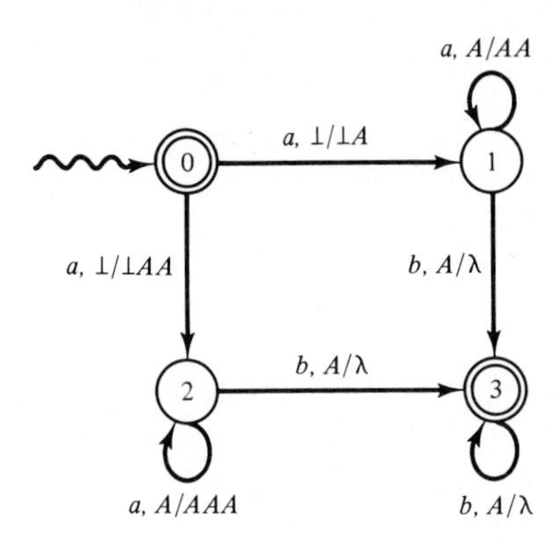

Figure 5.2.1

This language cannot be accepted by any *DPDA* or λ-*DPDA*, that is, $L(M)$ is not in $\boldsymbol{L}_{DCF}$; see Section 8.2.

We now demonstrate that every λ-free *CFL* is accepted by some *NPDA* by way of:

Construction 5.2.1 *CFG to NPDA.*
Given a CFG, $G = (N,\Sigma,P,S)$, *in Greibach normal form construct an NPDA,* $M = (\{s,f\},\Sigma,N \cup \{\perp\},\delta,s,\{f\})$, *from G as follows.*

For all a in Σ,

(i) $(s,a,\perp,f,\perp\alpha^R)$ is in δ, if $S \to a\alpha$ is in P, for some α in N^*.

(ii) For all A in N, (f,a,A,f,α^R) is in δ, if $A \to a\alpha$ is in P, for some α in N^*.

Observe that state s never reappears in any configuration sequence of M; we use this fact in Corollary 5.2.2 below. ■

Example 5.2.2 Let G be the *CFG*

$$S \to [T \mid [ST \mid [STS \mid [TS$$
$$T \to]$$

generating all nested parenthesized expressions. Then an *NPDA* M can be obtained by Construction 5.2.1; see Figure 5.2.2. In more compact form, allowing pushdown word sets as well as pushdown words we have Figure 5.2.3. Consider input x = [[ ][ ]]; then

$$\bot s[[\ ][\ ]] \vdash \bot TSf[[\ ][\ ]] \vdash \bot TSTf][\ ]] \vdash \bot TSf[]]$$
$$\vdash \bot TTf]] \vdash \bot Tf] \vdash \bot f$$

is one possible configuration sequence. Hence, [[ ][ ]] is accepted as anticipated. Moreover, observe the correspondence between this configuration sequence and the following left derivation of [[ ][ ]] in G

$$S \overset{L}{\Rightarrow} [ST \overset{L}{\Rightarrow} [[TST \overset{L}{\Rightarrow} [[\ ]ST \overset{L}{\Rightarrow} [[\ ][TT \overset{L}{\Rightarrow} [[\ ][\ ]T \overset{L}{\Rightarrow} [[\ ][\ ]]$$

But since there are four possible transitions with $(s,[,\bot)$ we also have nonaccepting configuration sequences, for example,

$$\bot s[[\ ][\ ]] \vdash \bot Tf[\ ][\ ]]$$

and M grinds to a halt, since no further move is possible.

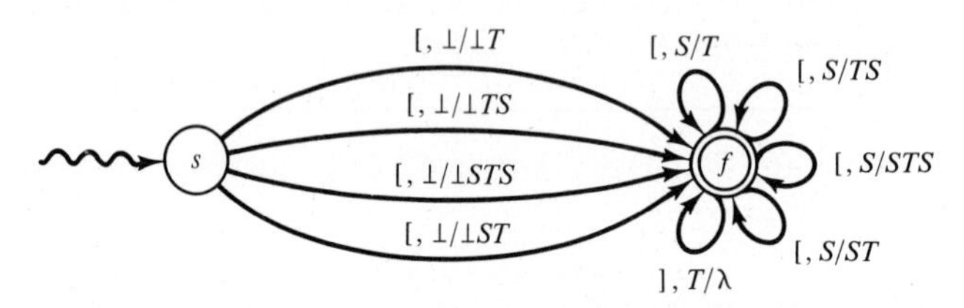

Figure 5.2.2

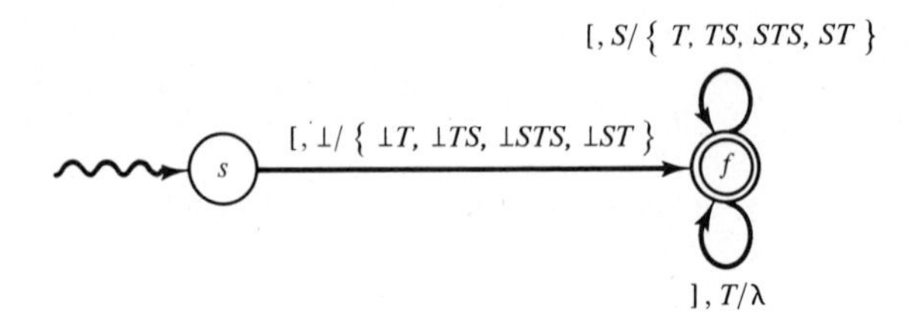

Figure 5.2.3

Lemma 5.2.1 *Let $G = (N,\Sigma,P,S)$ be a CFG in Greibach normal form and $M = (\{s,f\},\Sigma,N\cup\{\bot\},\delta,s,\{f\})$ be the NPDA obtained from G with Construction 5.2.1. Then $L(M) = L(G)$.*

Proof: By induction on the length of derivations and configuration sequences, see the Exercises. ■

Theorem 5.2.2 $\boldsymbol{L}_{CF} \subseteq \boldsymbol{L}_{NPDA}$.

Proof: Let L be an arbitrary *CFL*. Now $L' = L - \{\lambda\}$ is in $\boldsymbol{L}_{NPDA}$ by Lemma 5.2.1, since $L' = L(G')$ for some *CFG* G' in Greibach normal form, by Theorem 4.2.16. If $L = L'$, then L is also in $\boldsymbol{L}_{NPDA}$. Otherwise $L = L' \cup \{\lambda\}$ and in this case consider the *NPDA*, M, of Construction 5.2.1 for G'. Modify M, giving M', by letting both s and f be final states. Then $L(M') = L(M) \cup \{\lambda\} = L$ and, hence, L is in $\boldsymbol{L}_{NPDA}$. ■

Construction 5.2.1 provides the basis for implementing a nondeterministic membership tester (or a nondeterministic top-down parser; see Section 7.3) for a given *CFG*. Each configuration sequence corresponds to a partial left derivation in the underlying *CFG*; thus, an input word x is accepted by M exactly when it can mimic a corresponding left derivation of G generating x.

In order to prove the converse of the above theorem we proceed in two stages. First, assume we are given an *NPDA*, M, of the form resulting from Construction 5.2.1. Such an M has two states s and f, where s is the start state, f is the final state, s only appears in the initial configuration, and $\Gamma \cap \Sigma = \varnothing$. We say that such an *NPDA* is in *2-normal form*, and we write that M is a 2-*NPDA*. Given such a 2-*NPDA*, M, we apply the inverse of Construction 5.2.1 to give a *CFG*, G, in Greibach normal form satisfying $L(G) = L(M)$.

Second, we demonstrate that for every *NPDA*, M, there is a 2-*NPDA*, M', with $L(M') = L(M) - \{\lambda\}$, thus completing the proof that every *NPDA* language is a *CFL*. In order to prove this result we require that the 2-*NPDA*, M', simulate the original *NPDA*, M, by recalling at each step of M' the state that M would have been in. To perform this simulation M' must store, in some way, the future states of M. The only place these can be stored is in the pushdown of M'.

We first define the inverse of Construction 5.2.1.

Construction 5.2.2 *2-NPDA to CFG.*
Given a 2-*NPDA*, $M = (\{s,f\},\Sigma,\Gamma,\delta,s,\{f\})$, we construct a context-free grammar, $G = (\Gamma,\Sigma,P,\perp)$, from M as follows

For all a in Σ,

(i) $\perp \rightarrow ag^R$ is in P if $(s,a,\perp,f,g)$ is in δ, for some g in Γ^*.

(ii) For all Z in Γ, $Z \rightarrow ag^R$ is in P if (f,a,Z,f,g) is in δ, for some g in Γ^*. ■

Immediately, we have the following.

Lemma 5.2.3 *Let M be a 2-NPDA and G be a CFG obtained from M by Construction 5.2.2. Then G is in Greibach normal form and $L(G) = L(M)$.*

Proof: Clearly G is in Greibach normal form since each production is of the form $A \rightarrow ag$ where A is in Γ, a is in Σ, g is in Γ^* and Γ has no symbols in common with Σ. That $L(G) = L(M)$ follows by similar arguments to those used in the proof of Lemma 5.2.1; see the Exercises. ■

If we have an *NPDA*, M, which is a 2-*NPDA* except that $\Gamma \cap \Sigma \neq \varnothing$, then the symbols in Γ can always be renamed to obtain $\Gamma \cap \Sigma = \varnothing$. See the remarks on renaming for *DFAs* in Section 2.2.

We now consider the more difficult stage of the proof that every *NPDA* language is a *CFL*. We first give a construction that produces an equivalent 2-*NPDA* from an *NPDA* which has one final state and this is distinct from the start state.

Construction 5.2.3 *NPDA to 2-NPDA.*
Given an *NPDA*, $M = (Q,\Sigma,\Gamma,\delta,s,\{f\})$, where $s \neq f$, construct a 2-*NPDA*, $M' = (\{s',f'\},\Sigma,\Gamma',\delta',s',\{f'\})$, from M as follows.

Let $\Gamma' = \{[rZt] : r, t \text{ in } Q, \text{ and } Z \text{ in } \Gamma\}$ be a set of new pushdown symbols and define δ' by

For all a in Σ

(i) take $(s',a,\bot,f', \bot[s_2Z_2s_1] \cdots [s_n Z_n s_{n-1}])$ into δ', for all sequences $s_1, \ldots, s_n$ of states from Q, for which $(s,a,\bot,s_n,Z_1 \cdots Z_n)$ is in δ, where Z_i is in Γ, $1 \leq i \leq n$, and $Z_1 = \bot$.

(ii) For all Z in $\Gamma - \{\bot\}$, take the transition $(f',a,[s_{n+1}Z s_0], f',[s_1Z_1s_0][s_2Z_2s_1] \cdots [s_n Z_n s_{n-1}])$ into δ' for all sequences $s_0,s_1, \ldots, s_{n+1}$ of states from Q for which $(s_{n+1},a,Z,s_n,Z_1 \cdots Z_n)$ is in δ, where Z_i is in Γ, $1 \leq i \leq n$. ■

Informally, a new pushdown symbol $[rZt]$ represents the set of words x in Σ^* which satisfy $\bot Zrx \vdash^+ \bot t$, that is, M is initialized to state r and the pushdown is initialized to $\bot Z$. But this occurs as a part of a computation whenever $\bot s x \vdash^* g Zr y \vdash^* g t z$, where $x = x_1x_2z$ and $y = x_2z$. Hence, the new symbols are used to predict the future state of M as desired. Furthermore a configuration $gZry$, for y in Σ^+, leads in one step of M to a configuration $gZ_1 \cdots Z_nt_ny'$ where $(r,a,Z,t_n,Z_1 \cdots Z_n)$ is in δ and $y = ay'$. Eventually Z_n and the symbols it gives rise to will all be erased from the pushdown, giving $gZ_1 \cdots Z_{n-1}t_{n-1}y_n$. Similarly, Z_{n-1} will be erased, giving $gZ_1 \cdots Z_{n-2}t_{n-2}y_{n-1}$, and so on, that is,

$$\begin{aligned} gZry &\vdash gZ_1 \cdots Z_nt_ny' \\ &\vdash^+ gZ_1 \cdots Z_{n-1}t_{n-1}y_n \\ &\vdash^+ gZ_1 \cdots Z_{n-2}t_{n-2}y_{n-1} \\ &\;\;\vdots \end{aligned}$$

$$\vdash^+ gZ_1t_1y_2$$
$$\vdash^+ gtz.$$

This configuration sequence for M is shown in Figure 5.2.4, while the corresponding sequence for M' is shown in Figure 5.2.5.

It is this behavior that is captured by the symbol $[r\,Z\,t]$ and its transition $(f',a,[rZt],f',[t_1Z_1t][t_2Z_2t_1]\cdots[t_nZ_nt_{n-1}])$. Before proving that the construction preserves the language of M, consider a small example.

Example 5.2.3 Consider the *NPDA* of Example 5.2.1 in which 0 is no longer a final state. Construction 5.2.3 gives transitions

$$\{(s',a,\bot,f',\bot[1Ai]) : 0 \le i \le 3\}$$

and

$$\{(s',a,\bot,f',\bot[iAj][2Ai]) : 0 \le i,j \le 3\}$$

from s'. Now, noting that in the original *NPDA* only states 1 and 3 are reachable from 1, states 2 and 3 from 2, and state 3 from 3, we may restrict i and j accordingly. In other words, the only transitions from s' which can lead to acceptance are

$$\{(s',a,\bot,f',\bot[1Ai]) : i = 1,3\}$$
$$\{(s',a,\bot,f',\bot[iAj][2Ai]) : (i,j) = (2,2),(2,3),(3,3)\}$$

We follow the first two leaving the others for the Exercises. For pushdown symbols $[1A1]$ and $[1A3]$ we obtain

$$\{(f',a,[1A1],f',[iA1][1Ai]) : 0 \le i \le 3\}$$
$$\{(f',a,[1A3],f',[iA3][1Ai]) : 0 \le i \le 3\}$$
$$\{(f',b,[1A3],f',[iA3][3Ai]) : 0 \le i \le 3\}$$

since no transition with b and $[1A1]$ is possible. Simplifying once more we obtain

$$\{(f',a,[1A1],f',[1A1][1A1])\}$$
$$\{(f',a,[1A3],f',[iA3][1Ai]) : i = 1,3\}$$
$$\{(f',b,[1A3],f',[3A3][3A3])\}$$

Now $[3A3]$ gives

$$\{(f',b,[3A3],f',\lambda)\}$$

and we now have a partial pushdown state diagram with only pushdown symbols involving 2 omitted; see Figure 5.2.6. The reader is invited to complete this 2-*NPDA* and test it in the Exercises.

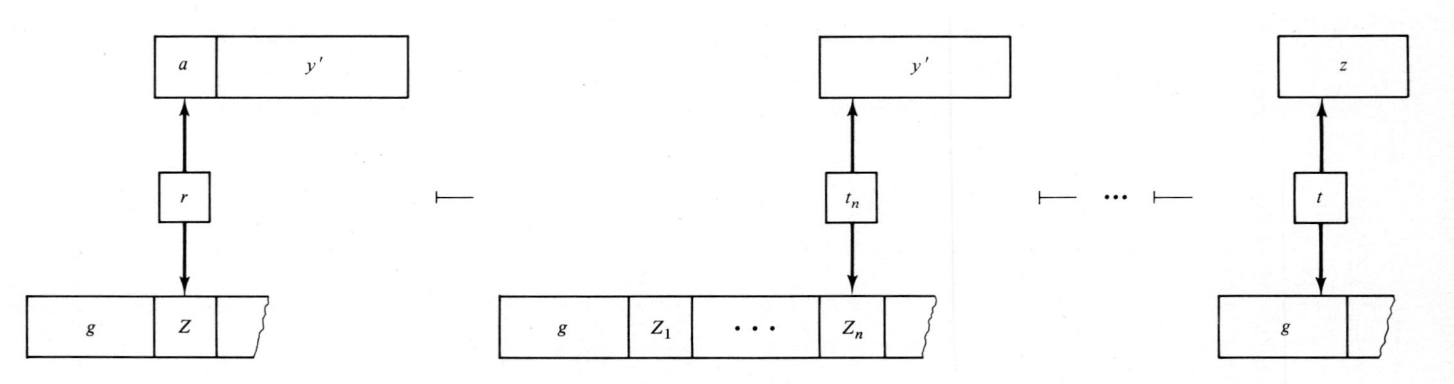

Figure 5.2.4

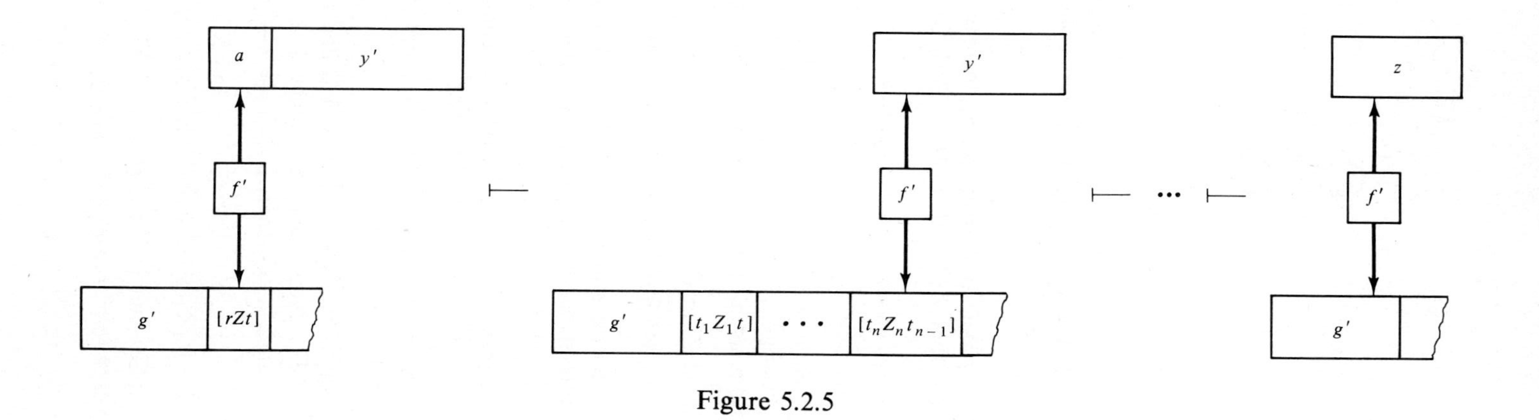

Figure 5.2.5

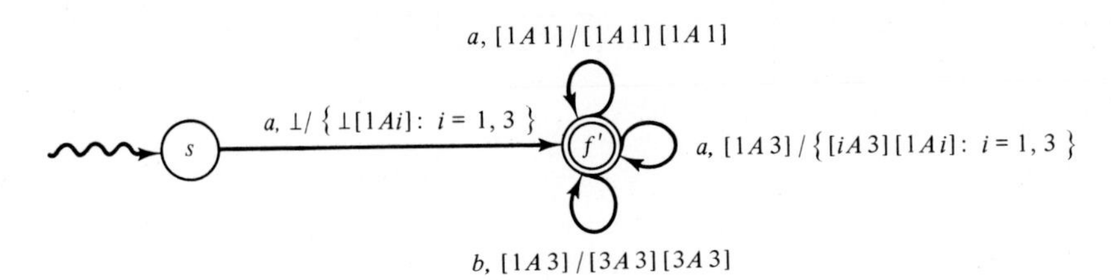

Figure 5.2.6

Theorem 5.2.4 *Given the NPDA, $M = (Q,\Sigma,\Gamma,\delta,s,\{f\})$, from which we obtain the 2-NPDA, $M' = (\{s',f'\},\Sigma,\Gamma',\delta',s',\{f'\})$, using Construction 5.2.3, we have* $L(M') = L(M)$.

Proof: We split the proof into two parts, namely, we first prove that $L(M') \subseteq L(M)$ and then that $L(M) \subseteq L(M')$. Both parts are proved by induction.

I. $L(M') \subseteq L(M)$.

Now $L(M') \subseteq L(M)$ if each word x accepted by M' is also accepted by M. In other words, if the configuration sequence $\perp s'x \vdash^+ \perp f'$ in M' implies $\perp sx \vdash^+ \perp f$ in M. As is usual in inductive proofs we strengthen this to

$$\text{For all } r, t \text{ in } Q, \text{ for all } Z \text{ in } \Gamma, \text{ for all } x \text{ in } \Sigma^+,\quad [rZt]f'x \vdash^+ f' \text{ in } M' \text{ implies } Zrx \vdash^+ t \text{ in } M, \tag{5.2.1}$$

while at the same time ignoring state s', since it can only appear in the initial configuration. We prove this stronger statement by induction on the length of the input word or, equivalently on the length of the configuration sequence.

Basis: $|x| = 1$. Then $[rZt]f'x \vdash f'$ in M' if δ' contains $(f', x, [rZt],f',\lambda)$, that is, if δ contains (r,x,Z,t,λ) which implies $Zrx \vdash t$ as desired.

Induction Hypothesis: Assume equation (5.2.1) holds for all x in Σ^+ with $|x| \leq k$, for some $k \geq 1$.

Induction Step: Let x in Σ^+ satisfy $|x| = k+1$, and assume $[rZt]f'x \vdash^+ f'$ in M', for some $[rZt]$ in Γ'. Now $|x| \geq 2$, since $k \geq 1$, we have $x = ay$ for some a in Σ and some y in Σ^+. Further, $[rZt]f'ay \vdash^+ f'$ in M' implies δ' contains a transition $(f', a, [rZt],f',[t_1Z_1t][t_2Z_2t_1] \cdots [t_nZ_nt_{n-1}])$, satisfying

$$[rZt]f'ay \vdash [t_1Z_1t][t_2Z_2t_1] \cdots [t_nZ_nt_{n-1}]f'y \vdash^+ f'$$

in M'. This, in turn, implies that $[t_iZ_it_{i-1}]f'z_i \vdash^+ f'$ in M', for some z_i in Σ^+, for all i, $1 \leq i \leq n$, where $t_0 = t$ and $y = z_nz_{n-1} \cdots z_1$.

Finally, $[t_i Z_i t_{i-1}] f' z_i \vdash^+ f'$ in M' implies $Z_i t_i z_i \vdash^+ t_{i-1}$ in M by the induction hypothesis since $|z_i| \leq |y| < |x|$. But this means that

$$Zray \vdash Z_1 \cdots Z_n t_n y$$

in M, since the transition $(f', a, [rZt], f', [t_1 Z_1 t] \cdots [t_n Z_n t_{n-1}])$ is in δ' only if the transition $(r, a, Z, t_n, Z_1 \cdots Z_n)$ is in δ, and

$$\begin{aligned} Z_1 \cdots Z_n t_n y &\vdash^+ Z_1 \cdots Z_{n-1} t_{n-1} z_{n-1} \\ &\vdash^+ \cdots \vdash^+ Z_1 Z_2 t_2 z_2 \vdash^+ Z_1 t_1 z_1 \vdash^+ t \end{aligned}$$

in M. But this ensures that $Zrx \vdash^+ t$ as required, completing the induction step.

To complete the proof of part I observe that $\perp s'x \vdash^+ \perp f'$ if and only if either x is in Σ and $(s, x, \perp, f, \perp)$ is in δ or $x = ay$, where a is in Σ and y is in Σ^+. In this latter case $\perp s'ay \vdash^+ \perp f'$ in M' if and only if

$$\perp [t_2 Z_2 t_1] \cdots [t_n Z_n t_{n-1}] f' y \vdash^+ f'$$

in M', for some transition $(s', a, btm, f', \perp [t_2 Z_2 t_1] \cdots [t_n Z_n t_{n-1}])$ in δ'. Now by the construction this transition is in δ' only if there is a corresponding transition $(s, a, \perp, t_n, Z_1 \cdots Z_n)$ in δ with $Z_1 = \perp$. This together with equation (5.2.1) implies there is a configuration sequence

$$\begin{aligned} \perp sx &\vdash \perp Z_2 \cdots Z_n t_n y \\ &\vdash^+ \perp Z_2 \cdots Z_{n-1} t_{n-1} z_{n-1} \\ &\vdash^+ \cdots \vdash^+ \perp Z_2 t_2 z_2 \vdash^+ \perp z_1 \vdash^+ \perp f \end{aligned}$$

in M, for some $z_1, \ldots, z_{n-1}$. Thus, in both cases x is in $L(M)$, establishing part I.

II. $L(M) \subseteq L(M')$.

The proof of this result is similar to that of part I and is, therefore, left to the Exercises. ■

To remove the restrictions on the input *NPDA* in Construction 5.2.3 we need to show that

(1) Every *NPDA*, M, in which the start state is also a final state can be replaced by an *NPDA*, M', in which the start state is not a final state such that $L(M') = L(M) - \{\lambda\}$.
(2) Construction 5.2.3 can be extended to the case that the given *NPDA* has more than one final state, none of which are s.

We deal with (1) first. Let $M = (Q, \Sigma, \Gamma, \delta, s, F)$, where s is in F, be the given *NPDA*. Modify M to give $M' = (Q', \Sigma, \Gamma, \delta', s, (F - \{s\}) \cup \{s'\})$, by adding a new state s', a copy of s, to Q to give $Q' = Q \cup \{s'\}$. The new state s' becomes the final state in M', replacing the final state s in M, and δ' is the same as δ except that we modify the transitions in which s appears

(a) For all (s,a,Z,q,g) in $\delta, q \neq s$, (s,a,Z,q,g) and (s',a,Z,q,g) are in δ'.
(b) For all (q,a,Z,s,g) in $\delta, q \neq s$, (q,a,Z,s',g) is in δ'.
(c) For all (s,a,Z,s,g) in δ, (s,a,Z,s',g) and (s',a,Z,s',g) are in δ'.

We leave to the Exercises the simple proof that $L(M') = L(M) - \{\lambda\}$.

Now although Construction 5.2.3 can be modified to include *NPDAs*, $M = (Q,\Sigma,\Gamma,\delta,s,F)$ with $\#F > 1$ and s not in F, it suffices to observe that $L(M) = \cup_{f \text{ in} F} L(M_f)$, where $M_f = (Q,\Sigma,\Gamma,\delta,s,\{f\})$. Now M_f satisfies the conditions of Construction 5.2.3 and, since we prove in Section 9.1 that $\boldsymbol{L}_{CF}$ is closed under finite union, $L(M)$ is in $\boldsymbol{L}_{CF}$, also.

Since $\{\lambda\}$ is also in $\boldsymbol{L}_{CF}$, we have the following.

Theorem 5.2.5 $\boldsymbol{L}_{NPDA} = \boldsymbol{L}_{CF}$.

As for *DPDAs* it is often useful to allow an *NPDA* to have λ-transitions in which case we call it a λ-*NPDA*. The results of this section can be extended to include this possibility, proving that λ-*NPDAs* also characterize the family of *CFLs*. This extension is left to the Exercises.

5.3* COUNTER AUTOMATA

A restricted version of *NPDAs* which is central to a result on Turing machines in Section 6.4 is the nondeterministic counter automaton.

Definition A *nondeterministic counter automaton*, or *NCA*, M is an *NPDA*, $M = (Q,\Sigma,\{1,\perp\},\delta,s,F)$. It is called a counter automaton because a string of 1's, on the pushdown, represents a nonnegative integer in unary notation.

Example 5.3.1 Let $L = \{a^i b^i : i \geq 1\}$. Then L is accepted by the *NCA* shown in Figure 5.3.1.

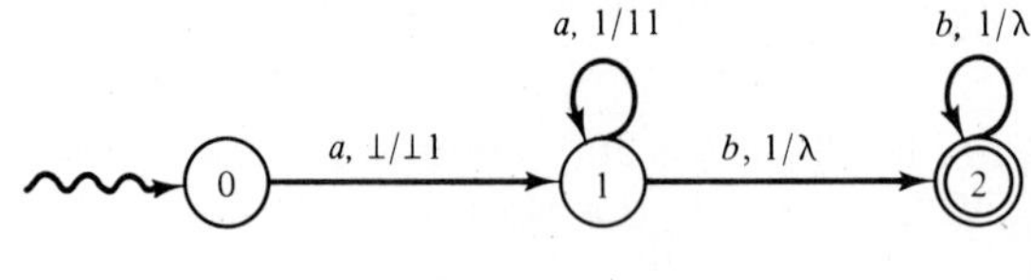

Figure 5.3.1

However, there are *CFLs* that are not accepted by any *NCA*. Although we do not prove this result we give a language and explain why it is not accepted by any *NCA*.

Example 5.3.2 Let $L = \{a^i b^j c^j d^i : i,j \geq 1\}$. Then L is not accepted by any *NCA*. The intuitive reasons for this are that in order to check that there are as many d's as a's it is necessary to keep the count of the number of a's in the pushdown. A similar observation holds for the b's and c's. But this means that by the time all the a's and b's have been read the pushdown must contain their separate counts. But with a single pushdown symbol this separation is not possible.

Note that the languages $L_1 = \{a^i b^j c^k d^i : i,j,k \geq 1\}$ and $L_2 = \{a^i b^j c^k d^l : i,j \geq 1,\ k+l = i+j\}$ are both accepted by some *NCA*.

5.4* TWO-WAY DETERMINISTIC PUSHDOWN AUTOMATA

In Chapters 2 and 3 we have discussed the basic operation found in text editors, namely, pattern matching. Given a text t over an alphabet Σ and a pattern p over Σ the pattern-matching problem, at its simplest, can be formulated as: Does p occur in t, that is, are there some t_1 and t_2 over Σ such that $t = t_1 p t_2$? This can be formulated as a language question as follows. Given Σ and a symbol c not in Σ, let L_Σ be the set of all words pct, such that p is a subword of t. Formally,

$$L_\Sigma = \{pct : p,t \text{ are in } \Sigma^* \text{ and } t = t_1 p t_2 \text{ for some } t_1 \text{ and } t_2\}$$

Now, given a specific pattern p and a specific text t, determining whether or not p occurs in t is equivalent to determining whether or not pct is in L_Σ. Hence, if we can obtain a deterministic automaton M_Σ with $L(M_\Sigma) = L_\Sigma$, then M_Σ may be used as the basis of a pattern-matching algorithm. Although, in general, L_Σ cannot be accepted by any *DPDA* (this requires a nontrivial proof; see the Exercises) it can be accepted by a two-way *DPDA*, that is, a *DPDA* in which the read head is allowed to move left and right. Formally, we have the following.

Definition A *two-way deterministic pushdown automaton*, or 2*DPDA*, M is specified by a sextuple $M = (Q,\Sigma,\Gamma,\delta,s,F)$, where Q,Σ,s, and F are unchanged from their specification in a *DPDA*. However, δ, the *transition function*, maps $Q \times (\Sigma \cup \{\boldsymbol{l},\boldsymbol{r}\}) \times \Gamma$ to $Q \times \Gamma^* \times \{L,R\}$, where $\boldsymbol{l}$ and $\boldsymbol{r}$ are *left* and *right endmarkers*, respectively, for the input, $\boldsymbol{l}$ and $\boldsymbol{r}$ are not in Σ, and L and R denote a *left* and *right* movement, respectively, of the read head.

The initial configuration of such a 2-*DPDA* is shown in Figure 5.4.1. The read head is positioned over the leftmost input symbol rather than over the left endmarker $\boldsymbol{l}$. When the input word is the empty word, however, the read head is over the right endmarker $\boldsymbol{r}$. The two-way movement of the read head implies that

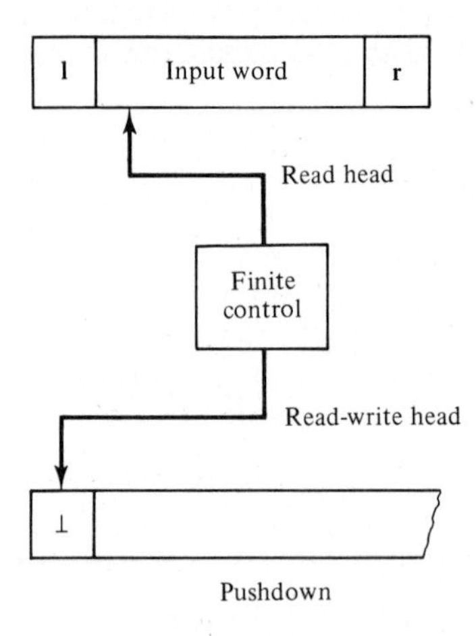

Figure 5.4.1

we cannot ignore the input symbols to the left of the read head, since they can be scanned a second time. We extend configurations to allow for this possibility as follows.

A *configuration* of a $2DPDA$, M, is a word gx_1qx_2 satisfying g is in Γ^*, q is in Q, and x_1x_2 is in $\{\boldsymbol{l}\}\Sigma^*\{\boldsymbol{r}\}$ with $x_2 \neq \lambda$. It denotes that M is in state q, its input is x_1x_2, including endmarkers, g is on the pushdown, and the read head is over the leftmost symbol of x_2. An initial configuration has the form $\perp \boldsymbol{l}sx\boldsymbol{r}$.

Again, because of the two-way movement of the read head, we modify the conditions for acceptance to be an empty pushdown and a final state. The read head may be anywhere when acceptance occurs. Hence, a final or accepting configuration is $\perp x_1fx_2$, for f in F and x_1x_2 in $\{\boldsymbol{l}\}\Sigma^*\{\boldsymbol{r}\}$.

A straightforward modification of the definitions of $\vdash$, $\vdash^i$, $\vdash^+$, and $\vdash^*$ for a $DPDA$, which takes into account the movement of the read head, enables us to define $L(M)$, the *language accepted by a 2DPDA*, as

$$L(M) = \{x : x \text{ is in } \Sigma^*, \perp \boldsymbol{l}sx\boldsymbol{r} \vdash^* \perp \boldsymbol{l}x_1fx_2\boldsymbol{r} \text{ in } M, \text{ and } f \text{ is in } F\}$$

Consider a pattern-matching example.

Example 5.4.1 Let $\Sigma = \{a,b,c\}$, where patterns and texts only contain the symbols a and b. Let $L = \{pct : p \text{ and } t \text{ are in } \{a,b\}^* \text{ and } p \text{ occurs in } t\}$. We describe the actions of a $2DPDA$, M, for which $L(M) = L$.

Let $x = a_1 \cdots a_n$ be a word over Σ, $a_0 = \boldsymbol{l}$, and $a_{n+1} = \boldsymbol{r}$.

Step 1: M scans $a_1 \cdots a_{n+1}$ until either an $a_i = c$ is found, in which case M continues with Step 2, or no such a_i is found, that is, $\boldsymbol{r}$ is met, in which case M halts in a nonfinal state.

Step 2: M scans $a_{i+1} \cdots a_{n+1}$ until either an $a_j = c$ is found, in which case, M halts in a nonfinal state, or $\boldsymbol{r}$ is met, in which case M continues with Step 3.

Step 3: M scans $a_n \cdots a_i$ copying each a_j onto the pushdown until a_i is met when M continued with Step 4.

Step 4: M scans $a_i \cdots a_0$ until a_0 is met, when it moves right to a_1. Letting $k = i+1$ M continues with Step 5.

Step 5: M now scans the symbols $a_1, \ldots, a_i$, comparing them with the topmost symbols of the pushdown $a_k, a_{k+1}, \ldots, a_n$ until a symbol $a_j \neq a_{i+j}$ is reached, in which case M performs Step 6; or $a_i = c$ is reached, in which case x is in L and M moves into the final state; or the pushdown is empty, that is, x is not in L and M halts in a nonfinal state.

Step 6: M restores the pushdown to its value on entering Step 5. This it can do because $a_1 \cdots a_{j-1} = a_k \cdots a_{k+j-2}$. M scans $a_{j-1} \cdots a_0$, copying the symbols into the pushdown until a_0 is reached. It then erases a_k from the pushdown, letting let $k = k-1$, and returns to Step 5.

M simulates the action of the simplest pattern-matching algorithm, that is, it compares $a_1, \ldots, a_i$, with $a_k, \ldots, a_n$, until either a complete match is found or a mismatch of two symbols occurs, when the process is repeated with $a_1, \ldots, a_i$, against $a_{k+1}, \ldots, a_n$. We have used a counting variable k throughout the above algorithm; however, this is included only for clarity. It is unessential because of the remark in Step 6 of the algorithm. This algorithm and the *2DPDA* described above require at most $d \cdot |p| \cdot |t|$ steps, for some constant $d \geq 1$, that is, $O(|x|^2)$ steps, since $|p| \cdot |t| = O(|x|^2)$. Fortunately, the behavior of an arbitrary *2DPDA* can be simulated in $O(|x|)$ steps, using a technique found in Section 8.3 for simulating the action of an *NPDA*. See the Exercises at the end of Chapter 8 for further details.

5.5 ADDITIONAL REMARKS

5.5.1 Summary

We have presented PUSHDOWN AUTOMATA in their deterministic and nondeterministic forms and have demonstrated that $\boldsymbol{L}_{CF} = \boldsymbol{L}_{NPDA}$. To prove this result we have shown how leftmost derivations in a *CFG* in Greibach normal form can be simulated by an *NPDA*. This means that *CFG* MEMBERSHIP can be implemented by *NPDAs*. In Section 7.3 we will show how this construction can be modified to solve *CFG* PARSING.

Finally, we have introduced one restricted class of *NPDAs*, NONDETERMINISTIC COUNTER AUTOMATA, in which there is, essentially, only one pushdown symbol and introduced one extended class of *NPDAs*, TWO-WAY DETERMINISTIC PUSHDOWN AUTOMATA. Counter automata are surprisingly powerful as we shall see in Section 6.4, while 2-way *DPDAs* provide a general method of modeling pattern matching and at the same time lead, via a technique used in Section 8.2, to efficient implementations.

5.5.2 History

NPDAs were first formalized by Chomsky (1962) and Evey (1963), while Schutzenberger (1963) is responsible for one notion of *DPDAs*. However, the notion of a stack or pushdown has been in use from the earliest days of computing; see Burks et al. (1954) and Newell and Shaw (1957), for example. Membership testing and parsing of *CFG*s has been their major application; see Oettinger (1961), Aho and Ullman (1972), and Gries (1971), for example. Simple deterministic grammars were introduced in Korenjak and Hopcroft (1966).

5.6 SPRINGBOARD

5.6.1 Data Structure Automata

Choose your favorite data structure — a queue, a dequeue, a tree — and define a class of automata based on it. Can you come up with any reasonable applications of your model?

5.6.2 Preset Pushdown Automata

In this variant of *PDAs* a preset pushdown automaton nondeterministically decides on some positive integer bound on the size of the pushdown throughout the computation. As long as the pushdown has size less than the threshold it computes as usual, but when the pushdown reaches the predetermined size it uses special transitions. If the pushdown's size goes above the threshold the automaton halts. Acceptance, as before, is by empty input, empty pushdown, and final state. See van Leeuwen (1974b) and Engelfriet et al. (1980).

5.6.3 Multihead and Multipushdown Automata

As we shall see in Section 6.4 nondeterministic multipushdown automata are of little interest, since two-pushdown automata have the same power as Turing machines. But how about multiple heads? Formally define these pushdown automata, study their relationships, and investigate possible applications of these models.

5.6.4 Finite-Buffer *PDAs* and Lazy *PDAs*

The reason that the *PDAs* (*FAs* and *TMs*) read one input symbol, at most, at each step is that this is sufficient to model most situations and is simple. However, one can think of *PDAs* being extended by adding a fixed-length buffer of predetermined size between the input tape and the finite state control; see Figure 5.6.1. If the buffer is of size $n \geq 1$, we call such a *PDA* an *n-BPDA*, for *size-n-buffer PDA*. The buffer is initialized to blank, denoted by $\boldsymbol{B}$, and then the first n symbols of the input are transferred to the buffer.

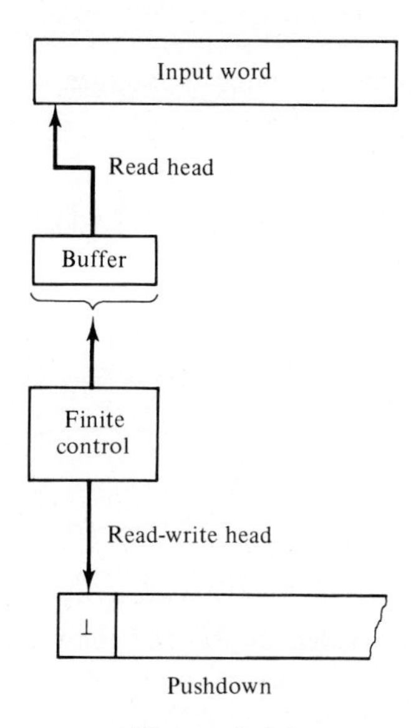

Figure 5.6.1

At each step the *n-BPDA* transition depends not on one input symbol but on the whole buffer. However, the effect of the transition is to either leave the buffer intact or transfer the next symbol from the input, causing the symbols in the buffer to slide up to make room for the new symbol and, in so doing, causing the first symbol to be lost. In other words, an *n-BPDA* simulates a look-ahead of n symbols. Clearly, if $n = 1$ it is a *PDA*, but what about $n > 1$? Such a model is useful for deterministic top-down membership testing; see Section 11.4, Rosenkrantz and Stearns (1970), and Lewis et al. (1976). Investigate the nondeterministic variant.

Another variant is to allow the *PDA* to read a word from the input tape at each step — similar to the lazy *FA*. Does this add any power to the *PDA*? See Exercise 2.11.

5.6.5 Small *PDAs*

What is a reasonable measure of the size of a *PDA*? Having decided on a size measure can you exhibit languages that require a larger *DPDA* to accept them than *NPDA*? Are there *FA* languages or regular languages that can be accepted by a smaller *PDA* than *FA*? Can you always find the smallest *DPDA* (or *NPDA*) for a given language?

EXERCISES

1.1 Construct *DPDAs* for the following languages.

(i) $\{a^{2i}cb^i : i \geq 0\}$.
(ii) $\{a^{2i+1}b^i : i \geq 0\}$.
(iii) $\{x : x \text{ is in } \{a,b\}^* \text{ and } |x|_a = |x|_b\}$.
(iv) $\{xcx^R : x \text{ is in } \{a,b\}^*\}$.

1.2 Given the following *DPDA*, M,

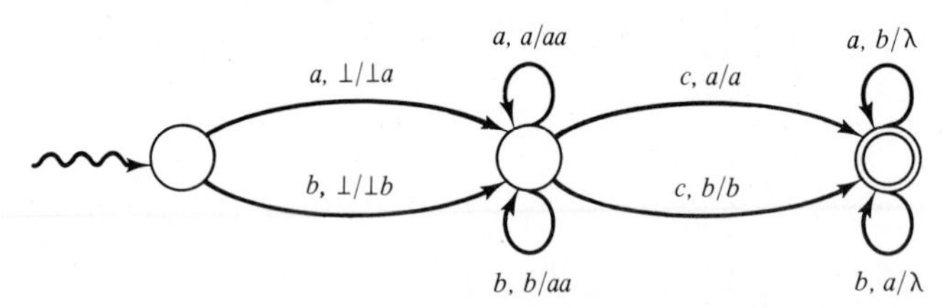

(i) Give four words accepted by M and their configuration sequences.
(ii) Give four words rejected by M and their configuration sequences.
(iii) Give the language of M.

1.3 Prove Theorem 5.1.1.

1.4 Given the language $\{a^ib^jc^ka^i : i \geq 1 \text{ and } j \geq k \geq 1\}$ construct a λ-*DPDA* that accepts it. Can this language be accepted by a *DPDA*? Justify your answer.

1.5 Let $M = (Q,\Sigma,\Gamma,\delta,s,F)$ be a *DPDA* or λ-*DPDA*. Let $L_e(M)$ denote the set $\{x : \perp sx \vdash^* \perp p, \text{ for some } p \text{ in } Q\}$ and $L_f(M)$ denote the set $\{x : \perp sx \vdash^* \alpha f, \text{ for some } \alpha \text{ in } \Gamma^+ \text{ and } f \text{ in } F\}$.

(i) Prove that $\{a^ib^jc : i \geq j \geq 0\}$ is $L_f(M)$, for some *DPDA*, M.
(ii) Prove that $\{a^ib^jc : i \geq j \geq 0\}$ is $L_e(M)$, for some λ-*DPDA*, M.
(iii) Prove that $\{a^ib^jc : i \geq j \geq 0\} = L_e(M)$, for any *DPDA*, M.

1.6 Prove the following

(i) If $L = L_e(M)$, for some λ-*DPDA*, M, then there is a λ-*DPDA*, M', with $L(M') = L$.
(ii) If $L = L_f(M)$, for some λ-*DPDA*, M, then there is a λ-*DPDA*, M', with $L(M') = L$.

2.1 Construct *NPDAs* for the following languages.

(i) $\{a^ib^jc^k : i,j,k \geq 0 \text{ and } i = j \text{ or } j = k\}$.
(ii) $\{xx^R : x \text{ is in } \{a,b\}^*\}$.
(iii) $\{x : x \text{ is in } \{a,b\}^* \text{ and either } |x|_a = |x|_b \text{ or } |x|_a = 2|x|_b\}$.
(iv) $\{x : x \text{ is in } \{a,b\}^* \text{ and } x = x^R\}$.

2.2 Given the following *NPDA*, M,

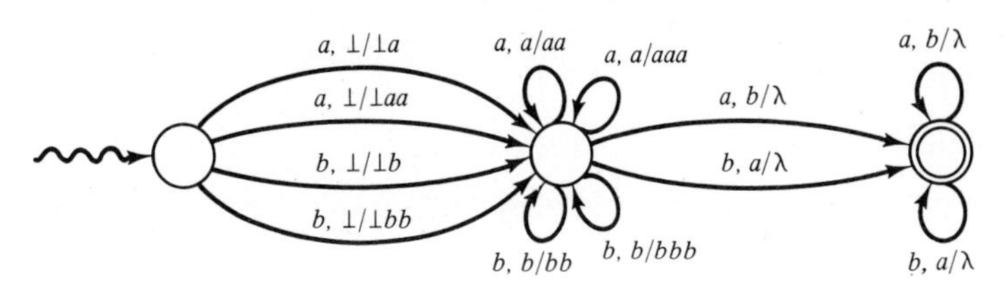

(i) Give four words accepted by M.
(ii) Give four words rejected by M.

In both cases justify your answers.

2.3 Given the following *CFGs* in *GNF* construct equivalent *NPDAs*.

(i) $S \to cSS \mid bSA \mid aSb \mid aB \mid bA;\ A \to a;\ B \to b.$
(ii) $S \to aAB\ ;\ A \to aAb \mid b\ ;\ B \to bBa \mid a.$

2.4 We have demonstrated how a *CFG* in *GNF* can be converted into an equivalent *NPDA*. If we allow λ-*NPDAs* rather than *NPDAs*, then this construction can be generalized to include arbitrary *CFGs*.

(i) Define a reasonable generalization.
(ii) In the original construction of an *NPDA*, since an input symbol had to be read at every transition, all configuration sequences were finite. Does this remain valid in the λ-*NPDA* given by your generalized construction?

2.5 Given the following 2-*NPDAs* construct equivalent *CFGs*.

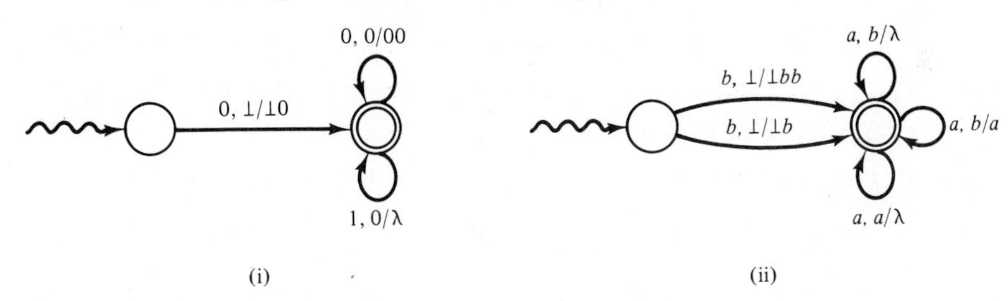

2.6 We can construct a λ-*NPDA*, M, from a *CFG*, $G = (N,\Sigma,P,S)$, as follows. Let M have two states s and f, and include all transitions of the forms

(i) $(s,\lambda,\perp,\alpha^R,f)$ if $S \to \alpha$ is in P;
(ii) (f,λ,A,α^R,f) if $A \to \alpha$ is in P; and
(iii) (f,a,a,λ,f) if a is in Σ.

Prove that $L(M) = L(G)$ and describe the relationship of configuration sequences in M to derivation sequences in G.

2.7 Prove Lemma 5.2.1.

2.8 Prove Lemma 5.2.3.

2.9 Complete the construction of a 2-*NPDA* in Example 5.2.3.

2.10 Complete the proof of Theorem 5.2.4.

2.11 We can generalize *NPDAs* even further as follows.

(i) Let an *NPDA* read a word, including λ, from the input at each transition.
(ii) Let an *NPDA* read a word, including λ, from the pushdown at each transition.
(iii) Let an *NPDA* read a word from a given regular language from the input at each transition.

Prove that each of these generalizations does not add to the power of *NPDAs*.

2.12 Given some *NPDA*, $M = (Q,\Sigma,\Gamma,\delta,s,F)$, an accepting configuration sequence for any input word defines a sequence of pop and push operations on the pushdown. Letting $\overline{\Gamma} = \{\overline{c} : c \text{ is in } \Gamma\}$ we can write such a sequence as a word $\perp\overline{\perp}x_1\overline{X}_1x_2\overline{X}_2 \cdots \overline{X}_nx_{n+1}$ in $(\Gamma \cup \overline{\Gamma})^*$, where the barred symbols correspond to pops and the unbarred to pushes. Is the language of such words context-free? Justify your answer.

3.1 Which of the *PDAs*, if any, of Exercises 1.1 and 2.1 are counter automata?

3.2 Let L be an arbitrary regular language. Is L accepted by some counter automata? Prove your answer.

4.1 Given the following two-way *DPDA*

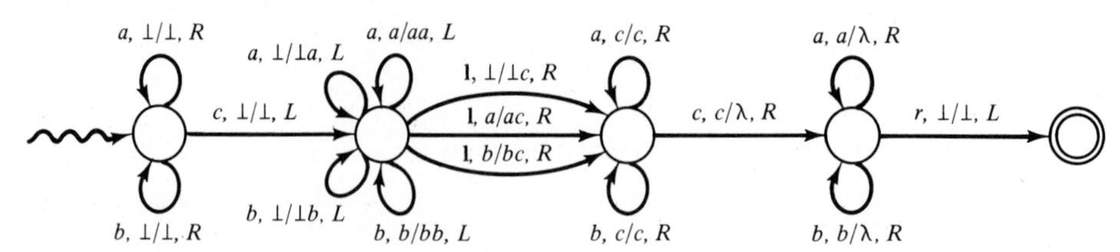

(i) Give four words accepted by it.
(ii) Give four words rejected by it.
(iii) What is the accepted language?

In each case justify your answers.

PROGRAMMING PROJECTS

P1.1 Write a program to display the action of a *DPDA* on an input word.

P1.2 Write a program that, for a given *DPDA* or *NPDA*, produces samples of accepted words.

P2.1 Write a program that constructs an *NPDA* from a given *CFG* in *GNF*.

P2.2 Write a program that constructs a λ-*NPDA* from a given *CFG* (see Exercise 2.4).

P2.3 Write a program that constructs a *CFG* from a given *NPDA*, then extend it to include λ-*NPDA*.

Chapter 6

Turing Machines

6.1 THE TURING MACHINE

We introduce, in this chapter, the most general automaton. Turing machines are, however, much more than this. Originally introduced in the mid-thirties to provide a method for specifying algorithms, they have since become the universally accepted basic computational model for algorithms, that is, the touchstone. The Turing machine provides a theoretical model of computers which has withstood the test of time, almost 50 years in production and still going strong. It's the ultimate personal computer since only pencil and paper are needed, yet at the same time, it is as powerful as any real machine.

The Turing machine is, in many respects, similar to the finite automaton. It has a finite number of states and symbols and has transitions which depend on a symbol and state. But the similarity ends here since rewriting of the input symbols is permitted as is the writing of additional symbols. For this reason the Turing machine has a read-write head which is allowed a two-way movement. To provide for the writing of additional symbols we assume the Turing machine has a semi-infinite tape or file of cells. Pictorially we have Figure 6.1.1.

We will soon discover the usefulness of additional symbols which cannot appear as input symbols when the Turing machine is started up. Moreover, we need to assume that the tape is initially filled with a filler or blank symbol. We use ***B*** throughout for this purpose. We now give the formal definition of the Turing machine.

Definition A *deterministic Turing machine* (*DTM*) M is specified by a sextuple $(Q,\Sigma,\Gamma,\delta,s,f)$, where

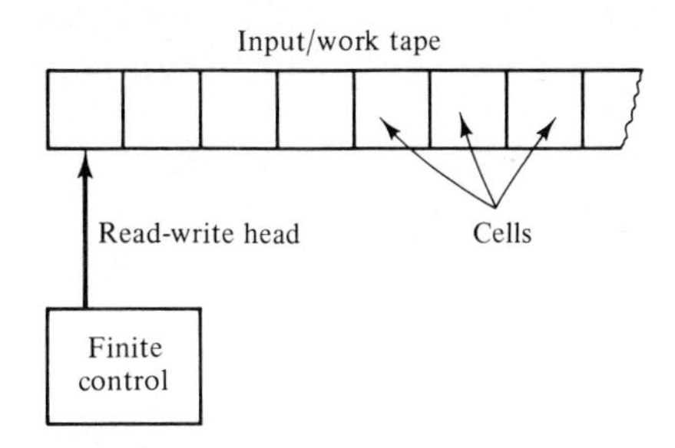

Figure 6.1.1

Q is a finite set of *states*;
Σ is an alphabet of *input* symbols;
Γ is an alphabet of *tape* symbols, where $(\Sigma \cup \{\boldsymbol{B}\}) \subseteq \Gamma$;
δ is a *transition function* $\delta : Q \times \Gamma \rightarrow Q \times \Gamma \times \{L,R,\lambda\}$, where L,R, and λ denote move left, move right, and do not move, respectively;
s in Q is a *start state*; and
f in Q is a *final state*.

A *DTM* is initialized by

(i) resetting the read-write head over the first cell of the tape;
(ii) letting the state of the *DTM* be s;
(iii) placing an input word x over Σ into the first $|x|$ cells, one symbol to a cell; and
(iv) assuming $\boldsymbol{B}$ is in all other cells.

We capture all information about the *DTMs* current situation with the notion of a *configuration*, as we did with *FAs*. A configuration is a word in $\Gamma^*Q\Gamma^*$, which denotes the current state, the position of the read-write head, and also the currently written portion of the tape. In a configuration g_1pg_2, the read-write head is assumed to be over the first symbol of g_2 if $g_2 \neq \lambda$ and over a blank cell otherwise. Strictly speaking, g_2 is either λ or a word in $\Gamma^*(\Gamma - \{\boldsymbol{B}\})$; there are no trailing blanks. If Γ and Q have elements in common, it is possible to have configurations that contain more than one state. Such a configuration does not identify a unique situation in the *DTM*, so we forbid this by assuming $\Gamma \cap Q = \varnothing$, throughout.

In any configuration g_1pg_2, M proceeds in one of three ways:

(a) If $p = f$, then M *halts* and, by definition, no move is possible. This is the reason we only require one final state.
(b) If $g_2 = \lambda$, then the read-write head is over a blank cell. In this case, if $\delta(p,\boldsymbol{B}) = (q,X,D)$, then M enters state q, rewriting the blank cell with X and moving one cell to the left or right or remaining where it is, depending on D.

(c) If $g_2 = ag_3$, where a is in Γ, g_3 is in Γ^*, and $\delta(p,a) = (q,X,D)$, then a is rewritten as X, M enters state q, and the read-write head moves according to D.

(d) In (b) and (c) if $\delta(p,\mathbf{B})$ or $\delta(p,a)$ is undefined, then M *hangs*, that is, no move is possible.

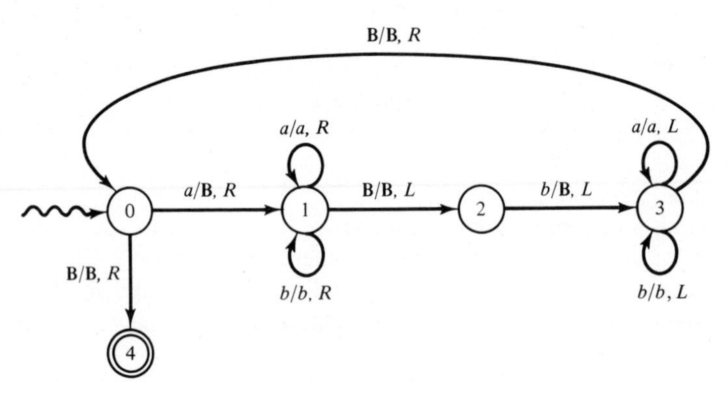

Figure 6.1.2

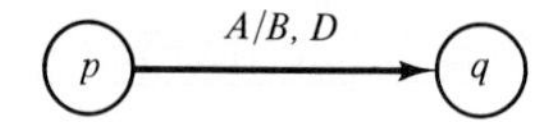

Figure 6.1.3

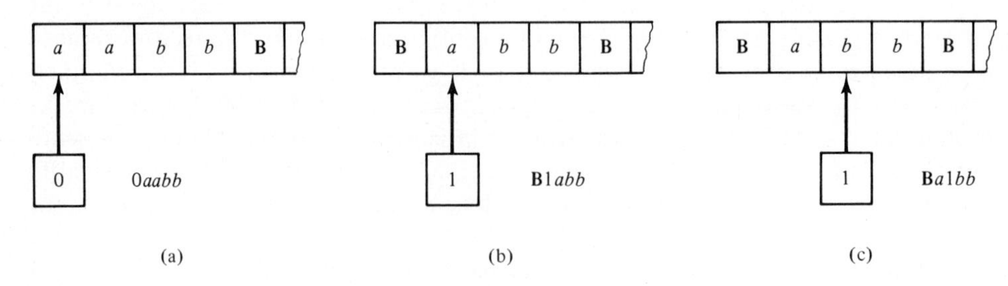

Figure 6.1.4

Example 6.1.1 Let M be specified by the Turing state diagram of Figure 6.1.2. We use the convention that the labeled directed edge shown in Figure 6.1.3 indicates that (q,B,D) is in $\delta(p,A)$, that is, move in direction D after rewriting A as B. A slash (/), is used to separate the current symbol from the next symbol and the direction of movement. In this example $Q = \{0,1,2,3,4\}$, $\Sigma = \{a,b\}$, $\Gamma = \{a,b,\boldsymbol{B}\}$, $s = 0$, and $f = 4$.

Let's see what happens with M when given $aabb$. Initially, we have the configuration $0aabb$ shown in Figure 6.1.4(a). Now $\delta(0,a) = (1,\boldsymbol{B},R)$ hence we obtain Figure 6.1.4(b), and $\delta(1,a) = (1,a,R)$, so we have Figure 6.1.4(c). Now $\delta(1,b) = (1,b,R)$ so we have two similar moves shown in Figures 6.1.5(a) and (b), but $\delta(1,\boldsymbol{B}) = (2,\boldsymbol{B},L)$, giving Figure 6.1.5(c). Continuing, $\delta(2,b) = (3,\boldsymbol{B},L)$, $\delta(3,b) = (3,b,L)$, and $\delta(3,a) = (3,a,L)$ as shown in Figures 6.1.6(a)-(c). At this stage $\delta(3,\boldsymbol{B}) = (0,\boldsymbol{B},R)$ and we have come full circle; see Figure 6.1.7(a). Since $\delta(0,a) = (1,\boldsymbol{B},R)$ we obtain Figure 6.1.7(b).

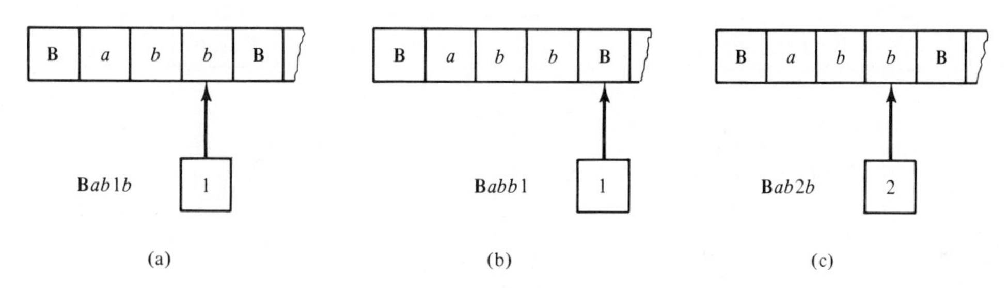

Figure 6.1.5

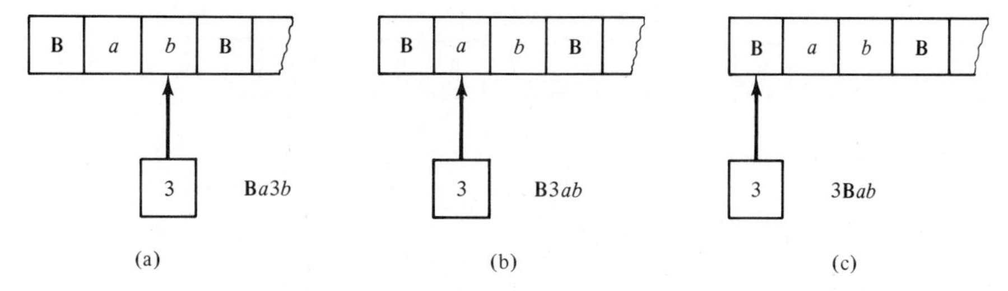

Figure 6.1.6

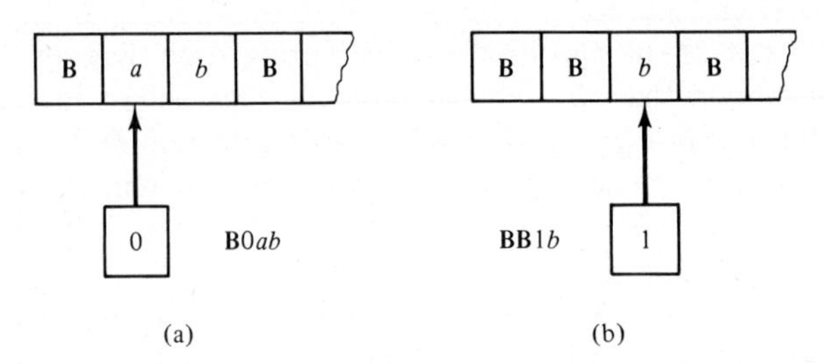

Figure 6.1.7

Now $\delta(1,b) = (1,b,R)$, $\delta(1,\boldsymbol{B}) = (2,\boldsymbol{B},L)$, and $\delta(2,b) = (3,\boldsymbol{B},L)$; see Figures 6.1.8(a)-(c). Finally, $\delta(3,\boldsymbol{B}) = (0,\boldsymbol{B},R)$, and now $\delta(0,\boldsymbol{B}) = (4,\boldsymbol{B},R)$ for the first time and, since 4 is the final state, M halts; see Figures 6.1.9(a) and (b). Hence, *aabb* is accepted.

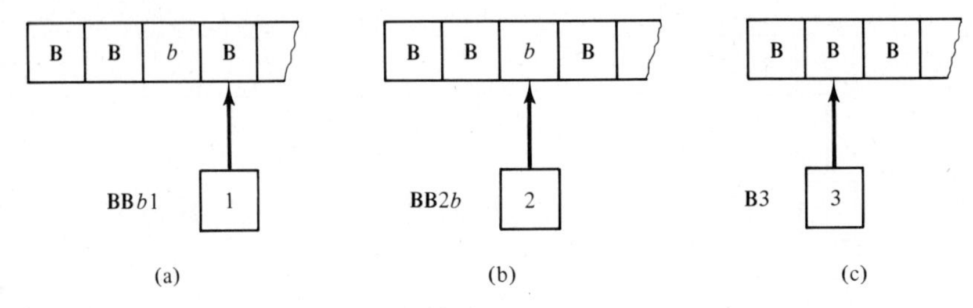

Figure 6.1.8

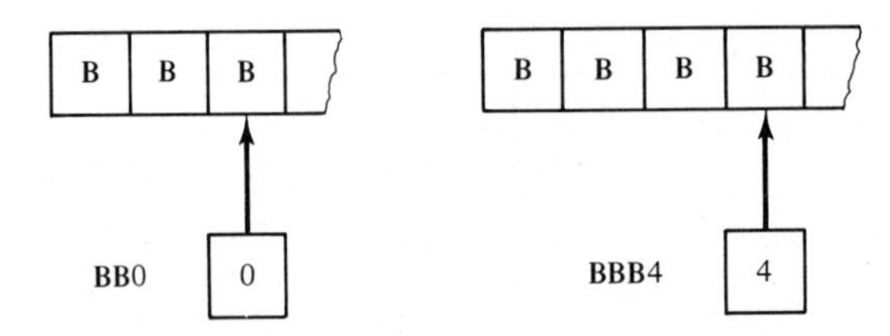

Figure 6.1.9

But with input aa we obtain Figures 6.1.10(a)-(d), and since $\delta(2,a)$ is undefined, M hangs and aa is not accepted.

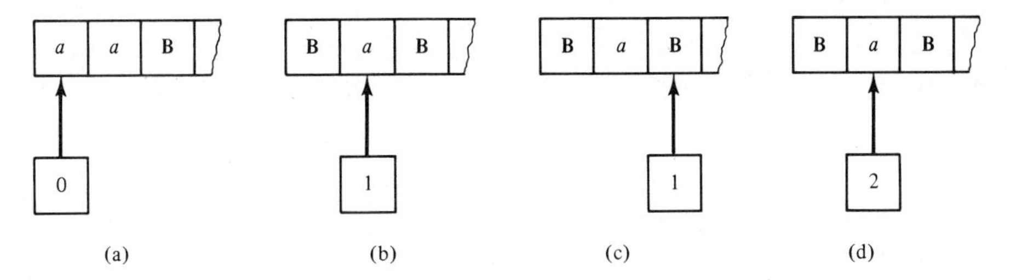

Figure 6.1.10

In the above example, we have introduced intuitively the *DTM* notions of *configuration sequence* and *acceptance*. We now give precise definitions of these notions before introducing Turing machines in their more general rôle as information processors, function computers, and decision makers.

Definition Let $M = (Q,\Sigma,\Gamma,\delta,s,f)$ be a *DTM*. Then for two configurations gph and $g'p'h'$ we write $gph \vdash g'p'h'$ if

(i) either $h = Ah_1$, for some A in Γ and h_1 in Γ^*, or $h = \lambda$ and by convention $A = \boldsymbol{B}$ and $h_1 = \lambda$;

(ii) $\delta(p,A)$ is defined, and $p \neq f$;

(iii) $\delta(p,A) = (p',B,D)$ and one of the following holds:

(a) $D = L$, then $g = g'C$ for some C in Γ, and $h' = CBh_1$;

(b) $D = \lambda$, then $g' = g$ and $h' = Bh_1$;

(c) $D = R$, than $g' = gB$ and $h' = h_1$.

Note that condition (iii)(a) implies that $g \neq \lambda$, since a move left is impossible in this case; in other words, M hangs once more.

We also write $gph \vdash^i g'p'h'$, for some $i > 0$, if either $i = 1$ and $gph \vdash g'p'h'$ or $i > 1$, $gph \vdash g''p''h''$, for some $g''h''$ in Γ^* and p'' in Q, and $g''p''h'' \vdash^{i-1} g'p'h'$. Finally, we write $gph \vdash^+ g'p'h'$ if $gph \vdash^i g'p'h'$ for some $i > 0$ and we write $gph \vdash^* g'p'h'$ if either $gph = g'p'h$ or $gph \vdash^+ g'p'h'$. These are both called *configuration sequences.*

In an *FA* the finiteness of the input together with the absence of rewriting ensures that the reading head will eventually fall off the tape to the right. However, it is possible that a *DTM* may neither hang nor halt.

Example 6.1.2 Let M be defined by the diagram of Figure 6.1.11. If the input configuration is $0a$, then we find that

$$0a \vdash a1 \vdash aa3 \vdash aaa3 \vdash aaaa3 \vdash \cdots$$

On the other hand, if the input configuration is $0aa$ we have

$$0aa \vdash a1a \vdash aa2$$

In the first case 3 is nonfinal and in the second 2 is final. We have avoided the difficulty of nonhalting on reaching a final state by specifying that a *DTM* always halts on entering a final state. This is the reason for having only *one* final state.

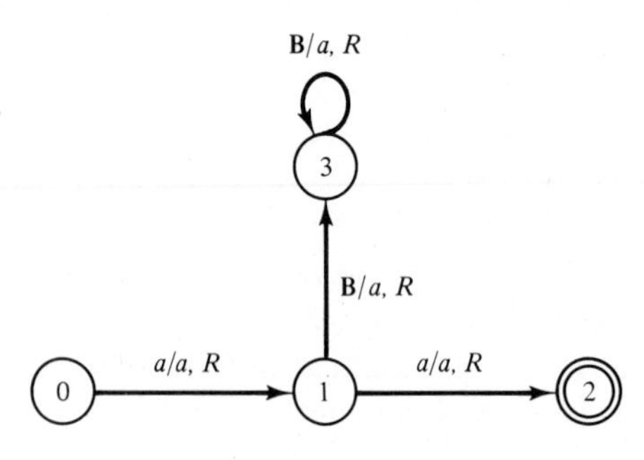

Figure 6.1.11

Definition Let $M = (Q,\Sigma,\Gamma,\delta,s,f)$ be a *DTM*. A word x in Σ^* is *accepted* by M if M terminates in a configuration gfh, when given x as input, that is,

$$sx \vdash^* gfh$$

The corresponding configuration sequence is called an *accepting configuration sequence*. Otherwise, either x causes M to hang or M never terminates.

The *language accepted by* M, denoted $L(M)$, is defined by

$$L(M) = \{x : x \text{ is in } \Sigma^* \text{ and } sx \vdash^* gfh\}.$$

We say two *DTMs*, M_1 and M_2, are *equivalent* if $L(M_1) = L(M_2)$. *The family of languages specified by all DTMs* is denoted by $\boldsymbol{L}_{DTM}$ and defined by

$$\boldsymbol{L}_{DTM} = \{ L : L = L(M) \text{ for some } DTM\ M\}.$$

A language in $\boldsymbol{L}_{DTM}$ is said to be a *deterministic Turing machine language* or a *DTML*.

Since *DTMs* were introduced as a general model for computation it is worthwhile seeing how we might program them to carry out some basic operations, rather than using them only as language recognizers.

Example 6.1.3 Program a *DTM* to shift its input word right by one cell, placing a blank in the leftmost cell.

So the input of Figure 6.1.12(a) is transformed as shown in Figure 6.1.12(b). We assume the input alphabet is $\{a, b\}$.

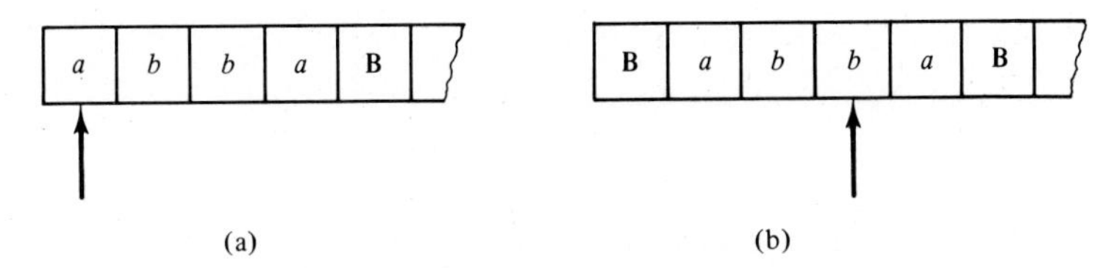

Figure 6.1.12

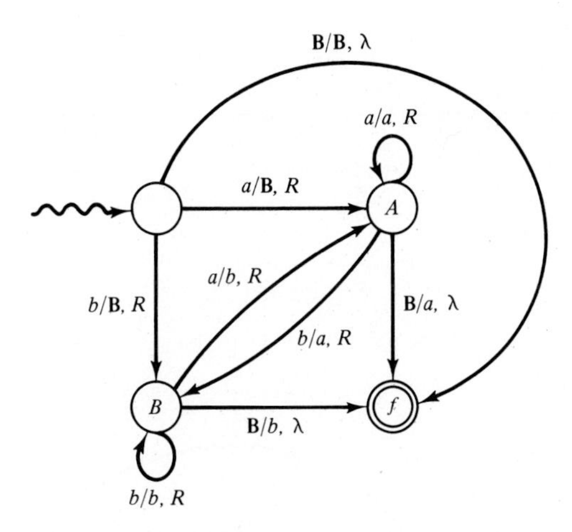

Figure 6.1.13

One possible solution is given in Figure 6.1.13. The *DTM* repeatedly reads a symbol, remembering it by moving to state A if it is an a and state B if it is a b. At the same time it overwrites the current symbol with the previous symbol, again by referring to the current state. Consider how a left shift that destroys the first symbol can be implemented.

Example 6.1.4 Program a *DTM* to shift its input word cyclically to the right by one position, that is, the input of Figure 6.1.14(a) should become the tape of Figure 6.1.14(b).

We do this by, first, shifting the input word to the right as in Figure 6.1.14(c) and, second, transferring the rightmost symbol to the first cell to give the required result. In breaking the problem down into these two simpler steps we can use the

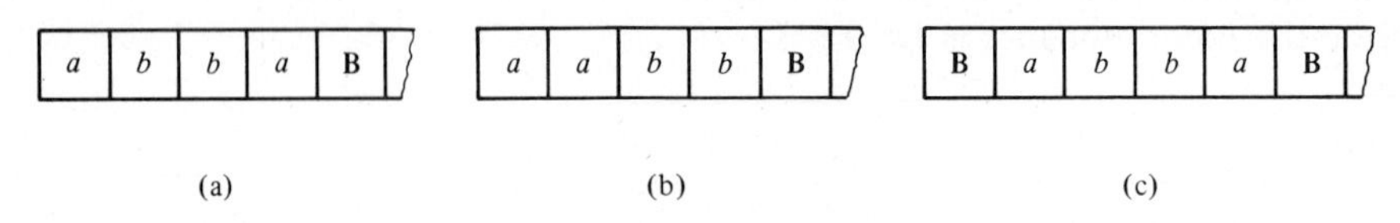

Figure 6.1.14

DTM of Example 6.1.3 as a subprogram of the present *DTM*. The right shift *DTM* leaves itself over the rightmost symbol as shown in Figure 6.1.15(a), unless the input word is λ, when it leaves itself over the first cell; see Figure 6.1.15(b). Therefore, we construct a transfer *DTM* which takes account of both situations; see Figure 6.1.16.

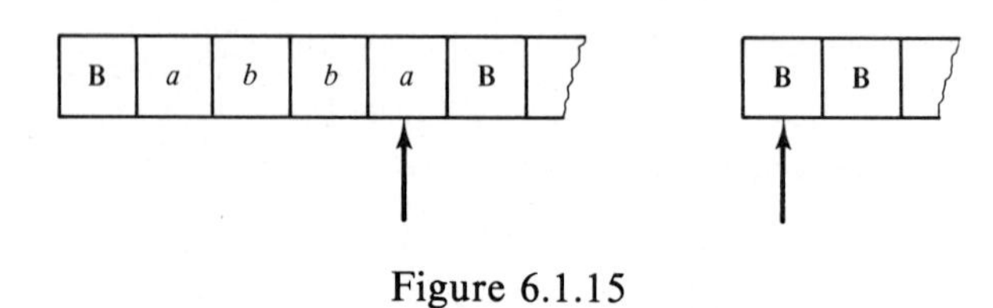

Figure 6.1.15

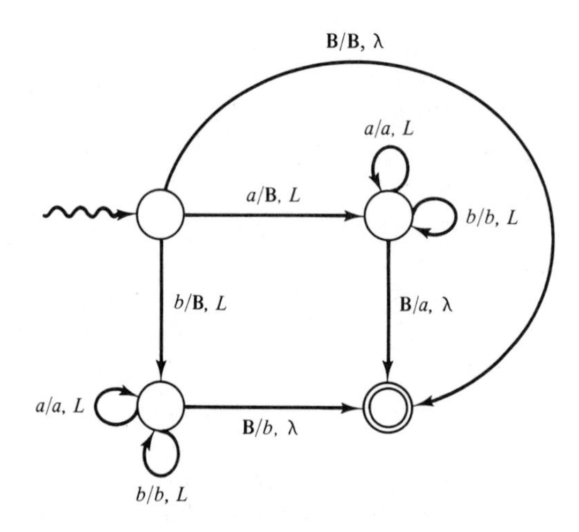

Figure 6.1.16

This DTM ends up over the leftmost cell once more. Putting them together we obtain the DTM of Figure 6.1.17.

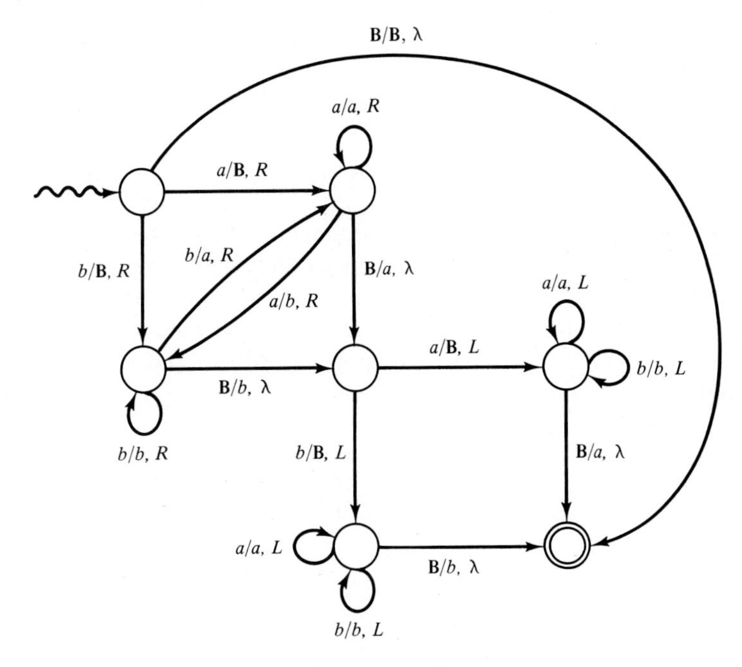

Figure 6.1.17

We now consider how $DTMs$ may be used to compute functions. We only consider one-place functions from Σ^* to $(\Gamma - \{\boldsymbol{B}\})^*$. The extension to more general functions is accomplished in Section 7.4.

Definition Let $M = (Q,\Sigma,\Gamma,\delta,s,f)$ be a DTM. Then M computes the function $f_M : \Sigma^* \to (\Gamma - \{\boldsymbol{B}\})^*$ that is defined by

For all x in Σ^*,
$f_M(x) = y$ in $(\Gamma - \{\boldsymbol{B}\})^*$ iff $sx \vdash^* y_1 f y_2$, where $y = y_1 y_2$

Example 6.1.5 Program a DTM to produce the reversal of its input word. In other words, with input *abbab* we require *babba* as output. We assume the input alphabet $\Sigma = \{a,b\}$ again, so for a DTM M we require that the function $f_M : \Sigma^* \to \Sigma^*$ satisfies $f_M(x) = x^R$, for all x in Σ^*.

The DTM we obtain mimics the PASCAL-like procedure

procedure *reversal*(A,i,j);
$\{A$ is a one-dimensional array and i and j are valid indexes into $A\}$

begin
  if $i > j$ **then** {nothing to reverse} **else**
  if $i = j$ **then** {reverse a single element} **else**
  begin $\{i < j\}$ Swap $A[i]$ and $A[j]$; *reversal*$(A, i+1, j-1)$ **end**
end

To implement swapping in a *DTM* we need to sweep back and forth along the input. This implies that the *DTM* needs to mark symbols as already swapped so it doesn't swap them again, and it needs to be able to detect the left end of the tape, so it doesn't hang.

We deal with the first problem by introducing new symbols a' and b' for this purpose. We deal with the second problem by shifting the input right one cell to leave a blank in the first cell.

Using the *DTM* of Example 6.1.3, we obtain the output of Figure 6.1.18(b) from the input of Figure 6.1.18(a). The empty word as input leads immediately to the final state, corresponding to $f_M(\lambda) = \lambda$.

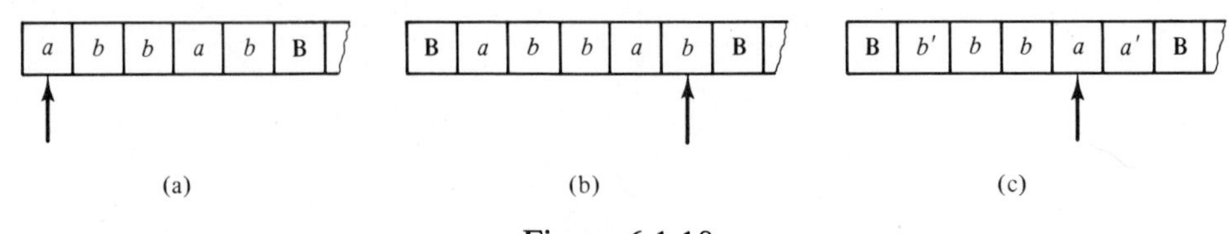

Figure 6.1.18

We first give a *DTM* to produce the output shown in Figure 6.1.18(c), namely, it swaps the outer two letters and marks them; see Figure 6.1.19. The top line of states remembers a rightmost a while scanning left and a leftmost a when scanning right. The bottom line performs similarly for b's. Observe that it treats the reversal of one letter correctly.

After swapping and marking the two outer letters we continue with a second *DTM* to swap the outermost unmarked letters. It is almost identical to the first *DTM*, see Figure 6.1.20. This *DTM* repeatedly swaps and marks pairs of letters until there are none left. It halts at two different points depending on there being an even or odd number of symbols.

After the reversal is completed the *DTM* still needs to replace the marked letters by their unmarked versions and shift the input left. We give a *DTM* to produce unmarked letters and, finally, one to shift the output left.

At this stage the read-write head is at the middle of the input; see Figure 6.1.21. First, move it to the leftmost cell, and, second, perform the replacements. We obtain the *DTM* of Figure 6.1.22. Finally, we have the left shift *DTM* of Figure 6.1.23 which is symmetric to the right shift *DTM* of Figure 6.1.13.

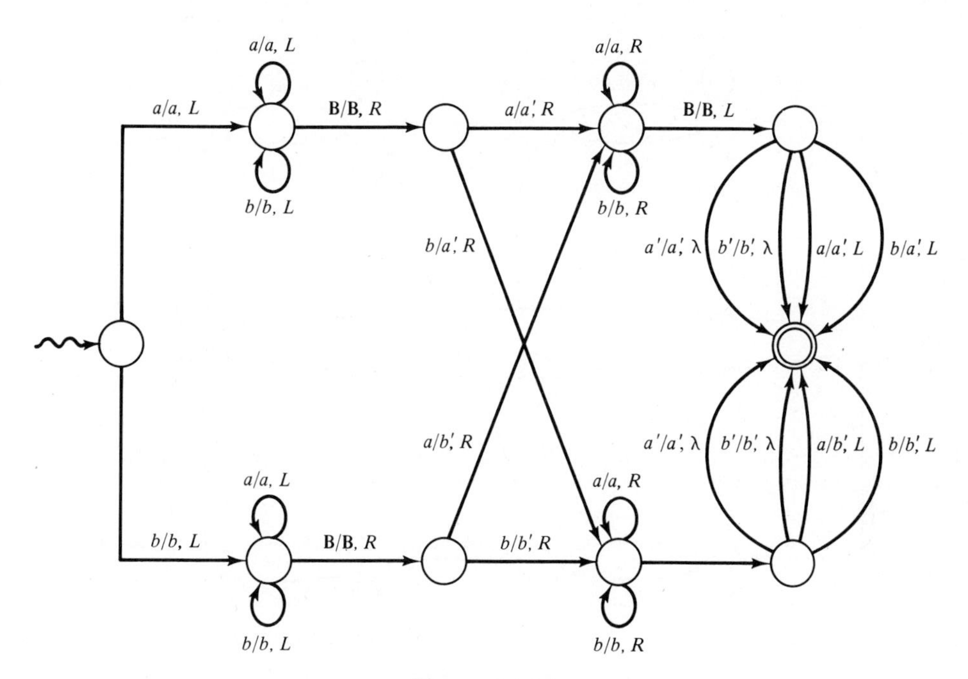

Figure 6.1.19

Example 6.1.6 Let $\Sigma = \{0,1\}$ and let w in $\{0,1\}^+$ represent a nonnegative integer in binary, where w may contain leading zeros. This means 11,011,0011, . . . , are all valid representations of 3.

Give a *DTM* which adds one to such an input word that represents a nonnegative integer in binary. For example, the input 0011 should give 0100 and the input 11 should give 100.

When w consists solely of 1's we need to shift w to the right by one cell to accommodate the leading one after the addition. So we shift w to the right whatever its value. In other words, the input of Figure 6.1.24(a) becomes the output of Figure 6.1.24(b) using the right shift *DTM* of Figure 6.1.13. We now add 1 using the *DTM* of Figure 6.1.25, after which a left shift is performed, if necessary.

The final example of this section demonstrates the use of Turing machines as decision makers.

Definition Let $\boldsymbol{y}$ and $\boldsymbol{n}$ be tape symbols representing yes and no. Then a *DTM* $M = (Q,\Sigma,\Gamma,\delta,s,f)$ is said to be a *decision-making Turing machine* if $\boldsymbol{y}$ and $\boldsymbol{n}$ are in Γ and not in Σ, and

For all x in Σ^*, either $sx \vdash^* f\boldsymbol{y}$ or $sx \vdash^* f\boldsymbol{n}$ in M.

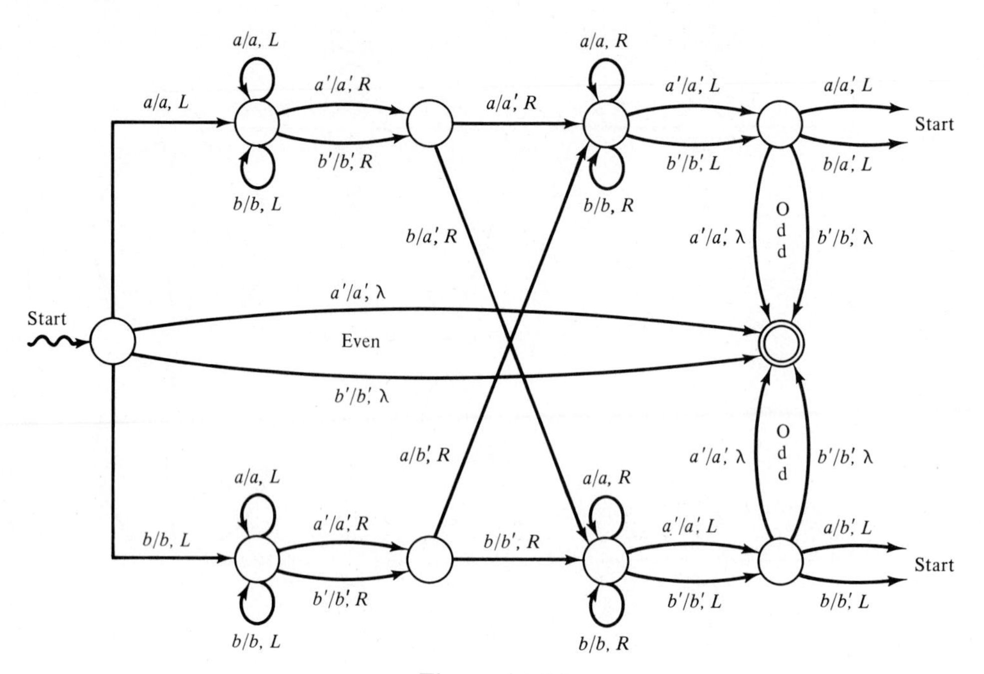

Figure 6.1.20

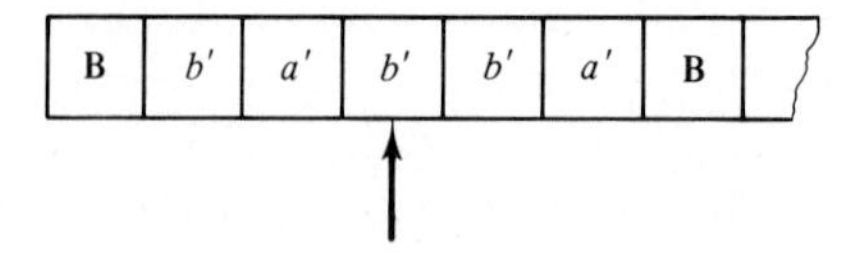

Figure 6.1.21

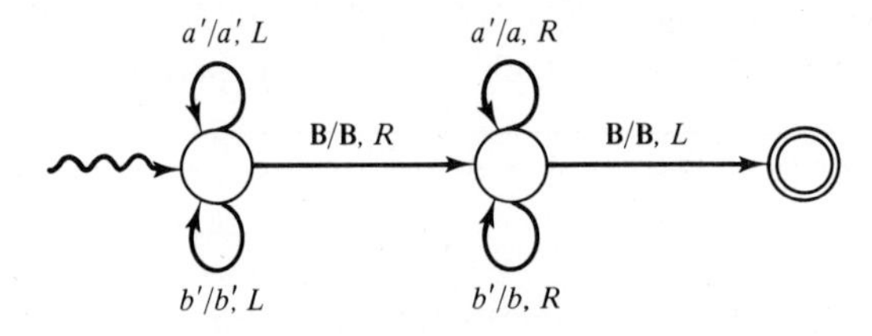

Figure 6.1.22

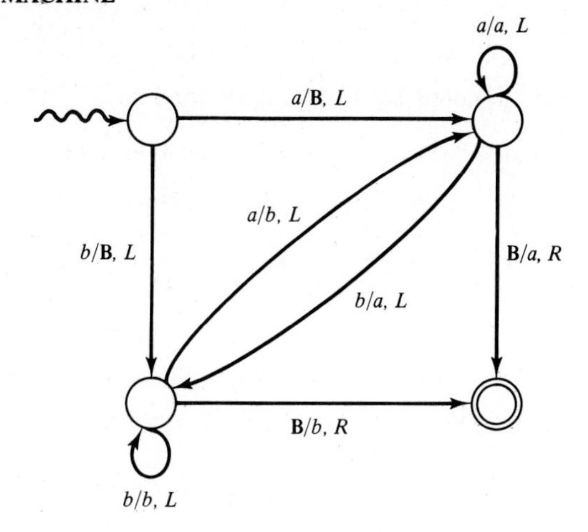

Figure 6.1.23

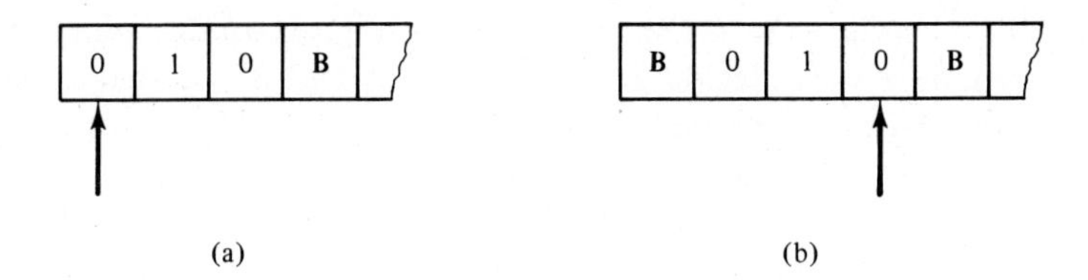

Figure 6.1.24

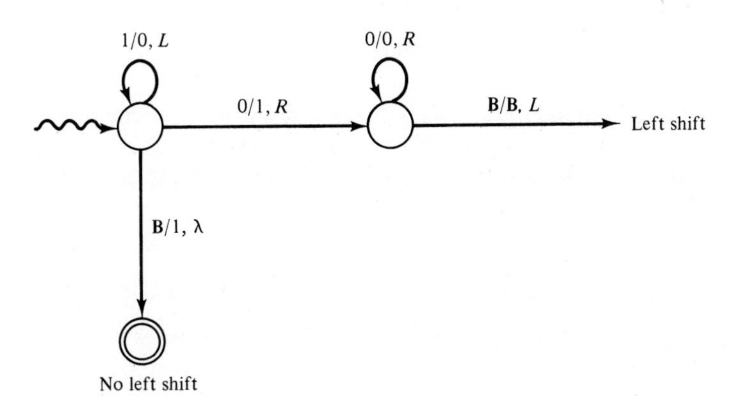

Figure 6.1.25

The *yes language of M*, denoted by $Y(M)$, is defined as

$$Y(M) = \{x : x \text{ is in } \Sigma^* \text{ and } sx \vdash^* f\boldsymbol{y}\}$$

and the *no language of M*, denoted by $N(M)$, is defined as

$$N(M) = \{x : x \text{ is in } \Sigma^* \text{ and } sx \vdash^* f\boldsymbol{n}\}$$

Such a *DTM* is a decision maker since for each word in Σ^* it always halts and gives either a yes or a no answer. Note that $L(M) = Y(M) \cup N(M) = \Sigma^*$ and $Y(M) \cap N(M) = \varnothing$. Decision-making machines will be used to model decision algorithms in Chapter 10.

Definition Given an alphabet Σ and a language $L \subseteq \Sigma^*$ we say that L is *decidable* if there is a decision-making Turing machine $M = (Q,\Sigma,\Gamma,\delta,s,f)$ with $Y(M) = L$.

Example 6.1.7 Let $\Sigma = \{a,b\}$ and $L = \{ba^ib : i \geq 0\}$. We demonstrate that L is decidable. Consider the *DTM*, M, given in Figure 6.1.26, which accepts L, but does not decide L. To decide L we have to extend M so that it halts on all input words and rewrites the tape with only a yes or no. This gives M' shown in Figure 6.1.27.

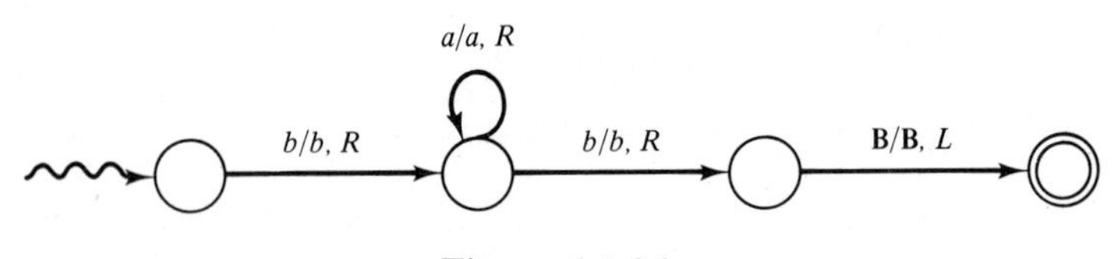

Figure 6.1.26

On the initial left-right scan we always leave a b in the first cell and replace every other b with an a. On the right-left tidy-up scan we replace every a with a blank and rewrite the only b with either $\boldsymbol{y}$ or $\boldsymbol{n}$. Thus, $Y(M') = L$ and we have $N(M') = \Sigma^* - L$ as desired.

The languages that are decidable by decision-making Turing machines form an important family — *the family of recursive languages*. This is denoted by $\boldsymbol{L}_{REC}$ and is defined by

$$\boldsymbol{L}_{REC} = \{Y(M) : M \text{ is a decision-making Turing machine}\}$$

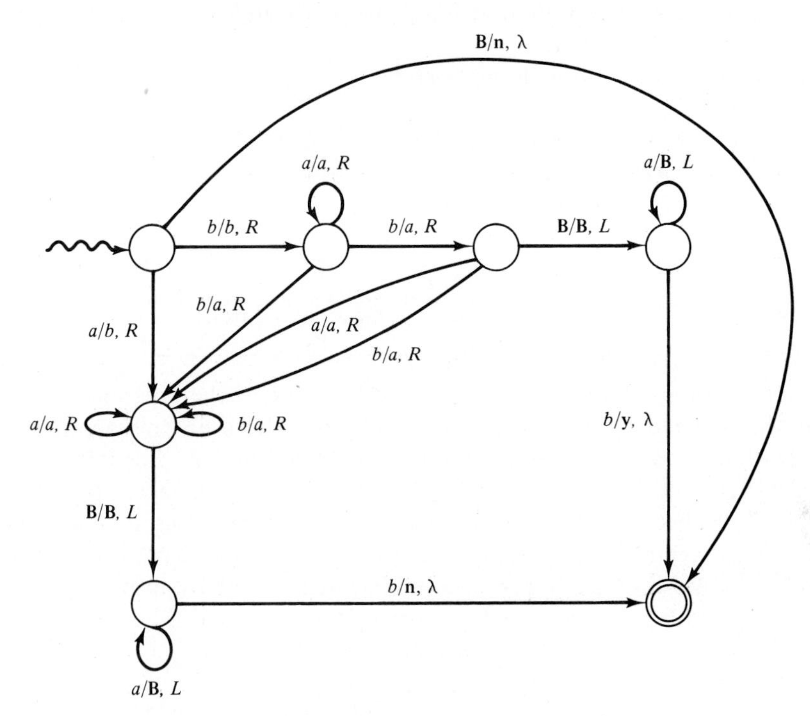

Figure 6.1.27

6.2 TURING MACHINE PROGRAMMING

Constructing the transition function of a *DTM* for a given problem is programming at its most primitive. Transitions are the machine language of *DTMs*. This is an observation which is independent of the use we have in mind, that is, language acceptance, function computation, or decision making. We will concentrate on language acceptance in this and the following sections, but all we say is equally applicable to function-computation and decision-making machines.

Constructing *DTMs* for complex examples requires discipline, otherwise the transition function will grow horrendously and opaquely. To avoid this we bring to bear some of the tools of a programmer's craft, namely, top-down design, high-level description, and modularity. For example, the state diagram or flow diagram of Example 6.1.1 could be described in the following high-level terms.

Example 6.2.1 A high-level description of the *DTM* of Example 6.1.1 can be given in pseudo-PASCAL follows.

{On entry *symbol* is the symbol in the first cell}
while *symbol* <> *blank* **do**
begin

```
    if symbol =a then rewrite it as blank and move right
                      else reject;
    while symbol <> blank do move right;
    move left;
    if symbol =b then rewrite it as blank and move left
                      else reject;
    while symbol <> blank do move left;
    move right
end {while}; accept;

{Whenever a move or rewrite is executed symbol is updated}
```

In the above example we check off, on each full scan of the tape, one a with one b. This is indicated by rewriting them as blanks. However, this approach is not always feasible — we often need to retain a nonblank symbol. For this purpose we use mnemonic symbols which consist of more than one character or mark. For example, when checking off an a we might do this by rewriting it as $a^{\dagger}$, where $a^{\dagger}$ is a new single symbol. Another approach to the same problem is to consider the tape to have more than one track. We might, with the above example, consider a two-track tape with the upper track containing the input and the lower track being used for checking off purposes; for example, see Figure 6.2.1.

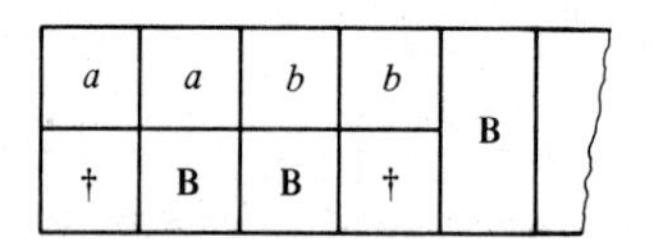

Figure 6.2.1

It must be emphasized that the idea of multitracking is conceptual rather than real. We introduce new symbols of the form $[a,\boldsymbol{B}]$ and $[a,\dagger]$ to "simulate" the twin tracking. Note that $[\boldsymbol{B},\boldsymbol{B}] = \boldsymbol{B}$, that is, blank two-track tape is just blank tape. In the following examples we illustrate these ideas. We begin with a two-track version of Example 6.1.1.

Example 6.2.2 Let $[a,\boldsymbol{B}],[b,\boldsymbol{B}]$, and $[\boldsymbol{B},\boldsymbol{B}]$ be two-track synonyms for a,b and $\boldsymbol{B}$ on a one-track tape. Introduce new symbols $[a,\dagger]$ and $[b,\dagger]$. We obtain the *DTM* of Figure 6.2.2. On input ab, M checks off the matching a and b and halts in the final state; see Figure 6.2.3.

Example 6.2.3 Given the language $L = \{a^i b^j : i \geq j \geq 1\}$, program a *DTM* to accept it.

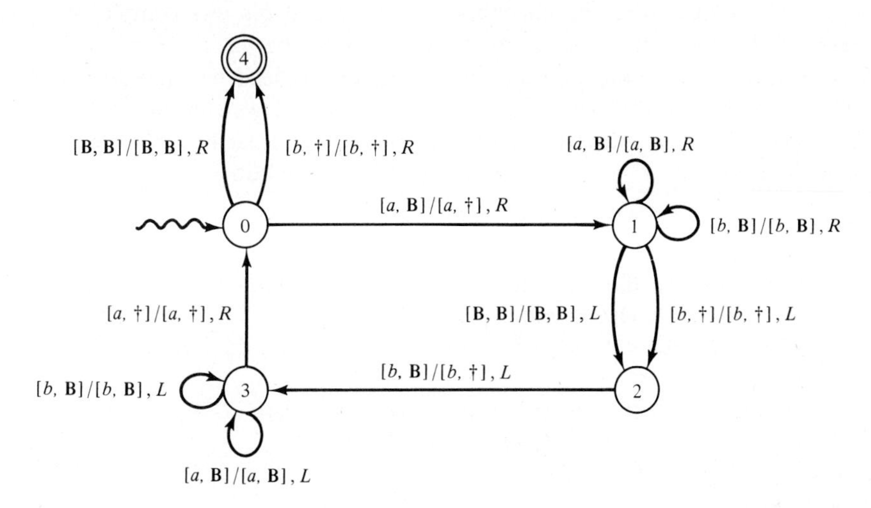

Figure 6.2.2

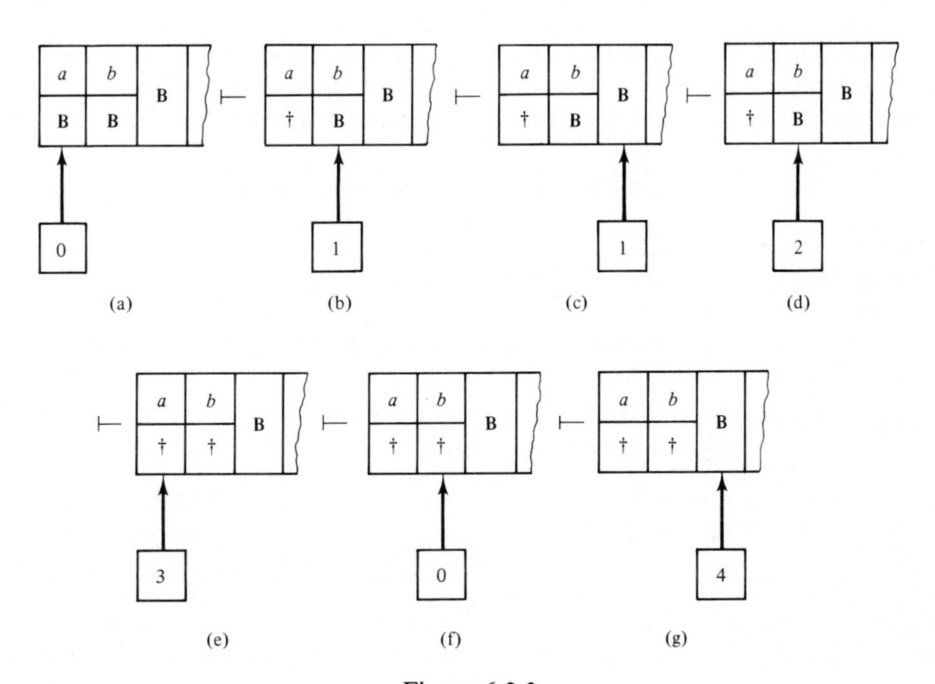

Figure 6.2.3

In Example 6.2.2 the checking off of a's and b's was completely symmetric. Here it is asymmetric: for every b there should be an a. Hence, after checking off all the a's there must be no unchecked b's. Indeed, we may check a's and b's in pairs as before with the slight adjustment that if we run out of b's the input is accepted. We also need to ensure that there is at least one b. We introduce two new tape symbols $a^{\dagger}$ and $b^{\dagger}$. So at high level we obtain

while (*symbol* <> *blank*) **and** (*symbol* <> $b^{\dagger}$) **do**
begin
    if *symbol* $=a$ **then** *rewrite* it as $a^{\dagger}$ and *move right*
                    else *reject*;
    while (*symbol* <> *blank*) **and** (*symbol* <> $b^{\dagger}$) **do** *move right*;
    move left;
    if *symbol* $=b$ **then** *rewrite* it as $b^{\dagger}$ and *move left*;
    while *symbol* <> $a^{\dagger}$ **do** *move left*;
    move right;
end {while};
if *symbol* $=b^{\dagger}$ **then** *accept* **else** *reject*;

Observe that all a's are checked even if there are no b's! Also, as in Examples 6.2.1 and 6.2.2, we carry out two distinct checks simultaneously on the input word. These are, first, to confirm that the input word belongs to a^*b^*, and, second, to ensure that there are more a's than b's. We could separate these concerns by first checking the input format and, then checking the numbers of a's and b's, see Exercise 6.2.2.

Example 6.2.4 Construct a *DTM* to accept the language $L = \{a^i b^j : i = kj$, for positive integers k and $j\}$.

Observe that if $a^i b^j$ is in L then $i \geq j \geq 1$. Thus, one approach to programming a *DTM* is to split the task into 2 stages

Stage 1: Check if the input word x is of the form $a^i b^j$ with $i \geq j \geq 1$.
Stage 2: Check that $i = kj$ for some $k \geq 1$.

Stage 1 is identical with the task of Example 6.2.3; thus we only consider Stage 2 in detail. Recall that multiplication is repeated addition and division is repeated subtraction. To check if $i = kj$ or, equivalently, i/j is an integer, we repeatedly check off j a's until the a's are exhausted. If this occurs in the middle of the checking, we reject, otherwise we accept. Since Stage 1 produces symbols $a^{\dagger}$ and $b^{\dagger}$ on the tape, this is the form of our input to Stage 2. We could check off by introducing new symbols $a^{\dagger\dagger}$ and $b^{\dagger\dagger}$, but we choose to do this by unchecking the checked symbols $a^{\dagger}$ and $b^{\dagger}$ to give a and b once more (avoiding the proliferation of † marks). Further, we need to know where the first cell is, so we further assume that the new symbols a_1 and $a_1^{\dagger}$ appear in Stage 1, but only in the first cell. Such a marking of the first cell is a standard trick. It enables us to either do a rewind to the first cell or detect that we are at the first cell. This could also be accomplished by shifting the input right initially.

Stage 1 successfully terminates with the *DTM* positioned over the leftmost $b^\dagger$. This will be the invariant position for the read-write head in the outer loop.

```
repeat  {Assume the DTM is at the leftmost b†; uncheck b's and a's}
    while (symbol <> blank) and (symbol <> a1) do
    begin
        while (symbol <> b†) and (symbol <> blank) do move right;
        if symbol = b† then
        begin rewrite it as b and move left;
            while (symbol <> a†) and (symbol <> a1†) do move left;
            {This may cause the DTM to fall off the left end of the tape
            implying i/j is not an integer}
            if symbol = a† then rewrite it as a and move right
                           else rewrite it as a1
                           {In the latter case we have exhausted
                           the a's}
        end {if}
    end {while};
    if symbol = a1 then {check that there are no checked b's}
    begin
        while (symbol <> blank) and (symbol <> b†) do move right;
        if symbol = blank then accept else reject
    end else {continue to uncheck a's}
    begin move left;
        while symbol = b do rewrite it as b† and move left;
        move right {reposition read-write head}
    end;
until forever
```

In the final example we use three tracks and checking.

Example 6.2.5 Construct a *DTM* to accept the language $L = \{a^i : i = n^2 \text{ for some } n \geq 0\}$. We assume the input is on the upper track, while on the middle track we have a candidate value of n, with $n^2 \leq i$, and on the lower track the value of n^2. These values are represented in unary as a sequence of 1's. The case $n = 0$ is dealt with separately as a special case. Now if the number of 1's in n^2 is the same as the number of a's in the input word we accept; if it is greater we reject; and otherwise we advance to the next value of n and n^2. For this purpose note that

$$(n+1)^2 = n^2+2n+1$$

so we only need add two 1's to n^2 for every 1 in n together with an extra 1, and also add an extra 1 to n. For $n = 2$, for example, we have Figure 6.2.4. Thus, our symbols are, conceptually, $[a,1,1], [a_1,1,1]$ $[a,\mathbf{B},1], [a,\mathbf{B},\mathbf{B}]$, and $[a,1^{\dagger},1]$, since we need to use checking for the addition of $2n+1$ to n. We obtain

```
if symbol = blank then accept else
    rewrite it as [a_1,1,1] and move right; {mark the first cell}
while symbol <> blank do
begin {increment n and n^2}
    {move to last 1 of n}
    while (symbol <> [a,1,1]) and (symbol <> [a_1,1,1]) do move left;
    {Now check each 1 of n in turn and add two 1's to n^2}
    while symbol <> [a_1,1†,1] do
    begin
        if symbol = [a,1,1] then rewrite it as [a,1†,1]
                                 else rewrite it as [a_1,1†,1];
        while (symbol <> blank) and (symbol <> [a,B,B]) do move right;
        if symbol = [a,B,B] then
        begin rewrite it as [a,B,1] and move right;
            if symbol = [a,B,B] then
                rewrite it as [a,B,1] and move left
                                  else reject
        end else reject;
        while (symbol <> [a,1,1]) and (symbol <> [a_1,1,1]);
            and (symbol <> [a_1,1†,1] ) do move left
    end {while};
    {Remove † marks}
    rewrite symbols as [a_1,1,1] and move right;
    while symbol = [a,1†,1] do
        rewrite it as [a,1,1] and move right;
    {symbol = [a,B,1] at this stage}
    {Having formed n^2 + 2n, now form n+1 and n^2 + 2n+1}
    rewrite it as [a,1,1] and move right;
    while (symbol <> [a,B,B]) and (symbol <> blank) do move right;
    if symbol = [a,B,B] then
        rewrite it as [a,B,1] and move right;
                      else reject
end {while}; accept.
```

In the Exercises you are asked to express as a function the number of steps taken by this Turing machine for each input a^i.

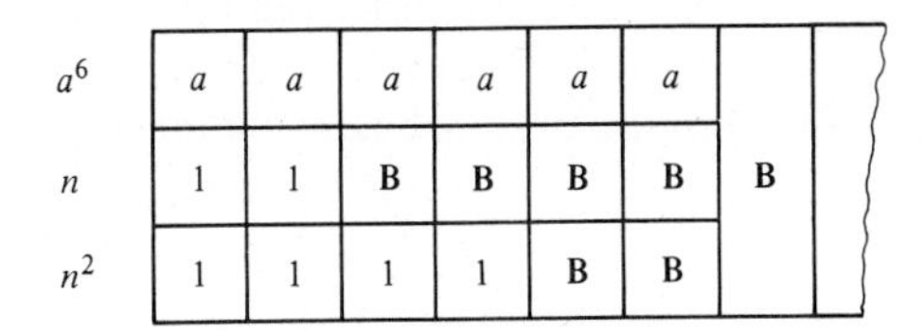

Figure 6.2.4

6.3* SIMPLIFICATIONS

In this section we consider some simplified versions of a *DTM*. We make use of the first two when constructing a universal Turing machine. The first simplification requires that the read-write head always moves; the second reduces the tape alphabet to $\Sigma \cup \{\boldsymbol{B}\}$; the third reduces the states to four; and the fourth simplification shows that every *DTM* can always be replaced by a *DTM* with read-only input and two additional highly restricted working tapes. The second and third simplifications are conflicting in that both cannot be attained simultaneously without restricting the languages accepted by *DTMs*; see the Exercises.

Theorem 6.3.1 No-Move Reduction
Let M be an arbitrary DTM. Then there exists an equivalent DTM, M', with a transition function δ' satisfying $\delta' : Q \times \Gamma \rightarrow Q \times \Gamma \times \{L, R\}$.

Proof: Split no-move transitions into two transitions with a new unique intermediate state. Let the first transition move right, the second move left. A detailed proof that the new *DTM* simulates the original *DTM* is left to the Exercises. ■

Theorem 6.3.2 Tape Symbol Reduction
Let $M = (Q,\Sigma,\Gamma,\delta,s,f)$ be an arbitrary DTM with $\#\Sigma \geq 2$. Then there exists an equivalent DTM, $M' = (Q',\Sigma,\Sigma \cup \{\boldsymbol{B}\},\delta',s',f')$.

Proof: We are given a *DTM*, $M = (Q,\Sigma,\Gamma,\delta,s,f)$, with $\#\Sigma \geq 2$. Let $2^{k-1} \leq \Gamma < 2^k$; then we can represent or encode each tape symbol as a string of k bits, that is, as a binary number. Since $\#\Sigma \geq 2$ we use two of the input symbols to represent 0 and 1. For simplicity of presentation assume 0 and 1 are actually in Σ. Let $code(a)$ be the k bit encoding of a in Γ.

Now since we have no restriction on the number of states we increase their number tremendously in order to simulate the original *DTM* M by M'. We only sketch the construction of M'; a worked example is to be found in the Exercises.

Initially, we have some input word $a_1 \cdots a_n$ in Σ^* on the tape of M'. We must first replace it by $code(a_1) \cdots code(a_n)$. To do this we move all symbols apart from a_1 $k-1$ cells to the right. Then we move all symbols apart from a_1 and a_2 $k-1$ cells to the right, and so on. This shifting is carried out by remembering in the states the contents of the previous $k-1$ cells. A final scan is carried out to replace each a_i by $code(a_i)$, resetting the read-write head to the first cell and moving into a state corresponding to s.

The central part of the simulation now begins, namely, the repeated simulation of an individual transition in M. We assume, as an invariant of this simulation, that

(i) the read-write head is positioned on the leftmost bit of an encoding of some tape symbol a in Γ;
(ii) the state of M' corresponds to the appropriate state p of M; and
(iii) the position of the read-write head over $code(a)$, corresponds to the position of the read-write head over a on the tape of M (divided by k plus 1).

Consider a generic transition of M, $\delta(p,a) = (q,b,X)$. We now demonstrate how this transition is simulated in M'. Without loss of generality we can assume X is either L or R by Theorem 6.3.1. In M' we have:

(a) read the first $k-1$ bits of $code(a)$ on the tape, remembering both p and the first $k-1$ bits of $code(a)$ in its state.
(b) read the kth bit of $code(a)$, identifying the bit sequence as $code(a)$, and simulate the transition:
    (i) remember q and X;
    (ii) replace $code(a)$ with $code(b)$ (from left to right);
    (iii) move the read-write head $2k$ cells to the left if $X = L$; otherwise it is correctly positioned;
    (iv) ensure that the new state of M' corresponds to q.

M' is now in a position to continue the simulation of the next transition of M, unless $q = f$ when M' halts.

One final remark. It is possible that during the simulation of M that M' meets $\boldsymbol{B}$ rather than $code(\boldsymbol{B})$. This means that M' has moved over a cell which hasn't been previously written, that is, it is to the right of the encoding of the input word. Before M' continues, it first writes $code(\boldsymbol{B})$ in the k cells to the right including the one currently under the read-write head. Then it is repositioned on the leftmost of these and the simulation continues as before. ■

When $\#\Sigma = 1$ we can still carry out the above simulation by using one new symbol. Thus, we obtain:

Theorem 6.3.3 *Let $L \subseteq \Sigma^*$ be an arbitrary DTM language. Then there exists a DTM $M = (Q,\Sigma,\Gamma,\delta,s,f)$ with $L(M) = L$ such that either $\Gamma = \Sigma \cup \{\boldsymbol{B}\}$ if $\#\Sigma \geq 2$, or $\Gamma = \Sigma \cup \{c\} \cup \{\boldsymbol{B}\}$, for some new symbol c, otherwise.*

We now consider the more difficult task, namely, reducing the number of states.

Theorem 6.3.4 State Reduction
Let $M = (Q,\Sigma,\Gamma,\delta,s,f)$ be an arbitrary DTM. Then there exists an equivalent four-state DTM M', that is, each DTML can be accepted by some four-state DTM.

Proof: In Theorem 6.3.2 we traded tape symbols for states. In the present theorem we must implement the converse, that is, trade states for tape symbols. The replacement of states by tape symbols is, conceptually, more difficult than the converse. (At this point, consider any specific *DTM* with more than three states and attempt to find an equivalent state-reduced *DTM*.)

Let $M = (Q,\Sigma,\Gamma,\delta,1,n)$ be a *DTM*, where $Q = \{1, \ldots, n\}$, and a transition always causes a move to left or right. We wish to simulate M with a four-state *DTM* M'. In order to do this we need to store the current state of M in the current symbol on the tape of M', for example, as a pair $[a,q]$. However, the difficulty with this approach is that M' will, in order to simulate M, move to the left or right of this cell after rewriting it in this way. In other words the current state of M will no longer be accessible to M', see Figure 6.3.1. To resolve this difficulty we transfer the current state of M to the next cell using the *bouncing* technique. M' repeatedly subtracts one from the current cell and adds one to the next cell until the current cell contains zero. We have, for example, the situation in Figure 6.3.2(a) which gives Figure 6.3.2(b) in one bounce and after 6 bounces results in Figure 6.3.2(c).

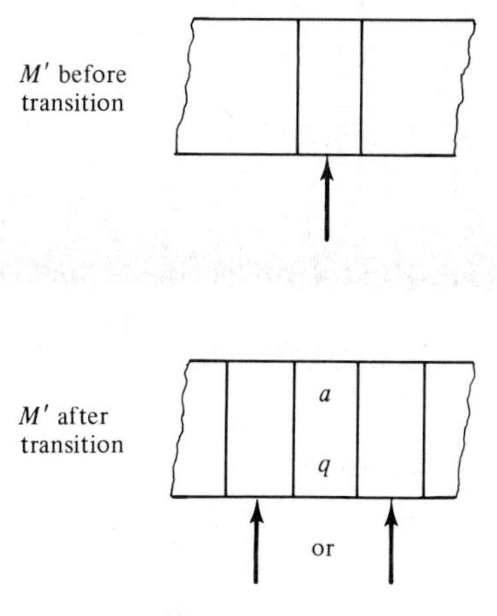

Figure 6.3.1

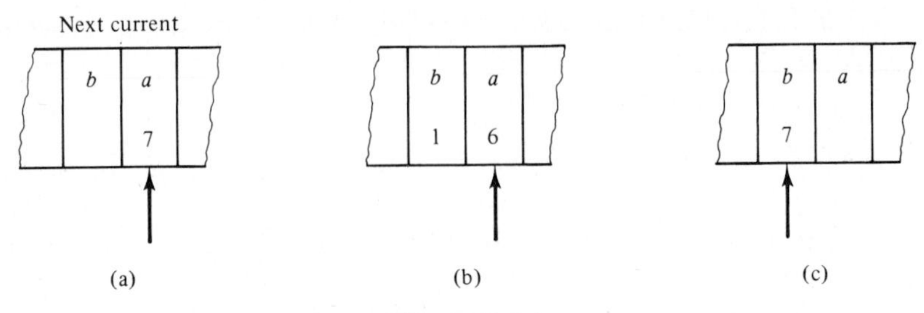

Figure 6.3.2

Because we want to keep the number of states as small as possible we also indicate in the current and next cell whether we want to add or subtract 1. So we have, for example, the situation displayed in Figure 6.3.3(a), which gives Figure 6.3.3(b), and ultimately Figure 6.3.3(c).

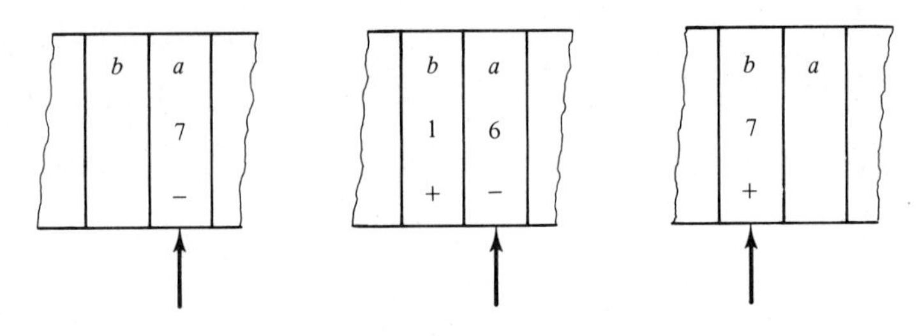

Figure 6.3.3

Thus, let $\Gamma' = \Gamma \cup \{[a,i,+],[a,i,-] : a \text{ in } \Gamma \text{ and } 0 \leq i \leq n\}$ and let us introduce four states, f, q_L, q_R, and s. f is the final state, q_L and q_R are bouncing states, bouncing to the left and to the right, respectively, and s is the start state and the state in which the simulation of a transition is commenced. Define δ' by

(1) A transition from the start state of M.
For all a in Γ, $\delta'(s,a) = (q_X,[b,i,-],X)$,
if $\delta(1,a) = (i,b,X)$ for $X = L$ or R and $i \neq n$.

(2) Start bouncing back.
For all a in Γ, $\delta'(q_X,a) = (q_X,[a,1,+],\overline{X})$,
where $\overline{X} = L$ if $X = R$ and R otherwise.

(3) Continue bouncing.
For all a in Γ, for all i, $1 \leq i \leq n$,
$\delta'(q_X,[a,i,-]) = (q_X,[a,i-1,-],X)$.

(4) Continue bouncing back.
For all a in Γ, for all i, $1 \le i < n$,
$\delta'(q_X,[a,i,+]) = (q_X,[a,i+1,+],\overline{X})$.

(5) Stop bouncing.
For all a in Γ, $\delta'(q_X,[a,0,-]) = (s,a,X)$.

(6) Begin next transition of M.
For all a in Γ, for all i, $1 \le i < n$, $\delta'(s,[a,i,+]) = (q_X,[b,j,-],X)$,
if $\delta(i,a) = (j,b,X)$ and $j \ne n$.

(7) Accept from start state of M.
For all a in Γ, $\delta'(s,a) = (f,b,X)$, if $\delta(1,a) = (n,b,X)$.

(8) Accept from nonstart state of M.
For all a in Γ, for all i, $1 \le i < n$, $\delta'(s,[a,i,+]) = (f,b,X)$,
if $\delta(i,a) = (n,b,X)$.

The transitions in 1 and 6 correspond to beginning the simulation of a single transition of M, in 1 the initial transition and in 6 the continuing transitions. The transitions in 2 and 4 define the bouncing actions to transfer the current state to the next cell, while those in 2 and 5 begin and terminate the bouncing. Finally, the transitions in 7 and 8 are accepting transitions and therefore are dealt with separately.

The detailed proof that $L(M') = L(M)$ is left to the Exercises. ■

The fourth simplification involves at the same time, the following extension.

Definition An *off-line DTM* $M = (Q,\Sigma,\Gamma,\delta,s,f)$ has two tapes, a *read-only input tape* and a *work tape*. The reading head on the read-only tape and the read-write head on the work tape operate simultaneously, but independently. The transition function δ satisfies

$$\delta : Q \times (\Sigma \cup \{\boldsymbol{l},\boldsymbol{r}\}) \times \Gamma \to Q \times \{L,R,\lambda\} \times \Gamma \times \{L,R,\lambda\}.$$

where $\boldsymbol{l}$ and $\boldsymbol{r}$ mark the leftmost and rightmost ends of the input, which is placed on the read-only tape. Because the off-line *DTM* is not allowed to write on the read-only tape the endmarkers are needed to enable the *DTM* to detect the ends of the input.

An off-line *DTM* can be simulated easily by a four-track *DTM* (see, for comparison, the simulation of multitape *DTMs* in Section 6.4.4) and, conversely, given a *DTM* an equivalent off-line *DTM* first copies the input to its work tape and then operates identically to the original *DTM* with the worktape as the *DTMs* tape. Thus, we have:

Theorem 6.3.5 *Let M be an off-line DTM. Then an equivalent DTM, M', can be constructed, and conversely.*

But why introduce such a variant in the first place? The answer is three-fold. First, an off-line *DTM* models a typical computing situation, namely, a file of data is referenced during a computation, but it isn't changed by it, for example, running a spell program against a text file. Second, we need it to obtain the promised simplification and, third, it separates the input data from the working data. This latter observation means that the working space can, in fact, be smaller than the input space, which models the situation when the input is on unlimited secondary storage, but the work space is in central storage, which is a limited resource.

We now introduce two kinds of work tapes, classified by how they are to be used.

Definition A work tape of an off-line *TM* is a *pushdown tape* if it contains the designated tape symbol $\perp$ in the leftmost cell and this cannot appear elsewhere on the tape. Moreover, a move left on the tape enforces rewriting with a blank. A work tape of an off-line *TM* is a *counter tape* if it is a pushdown tape with the additional restriction that apart from $\perp$ the remainder of the tape is blank, that is, a move right and a no move must only write blanks as well.

Definition A *two-pushdown off-line DTM* or *2-PDTM* has three tapes, an input tape and two pushdown tapes. The transition function δ satisfies

$$\delta: Q \times (\Sigma \cup \{\boldsymbol{l},\boldsymbol{r}\}) \times \Gamma \times \Gamma \rightarrow Q \times \{L,R,\lambda\} \times \Gamma \times \{L,R\} \times \Gamma \times \{L,R\}$$

A *2-PDTM* can always be simulated by an equivalent (6-track) *DTM*. We now prove that the converse also holds, that given an arbitrary *DTM*, $M = (Q,\Sigma,\Gamma,\delta,s,f)$, we can construct an equivalent *2-PDTM*, M'. M' first copies the input word, apart from its first symbol, onto the second pushdown tape in reverse and then copies the first symbol to the first pushdown tape. This transformation is illustrated in Figure 6.3.4. From hereon in M' ignores the input tape. Now M' begins to simulate the transitions of M. Observe that initially the first symbol of the input is on the top of the first pushdown tape, while the second is on the top of the second pushdown tape. The simulation is so arranged that the symbol on the top of the first pushdown is always the symbol under the read-write head of M at the corresponding move and, furthermore, the symbol on the top of the second pushdown is the symbol immediately to its right, where $\perp$ corresponds to blank in this instance. This is the invariant of the simulation. Figure 6.3.5(a) gives an example of a configuration of M, while Figure 6.3.5(b) gives the corresponding configuration of M'. It only remains to demonstrate how the invariant can be maintained at each simulated step of M.

There are two cases to consider, as follows:

Case 1: $\delta(p,c_i) = (q,c,L)$.

In M' we have

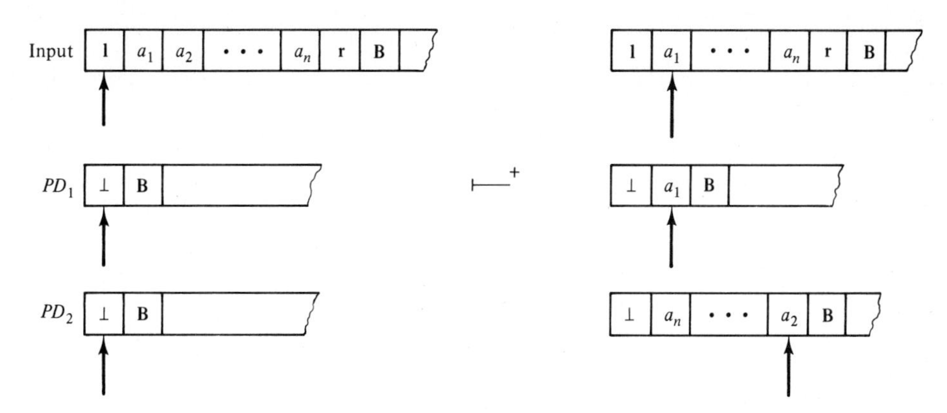

Figure 6.3.4

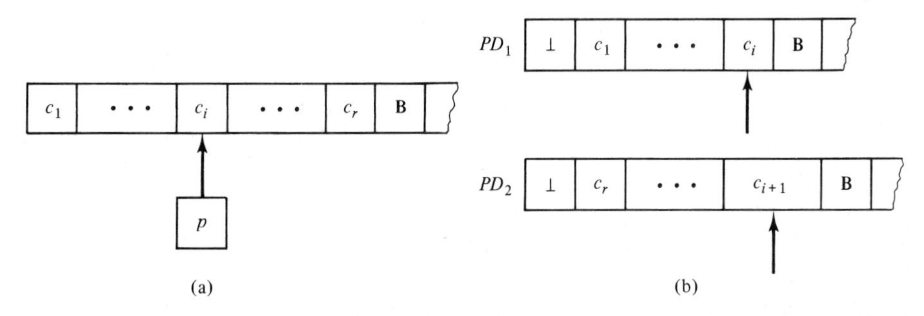

Figure 6.3.5

$$\delta'(p,a_1,c_i,c_{i+1}) = ([p,c_i],\lambda,\boldsymbol{B},L,c_{i+1},R)$$
$$\delta'([p,c_i],a_1,c_{i-1},\boldsymbol{B}) = (q,\lambda,\boldsymbol{B},\lambda,c,\lambda)$$

where $[p,c_i]$ is some new state not in Q. This results in Figure 6.3.6.

Case 2: $\delta(p,c_i) = (q,c,R)$.

In M' we have, if $i \neq r$,

$$\delta'(p,a_1,c_i,c_{i+1}) = ([p,c_i],\lambda,c,R,\boldsymbol{B},L)$$
$$\delta'([p,c_i],a_1,\boldsymbol{B},c_{i+2}) = (q,\lambda,c_{i+1},\lambda,c_{i+2},\lambda)$$

where $[p,c_i]$ is some new state not in Q. This results in Figure 6.3.7. If $i = r$, then in M' $c_{i+1} = \perp$ rather than blank, hence we obtain

$$\delta'(p,a_1,c_i,\perp) = (q,\lambda,c,R,\perp,\lambda)$$

This results in Figure 6.3.8.

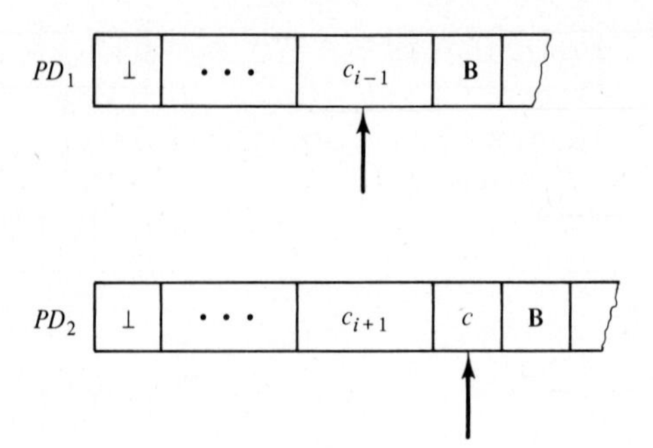

Figure 6.3.6

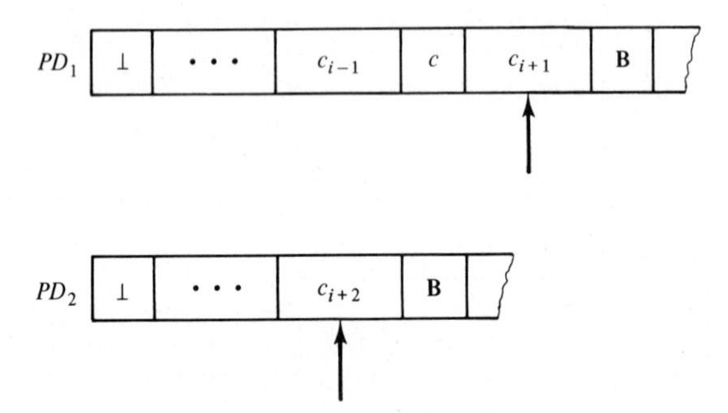

Figure 6.3.7

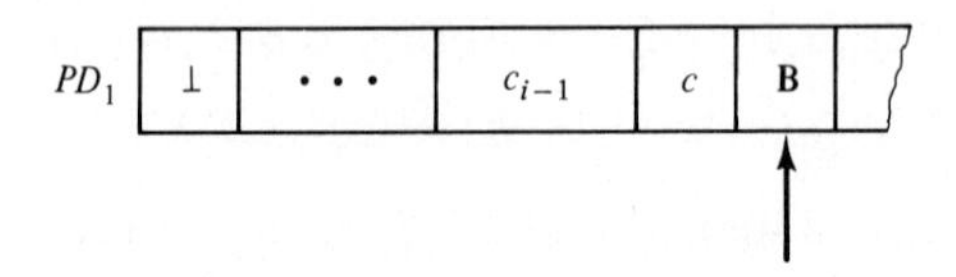

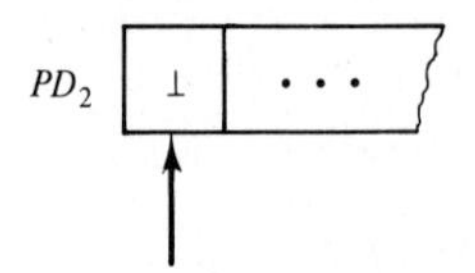

Figure 6.3.8

Note that we indeed ignore the input tape by remaining over the first symbol of it; that we prevent moving left at the bottom of both pushdowns since $\perp$ is not in the tape alphabet of M; and that we are able to keep the same states as M, and simulate a transition of M by, at most, two transitions of M', apart from the initialization phase of M'. It is also necessary to deal separately with the special case that the input word is empty. In this case the initialization of M' is as shown in Figure 6.3.9.

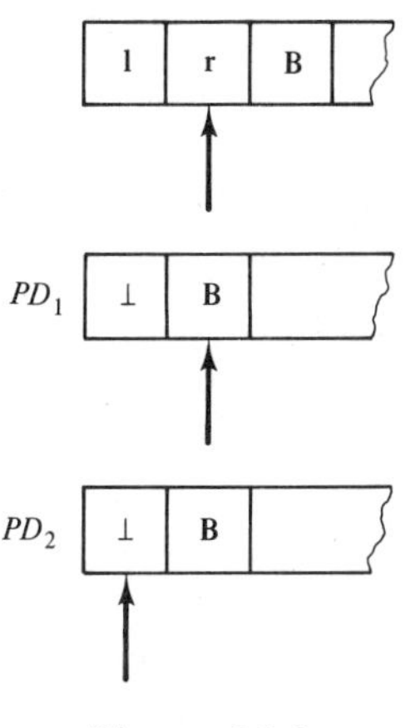

Figure 6.3.9

We have demonstrated

Theorem 6.3.6 *Let M be a DTM. Then an equivalent 2-PDTM M' can be constructed from M, and conversely.*

We now sketch how a *DTM* can be simulated by a 4-*CDTM*, an off-line *DTM* with five tapes, an input tape and four counter tapes. The resulting 4-*CDTM* simulates one pushdown of a 2-*PDTM* with two counter tapes.

Let $M = (Q,\Sigma,\Gamma,\delta,s,f)$ be a 2-*PDTM*, where $\#\Gamma = k$. The basic idea in the counter simulation is that, first, a counter tape represents an integer expressed in unary, that is, Figure 6.3.10(a) represents $n \geq 1$, while Figure 6.3.10(b) represents zero. Second, a pushdown tape also represents an integer, but expressed in base k; rather than in unary. Simply consider the k symbols of Γ to represent the k-ary digits $0, \ldots, k-1$. Immediately, any word over Γ represents a base k integer, for example,

$$A_m \cdots A_1$$

represents the integer $d(A_m) \times k^{m-1} + \cdots + d(A_1) \times k^0$, which we denote by $repr(A_m \cdots A_1)$, where $d(A)$ is the digit that is represented by A in Γ. For our purposes the pushdown tape C_1 of Figure 6.3.11(a) represents the integer

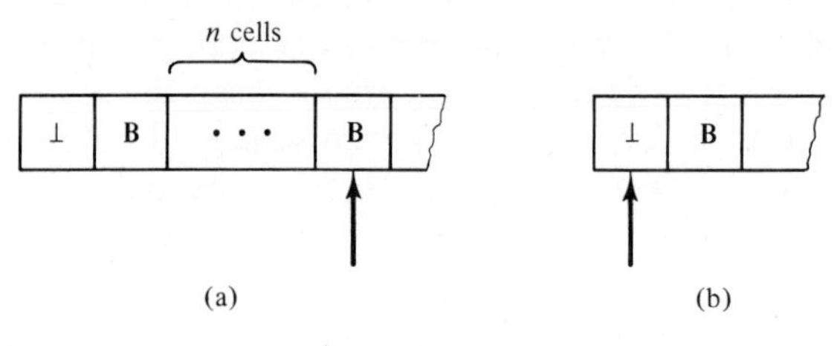

Figure 6.3.10

repr$(A_m \cdots A_1)$, where $A_m = \perp$. Thus, the least significant digit is at the top of the pushdown, the most significant, that is, $\perp$, is at the bottom of the pushdown tape. The pushdown tape of Figure 6.3.11(a) is represented by the counter tape C_1 of Figure 6.3.11(b).

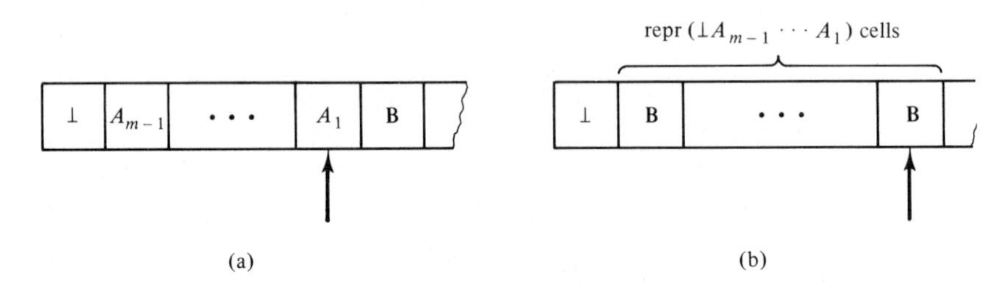

Figure 6.3.11

Now moving left on the pushdown, that is, erasing A_1, corresponds to dividing *repr*$(\perp \cdots A_1)$ by k, since *repr*$(\perp \cdots A_2) \times k + d(A_1) =$ *repr*$(\perp \cdots A_1)$. To divide C_1 by k we need a second counter tape C_2; see Figure 6.3.12(a). Now repeatedly move left on C_1 by k cells. As long as this is successful move right one cell on C_2. When a move left by k cells on C_1 is no longer possible, that is, $\perp$ is met, C_1 contains $d(A_1)$; see Figure 6.3.12(b). At this stage C_2 contains *repr*$(\perp \cdots A_2)$. Of course, to discover that only $d(A_1)$ cells remain on C_1 means that we must move left to $\perp$ and, hence, lose $d(A_1)$. However, we can retain this value in the state. So on termination we have both A_1 and the new C_1. We need A_1 when deciding which transition of M must be simulated.

Moving right on the pushdown corresponds to the case of multiplying the integer *repr*$(\perp \cdots A_1)$ by k, that is, moving right k cells on C_2 for every cell erased on C_1. To simulate a move right we must also simulate the writing of a new symbol A_0 in M [see Figure 6.3.12(c)], but this corresponds to copying C_2 to C_1 and moving $d(A_0)$ cells to the right on C_2.

We leave to the exercises the details of the simulation and the proof that:

Theorem 6.3.7 *Let M be a DTM. Then an equivalent 4-CDTM can be constructed, and conversely.*

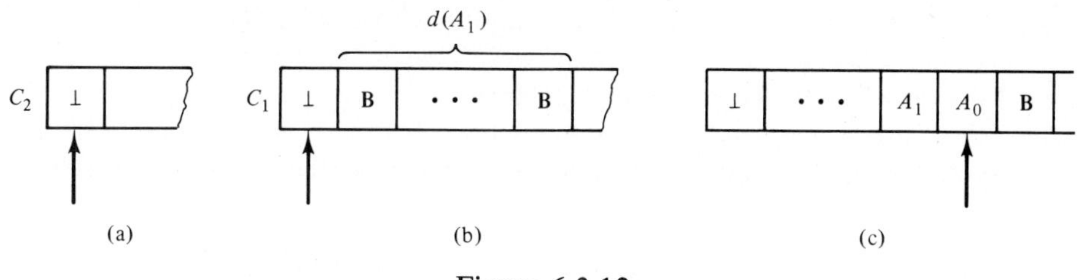

Figure 6.3.12

Finally, we sketch how a *DTM* can be simulated by a 2-*CDTM*.

In the 4-*CDTM* simulation of a *DTM* the only basic operations performed on the counters are

subtract one	(move left),
add one	(move right), and
test for zero	(over $\perp$).

These are combined to obtain integer division and multiplication, that is, the three PASCAL operations **div**, **mod**, and $*$.

Now, four counter tapes containing integers k, l, m, n can be represented by the single integer $i = 2^k 3^l 5^m 7^n$. The three basic operations can be simulated by

(1) dividing by the corresponding prime;
(2) multiplying by the corresponding prime; and
(3) attempting to divide by the corresponding prime.

The integer i can be represented on one counter tape, while the other is used while carrying out these operations.

Leaving the details once more to the Exercises, the following theorem can be obtained.

Theorem 6.3.8 *Let M be a DTM. Then an equivalent 2-CDTM can be constructed, and conversely.*

6.4 EXTENSIONS

We consider four extensions of deterministic Turing machines. Each one of them can be shown to be equivalent in accepting power to the basic model, although different in descriptive power. It is this latter property that is, useful when demonstrating the existence of a *DTM*, since we need only construct one of the extended models. If we actually require a *DTM*, then we can use a generic construction to obtain one.

In the following, we only sketch the required constructions, leaving the proofs of their correctness to the Exercises. In general, we only show how to simulate an extended *TM* by a *DTM*, since the converse simulation is straightforward.

Furthermore, there are many extensions which we leave to the Exercises; for example, nondeterministic multihead *TMs* and multidimensional *TMs*.

6.4.1 Two-Way Infinite Tape Turing Machines

All of the Turing machines considered in this chapter have a semi-infinite tape. This asymmetry is, to some extent, unpleasing, but nevertheless it is not restrictive, as we now demonstrate.

A *two-way-infinite-tape DTM*, or *2-DTM*, M, is specified a sextuple by $(Q,\Sigma,\Gamma,\delta,s,f)$ as for a *TM*. However, we assume the tape is infinite in both directions; see Figure 6.4.1. Initially, the input is placed in some contiguous cells, the remaining cells are blank, the read-write head is positioned on the leftmost symbol of the input, and the state is s. The only major difference, from our point of view, is that moving left on the leftmost symbol does not cause the 2-*DTM* to halt. 2-*DTMs* are used in the Busy Beaver Game; see Section 6.8.1.

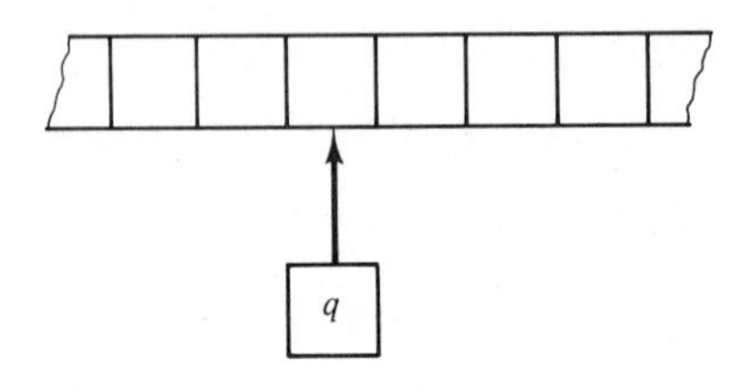

Figure 6.4.1

A *DTM* can easily be simulated by a 2-*DTM* which first marks the cell to its left with a new symbol, before initiating the original *DTM*. Now a further move left to this cell will cause the 2-*DTM* to halt, since no transitions have been added for it.

To see that the converse also holds, consider a 2-*DTM*, M, where for convenience the cells have been numbered as in Figure 6.4.2(a) and the cell containing the leftmost symbol of the input word is numbered 0. Then, conceptually, we fold the tape to give the two-track tape of Figure 6.4.2(b) and mark the lower track of cell 0 with a new symbol **l**, to indicate this is the left end of the tape. Now as long as the 2-*DTM* remains in the nonnegative cells the *DTM* remains on the upper track, and when the 2-*DTM* is in the negative cells the *DTM* is on the lower track. Observe that moving left and right in negative cells is reversed in the *DTM*. Apart from this the only tricky part of the simulation is when the 2-*DTM* moves from cell 0 to cell -1, or conversely. When the *DTM* is at cell 0 it must move right rather than left to get to cell -1. The symbol **l** only appears in cell 0 so that the *DTM* can decide if it is at cell 0. On the other hand, if the *DTM* is at cell -1 and is to move to cell 0, then this is accomplished by moving left, rather than right. But this change of direction has been incorporated into the *DTM* already.

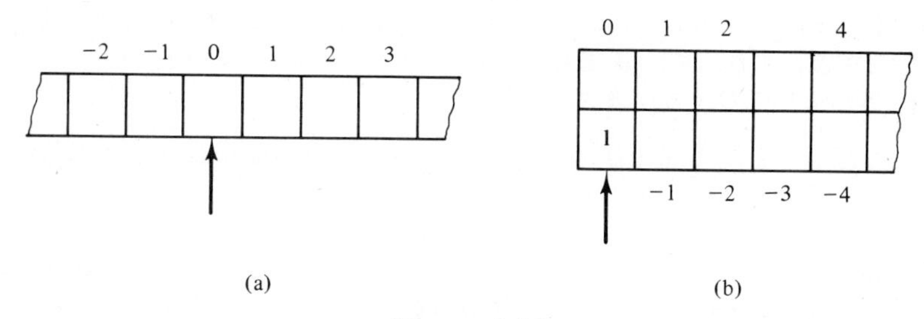

Figure 6.4.2

In more detail the *DTM*, M', simulating a 2-*DTM*, $M = (Q,\Sigma,\Gamma,\delta,s,f)$, consists of three phases

(i) Preprocess the input word x to obtain a 2-track version of it;
(ii) Simulate the configuration sequence of M on the 2-track input;
(iii) Postprocess the resulting 2-track output to give single-track output.

Let the tape symbols of M' be $[a,b]$, for all a, b in Γ, together with all symbols in Γ, and other necessary symbols. The first component is an upper-track symbol and the second a lower-track symbol. We design a *DTM* for each of these phases.

The *preprocessor DTM* consists of two simple while loops

{On entry *symbol* is the symbol in the first cell}
rewrite symbol as [symbol, ***l***] and *move right*;
while *symbol* <> *blank* **do**
    rewrite symbol as [symbol, ***B***] and *move right*;
while *symbol* <> [*?*, ***l***] **do** *move left*;

This conversion always modifies the leftmost cell, even if the input word is empty, and resets the read-write head over the first cell.

The *simulator DTM* requires that its states be structured as $[p,T]$ and $[p,B]$, where p is a state of M, T indicates M' is on the top track, and B indicates it is on the bottom track. On entry the state is $[s,T]$ and the read-write head is over the leftmost cell. The simulating transitions of M' are

(i) For all p in Q, for all a and c in Γ,
$\delta'([p,T],[a,c]) = ([q,T],[b,c],X)$, if $\delta(p,a) = (q,b,X)$.
(ii) For all p in Q, for all a and c in Γ,
$\delta'([p,B],[c,a]) = ([q,B],[c,b],\overline{X})$, if $\delta(p,a) = (q,b,X)$,
where $\overline{X} = L$ if $X = R$, $\overline{X} = R$ if $X = L$, and $\overline{X} = \lambda$ otherwise.
(iii) For all p in Q, for all a in Γ, $\delta'([p,T],[a,\boldsymbol{l}]) = ([q,X],[b,\boldsymbol{l}],Y)$,
if $\delta(p,a) = (q,b,Z)$, where $X = T$ and $Y = Z$, if $Z = \lambda$ or R, and $X = B$ and $Y = R$ if $Z = L$.

(iv) For all p in Q, for all a in Γ, $\delta'([p,B],[a,\boldsymbol{l}]) = ([p,T],[a,\boldsymbol{l}],\lambda)$

(v) For all p in Q,
$\delta'([p,T],\boldsymbol{B}) = ([p,T],[\boldsymbol{B},\boldsymbol{B}],\lambda)$
$\delta'([p,B],bb) = ([p,B],[\boldsymbol{B},\boldsymbol{B}],\lambda)$

Transitions of types (i) and (ii) are straightforward simulations of the original transitions in M, except that on the bottom track a move left becomes a move right and a move right becomes a move left. Transitions of type (iii) detect the end of the tape, and a move left becomes a move right and a shift to the bottom track. Type (iv) transitions carry out a shift from the bottom to top tracks. The only way that M' can be over the first cell reading $\boldsymbol{l}$ is when the previous transition was a move left from cell -1 on the bottom track (originally a move right). Type (v) transitions deal with the problem of entering blank cells on the negative and positive sides of the original tape. They convert a single blank into a twin-track blank without changing state.

The simulation ends when either $[f,T]$ or $[f,B]$ is entered.

The *postprocessor DTM* is not needed when only considering language acceptance. However, when computing functions we require that the output be in single-track form. We consider the case when the output is a single word. For example, as the output of the simulation we have Figure 6.4.3(a), which corresponds to Figure 6.4.3(b) except that some of the $a_{-m}, a_{-m+1}, \ldots, a_m$, may be $\boldsymbol{B}$. This implies that it is insufficient to produce Figure 6.4.3(c) as the transformed output; we must also eliminate leading blanks.

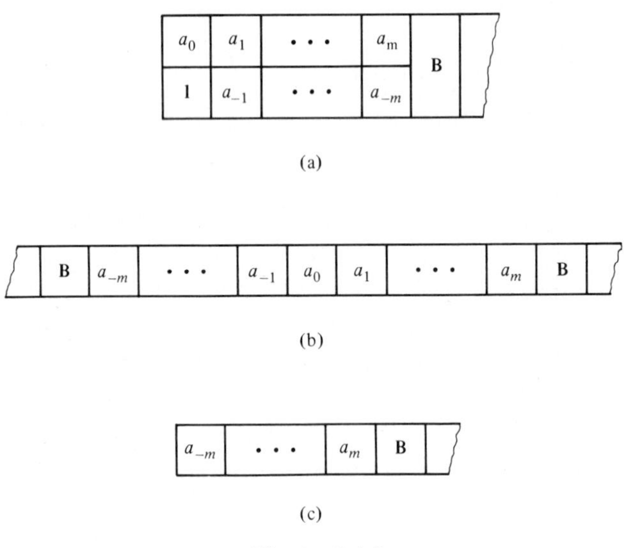

Figure 6.4.3

The transformation can be accomplished by using some basic *DTM* functions. First, call the *reversal DTM* that reverses the cells $-m$ to -1, yielding Figure 6.4.4(a) Second, call the *copy DTM* which produces Figure 6.4.4(b), that is, it squares its input word. Third, call the *transliterator DTM*, which replaces the first $m+1$ symbols by their bottom track components and the second $m+1$ symbols by their top track components. This can be viewed as two while loops

{On entry the read-write head is over the leftmost cell}
rewrite symbol as ***l*** and *move right*;
while *symbol* <> [?,***l***] **do**
    rewrite symbol as its bottom component and *move right*;
while *symbol* <> ***B*** **do**
    rewrite symbol as its top component and *move right*;
rewrite symbol as ***r***; {to mark the right end}

This yields Figure 6.4.4(c). Fourth, if the second cell contains ***B***, perform a left shift. Repeat this until a nonblank symbol is obtained in the second cell. This may be ***r***. Finally, perform a further left shift to remove ***l*** and follow this by replacing ***r*** with ***B***.

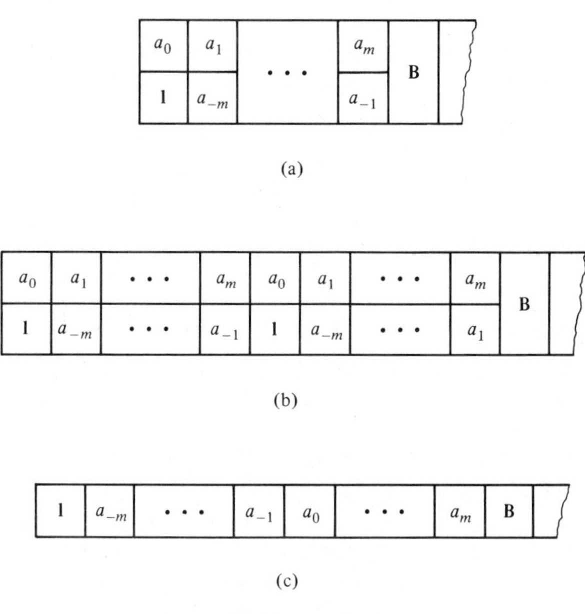

Figure 6.4.4

The operations left shift and reversal have been treated earlier and we leave the copy *DTM* to the Exercises. The remaining two operations are straightforward to implement. Thus, we have sketched the proof of:

Theorem 6.4.1 *For every 2-DTM, M, there is an equivalent DTM, M′, and conversely.*

6.4.2 Nondeterministic Turing Machines

A *nondeterministic TM* or *NTM*, $M = (Q,\Sigma,\Gamma,\delta,s,f)$, is similar to a *DTM* except that δ is a *finite transition relation* as in an *NFA*, that is, δ is a finite subset of $Q \times \Gamma \times Q \times \Gamma \times \{L,R,\lambda\}$. We may also consider δ to be a function $\delta : Q \times \Gamma \to 2^{Q \times \Gamma \times \{L,R,\lambda\}}$. Clearly an *NTM*, M, with $\#\,\delta(q,A) \leq 1$, for all q in Q and all A in Γ, is a *DTM*. As for *DTMs* we can define $\vdash, \vdash^+, \vdash^*$ between configurations, and as for *NFAs* we can define acceptance as the existence of an accepting configuration sequence. In other words, for x in Σ^*, x is accepted if and only if there is a configuration sequence of M leading to the final state f, that is, $sx \vdash^* y_1 f y_2$ in M. *NTMs* are often useful for demonstrating indirectly that, as a result of the theorem below, there is a *DTM* to accept a given language. This methodology is analogous to that used with *NFAs* and *DFA*s; see Chapter 2. To illustrate this idea, we consider an example.

Example 6.4.1 Let

$$L = \{ca^{i_1}ba^{i_2} \cdots ba^{i_n}c : n \geq 2, i_j \geq 1, 1 \leq j \leq n, \text{ and } i_j = i_{j+1} \text{ for some } j, 1 \leq j < n\}.$$

An *NTM*, M, which accepts L can be constructed as follows. We only give a high-level algorithm.

Step 1: (Left-Right Scan)
Test that the input word is of the form $ca^{i_1}ba^{i_2}b \cdots ba^{i_n}c$, for some $n \geq 2$ and $i_j \geq 1$, $1 \leq j \leq n$. This test can be carried out by a *DFA* and therefore by a *DTM* or an *NTM*.

Step 2: (Right-Left Scan)
Nondeterministically position the read-write head over some b.

Step 3: (We are now looking at a^iba^j, for some $i, j \geq 1$)
In this step decide whether $i = j$. If it does, then accept the input word. To test if $i = j$, use a marking or checking-off algorithm. Repeatedly check-off an a to the left of b and an a to the right of b, until this process cannot continue. This is because (i) there are more a's to the left than to the right; (ii) more a's to right than to the left; or (iii) an equal number of a's. Distinguishing between these three termination possibilities is straightforward.

The reader is invited to compare this nondeterministic algorithm with a deterministic one. In the latter case all adjacent pairs need to be tested.

Using the analogy of *NFAs* and *DFAs* to *NTMs* and *DTMs* we are tempted to show that every *NTM* can be transformed into an equivalent *DTM* by using the subset construction. Unfortunately, the subset construction does not work here. This is because both the rewriting of a tape symbol and the read-write head movement have to be taken into account. Notwithstanding this difficulty, an *NTM* can be simulated by a *DTM* by exploring all possible configuration sequences of a *NTM*. Since an *NTM*, M, is finitely specified, to each pair (p,a) in $Q \times \Gamma$, there corresponds a finite number of triples (q,b,X) in $Q \times \Gamma \times \{L,R,\lambda\}$. Note that every configuration sequence beginning with sx, for x in Σ^*, is uniquely determined by the sequence of triples (q,b,X) used in the configuration sequence.

Let $r \geq 1$ be the maximum number of these triples over all pairs in $Q \times \Gamma$. For each pair in $Q \times \Gamma$ number its triples from 0 to, at most, $r-1$. Now every configuration sequence beginning with sx is uniquely determined by an integer sequence, whose values are chosen from the range 0 to $r-1$. Note that the converse isn't necessarily so, since given a sequence $i_1, i_2, \ldots, a_n$, one or more values in the sequence may not correspond to any triple since not all pairs (p,a) have r triples.

To construct a *DTM*, M', to simulate M we assume M' has a 3-track tape. As in the simulation of two-way-infinite tape *DTMs* we must preprocess the standard input to obtain input in 3-track form and, after the simulation is complete, we should postprocess the 3-track form to yield single-track output. The postprocessing phase can be omitted if we are only concerned with language acceptance. The first track contains the input, the second track contains a sequence $i_1, i_2, \ldots, i_t$ of integers, each one chosen from the values 0 to $r-1$, and the third track is used to simulate M's action on the input with the sequence on the second track.

In more detail, M' first initializes track 2 to 0; see Figure 6.4.5. Then M' copies x to track 3, and simulates the one-step computation of M, given by 0, on x. If the resulting state of M is not final, then track 3 is erased, the next sequence is placed on track 2, and the simulation is repeated.

To see what is meant by *next sequence* consider a sequence to be a base r number written backward. If $r = 13$ then we could have

$$12,\ 0,\ 1,\ 4$$

which is a 4 13-ary-digit number. But as it is to be read from right-to-left rather than left-to-right, the next number in sequence is

$$0,\ 1,\ 1,\ 4$$

since $12 + 1 = 13$ which is 0 and carry 1. Clearly all possible sequences of r-ary digit numbers will be explored if we obtain the next sequence by adding one to the present sequence in this manner. We have chosen a backward representation to ensure that the number grows to the right, that is, where there is blank tape.

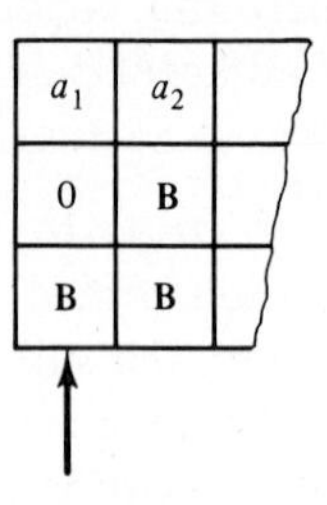

Figure 6.4.5

Finally, if x is accepted by M, then there is an integer sequence $i_1, \ldots, i_t$ which defines an accepting configuration sequence $sx \vdash^+ y_1 f y_2$. This sequence will eventually be examined by M' and hence x is accepted by M', also. Conversely if x is accepted by M' it is because there is such a sequence and hence M also accepts x. So we have:

Theorem 6.4.2 *Let M be a NTM. Then there is an equivalent DTM, M'.*

The time bounds or number of moves of M' on input x are exponential in the time bounds of M on input x. It is a long-standing open problem of whether or not there a simulation which is only polynomial in the time bounds of M. This is the well known $P = NP$ problem (see Section 6.6 and Chapter 11).

6.4.3 Multihead Turing Machines

A *k-head DTM*, $k \geq 1$, $M = (Q, \Sigma, \Gamma, \delta, s, f)$ is the same as a *DTM* except that there are k independently moving heads. This means that the transition function satisfies $\delta : Q \times \Gamma^k \rightarrow Q \times (\Gamma \times \{L, R, \lambda\})^k$ For example, for $k = 3$; see Figure 6.4.6.

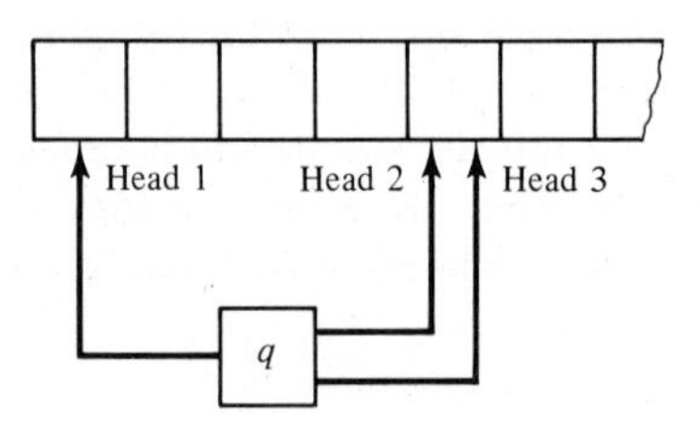

Figure 6.4.6

The extra heads enable us to keep a finger on one particular cell while other heads move around the tape. For example, with $L = \{a^i b^i : i \geq 1\}$ we can use a 2-head *DTM*, in which one head immediately goes to the rightmost b, and then the two heads advance, in synchrony, one cell at a time to the center. If they meet over ab, then the input is accepted, otherwise the input is rejected. Compare this DTM with the one in Example 6.1.1. If two or more heads are over the same cell in some configuration, then the next configuration is well defined only if each of these heads writes the same symbol in the same cell.

Now given a k-head DTM, $M = (Q,\Sigma,\Gamma,\delta,s,f)$, we can construct a 1-head DTM, M', which accepts the same language as M. We again sketch the construction, leaving the details to the Exercises. M' has $k+1$ tracks, the first corresponding to the original tape, the others corresponding to the k heads; see Figure 6.4.7. The ith head track is all blank apart from one cell which contains $\uparrow$, the position of the ith head. To simulate one transition of M, M' sweeps from the leftmost cell building up and retaining in its state the k symbols which have been found, the tape is scanned from right to left carrying the appropriate rewriting of tape symbols, and moving the $\uparrow$'s left, right or not at all. To do this the state retains the corresponding k-tuple of rewrite symbols and movements. Initially, all $\uparrow$'s appear in the leftmost positions of their tracks.

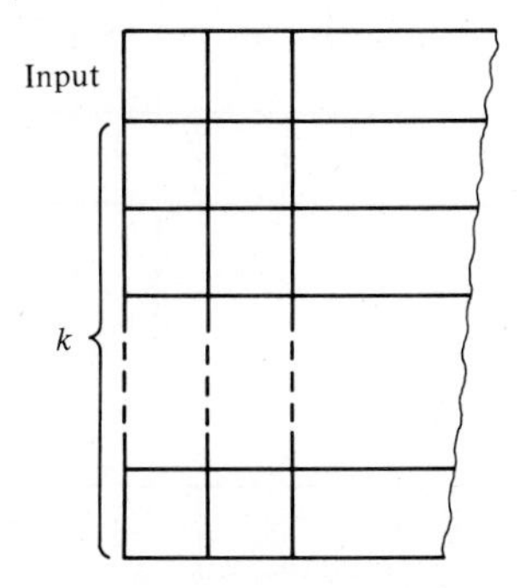

Figure 6.4.7

Not only does $L(M') = L(M)$, but also the running time of M' is polynomial if M runs in polynomial time. To see this latter claim, observe that to simulate one transition of M, M' scans the written tape twice. Now letting $T(n)$ be the worst-case running time for M on an input of size n implies that the number of written cells is at most $T(n)$. Since M' uses exactly the same space as M for the same input, one transition of M can be simulated in, at worst, $T(n)$ steps in M'.

In other words $T(n)$ steps in M can be simulated in M' in $T^2(n)$ steps. If $T(n)$ is a polynomial, then $T^2(n)$ is also a polynomial. So, to summarize, we have:

Theorem 6.4.3 *Let M be a k-head TM. Then there is an equivalent TM, M', which is polynomially space bounded if M is, and polynomially time bounded if M is.*

6.4.4 Multitape Turing Machines

A *k-tape TM*, $k \geq 1$, $M = (Q,\Sigma,\Gamma,\delta,s,f)$ is similar to a *DTM* except that there are k independent tapes each with its own read-write head. By convention the first tape is the input tape; initially all others are blank. The transition function satisfies $\delta : Q \times \Gamma^k \rightarrow Q \times (\Gamma \times \{L,R,\lambda\})^k$ which formally appears similar to the transition function for k-head *DTMs*, but pictorially is distinct; see Figure 6.4.8. However, since the k tapes can be viewed as k tracks on one tape and the k heads also as k tracks, we immediately have:

Theorem 6.4.4 *Let M be a k-tape DTM. Then there is an equivalent DTM, M', which is polynomially space bounded if M is, and polynomially time bounded if M is.*

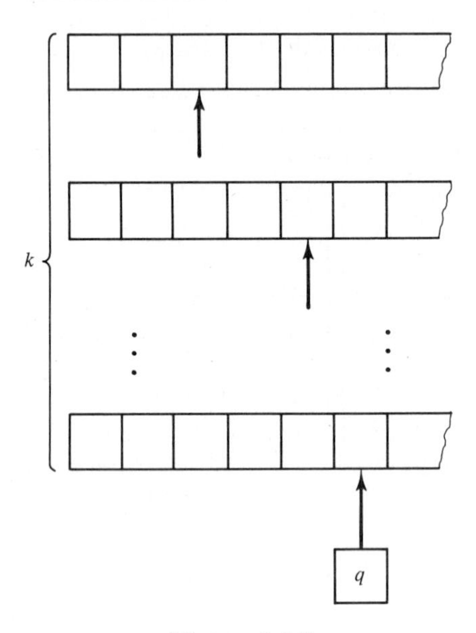

Figure 6.4.8

6.5 UNIVERSAL TURING MACHINES

The Turing machines considered so far are similar to special-purpose computers, machines built to execute a single program. We now demonstrate that just as there are programmable computers that are completely general purpose or universal, so there are programmable *TMs* that are universal — the ultimate personal computer! The basic idea is simple, just as we usually provide both a program and its data to a computer, we attempt to construct a *DTM*, U, which will take both the description of some *DTM*, M, and an input word x for M and simulate the action of M on x; see Figure 6.5.1.

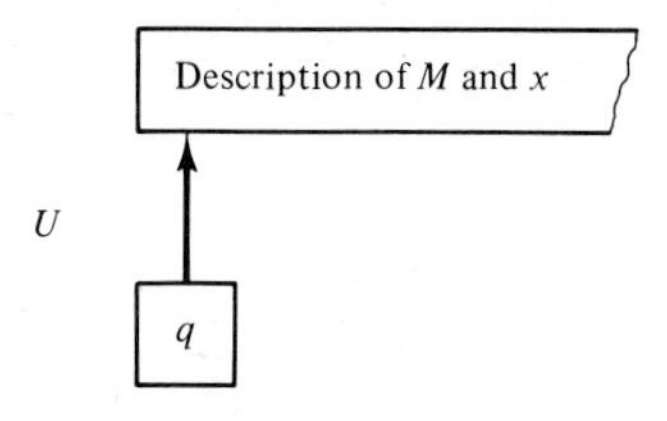

Figure 6.5.1

However, this simple idea does not mean that such U exist. (Indeed, there are no universal finite automata, for example.) We have to demonstrate their existence. To do this we need to consider, in somewhat more detail, what is meant by a description of M. Since a description appears on the tape of U it must, clearly, be a sequence of symbols, just as a PASCAL program must be transformed into a sequence of bits before it can be compiled and executed.

First, we note that U is a specific *TM* and, because of this, its tape alphabet is fixed in advance. This means that U cannot be programmed for all *DTMs*, but only for those *DTMs* whose input symbols are also input symbols of U.

Second, we recognize that every *DTM* with input alphabet $\{a,b\}$, say, is equivalent to some *DTM* with the same input alphabet and a tape alphabet equal to $\{a,b,\boldsymbol{B}\}$ by Theorem 6.3.2. We also assume $Q = \{0, \ldots, n\}$, where 0 is the start state and 1 is the final state. This is no loss of generality, since we can always rename the states to satisfy this condition. Now, such a *DTM*, $M = (Q,\{a,b\},\{a,b,\boldsymbol{B}\},\delta,0,1)$, can be described, without loss of generality, in coded form, as a word, denoted by <M>

$$[\delta(code(i_1),A_{i_1}) = (code(i_2),A_{i_2},D_{i_2}); \cdots] \tag{6.5.1}$$

where $code(i)$ is an encoding of state i; A_{i_1} and A_{i_2} are the symbols a,b, or B, where B denotes $\boldsymbol{B}$, since $\boldsymbol{B}$ cannot appear in an input word; and D_{i_2} is either L or R. Consider a state i, $0 \leq i \leq n$; then i can be encoded by

$$a^{i+1}\ ,\ i \geq 0 \tag{6.5.2}$$

This method of encoding or describing a *DTM* provides for unambiguous decoding. Clearly, there are words over $\{[,],a,b,B,L,R,;,(,),\delta,=\}$ which are not <M> for any Turing machine M. It is instructive to construct a *DTM* to check whether or not a word over this alphabet is or is not the encoding of some *DTM*.

Example 6.5.1A Let $M = (\{0,1,2\},\{a,b\},\{a,b,\boldsymbol{B}\},\delta,0,1)$ where

$$\delta(0,\boldsymbol{B}) = (2,\boldsymbol{B},R) \qquad \delta(2,\boldsymbol{B}) = (2,\boldsymbol{B},R)$$

$$\delta(0,a) = (1,b,R) \qquad \delta(2,a) = (2,b,R)$$
$$\delta(0,b) = (1,b,R) \qquad \delta(2,b) = (2,b,R)$$

Then $\delta(0,a) = (1,b,R)$ becomes, in encoded form,

$$\delta(a,a) = (aa,b,R)$$

and $\delta(2,\boldsymbol{B}) = (2,\boldsymbol{B},R)$ becomes

$$\delta(aaa,B) = (aaa,B,R)$$

Thus, M can be encoded as

$$[\delta(a,B) = (aaa,B,R);\ \delta(a,a) = (aa,b,R);$$
$$\delta(a,b) = (aa,b,R);\ \delta(aaa,B) = (aaa,B,R);$$
$$\delta(aaa,a) = (aaa,b,R);\ \delta(aaa,b) = (aaa,b,R)]$$

or as

$$[\delta(a,b) = (aa,b,R);\ \delta(aaa,a) = (aaa,b,R);$$
$$\delta(a,B) = (aaa,B,R);\ \delta(aaa,B) = (aaa,B,R);$$
$$\delta(aaa,b) = (aaa,b,R);\ \delta(a,a) = (aa,b,R)]$$

depending on which choice of an ordering of the transitions is made.

Given a *DTM*, $M = (Q,\{a,b\},\{a,b,\boldsymbol{B}\},\delta,0,1)$, and a word x in $\{a,b\}^*$, then by $<M,x>$ we denote the word consisting of some $<M>$ followed by x. Again, because there are many different encodings of M, there are many different encodings of M and x, and we use $<M,x>$ to denote any one of them as we do for $<M>$. Since $<M>$ ends with], which appears nowhere else, it is easy to find the first symbol of x.

We are now in a position to describe a universal *DTM*, U, since the input to U will be $<M,x>$ for some *DTM*, M, and word x satisfying the conditions laid down above. We describe a three-tape *DTM* for this purpose, which by Theorem 6.4.4 can then be replaced by an equivalent *DTM*.

The first tape contains the input word x, which together with the first head is used to simulate the tape of M. The second tape contains $<M>$ and the second head is used to find the corresponding $\delta(i,A)$, if there is one, when given a state i and tape symbol A. The third tape contains the current state in encoded form in the second cell onward. The first cell contains the left endmarker $\boldsymbol{l}$.

U's first task is to transform its input word w — see Figure 6.5.2 — so that it is in the desired three-tape form, if w is indeed $<M,x>$ for some *DTM* M and word x in $\{a,b\}^*$; see Figure 6.5.3. The most difficult step is determining that $<M>$ is the encoding of a *deterministic* rather than a *nondeterministic* *TM*. The details of this task, as with the following tasks, we leave to the Exercises.

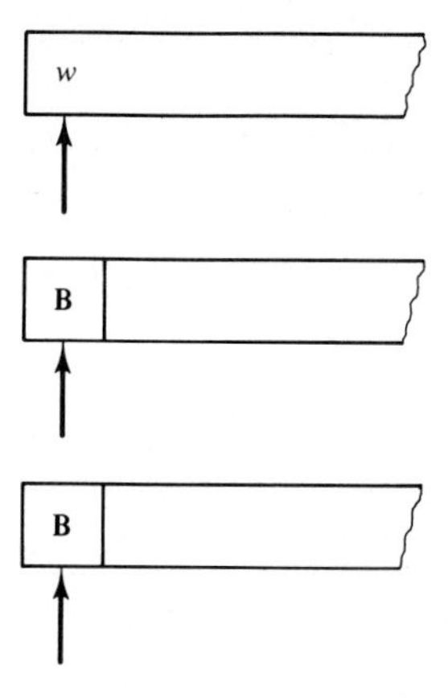

Figure 6.5.2

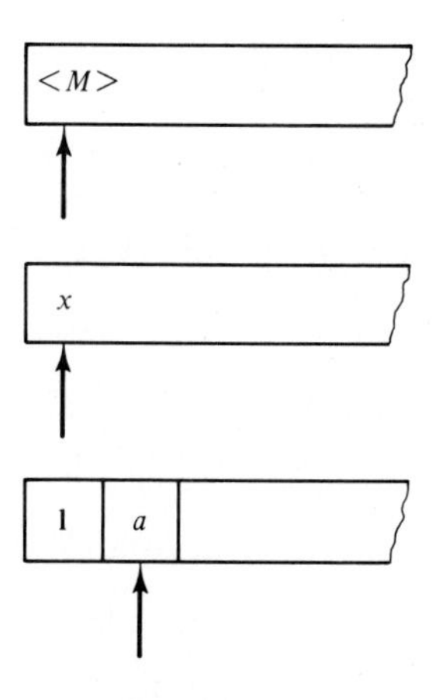

Figure 6.5.3

To simulate M on input x we maintain the invariant displayed in Figure 6.5.4. Namely, after the simulation of the ith move of M, y_1jy_2 is the configuration of M after the ith move, where $y = y_1y_2$, head 1 is over the first symbol of $y_2\boldsymbol{B}$, head 1 is over the first cell, and head 3 is over the second cell of tape 3, where tape 3 contains, in encoded form, the state j.

After the simulation of the ith move of M, track 3 is examined to see whether or not it contains aa, the final state. If not, the $(i+1)$st step of M is simulated.

Consider the simulation of one transition of M. We use *symbol* i, $i = 1,2,3$, to denote the current symbol on the ith tape, while *move* i *left*, *move* i *right*, and *rewrite* i *with*, $i = 1,2,3$ have the obvious meanings.

{Simulating one transition of M}

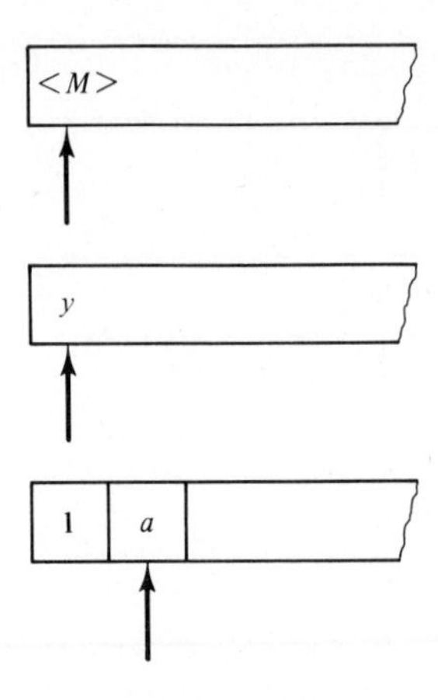

Figure 6.5.4

```
while symbol 2 <> [ do
begin move 2 right 3 cells; {skip "δ("}
        {Now compare states}
    while (symbol 2 <> ,) and (symbol 3 <> B) do
        move 2 and 3 right;
    if (symbol 2 = ,) and (symbol 3 = B) then
    {Found state — now check input symbol}
    begin move 2 right;
        if symbol 2 = symbol 1 or (symbol 2 = B and symbol 1 = B) then
        {Found match — execute transition}
        begin move 2 right 4 cells;
            reset and erase 3; {Move head 3 to cell 2 and erase a's}
            {Copy new state to tape 3}
            while symbol 2 <> , do
            begin rewrite 3 with symbol 2;
                move 2 and 3 right
            end;
            move 2 right;
            rewrite 1 with symbol 2;
            {If symbol 2 = B then symbol 1 becomes B}
            move 2 right 2 cells;
            {Move head 1}
            if symbol2 = L then move 1 left else move 1 right;
            {Transition has been simulated — maintain invariant}
            reset 3; {to second cell}
            reset 2; {to first cell}
        end {if}
    end {if};
```

```
        if symbol 2 <> ] then
        begin {Move head 2 to next transition}
            while (symbol 2 <> ;) and (symbol 2 <> ] do
                    move 2 right;
            if symbol 2 = ; then {Continue search for transition}
                reset 3 {to second cell}
                        else {No such transition — abort}
                begin reset 2; move 2 left end
        end
    end {while}
```

Note that if M ever moves left on the first cell of its first tape, then head 1 of U also does, and therefore the simulation is faithful. To complete the simulation of one step it is necessary to check whether or not the current state is final. Thus, the complete simulation can be summarized by

```
while nonfinal state do
begin
    {Simulate one transition of M, as above}
end {while};
```

Letting $U = (P,\Delta,\Delta \cup \{\boldsymbol{B}\},\delta_u,0,1)$, where $\Delta = \{a,b,B,L,R,[,],(,),\delta,=,;,,\}$, we have sketched the proof of:

Theorem 6.5.1 *Let U be the DTM defined as above. Then for all DTMs $M = (Q,\{a,b\},\{a,b,\boldsymbol{B}\},\delta,0,1)$ and words x in $\{a,b\}^*$*

x is accepted by M iff $<M,x>$ is accepted by U,

that is, U is universal with respect to $\{a,b\}$. In other words, x is in $L(M)$ if and only if $<M,x>$ is in $L(U)$.

Finally, observe that we may encode a *DTM* solely with the alphabet $\{a,b\}$. For example, let

[and] be encoded as bbb,
i as $a^{i+1}b$,
a,b, and $\boldsymbol{B}$ as ab,aab, and $aaab$, respectively,
L and R as ab and aab,
; as b, and
all other symbols as λ.

The operation of U needs to be modified slightly to accommodate these changes, with the end result that U has a tape alphabet $\{a,b,\boldsymbol{B}\}$, also. The details of the modifications are left to the Exercises. We let $<<M>>$ denote the $\{a,b\}$-encoding of M, and $<<M,x>>$ denote the $\{a,b\}$-encoding of M followed by x, respectively. This means that $<<U,x>>$ becomes a valid input for U itself, which fact we make use of in Chapters 8 and 10.

Example 6.5.1B Under this encoding scheme

$$\delta(0,\boldsymbol{B}) = (2,\boldsymbol{B},R)$$

becomes

abaaabaaabaaabaab

and one encoding $<\!\!<M>\!\!>$ of M is

bbbabaaabaaabaaabaabbababaabaabaabbabaabaabaabaab

baaabaaabaabaabbaaababaabaabaabbaaabaabaaababaabbbb

It begins with *bbb* and ends with *bbbb* and two encoded transitions are separated by *bb*. Therefore, it is straightforward to find the beginning and end of $<\!\!<M>\!\!>$ and to find the next transition.

Similarly, for any word x in $\{a,b\}^*$ the separation point of $<\!\!<M>\!\!>$ from x in $<\!\!<M,x>\!\!>$ is uniquely defined as *bbbb* when scanning it from the left.

6.6 RESOURCE-BOUNDED TURING MACHINES

The focus of this section is Turing machines which have a prespecified bound on the space and/or time available for each computation. This enables us to study the computational complexity of languages and the computational complexity of functions.

Two natural measures of the difficulty of accepting or rejecting a word on a given *TM*, M, are: (i) the time taken to accept the word and (ii) the space required during the computation. These correspond to the basic measures used for evaluating the performance of any program. For a *DTM* an abstract measure of time is the number of steps in the accepting or rejecting configuration sequence, while an abstract measure of space is the number of distinct cells visited in the configuration sequence. Since, in an *NTM*, each input word can give rise to many different configuration sequences we take the length of an accepting configuration sequence to be the number of steps and the maximum number of distinct cells visited in any of the configuration sequences to be the space needed.

We could express the time and space required by a given *TM*, M, for a given input word x as a function from words to integers. However, following tradition we express it as a function from the size of words to integers, by taking the maximum amount of the resource needed over all input words of the same size. This is called a *worst-case* measure of complexity. We now formalize these ideas, basing them on off-line *TMs* since this allows us to obtain sublinear space bounds. For off-line *TMs* the measure of the space required by an input word is the maximum number of distinct cells *of the work tape* occupied or visited over all computation sequences. This reflects the possible use of secondary storage for the input data of a program while using central storage for the actual computation.

Given an off-line *DTM*, *M*, for which each accepted word of length n causes M to visit at most $S(n)$ distinct cells, then M is said to be an $S(n)$ *space-bounded DTM*, or of *space complexity* $S(n)$. Further, a language L is said to be of *deterministic-space complexity* $S(n)$ if there is a *DTM*, *M*, such that M accepts L and M is of space complexity $S(n)$.

Similarly, if for each accepted word of length n, M makes at most $T(n)$ moves before accepting, then M is said to be a $T(n)$ *time-bounded DTM*, or of *time complexity* $T(n)$. Again a language L is said to be of *deterministic-time complexity* $T(n)$ if there is a *DTM*, *M*, such that M accepts L and M has time complexity $T(n)$.

These concepts apply to *NTMs* also. We say an *NTM*, *M*, is of space complexity $S(n)$ if each accepting configuration sequence visits no more than $S(n)$ distinct cells. Similarly, it is of time complexity $T(n)$ if each accepting configuration sequence causes it to make no more than $T(n)$ moves. We apply these concepts to languages to obtain the *nondeterministic-space* and *nondeterministic-time complexity* of a language.

Definition $\boldsymbol{L}_{DSPACE(S(n))}$ denotes the family of languages of deterministic-space complexity $S(n)$, and $\boldsymbol{L}_{DTIME(T(n))}$ the family of languages of deterministic-time complexity $T(n)$. Similarly, we obtain $\boldsymbol{L}_{NSPACE(S(n))}$ and $\boldsymbol{L}_{NTIME(T(n))}$. Of particular importance are the following four families

$$\boldsymbol{L}_{DPSPACE} = \bigcup \boldsymbol{L}_{DSPACE(S(n))}$$

$$\boldsymbol{L}_{NPSPACE} = \bigcup \boldsymbol{L}_{NSPACE(S(n))}$$

where the union in each case is over all polynomials $S(n)$ in n, and

$$\boldsymbol{L}_{DPTIME} = \bigcup \boldsymbol{L}_{DTIME(T(n))}$$

$$\boldsymbol{L}_{NPTIME} = \bigcup \boldsymbol{L}_{NTIME(T(n))}$$

where the union is once more over all polynomials $T(n)$ in n. These are called the families of *deterministic polynomial-space-bounded* languages, *nondeterministic polynomial-space-bounded* languages, *deterministic polynomial-time-bounded* languages, and *nondeterministic polynomial-time-bounded* languages, respectively.

Now, $\boldsymbol{L}_{DPSPACE} \subseteq \boldsymbol{L}_{NPSPACE}$ and $\boldsymbol{L}_{DPTIME} \subseteq \boldsymbol{L}_{NPTIME}$ by definition, since every *DTM* is a degenerate *NTM*. Furthermore, $\boldsymbol{L}_{NPTIME} \subseteq \boldsymbol{L}_{REC}$, and $\boldsymbol{L}_{NPSPACE} \subseteq \boldsymbol{L}_{REC}$, since each language in $\boldsymbol{L}_{NPTIME}$ or in $\boldsymbol{L}_{NPSPACE}$ can be shown to be accepted by an *NTM* which always terminates; see Section 8.3. Indeed, $\boldsymbol{L}_{DPTIME} \subseteq \boldsymbol{L}_{NPTIME} \subseteq \boldsymbol{L}_{DPSPACE} = \boldsymbol{L}_{NPSPACE} \subset \boldsymbol{L}_{REC}$ as we demonstrate in Section 8.3. The family $\boldsymbol{L}_{DPTIME}$ contains those languages that can be recognized efficiently, since it is generally agreed that a polynomial time bound on the input size captures the notion of efficiency appropriately. $\boldsymbol{L}_{DPTIME}$ is said, for this reason, to be the family of *tractable* languages, whereas $\boldsymbol{L}_{NPTIME}$ appears to contain *intractable languages*. It is one of the major open problems whether or not

$\boldsymbol{L}_{NPTIME} = \boldsymbol{L}_{DPTIME}$, the so-called $P = NP$ problem, since although the classes appear to deal only with languages, this is not so, as we now demonstrate.

Given a programming problem, we can always turn it into a decision problem (see the discussion in Chapter 1), and every decision problem can be encoded as a language recognition problem. For example, consider once more the optimization problem known as the Traveling Salesman Problem; see Section 1.4. In this problem, n cities $C_1, \ldots, C_n$, are given together with their positive integer distances apart $d(C_i, C_j)$, $1 \leq i,j \leq n$. Beginning at C_1 a traveling salesman wishes to visit each of the other cities once and only once before returning to C_1, that is, make a *tour*, and more importantly minimize the distance traveled. A tour is given by any nonrepeating sequence of the cities, since if it doesn't begin with C_1 it can be broken at C_1 and the initial sequence moved to the end. Thus, the problem is to find a minimum-cost permutation of the cities. More formally

TRAVELING SALESMAN
INSTANCE: n distinct cities $C_1, \ldots, C_n$ and positive integral distances between every pair C_i, C_j denoted by $d(C_i, C_j)$.
QUESTION: Which tour $C_{i_1}, \ldots, C_{i_n}$ of the cities minimizes

$$COST(i_1, \ldots, i_n) = \sum_{j=1}^{n-1} d(C_{i_j}, C_{i_{j+1}}) + d(C_{i_n}, C_{i_1})?$$

This optimization problem can be converted into a decision problem, that is, a problem with a yes or no answer, by specifying, additionally, a bound B and asking if there is a tour $C_{i_1}, \ldots, C_{i_n}$ with $COST(i_1, \ldots, i_n) \leq B$. We obtain

TRAVELING SALESMAN CHECKING
INSTANCE: n distinct cities $C_1, \ldots, C_n$, positive integral distancces between every pair C_i, C_j, denoted by $d(C_i, C_j)$, and a positive integral bound B.
QUESTION: Is there a tour $C_{i_1}, \ldots, C_{i_n}$ of the cities such that $COST(i_1, \ldots, i_n) \leq B$?

If B is less than the minimal cost tour the answer is clearly no and otherwise it is yes.

Now, transforming problems to decision problems may appear to be unduly restrictive. Fortunately this is not so, at least from the viewpoint of time complexity. Indeed the decision problem given above is no more difficult, from a time-complexity viewpoint, than the corresponding optimization problem. This is simply because an algorithm for the decision problem can be constructed as follows. First execute the algorithm for the optimization problem and, second, test whether or not the cost of the solution is less than B. Assuming that comparing two integers is an inexpensive operation, as it usually is, the claim follows. Moreover, TRAVELING SALESMAN and many similar problems can also be shown to be no more difficult, up to a polynomial relationship, than the corresponding decision problems. We demonstrate this for TRAVELING SALESMAN.

Given an instance of TRAVELING SALESMAN (TS) and an algorithm for TRAVELING SALESMAN CHECKING (TSC) we demonstrate how to construct an algorithm for TS. We then show that the two algorithms have similar time bounds.

Define $d_{\max}$ to be $\max(\{d(C_i,C_j) : 1 \leq i,j \leq n\})$ and B^* to be the cost of a minimal cost tour. Observe that $n \leq B^* \leq nd_{\max}$, since each intercity distance lies between 1 and $d_{\max}$. The TS algorithm has two phases. In the first phase the value of B^* is found and in the second a tour with cost B^* is found.

To search for B^* use divide and conquer. Let $i = 1$, $lower = n$, and $upper = nd_{\max}$. Let $B_i = (lower + upper)/2$ and using the TSC algorithm test if there is a tour with cost $\leq B_i$. If there is such a tour, then i is replaced by $i+1$ and $upper$ by B_{i-1} and the search is repeated. Otherwise i is replaced by $i+1$ and $lower$ by B_{i-1} and the search is repeated.

If at any time $lower$ equals $upper$, then the search terminates with $B^* = lower = upper$. The search for B^* requires $O(\log(nd_{\max}))$ calls of the TSC algorithm, that is, $O(\log n + \log d_{\max})$ calls.

In the second phase we iteratively construct a tour of cost B^*. The method is quite simple. Assuming the tour is to begin at city C_1, we replace the value of $d(C_1,C_2)$ by $d(C_1,C_2)+1$ and call the TSC algorithm with B^* and the modified d. This change of value forces a tour with cost B^* to go from C_1 to some city $C_3, \ldots, C_n$. It cannot go from C_1 to C_2. The reason for this is that $d(C_1,C_2)$ is one more than it was originally. If there is a tour of cost B^* in which C_1 is followed by C_2, then with the original values there must be a tour with cost B^*-1, a contradiction.

If the algorithm says there is no such tour, then a minimal cost tour must go to city C_2 after C_1. So we have identified a portion of the tour. If there is still a tour with cost B^*, then we also change $d(C_1,C_3)$ to $d(C_1,C_3)+1$ and repeat the process.

Eventually after modifying $d(C_1,C_i)$, for some i, the TSC algorithm must report that no tour of cost B^* exists. Hence, we have identified C_1,C_i as a part of a minimal-cost tour, where $d(C_1,C_i)$ is the last value modified. Reset this distance to its original value. Once the second city has been found we carry out a similar process to identify the third city in a minimal cost tour. After at most $(n-1)+(n-2)+ \cdots +2$ calls $= n(n-1)/2+1$ calls of the TSC algorithm we will have identified a minimal cost tour.

Overall there are $O(n^2+\log d_{\max})$ calls of the TSC algorithm. Hence, the time complexity of the TS algorithm is $O((n^2+\log d_{\max})T(m))$, where $T(m)$ is the time complexity of the TSC algorithm. Note that m, the size of the input, is $O(n^2 \log d_{\max})$ if the integer distances are represented in binary. We may assume B satisfies $n \leq B \leq Bd_{\max}$ and, therefore, it requires $\log B$ cells in binary, that is, $O(\log n + \log d_{\max})$ cells.

Now $n^2 + \log d_{\max}$ is $O(m)$, therefore $T'(m)$, the time complexity of our TS algorithm is $O(m\, T(m))$, that is, it is polynomial if $T(m)$ is polynomial. Moreover, it is only a linear multiple of the input size worse than the time complexity of the TSC algorithm.

We now recall how decision problems can be encoded as language recognition problems. Again we use TRAVELING SALESMAN CHECKING to illustrate this.

An *instance* I, of *TSC* is determined by a specific assignment of positive integer values to n, B, and $d(C_i, C_j)$, $1 \leq i < j \leq n$. The collection of *yes instances Y* consists of those instances giving the answer yes, and the collection of *no instances N* those instances for which the answer is no. We encode an instance as a word over the alphabet $\Delta = \{\mathbf{0}, \mathbf{1}, [,], (,), ,\}$ as follows. The cities are numbered from 1 to n and they are encoded by the words $[binary(i)]$, $1 \leq i < n$, where $binary(i)$ is the binary encoding over $\{\mathbf{0}, \mathbf{1}\}$ of the integer i. The distances are encoded by the words

$$([binary(i)], [binary(j)], [binary(d_{ij})])$$

where $d_{ij} = d(C_i, C_j)$, $1 \leq i < j \leq n$. The instance can be encoded as a list of the cities, a list of distances, and the bound B,

$$(([binary(1)], \ldots, [binary(n)]),$$
$$(([binary(1)], [binary(2)], [binary(d_{12})]), \ldots,$$
$$([binary(n-1)], [binary(n)], [binary(d_{n-1,n})])),$$
$$[binary(B)])$$

where parentheses are used to group related objects together. Letting $code(I)$ denote the above encoding of instance I, then $code(I)$ is clearly a word over the alphabet Δ. Note that not all words over Δ are encodings of instances. Finally, let

$$L[TSC, code] = \{x : x \text{ is in } \Delta^* \text{ and } x = code(I) \text{ for some yes instance of } TSC\}$$

Thus, deciding whether or not an instance is a yes instance is equivalent to deciding whether or not its encoding belongs to $L[TSC, code]$.

Now, if we have a polynomial-time DTM, M, for a decision problem Π with $L(M) = L[\Pi, code]$, then M provides a deterministic-polynomial-time algorithm for Π. Hence, $\boldsymbol{L}_{DPTIME}$, which is a family of languages, is often thought of, because of this encoding technique, as a family of decision problems having deterministic-polynomial-time algorithms.

In a similar manner $\boldsymbol{L}_{NPTIME}$ can be said to consist of those problems which have polynomial-time nondeterministic algorithms. TRAVELING SALESMAN falls into this class and is in $\boldsymbol{L}_{DPTIME}$ if and only if $\boldsymbol{L}_{DPTIME} = \boldsymbol{L}_{NPTIME}$. All known problems of this kind have a nondeterministic algorithm of the form

guess a candidate solution;
verify if the candidate is indeed a solution.

For example, TRAVELING SALESMAN can be solved by guessing a permutation and verifying if its cost is small enough. Verification is, in all these cases, polynomially time bounded.

An in-depth discussion of the topic of tractable problems, that is, problems in $\boldsymbol{L}_{DPTIME}$, and intractable problems, that is, problems in $\boldsymbol{L}_{NPTIME}$ and above, is beyond the scope of this text; however, we discuss them in more detail in Sections 11.1 and 11.5.

6.7 ADDITIONAL REMARKS

6.7.1 Summary

We have introduced the most general class of automata DETERMINISTIC TURING MACHINES. We have illustrated how these may be programmed and shown that there is a UNIVERSAL TURING MACHINE.

We have considered a number of reduction results, the most important being NO-MOVE REDUCTION and TAPE SYMBOL REDUCTION. We have demonstrated that 2-COUNTER DETERMINISTIC TURING MACHINES are equivalent in power to *DTMs*, while MULTITAPE TURING MACHINES are no more powerful than *DTMs*. Finally, RESOURCE-BOUNDED TURING MACHINES have been introduced, which lead in a natural way to a classification of languages into TRACTABLE and INTRACTABLE LANGUAGES, and problems into TRACTABLE and INTRACTABLE PROBLEMS.

6.7.1 History

The Turing machine was invented by Alan M. Turing (1936). Consult Hodges (1983) for a readable biography of Turing and read the *Scientific American* articles of Hopcroft (1984) and Dewdney (1984). Off-line Turing machines were first discussed in Hartmanis et al. (1965). Shannon (1956) was the first to prove that 3-state Turing machines are as powerful as unrestricted Turing machines. A universal Turing machine was demonstrated in Turing (1936). The serious study of resource-bounded Turing machines was initiated in a series of three papers, Hartmanis and Stearns (1965), Hartmanis et al. (1965), and Lewis et al. (1965), although Cobham (1964) was the first to study $\boldsymbol{L}_{DPTIME}$.

6.8 SPRINGBOARD

6.8.1 The Busy Beaver Game

Consider a Turing machine with two-way infinite tape, an input alphabet equal to $\{1\}$, and a tape alphabet equal to $\{1,\mathbf{b}\}$. Tibor Rado (see Rado, 1962) asked the following question of such a machine with n states (apart from f, the only final state):

How many 1's can there be when the empty word is accepted.

Since 1 is the only nonblank symbol, this is equivalent to asking: if it halts when given the blank tape as input, then how many nonblank cells can there be? Rado called this the Busy Beaver Game, presumably because of the similarity of 1's to twigs and of the activity of the machine to the industrious activity of beavers.

The maximum number of nonblank cells that can be obtained by such an n-state Turing machine is denoted by $\Sigma(n)$. A n-state machine that produces $\Sigma(n)$ nonblank cells is called a *busy beaver*. For $n = 1, 2, 3$, and 4, $\Sigma(n) = 1, 4, 6$, and 13, respectively. For example, for $n = 1$ we have the busy beaver in Figure 6.8.1, where $D = L, R$, or λ. This is the only possibility, since a loop from s to itself will prevent halting. A 2-state busy beaver is displayed in Figure 6.8.2. At this stage we already encounter the major difficulty, namely, how do we know the Turing machine of Figure 6.8.2 is, indeed, a 2-state busy beaver?

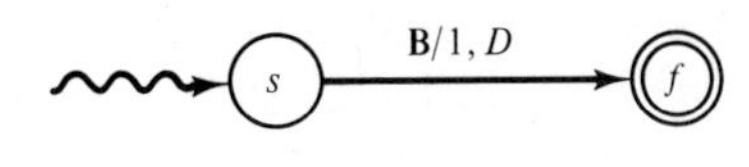

Figure 6.8.1

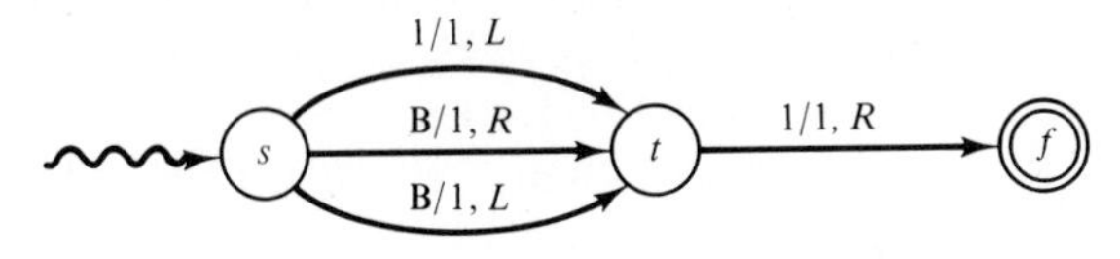

Figure 6.8.2

At present there is no general theory about the structure of busy beavers, nor for that matter for n-state busy beavers, for particular values of n. The only available technique for finding them is to perform an exhaustive search over all n-state Turing machines. What is the value of $\Sigma(5)$? It has been shown recently that it is at least 1,915; see Dewdney (1984, 1985).

6.8.2 The Turing Micros

An early puzzle that resulted from the demonstration of the existence of a universal Turing machine was: What is the smallest universal Turing machine? We call candidates for this honor *Turing micros*, since they are small computers indeed.

Because we can simulate states with tape symbols, and vice versa, a measure of size must be chosen carefully. The number of states or the number of tape symbols is insufficient. A more meaningful measure is the number of transitions. We approximate this with the product of the number of tape symbols and number of states.

The first universal Turing machine worthy of the name Turing micro was the six-symbol, ten-state machine of Ikeno (1958). This was followed by the six-symbol, eight-state micro of Watanabe (1960), the six-symbol, seven-state micro of Minsky (1960), the five-symbol, eight-state micro of Watanabe (1961), the six-symbol, six-state micro of Minsky (1962), and, finally, the four-symbol, seven-state micro of Minsky (1962). Over a period of four years from 1958 to 1962 the size of the Turing micro went from 60 down to 28. Minsky also conjectured that one of the three-symbol, six-state Turing machines was universal, but no one has ever proved this to be the case. Surprisingly, in the succeeding 23 years no smaller micro has appeared! Is there a three-symbol, nine-state micro or a four-symbol, six-state micro out there waiting to be discovered?

The only new results in this area are for universal two-dimensional Turing machines; see Kleine Buning and Ottmann (1977) and Priese (1979).

6.8.3 Computable Functions

This chapter has been primarily concerned with acceptance of languages and computation of Boolean functions by Turing machines. However, as we have indicated this can be extended to the computation of general functions; see Section 7.5. This gives rise to the class of *DTM-computable functions*, which are considered to be the class of functions computable by any computing device. Read the original paper of Turing (1936) and peruse Davis (1958).

6.8.4 *NP*-Completeness

Two languages, $L_1 \subseteq \Sigma_1^*$ and $L_2 \subseteq \Sigma_2^*$, are said to be polynomially related if there is a function $f : \Sigma_1^* \rightarrow \Sigma_2^*$ for which $f(L_1) = L_2$ and which can be computed in deterministic polynomial time. This relation leads to a finer classification of $\boldsymbol{L}_{NPTIME}$ and, in particular, its reflexive, transitive closure focuses attention on the set of hardest languages in $\boldsymbol{L}_{NPTIME}$. These languages are said to be *NPTIME-complete* or simply *NP-complete.* A similar approach to the languages in $\boldsymbol{L}_{DPSPACE}$ leads to *DPSPACE-completeness* or *PSPACE-completeness.* These notions can be extended, in a natural manner, to problems leading to the concepts of *NP-complete* and *PSPACE-complete problems.* An introduction to these topics is to be found in Section 11.1.

However, the excellent articles of Graham (1978), Pippenger (1978), and Stockmeyer and Chandra (1979) can also be consulted, before perusing the superb book by Garey and Johnson (1979).

6.8.5 Simulation Efficiency

Whenever a transformation of a *TM* is carried out, for example, transforming a multihead *TM* into an equivalent *TM*, there are always two questions of efficiency raised. First, how efficient is the transformation itself and, second, how efficient is the transformed *TM* compared with the initial *TM*. It is this second notion of efficiency that is, our main concern. Since the transformed *TM* simulates, in some way, the initial *TM*, this implies we are concerned with simulation efficiency. Given that the initial *TM* has time complexity $T(n)$, does the simulating *TM* have the same complexity, a polynomial (in n) multiple of $T(n)$, or some polynomial of $T(n)$ itself, for example.

For the simulations given in this chapter, compute their efficiencies. Can you come up with more efficient simulations? Consult Hopcroft and Ullman (1979).

6.8.6 Recursively Enumerable Languages

A language is said to be *recursively enumerable* if there is an algorithm that enumerates or lists all the words in the language. Since we have introduced Turing machines as a model for algorithms we can formalize this notion in the following way. Let $M = (Q,\Sigma,\Gamma,\delta,s,f)$ be a 2-tape *DTM* in which the first tape is considered to be the output tape and the second is the work tape. M is initialized with blanks on both tapes and the read-write heads over the leftmost cells. During its computation M only writes on the first tape and, moreover, the read-write head never moves left. Suppose M writes words from Σ^* separated by $\boldsymbol{B}$'s on the output tape. Then the collection of words that it writes or enumerates is said to be *the language enumerated by M* and is denoted by $E(M)$. Note that if M halts, then $E(M)$ is finite. $E(M)$ is said to be a recursively enumerable language. It can be proved that $\boldsymbol{L}_{RE}$, the family of recursively enumerable languages, equals $\boldsymbol{L}_{DTM}$. Prove this result yourself or consult Hopcroft and Ullman (1979).

6.8.7 Context-Sensitive Languages

First, consider an *NTM* $M = (Q,\Sigma,\Gamma,\delta,s,f)$ in which the number of cells that can be used during a computation on an input word x in Σ^* is exactly $|x|$. Such an *NTM* is called a *nondeterministic linear-bounded automaton.* If it is also a *DTM*, then it is called a *deterministic linear-bounded automaton.* We use the acronyms *DLBA* and *NLBA* and denote by $\boldsymbol{L}_{DLBA}$ and $\boldsymbol{L}_{NLBA}$, the families of *DLBA* and *NLBA* languages, respectively.

Second, consider a grammar $G = (N,\Sigma,P,S)$ in which the finite set of productions $P \subseteq N^+ \times (N \cup \Sigma)^+$. Moreover, each production $\alpha \rightarrow \beta$ in P satisfies $|\alpha| \leq |\beta|$. Such a grammar is said to be *context sensitive.* The family of context-sensitive languages is denoted by $\boldsymbol{L}_{CS}$.

It can be proved that $\boldsymbol{L}_{NLBA} = \boldsymbol{L}_{CS}$; see Hopcroft and Ullman (1979), Révész (1983), or Salomaa (1973). It is an longstanding open problem whether $\boldsymbol{L}_{DLBA}$ equals $\boldsymbol{L}_{NLBA}$; see Hopcroft and Ullman (1979).

EXERCISES

1.1 Given the following *DTM*

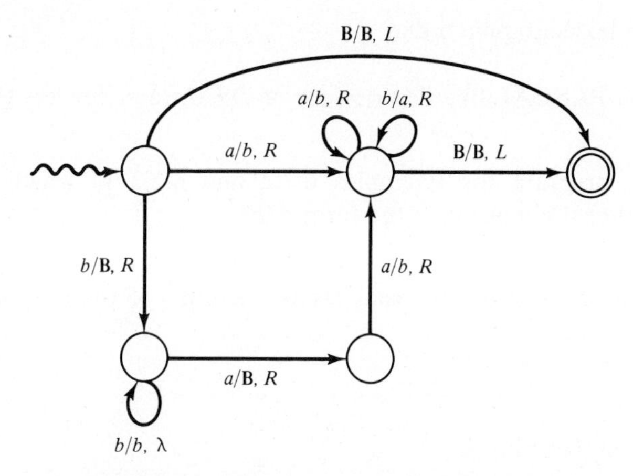

(i) Give four words accepted by the *DTM* together with their configuration sequences.
(ii) Give four words that are not accepted by the *DTM* and, in each case, explain why not.

1.2 Construct *DTMs* to carry out the following operations.

(i) A left shift of its input by one cell.
(ii) A cyclic left shift of its input by one cell.
(iii) Let c be in Σ. If the input word $x = x_1cx_2$, where x_1 is in $(\Sigma - \{c\})^*$, then produce cx_2 as the output.
(iv) A duplication of its input, that is, if the input is x, the output should be xx.
(v) Let 1 be in Σ. If the input is $x\mathbf{B}1^i$, for some $i \geq 0$, then the output should be x shifted left i cells. So input $x\mathbf{BB}$ is unchanged on output, while $x\mathbf{B}1$ is given by the solution to (i) above.
(vi) Let 1 be in Σ and the input be $x\mathbf{B}1^i$, for some $i \geq 0$. Then output x^i.

1.3 We may represent nonnegative integers in binary or unary (or indeed in any base). Implement *DTMs* to carry out the basic arithmetic operations on integers. Do this for both binary and unary input. The operations to be considered are addition, monus (x monus y equals x minus y if $x \geq y$, otherwise it equals 0), multiplication, integer division, exponentiation, integer square root.

1.4 Construct *DTMs* which, given input

$$x_1\mathbf{B}x_2\mathbf{B} \cdots \mathbf{B}x_m$$

where x_i is in $\{a,b\}^*$, $1 \leq i \leq m$, for some $m \geq 1$, will

(i) check if the x_i are in lexicographic order;

(ii) sort the x_i into lexicographic order.

2.1 Take any one of the PASCAL-like descriptions of *DTMs* and convert them into *DTMs*.

2.2 Construct a *DTM* program for Examples 6.2.2 and 6.2.3 in which the *DTM* first checks that the input is of the form a^*b^*.

2.3 For the *DTMs* of Examples 6.2.3, 6.2.4, and 6.2.5, express as a function the maximum number of steps taken by each *DTM* in terms of the size of its input.

3.1 Prove Theorem 6.3.1

3.2 Complete the proof of Theorem 6.3.2.

3.3 Let M be the sub-*DTM* of Example 6.1.4 which swaps the outermost symbols of an input word in $\{a,b\}^*$. Using the construction of Theorem 6.3.2 obtain a *DTM*, M', with tape alphabet $\{a,b,\mathbf{B}\}$.

3.4 Let M be the final *DTM* of Example 6.1.3. Using the bouncing technique of Theorem 6.3.4 reduce it to a 4-state *DTM*.

3.5 Complete the proof of Theorem 6.3.4.

3.6 A further simplification of *DTMs* is obtained by only allowing either a symbol to be rewritten or a move of the read-write head to be made. Formally, a *simple DTM* is a *DTM*, $M = (Q,\Sigma,\Gamma,\delta,s,f)$, in which the transition function satisfies $\delta: Q \times \Gamma \rightarrow Q \times (\Gamma \cup \{L,R,\lambda\})$, where L, R, and λ are not in Γ. Prove that the two models are equivalent in accepting power.

3.7 The simplifications given in this section are compared according to their accepting power; however, it is important to realize that this is not restrictive. The various simplifications compute the same functions and decide the same languages. Prove this for Theorems 6.3.1-6.3.5.

3.8 Prove Theorems 6.3.7 and 6.3.8.

4.1 Complete the proofs of Theorems 6.4.1-6.4.4.

5.1 We have provided an encoding of *DTMs* over the alphabet $\{a,b\}$. Design a *DTM* which given a word over $\{a,b\}$ decides whether or not it is the encoding of some *DTM*.

5.2 We encoded *DTMs* by assuming that there is a universal state set $\mathbb{N}_0$, while restricting the tape alphabet to $\{a,b,\mathbf{B}\}$. An alternative encoding scheme assumes there is a universal tape alphabet Γ_∞ whose symbols are $a_1, a_2, \ldots, a_i$.

(i) Letting a_1 be $\boldsymbol{B}$, $a_2 = a$, $a_3 = b$, extend the encoding scheme to tape symbols. The encoding should be a word in $\{a,b\}^*$.

(ii) Design a universal Turing machine which takes as input the encoding of a *DTM* followed by the encoding of an input word.

6.1 We defined a $T(n)$ time-bounded *TM* by restricting an accepting configuration sequence, for a word of length n,to have at most $T(n)$ steps. However, if a word of length n is not accepted, then its configuration sequence may be nonterminating.

(i) Prove that if a language L is of time complexity $T(n)$, then there is a *DTM* with time complexity $T(n)$ which decides L. In other words, L is in $\boldsymbol{L}_{REC}$, hence $\boldsymbol{L}_{DPTIME} \subseteq \boldsymbol{L}_{REC}$.

(ii) Prove that $\boldsymbol{L}_{NPTIME} \subseteq \boldsymbol{L}_{REC}$.

6.2 Prove that $\boldsymbol{L}_{DPSPACE} \subseteq \boldsymbol{L}_{REC}$.

6.3 Prove that $\boldsymbol{L}_{DPSPACE} = \boldsymbol{L}_{NPSPACE}$.

PROGRAMMING PROJECTS

P1.1 Design and implement a program to simulate an arbitrary deterministic Turing machine. A visual display of configuration sequences is essential.

P2.1 We used an informally defined programming language for *DTMs* — clearly a PASCAL dialect. Formalize its description and implement a compiler which given such a program produces a corresponding *DTM*.

P3.1 Implement a sequence of programs to carry out the transformation of a *DTM* into a *2CDTM*.

P5.1 Construct a universal-*DTM*-simulator program.

P5.2 First, construct a program to play the game of life. Second, construct a program which given a *DTM* produces a corresponding instance of life which simulates it; see Berlekamp, Conway, and Guy (1983).

Chapter 7

Functions, Relations, and Translations

7.1 INTRODUCTION

In the six previous chapters we have concentrated on the specification of languages or, alternatively, the specification of Boolean functions.

However, most computer programs produce much more substantial output; at least this is usually their intent! We can view a program as a black box which is functional, that is, given input data I, there are unique output data O associated with I; see Figure 7.1.1.

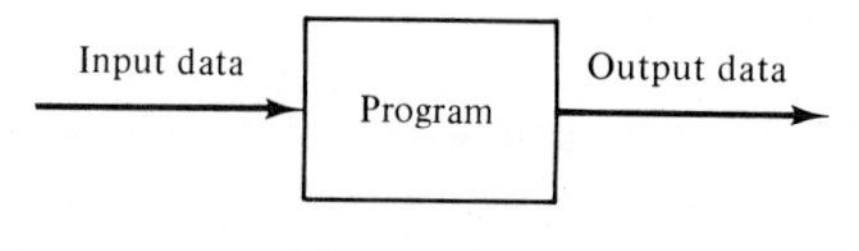

Figure 7.1.1

Input and output data can be represented as words over some alphabets, for example, as is done in Section 6.6. Therefore, letting Σ be the input data alphabet and Δ the output data alphabet we have the modified black box view displayed in Figure 7.1.2. In Figure 7.1.2 the program P is now a *partial* function $P : \Sigma^* \to \Delta^*$, since there may not be output data associated with all input data.

Because computation is the focus of our concerns it is necessary that we include models that mirror this view of programs, since we wish to be able to discuss what it means for a function to be computable, and if it is computable, then

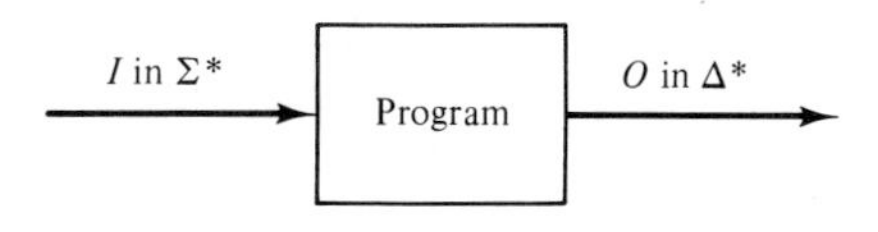

Figure 7.1.2

how efficiently it can be computed.

We consider in the following sections how functions and relations can be defined by finite automata, context-free grammars, pushdown automata, and Turing machines. As far as the first three models are concerned we quickly restrict our attention to functions of the form $f : \Sigma^* \to \Delta^*$ and binary relations $\Sigma^* \times \Delta^*$, known as *translations* or *transductions*, since this is the main reason for their existence. The associated automata are called *transducers*, while the associated grammars are called *translation grammars*. However, Turing machines are central to the fundamental issues of computable functions and, therefore, we treat them in full generality. Finally, we introduce recursive functions that are specified in a machine-independent manner. The class of recursive functions equals the class of Turing computable functions.

7.2 FINITE TRANSDUCERS

Consider a deterministic finite automaton M, having input alphabet Σ, which is extended by

(i) adding an *output alphabet* Δ and an *output tape*;
(ii) allowing it to write an *output word* on the output tape at each transition; and
(iii) adding $\boldsymbol{B}$ as an additional input and output symbol, to be used as a separator.

Pictorially we have Figure 7.2.1. Now an input word $x = x_1\boldsymbol{B}x_2 \cdots \boldsymbol{B}x_m$, where x_i is in Σ^*, $1 \leq i \leq m$, for $m \geq 1$, may be viewed as m arguments $x_1, \ldots, x_m$. If M accepts x, then it also produces as output a word y in Δ^*. We write $M(x) = y$.

Let $f_{m,n} : (\Sigma^*)^m \to (\Delta^*)^n$, $m \geq 1$, be an m,n-place function. If, for all $(x_1, \ldots, x_m)$ in $(\Sigma^*)^m$, $M(x_1\boldsymbol{B} \cdots \boldsymbol{B}x_m) = y_1\boldsymbol{B} \cdots \boldsymbol{B}y_n$ in Δ^* if and only if $f_{m,n}(x_1, \ldots, x_m) = (y_1, \ldots, y_n)$, then we say that M *computes* $f_{m,n}$.

Rather than considering functions of finite automata with this generality, we restrict our attention to functions having a single argument and a single result. These are called *transductions* or *translations*. In this case we do not need the separator symbol $\boldsymbol{B}$. Since we also allow nondeterministic automata we obtain, in general, binary relations which may or may not be functions.

The finite transducer is the simplest translation device; it is a finite automaton which is allowed to write an output word, on a separate output tape, at each transition. In terms of programming it is similar to a program which accesses a sequential input file and writes a sequential output file, that is, a typical PASCAL program. The input tape is read-only and the output tape is write-only.

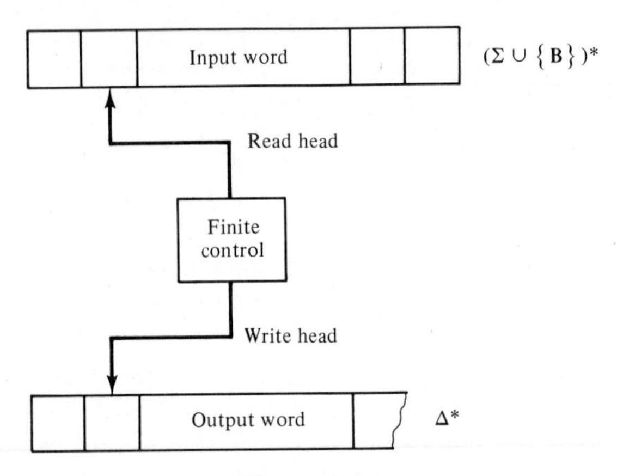

Figure 7.2.1

The finite transducer is based on *λ-NFAs* rather than on *NFAs* or *DFAs* since λ-transitions (on input) are important, both practically and theoretically.

Definition A *finite transducer*, or *FT*, M, is a sextuple $M = (Q,\Sigma,\Delta,\delta,s,F)$, where

Q is a finite set of *states*;
Σ is an *input alphabet*;
Δ is an *output alphabet*;
$\delta \subseteq Q \times (\Sigma \cup \{\lambda\}) \times Q \times \Delta^*$ is a *finite transition relation*;
s in Q is a *start state*; and
$F \subseteq Q$ is a set of *final states*.

We define δ as a relation for reasons of simplicity; however, it should be clear that it can also be viewed as a function from $Q \times (\Sigma \cup \{\lambda\})$ to finite subsets of $Q \times \Delta^*$.

A *configuration* of an *FT*, $M = (Q,\Sigma,\Delta,\delta,s,F)$, is a word yqx in $\Delta^* Q \Sigma^*$, which specifies the current state q, the remaining input x, and the output y generated so far. Note that we assume $\Delta \cap Q = \emptyset$ to avoid confusion in the interpretation of a configuration. An *initial configuration* is sx, where x in Σ^* is an input word. A *final configuration* is yf, where y is in Δ^* and f is in F. Given a configuration $ypax$, where a is in $\Sigma \cup \{\lambda\}$, we write

$$ypax \vdash yzqx$$

if (p,a,q,z) is in δ. As before we extend this to give $\vdash^i$, $\vdash^+$, and $\vdash^*$. Given a word x in Σ^* we say y *is an output for* x *with respect to* M if

$$sx \vdash^* yf$$

for some final configuration yf. For all x in Σ^* we denote by $M(x)$, *the set of all outputs for* x, that is,

$$M(x) = \{y : sx \vdash^* yf, \text{ for some final configuration } yf\}$$

$M(x) = \emptyset$ implies that no final configuration can be reached by x.

The *translation obtained from M* is denoted by $T(M)$ and is defined by

$$T(M) = \{(x,y) : x \text{ is in } \Sigma^* \text{ and } y \text{ is in } M(x)\}.$$

A subset T of $\Sigma^* \times \Delta^*$ is called a *finite transduction* (or *translation*), or an *FT*, if $T = T(M)$ for some finite transducer M. The *family of finite transductions* or *translations* is denoted by $\boldsymbol{T}_{FT}$ and is defined by

$$\boldsymbol{T}_{FT} = \{T : T \text{ is a finite transduction}\}$$

Let $T \subseteq \Sigma^* \times \Delta^*$; then *Domain*$(T)$ denotes the set of words $\{x : x$ is in Σ^* and (x,y) is in T for some y in $\Delta^*\}$ and *Range*(T) denotes $\{y : y$ is in Δ^* and (x,y) is in T for some x in $\Sigma^*\}$.

Observe that if $T \subseteq \Sigma^* \times \Delta^*$ is an *FT*, then *Domain*(T) is a *REGL*. More surprisingly *Range*(T) is also a *REGL*, as we now demonstrate.

Theorem 7.2.1 *Let $M = (Q,\Sigma,\Delta,\delta,s,F)$ be an FT. Then Range$(T(M))$ is a REGL.*

Proof: Recalling that lazy *FAs* have the same expressive power as *DFAs*, construct a lazy *FA*, M', from M as follows: Let $M' = (Q,\Delta,\delta',s,F)$, where $\delta' = \{(p,y,q) : (p,a,q,y)$ is in δ and y is in $\Delta^*\}$. Now, $L(M') =$ *Range*$(T(M))$, since for all words y in Δ^*,

y is in $L(M')$ iff

$sy_1 \cdots y_m \vdash p_1y_2 \cdots y_m \vdash \cdots \vdash p_{m-1}y_m \vdash p_m$ in M',
for some $y_1, \ldots, y_m$ in Δ^* with $y = y_1 \cdots y_m$ and
for some $p_1, \ldots, p_m$ in Q with p_m in F iff

$sx_1 \cdots x_m \vdash y_1p_1x_2 \cdots x_m \vdash \cdots \vdash y_1 \cdots y_mp_m$ in M,
for some $x_1, \ldots, x_m$ in $\Sigma \cup \{\lambda\}$ iff

$(x_1 \cdots x_m, y)$ is in $T(M)$ iff

y is in *Range*$(T(M))$. ■

Having defined *FTs* it remains to demonstrate their usefulness both practically and theoretically. Since they are used as a theoretical tool in Chapter 9 we only deal with practical applications in the remainder of this chapter through a series of examples.

Example 7.2.1 Encoding and Decoding

Let h_i be the code that replaces each letter of the English alphabet by its ith successor, so that h_0 is the identity code as is h_{26}. We can create a better encoding by using more than one h_i, as follows. First, the sender and receiver agree on a keyword, for example, *XMAS*. Now let $\#X$ be the number of letters preceding it in the alphabet, that is, 23, and similarly for the other letters. Since the keyword has four letters the sender encodes the first letter of the message using $h_{\#X}$, the second with $h_{\#M}$, the third with $h_{\#A}$, the fourth with $h_{\#S}$, the fifth with $h_{\#X}$ again, and so on. Note that $h_{\#X}(A) = X$, $h_{\#M}(A) = M$, etc. So

$$\overset{1}{H}\overset{2}{A}\overset{3}{P}\overset{4}{P}\overset{1}{Y}\ \overset{2}{N}\overset{3}{E}\overset{4}{W}\ \overset{1}{Y}\overset{2}{E}\overset{3}{A}\overset{4}{R}$$

becomes EMPHVZEOVQAJ under this keyword based encoding, where we ignore blanks.

A finite transducer realizing such an encoding, that is, an *encoder*, is straightforward to construct, see Figure 7.2.3. We use modified state diagrams once more in which a transition (p,a,y,q) is drawn as shown in Figure 7.2.2. Since all input sequences are valid, all states are final.

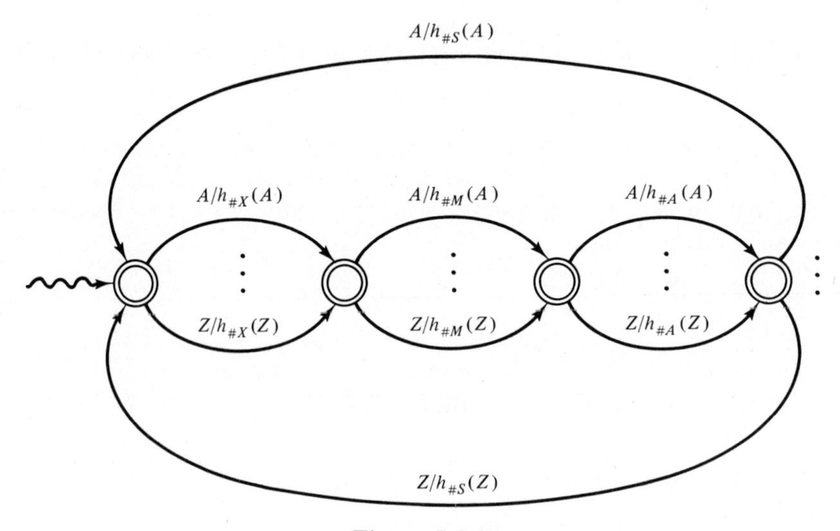

Figure 7.2.2

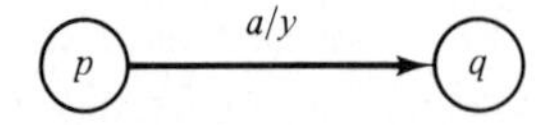

Figure 7.2.3

Conversely, the receiver constructs a decoding *FT* which is identical to the one above except that the rôles of input and output symbols are interchanged giving the *decoder* shown in Figure 7.2.4.

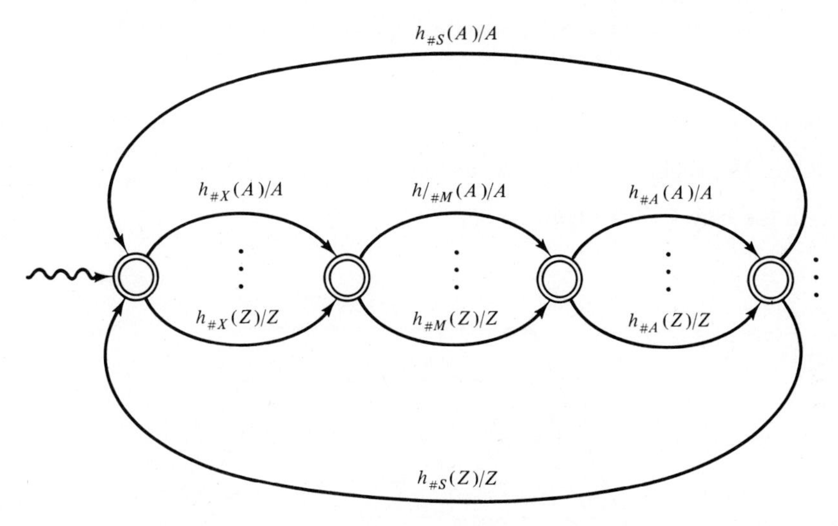

Figure 7.2.4

Both encoding and decoding are simple to implement in this scheme. Unfortunately given either the encoder or the decoder we can easily construct the other. For this reason this encoding scheme is useful when only low-level security is required.

Example 7.2.2 Lexical Analysis

Every compiler needs to partition the incoming sequence of characters, representing a program, into lexical units corresponding to the lexical symbols of the programming language being used. For example,

program x; **var** a, b : **integer**; **begin end.**

is partitioned into

program/x/;/**var**/a/,/b/:/**integer**/;/**begin**/**end**/./

Moreover, a compiler also speeds up subsequent processing by replacing variable-length lexical units with fixed-length internal values. The internal value of each lexical unit is uniquely chosen so that two or more appearances of the same unit have the same internal value and this value is different from that of other lexical units. Identifiers and constants are, typically, given unique internal values together with a reference to a symbol table in which the identifier or the constant is located. Moreover, the internal value of other symbols is often structured into a class or group designation, for example, an arithmetic operator, together with a unique internal value. Therefore, lexical analysis is, essentially, a decoding process, except for numeric constants, strings, and identifiers. Each lexical unit is decoded into its unique internal value. This is a translation which can be implemented with a finite transducer — if we ignore numeric constants, strings, and identifiers which are treated separately.

Given

if, then, else, begin, end, boolean, integer, $+$, $-$, **or**

we obtain the lazy *FT* of Figure 7.2.5.

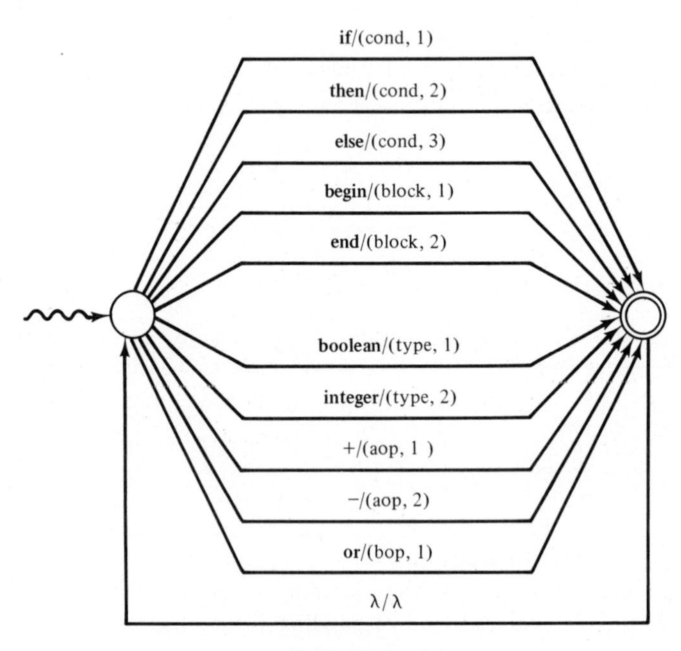

Figure 7.2.5

It is lazy since we allow words over the input alphabet in transitions. The proof that a lazy *FT* can always be transformed into an "equivalent" *FT* is left to the Exercises.

Example 7.2.3 Data Compression

There are many different methods of data compression in use; we discuss a method which replaces symbols by bit strings. Since every method of storing text in computer-readable form involves representing characters or symbols by bit strings, this can be applied to storing large files of data. Every computer system has at least one character code, usually of 6, 8, or 9 bits, which provides a one-to-one correspondence between characters on a keyboard and their internal representation. The codes are called *block codes* since they are fixed-length. Decoding is simply a matter of taking the next b bits, where b is the length of a block, and performing a table look-up to determine the corresponding external character.

Early in the development of communication theory it was observed that the block-encoding method is inefficient in its expected space usage when the characters being encoded do not occur with equal probability. If a text has m characters, then it always requires mb bits to represent it. But, for example, Morse code which is a variable-length code, see Table 7.2.1, does much better for typical English text. This is because the probabilities of occurrence of letters in typical English text varies tremendously, as displayed in Table 7.2.2. The five most frequent letters are E, T, A, O, and I, and the five least frequent are K, X, J, Q, and Z. Observe how the most frequent letters are given short code words and the infrequent ones long code words. In Figure 7.2.6 a sentence and its Morse code encoding is given.

The	cat	sat	on	the	mat
— •••• •	—•—• •— —	••• •— —	——— —•	— •••• •	—— •— —

Figure 7.2.6

Table 7.2.1

A	· —	N	— ·
B	— · · ·	O	— — —
C	— · — ·	P	· — — ·
D	— · ·	Q	— — · —
E	·	R	· — ·
F	· · — ·	S	· · ·
G	— — ·	T	—
H	· · · ·	U	· · —
I	· ·	V	· · · —
J	· — — —	W	· — —
K	— · —	X	— · · —
L	· — · ·	Y	— · — —
M	— —	Z	— — · ·

Table 7.2.2

A	.0805	N	.0710
B	.0153	O	.0760
C	.0310	P	.0202
D	.0397	Q	.0011
E	.1250	R	.0613
F	.0231	S	.0655
G	.0195	T	.0925
H	.0542	U	.0272
I	.0729	V	.0100
J	.0016	W	.0188
K	.0066	X	.0020
L	.0414	Y	.0172
M	.0254	Z	.0010

The encoding has a length of 35 bits rather than a length of $17 \times 5 = 85$ bits using a 5-bit block code (the minimum possible for 26 letters). There is an algorithm to construct codes for letters with known probabilities such that the constructed code has minimal expected length; see the Exercises.

Decoding variable-length codes is more difficult than decoding block codes, since it is not so clear how many bits should be taken next. Of course there should be no ambiguity. For example, Morse code as defined above is ambiguous, but ambiguity is avoided by adding a "pause" after each encoded letter. Another example is the code in Table 7.2.3 which is ambiguous since 001 can be decoded as c or as aab.

Table 7.2.3

a	0
b	1
c	001

One way to prevent this is to require that a variable-length code be *prefix-free*, that is, no encoding of a letter should be a proper prefix of the encoding of some other letter. For example, the code in Table 7.2.4 satisfies this requirement as does Morse code when a "pause" is added. Such a code, called a *prefix code*, can always be represented by a binary tree. For example, the tree in Figure 7.2.7 represents the code in Table 7.2.4. Observe how each codeword determines a unique path from the root to an external node, when 0 is interpreted as go left and 1 as go right.

Encoding with a prefix code is straightforward — we always have a single state FT as in Figure 7.2.8. Decoding, on the other hand, results in a treelike FT such as that in Figure 7.2.9 which is a simplified version of Figure 7.2.10. That is, it is based on the tree of Figure 7.2.10 with the root as the start and final state and λ-transitions connecting external nodes back to the root.

Table 7.2.4

a	0
b	10
c	11

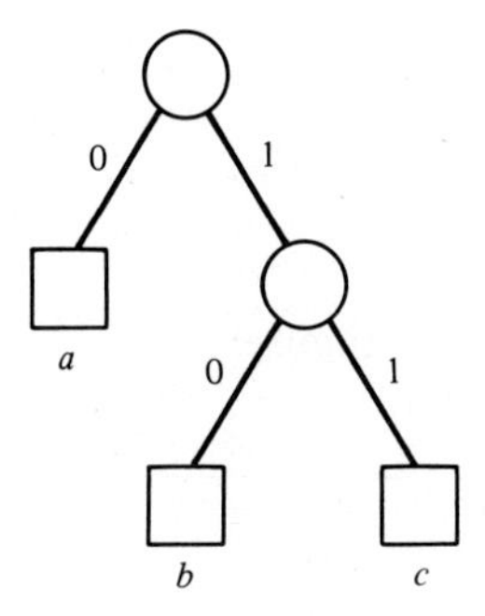

Figure 7.2.7

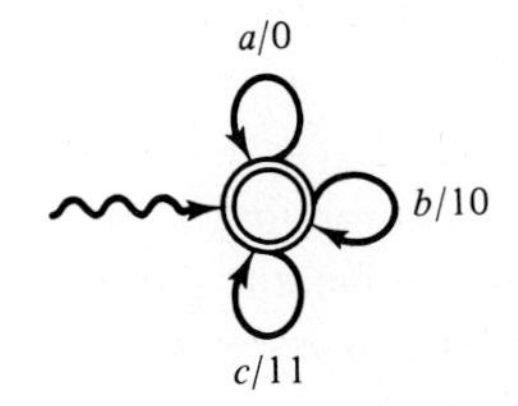

Figure 7.2.8

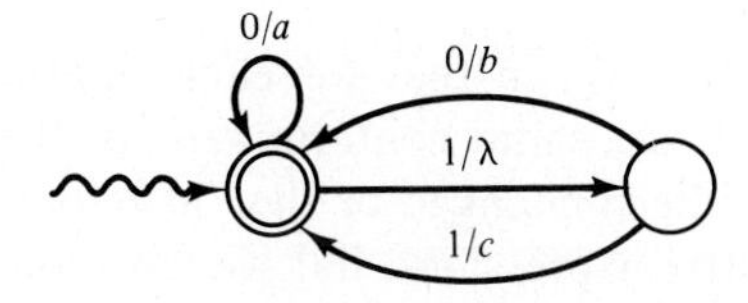

Figure 7.2.9

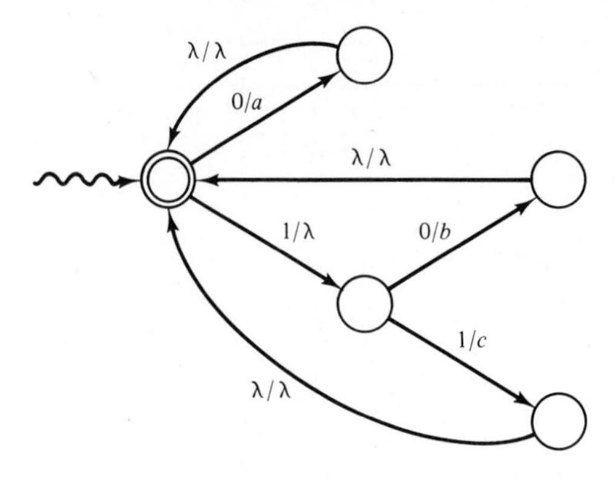

Figure 7.2.10

7.3 TRANSLATION GRAMMARS

Consider the following *CFG* G for infix arithmetic expressions

$$S \to S + T \mid T$$
$$T \to T * F \mid F$$
$$F \to (S) \mid a$$

It is standard practice, when compiling a programming language, to transform infix arithmetic expressions into equivalent parenthesis-free postfix expressions; for example, $a_1 * (a_2 + a_3)$ becomes $a_1 a_2 a_3 + *$. Grammatically, postfix expressions correspond to a *CFG*, G', specified by

$$S \to ST + \mid T$$
$$T \to TF * \mid F$$
$$F \to S \mid a$$

Observe how similar G' is to G, in that for each right-hand side in G there is a right-hand side in G' with the same nonterminals, in the same order. For terminals, either their order has been changed or they have been removed.

G' specifies only postfix expressions, but we wish to specify the transformation of infix expressions into postfix expressions. To this end we need to connect G and G' together in some way. One way to do this is as follows

$$S \to S + T\ ,\ ST + \mid T\ ,\ T$$
$$T \to T{*}F\ ,\ TF{*} \mid F\ ,\ F$$
$$F \to (S)\ ,\ S \mid a\ ,\ a$$

where a comma is used to separate the input right-hand side from the output right-hand side. Beginning with the pair (S,S) we obtain, for example

$$(S,S) \Rightarrow (T,T)$$

where $S \Rightarrow T$ in G and $S \Rightarrow T$ in G'. A pair represents an *input-output sentential form* or *translation form.* Continuing we obtain

$$\begin{aligned}(T,T) &\Rightarrow (T*F, TF*) \Rightarrow (F*F, FF*)\\ &\Rightarrow (a*F, aF*) \Rightarrow (a*(S), aS*)\\ &\Rightarrow (a*(S+T), aST+*) \Rightarrow (a*(T+T), aTT+*)\\ &\Rightarrow (a*(F+T), aFT+*) \Rightarrow (a*(a+T), aaT+*)\\ &\Rightarrow (a*(a+F), aaF+*) \Rightarrow (a*(a+a), aaa+*)\end{aligned}$$

that is, $(S,S) \Rightarrow^+ (a*(a+a), aaa+*)$, and, indeed, $aaa+*$ is the postfix version of $a*(a+a)$.

We say that the above construct is a *simple syntax-directed translation grammar*. We now give a formal definition of these grammars.

Definition A *simple syntax-directed translation grammar*, or *SSDTG*, G, is a quintuple $T = (N,\Sigma,\Delta,R,S)$, where

- N is a *nonterminal* alphabet;
- Σ is an *input* alphabet;
- Δ is an *output* alphabet;
- $R \subseteq N \times (N \cup \Sigma)^* \times (N \cup \Delta)^*$ is a finite set of *rules*; and
- S in N is a *start* symbol.

Moreover, for each rule (A,α,β) in R (usually written as $A \rightarrow \alpha,\beta$), the same nonterminals appear in both α and β and in the same order. Note that both $(N,\Sigma,\{A \rightarrow \alpha : A \rightarrow \alpha,\beta \text{ is in } R\},S)$ and $(N,\Delta, \{A \rightarrow \beta : A \rightarrow \alpha,\beta \text{ is in } R\},S)$ are *CFGs* as we would expect from the discussion above.

Given $G = (N,\Sigma,\Delta,R,S)$ we define a *translation form* recursively as

1. (S,S) is a translation form.
2. If (β_1,β_2) is a translation form with $\beta_1 = x_1 A \beta_1'$ and $\beta_2 = x_2 A \beta_2'$, where x_1 is in Σ^* and x_2 is in Δ^*, then $(x_1\alpha_1\beta_1', x_2\alpha_2\beta_2')$ is a translation form if $A \rightarrow \alpha_1,\alpha_2$ is in R.

We write $(\beta_1,\beta_2) \Rightarrow (x_1\alpha_1\beta_1', x_2\alpha_2\beta_2')$ in this case and, as for *CFGs*, we extend $\Rightarrow$ to $\Rightarrow^i$, $\Rightarrow^+$, and $\Rightarrow^*$ in an obvious way. Thus, the *translation obtained from* G is denoted by $T(G)$ and is the set of pairs

$$T(G) = \{(x,y) : (S,S) \Rightarrow^+ (x,y), x \text{ is in } \Sigma^*, \text{ and } y \text{ is in } \Delta^*\}.$$

We say that $T \subseteq \Sigma^* \times \Delta^*$ is a *simple syntax-directed translation*, or *SSDT*, if there exists an *SSDTG* G with $T = T(G)$.

This definition of a translation form mirrors the definition of left sentential forms, that is, sentential forms obtained under leftmost derivations. In the infix-postfix example we obtained

$$(S,S) \Rightarrow^+ (F{*}F,FF{*})$$

in which the leftmost *Fs* were both replaced at the next step, yielding

$$(F{*}\,F,FF{*}) \Rightarrow (a{*}F,aF{*})$$

This is crucial, since replacing the first F in the input and the second in the output yields

$$(F{*}F,FF{*}) \Rightarrow (a{*}F,Fa{*})$$

and now if the succeeding rules are applied as before we obtain

$$(a{*}(a+a),aa+a{*})$$

and these are no longer related expressions.

Just as we obtained a family of languages from *CFGs* we obtain

Definition The *family of SSDTs* is denoted by $\boldsymbol{T}_{SSDT}$ and is defined by $\boldsymbol{T}_{SSDT} = \{T : T \text{ is an } SSDT\}$.

In Section 7.4 we sketch the proof that $\boldsymbol{T}_{SSDT} = \boldsymbol{T}_{PDT}$, which demonstrates that pushdown transducers are a model for *SSDTGs* in the same way that pushdown automata are for *CFGs*. We close this section by briefly discussing a more general version of translation grammars. Other types of translation grammars are introduced in the Exercises.

Definition A *syntax-directed translation grammar*, or *SDTG*, G is a quintuple $G = (N,\Sigma,\Delta,R,S)$, as in an *SSDTG*, except that for each $A \to \alpha,\beta$ in R we only require the nonterminals in β to be a permutation of those in α.

We still require that at each derivation step the same nonterminal in the two words of a translation form be rewritten. Because the appearances may no longer be leftmost we need to keep track of which appearances are *associated* with each other in the form. Since there may be more than one appearance of a nonterminal in the input portion of a rule, we also need to specify the associations for each rule. The association in each rule can be indicated by adding superscripts. For example, let $G = (\{S\},\{a,b,(,)\},\{0,1,2\},R,S)$, where R is

$$S \to S^{(1)}aS^{(2)},\ 0S^{(2)}S^{(1)} \mid (S),1S \mid b,2$$

Then (S,S), the initial translation form, associates the two S's. We obtain the derivation

$$(S, S) \Rightarrow (S^{(1)}aS^{(2)}, 0S^{(2)}S^{(1)})$$

using the first rule, and this in turn yields

$$(S^{(1)}aS^{(2)}aS^{(3)}, 00S^{(3)}S^{(2)}S^{(1)})$$

applying the first rule once more to the $S^{(2)}$'s. We have renumbered the superscripts uniformly in the rule in order that the associations be correctly maintained. Otherwise there would be two appearances of $S^{(1)}$ in each half of the form which would cause confusion.

An *SDTG* gives rise to a *syntax-directed translation*, or *SDT*, and the *family of syntax-directed translations*, which is denoted by $\boldsymbol{T}_{SDT}$. Every *SSDT* is, clearly, an *SDT* but the converse does not hold. For example, let $G = (\{S\},\{(,)\},\{(,)\},R,S)$, where R is given by

$$S \rightarrow S^{(1)}S^{(2)},S^{(2)}S^{(1)} \mid (S),)S(\mid \lambda,\lambda$$

Then $T(G) = \{(x,x^R) : x \text{ is in } L(G_I)\}$, where G_I is the *CFG* specified by $S \rightarrow SS \mid (S) \mid \lambda$. It can be proved that $T(G)$ is not an *SSDT*.

We close this section with an example that relates *SSDTGs* to parsing.

Example 7.3.1 Let $G = (N,\Sigma,P,S)$ be a *CFG* in which the productions are numbered arbitrarily, but uniquely, from 1 to $\#P$. We refer to the ith production by writing

$$i : A_i \rightarrow \alpha_i$$

Consider a leftmost derivation sequence in G

$$\beta_0 \overset{L}{\Rightarrow} \beta_1 \overset{L}{\Rightarrow} \cdots \overset{L}{\Rightarrow} \beta_m$$

where $\beta_0 = S$, β_m is in Σ^*, and the i_jth production $i_j : A_{i_j} \rightarrow \alpha_{i_j}$, $1 \leq i_j \leq \#P$, is used at the jth step, $1 \leq j \leq m$. We say that the sequence

$$i_1 i_2 \cdots i_m$$

is a *left parse* of β_m in G. A left parse specifies the order of the productions applied to S to obtain β_m using a leftmost derivation. It is a compact representation of the corresponding syntax tree.

We can give an *SSDTG* which gives left parses as output as follows.

Let $d_\Sigma : (N \cup \Sigma)^* \rightarrow N^*$ be a morphism which erases Σ-letters and is the identity on N-letters. More formally,

$$\text{For all } X \text{ in } N \cup \Sigma,\ d_\Sigma(X) = \begin{cases} X, & \text{if } X \text{ is in } N \\ \lambda, & \text{otherwise} \end{cases}$$

Let $\Delta = \{i : 1 \leq i \leq \#P\}$ and construct, from G, the rules of an *SSDTG*

$G' = (N, \Sigma, \Delta, R, (S, S))$ as follows.

For all i, $1 \leq i \leq \#P$,

$A \rightarrow \alpha$, $id_{\Sigma}(\alpha)$ is in R, where $A \rightarrow \alpha$ is the ith production in P.

A simple induction proof establishes that (x,y) is in $T(G')$ iff y is a left parse of x in G.

7.4 PUSHDOWN TRANSDUCERS

The pushdown transducer is an extended version of a pushdown automaton in the same way a finite transducer is an extended version of a finite automaton. It has an additional tape, *the output tape*, and an additional write head; see Figure 7.4.1.

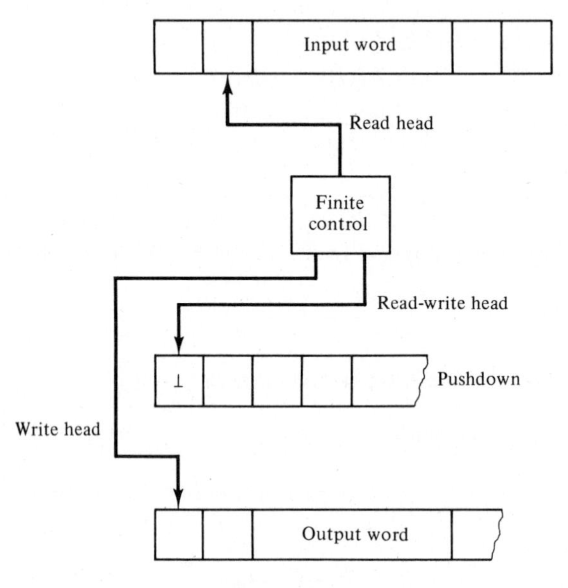

Figure 7.4.1

We consider pushdown transducers which are allowed to have λ-transitions, that is, a transition in which the read head is not advanced. Pushdown transducers with λ-transitions characterize the *SSDTs* and therefore they provide an appropriate implementation for *SSDTGs*. We sketch the proof of a restricted version of this characterization, again leaving the full theorem for the Exercises.

Definition A (*nondeterministic*) *pushdown transducer*, or *PDT*, M, is a septuple $M = (Q, \Sigma, \Delta, \Gamma, \delta, s, F)$, where

Q is a finite set of *states*;
Σ is an *input alphabet*;
Δ is an *output alphabet*;
Γ is a *pushdown alphabet*;
$\delta \subseteq Q \times (\Sigma \cup \{\lambda\}) \times \Gamma \times Q \times \Gamma^* \times \Delta^*$ is a *finite transition relation*;
s in Q is a *start state*; and
$F \subseteq Q$ is a set of *final states*.

A *configuration* of a *PDT* $M = (Q,\Sigma,\Delta,\Gamma,\delta,s,F)$ is a word $gqyqx$ in $\Gamma^*Q\Delta^*Q\Sigma^*$, where the state is repeated so that g and y can be distinguished. It specifies the current state q, the current pushdown g, the current output y, and the remaining input x. We assume $Q \cap \Delta = \varnothing$ to avoid confusion in the interpretation of a configuration. Initially, we have $\perp ssx$, where x is the input word. A *final configuration* is $\perp fyf$, where f is in F and y is in Γ^*.

Given a configuration $gZpypax$, where Z is in Γ and a is in $\Sigma \cup \{\lambda\}$, we write

$$gZpypax \vdash gg'qyy'qx$$

if (p,a,Z,q,g',y') is in δ. As before this is extended to give $\vdash^i$, $\vdash^+$, and $\vdash^*$. Given a word x in Σ^* we say y *is an output for* x *with respect to* M if

$$\perp ssx \vdash^* \perp fyf$$

for some final state f. For all x in Σ^* we denote by $M(x)$ *the set of all outputs for* x, that is,

$$M(x) = \{y : \perp ssx \vdash^* \perp fyf, \text{ for some final state } f\}$$

$M(x) = \varnothing$ implies no final configuration can be reached by x. Ignoring possible output this means that x is not accepted by M.

The *translation obtained from* M is denoted by $T(M)$ and is defined by

$$T(M) = \{(x, y) : x \text{ is in } \Sigma^* \text{ and } y \text{ is in } M(x)\}$$

A subset T of $\Sigma^* \times \Delta^*$ is called a *pushdown translation* (or *transduction*), or *PDT*, if $T = T(M)$ for some pushdown transducer M. The family of pushdown translations is denoted by $\boldsymbol{T}_{PDT}$ and is defined as

$$\boldsymbol{T}_{PDT} = \{T : T \text{ is a pushdown translation}\}$$

As for *FTs* we observe that both *Domain*(T) and *Range*(T) are *CFLs*; see the Exercises. We now sketch the proof that T is a *PDT* if and only if it is an *SSDT*. We will deal only with the special case that an *SSDT* is specified by an *SSDTG* in which the input grammar is in Greibach Normal Form. The general case is left to the Exercises.

Theorem 7.4.1 *Let* $G = (N,\Sigma,\Delta,R,S)$ *be an SSDT grammar in which the input grammar is in GNF. Then there is a PDT,* M, *such that* $T(M) = T(G)$.

Proof: This is by construction; it parallels the proof of Lemma 5.2.1. Construct a *PDT* $M = (\{s,f\},\Sigma,\Delta,N \cup \Delta \cup \{\perp\},\delta,s,\{f\})$ from G as follows.

(i) For all a in Σ, $(s,a,\perp,f,\perp\beta^R,\lambda)$ is in δ,
if $S \rightarrow a\alpha$, β is in R, for some α in $(N \cup \Sigma)^*$ and β in $(N \cup \Delta)^*$.
(ii) For all a in Σ and all A in N, $(f,a,A,f,\beta^R,\lambda)$ is in δ,
if $A \rightarrow a\alpha$, β is in R, for some α in $(N \cup \Sigma)^*$ and β in $(N \cup \Delta)^*$.
(iii) For all a in Δ, $(f,\lambda,a,f,\lambda,a)$ is in δ.

The only difference between this construction and that of Construction 5.2.1 is that we allow λ-transitions with respect to the input if a Δ-symbol appears on the push-down. This allows output to be generated in the correct position. A straightforward inductive proof that $T(M) = T(G)$ is left to the Exercises. ■

Conversely, given a *PDT*, M, an *SSDT* grammar G can be constructed from it such that $T(G) = T(M)$.

Theorem 7.4.2 *Let $M = (Q,\Sigma,\Delta,\Gamma,\delta,s,F)$ be a PDT. Then there exists an SSDT grammar $G = (N,\Sigma,\Delta,R,S)$ with $T(G) = T(M)$.*

Proof: A construction similar to Construction 5.2.3 is given which produces an *SSDTG* G from M.

Let $N = \{[pAq] : p,q$ are in Q and A is in $\Gamma\} \cup \{S\}$, where S is a completely new symbol, and define R by

(1) If $(p,a,A,r,Z_1 \cdots Z_k,y)$ is in δ, $k > 0$, then R contains the rules

$$[pAq_k] \rightarrow a[rX_1q_1][q_1X_2q_2] \cdots [q_{k-1}X_kq_k], y[rX_1q_1][q_1X_2q_2] \cdots [q_{k-1}X_kq_k]$$

for all sequences $q_1, \ldots, q_k$ of states from Q.
(2) If (p,a,A,r,λ,y) is in δ, then R contains the rule

$$[pAr] \rightarrow a,y$$

(3) R contains the rules

$$S \rightarrow [s\perp f],[s\perp f]$$

for all f in F.

To prove that $T(G) = T(M)$ it suffices to establish, by induction on m and n, that

For all A in Γ and for all p, q in Q
$([pAq],[pAq]) \Rightarrow^m (x,y)$ in G if and only if $Appx \vdash^n qyq$ in M

Since, in this case, $\perp ssx \vdash^n \perp fyf$, for some f in F, if and only if

$([s\perp f],[s\perp f]) \Rightarrow^m (x,y)$ in G, that is, if and only if $S \Rightarrow^{m+1} (x,y)$ in G. Hence, $T(G) = T(M)$ as desired. ■

7.5 TURING MACHINES AND COMPUTABLE FUNCTIONS

Our fourth and most important device with output is the Turing machine. In this case we do not introduce a special output tape, we simply use the given tape as an input, output, and work tape, as we did in Section 6.1; see Figure 7.5.1. However, to include the possibility that we have a different output alphabet we modify the definition slightly. Although this is not really necessary, it serves to distinguish the use of a Turing machine as a transducer from its use as a language recognizer or decision maker.

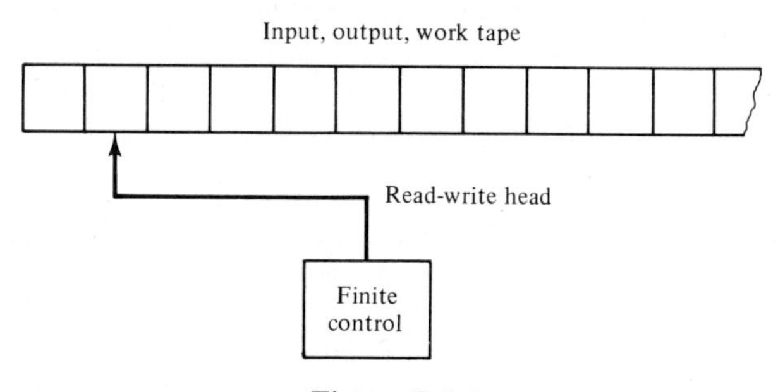

Figure 7.5.1

Definition A *deterministic Turing transducer*, or *DTT*, M, is a septuple $M = (Q,\Sigma,\Delta,\Gamma,\delta,s,f)$, where

- Q is a finite set of *states*;
- Σ is an *input alphabet*;
- Δ is an *output alphabet*;
- Γ is a *tape alphabet* with $\Sigma \cup \Delta \cup \{\boldsymbol{B}\} \subseteq \Gamma$, where $\boldsymbol{B}$ is the *blank* symbol;
- $\delta : Q \times \Gamma \rightarrow Q \times \Gamma \times \{L,R,\lambda\}$ is a *transition function*;
- s in Q is a *start state*; and
- f in Q is a *final state*.

Apart from the introduction of the output alphabet Δ the *DTT* is identical to the *DTM*. The definition of a configuration of a *DTT* is also identical to that of a *DTM*, as are configuration sequences, so as for *DTMs* we assume $\Gamma \cap Q = \varnothing$. The only other differences are in the definition of starting and final configurations. A starting configuration has the form sx, where $x = \boldsymbol{B}x_1\boldsymbol{B}x \cdots \boldsymbol{B}x_m$, for some x_i in Σ^*, $1 \leq i \leq m$, and for some $m \geq 1$. A final configuration has the form $f\boldsymbol{B}y_1\boldsymbol{B} \cdots \boldsymbol{B}y_n$, for some y_i in Δ^*, $1 \leq i \leq n$, and for some $n \geq 1$. Forcing the leftmost cell to contain a blank simplifies many constructions and forcing the final configuration to have the read-write head over the leftmost cell does not restrict the generality of the model.

DTTs are used to compute functions as follows.

Definition Let $M = (Q,\Sigma,\Delta,\Gamma,\delta,s,f)$ be a *DTT* and $g : (\Sigma^*)^m \rightarrow (\Delta^*)^n$ be an m,n-place function for some $m,n \geq 1$. We say M *computes* g if

For all $x_1, \ldots, x_m$ in Σ^* and for all $y_1, \ldots, y_n$ in Δ^*,

$$s\boldsymbol{B}x_1\boldsymbol{B} \cdots \boldsymbol{B}x_m \vdash^* f\boldsymbol{B}y_1\boldsymbol{B} \cdots \boldsymbol{B}y_n \text{ iff}$$
$$g(x_1, \ldots, x_m) = (y_1, \ldots, y_n)$$

We say that an m,n-place function $g : (\Sigma^*)^m \rightarrow (\Delta^*)^n$ is (*Turing-*) *computable* if there is a *DTT* M which computes g.

If $s\boldsymbol{B}x_1\boldsymbol{B} \cdots \boldsymbol{B}x_m \vdash^* fy$, where y contains a symbol in $\Gamma - (\{\boldsymbol{B}\} \cup \Delta)$, then $g(x_1, \ldots, x_m) =$ ***undef***. If the read-write head ends up over some cell other than the leftmost cell, then $g(x_1, \ldots, x_m) =$ ***undef***, as it is if M hangs or never halts on input $\boldsymbol{B}x_1 \cdots \boldsymbol{B}x_m$.

Examples 6.1.5 and 6.1.6 can easily be modified for the *DTT*. Let us examine two further examples.

Example 7.5.1 The function $add(m,n) = m + n$, for all nonnegative integers m and n, is Turing-computable.

We assume $\Sigma = \Delta = \{1\}$, that is, the input is given in unary notation with 1^i representing i, for all $i \geq 0$. On input the tape contents are as shown in Figure 7.5.2(a), while on output we have Figure 7.5.2(b). This means we should move the final 1 to close the gap between m and n, taking care of the case that $n = 0$. We obtain the *DTT* of Figure 7.5.3.

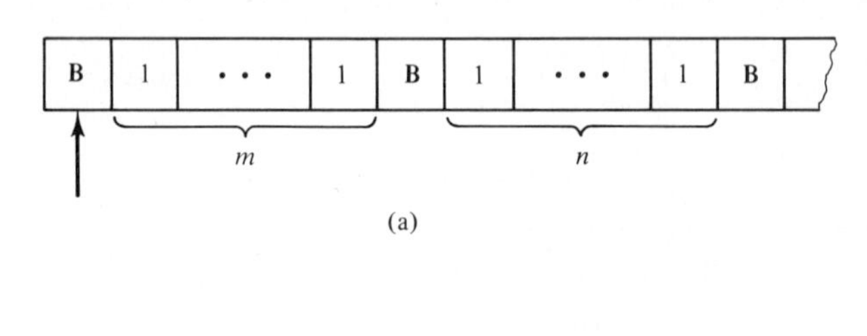

(a)

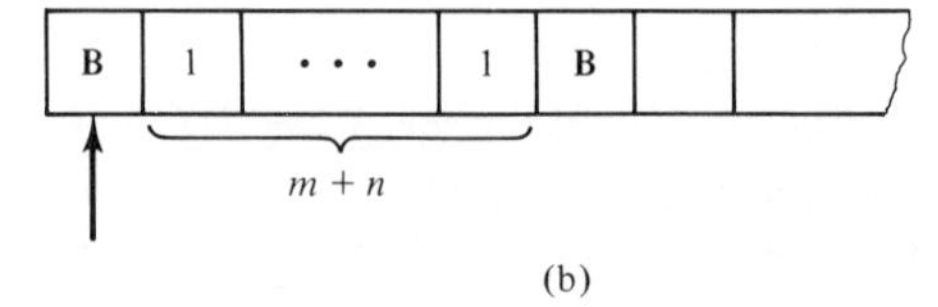

(b)

Figure 7.5.2

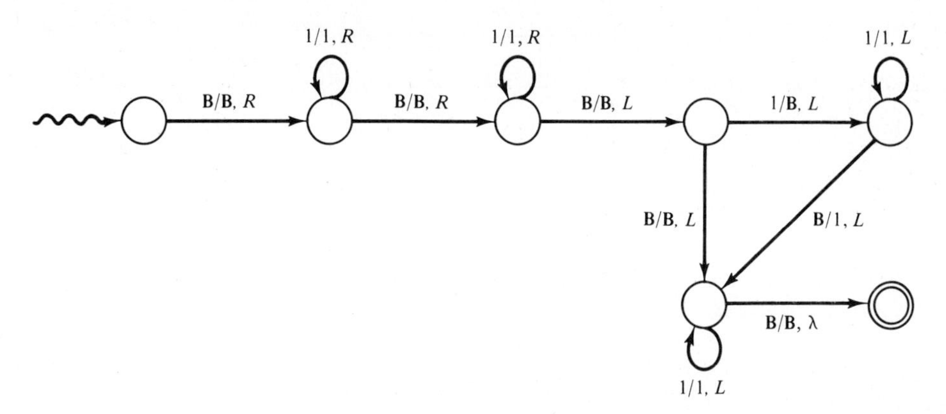

Figure 7.5.3

The various simplification and extension results proved, with respect to acceptance, in Sections 6.3 and 6.4 carry over to *DDTs* and computable functions. Thus, the functions computable by multitape *DTTs* are computable by *DTT*'s and vice versa. We make use of this result in our next example.

Example 7.5.2 The function $mult(m,n) = m \times n$, for all nonnegative integers m and n, is computable. We assume $\Sigma = \Delta = \{1\}$ once more.

We use a three-tape *DTT* to compute *mult*. We require the input to be transformed as shown in Figure 7.5.4, if both m and n are nonzero, otherwise the result is zero. Now the *DTT* makes one copy of tape 2 on tape 1, resets the read-write head on tape 2, and moves the read-write head on tape 3 one cell to the right. This process is repeated until the head on tape 3 is over a blank cell, when m will have been copied exactly n times to tape 1, that is, tape 1 contains $m \times n$. More formally,

```
while symbol 3 = 1 do
begin
    while symbol 2 = 1 do
    begin rewrite symbol 1 as 1;
        move 1 right;
        move 2 right
    end;
    move 2 left;
    while symbol 2 = 1 do move 2 left;
    move 2 right;
    move 3 right
end;
```

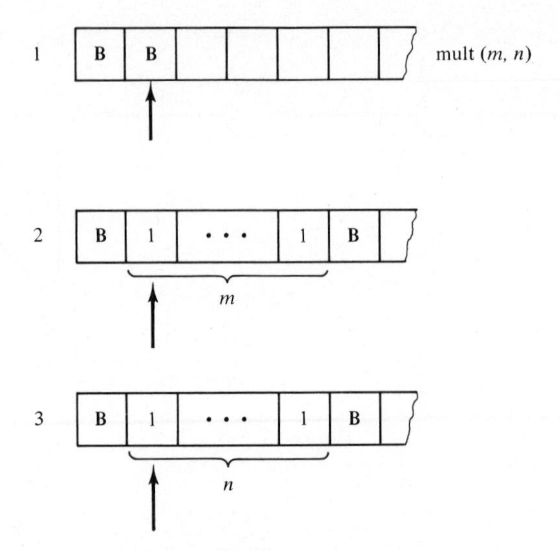

Figure 7.5.4

To transform the initial input we first perform

{Check for $m = 0$ or $n = 0$ } *move* 1 *right*;
if *symbol* 1 = 1 **then**
begin while *symbol* 1 = 1 **do** *move* 1 *right*;
    move 1 *right*;
    if *symbol* 1 = 1 **then**
    begin while *symbol* 1 = 1 **do** *move* 1 *right*;
        move 1 *left*
    end else
    begin {$n = 0$}
        move 1 *left*;
        move 1 *left*;
        while *symbol* 1 = 1 **do**
            rewrite symbol 1 as ***B*** and *move left*;
    end
end else
begin {$m = 0$}
    move 1 *right*;
    while *symbol* 1 = 1 **do** *move* 1 *right*;
    move 1 *left*;
    while *symbol* 1 = 1 **do**
        rewrite symbol 1 as ***B*** and *move left*;
    move 1 *left*;
    enter final state
end;

```
{On termination the read-write head is over the rightmost 1 if m ≠ 0 ≠ n, other-
wise it is over the leftmost cell of a completely blank tape.}
```

We may now transform the input as required.

```
{Transfer n to tape 3}
move 3 right;
while symbol 1 = 1 do
begin rewrite symbol 1 as B;
      move 1 left;
      rewrite symbol 3 as 1;
      move 3 right
end;
move 1 left;
move 3 left;
while symbol 3 = 1 do move 3 left;
move 3 right;
{Transferring m to tape 2 is identical, except that 3 is replaced by 2 throughout}
move 1 right {move to the second cell}
```

Not only is the class of computable functions larger than those defined by finite and pushdown transducers, but it includes all partial functions that can be computed by any formal machine, rewriting system, or programming language. For this reason it is hypothesized that whenever a new method of defining functions is introduced, these functions can always be computed by a Turing machine; in other words, the computable functions exhaust all functions which can be computed by an algorithm. This is known as Church's thesis and it implies that our intuitive notions of algorithm are captured by the computable functions. Clearly it cannot be proved, since this requires that an intuitive notion of an algorithm must be formalized, which gives another formal specification method and another Church's thesis, and so on. However, evidence in its favor is overwhelming. Many methods of specifying algorithms have been introduced, for example, Post and Markov systems and multitape and multihead Turing machines, but in each case they have been shown to define exactly the computable functions.

7.6 RECURSIVE FUNCTIONS

In this section the final models for computable functions are introduced, namely, the primitive recursive functions and the recursive functions. It can be shown that every computable function is a recursive function, and vice versa. However, this is beyond the scope of the text, so we only present their definition and some examples. We begin by defining primitive recursive functions which are then extended to give the recursive functions.

Throughout this section we assume all functions are m-place, for some $m \geq 1$, their domain is $\mathbb{N}_0^m$, and their range is $\mathbb{N}_0$. We will define a class of functions recursively based on two ways of combining functions to obtain new functions.

Given an m-place function f and m n-place functions $g_1, \ldots, g_m$ we obtain the n-place function h as follows. For all $x_1, \ldots, x_n$ in $\mathbb{N}_0$,

$$h(x_1, \ldots, x_n) = f(g_1(x_1, \ldots, x_n), \ldots, g_m(x_1, \ldots, x_n))$$

We say h is obtained by *composition* from f and $g_1, \ldots, g_m$. If f is a 1-place function, then the above definition reduces to the usual definition of composition. If $f, g_1, \ldots, g_m$ are total, then g is also total.

Example 7.6.1 Consider $f(x,y) = 3x^2 + y$, $g_1(x,y,z) = x + y + z$, and $g_2(x,y,z) = x - y - z$. Then

$$h(x,y,z) = f(g_1(x,y,z), g_2(x,y,z)) = 3(x+y+z)^2 + x - y - z.$$

The second method of combining functions involves recursion. Given a two-place total function g and an integer k a new one-place total function h can be defined by

$$h(i) = \begin{cases} k, & \text{if } i = 0 \\ g(i-1, h(i-1)), & \text{otherwise} \end{cases}$$

We say h is obtained from g by *(primitive) recursion*. More generally, let f be an m-place total function and g be an $(m+2)$-place total function, then a new $(m+1)$-place total function h can be defined by

$$h(x_1, \ldots, x_{m+1}) = \begin{cases} f(x_1, \ldots, x_m), \text{ if } x_{m+1} = 0 \\ g(x_{m+1}-1, h(x_1, \ldots, x_m, x_{m+1}-1), x_1, \ldots, x_m), \\ \quad \text{otherwise} \end{cases}$$

Again we say h is obtained from f and g by *(primitive) recursion*. Observe that when $m = 0$, this, essentially, reduces to the first definition.

Example 7.6.2 Let $k = 1$ and $f(x,y) = (x+1) \times y$; then

$$h(i) = \begin{cases} 1, & \text{if } i = 0 \\ f(i-1, h(i-1)), & \text{otherwise} \end{cases}$$

Now, we obtain $h(0) = 1$, $h(1) = 1 \times h(0) = 1$, $h(2) = 2 \times h(1) = 2$, and $h(3) = 3 \times h(2) = 6$, etc. In other words, $h(i) = i!$

Before defining our first class of functions we need some *initial functions*. The first and simplest is the *constant function zero*, defined by $zero(x) = 0$, for all x. Second, we need the *successor function succ*, defined by $succ(x) = x+1$, for all x. Third, we need the class of *projection functions* π_i^m, for all i, $1 \leq i \leq m$, and for all $m \geq 1$, defined by $\pi_i^m(x_1, \ldots, x_m) = x_i$, for all x_j, $1 \leq j \leq m$.

Example 7.6.3 Define the addition function $add(x,y)$ by $add(x,y) = x+y$. It satisfies the recurrence

$$add(x,y) = \begin{cases} x, & \text{if } y = 0 \\ add(x,y-1)+1, & \text{otherwise} \end{cases}$$

This in turn can be written as

$$add(x,y) = \begin{cases} \pi_1^1(x), & \text{if } y = 0 \\ g(y-1, add(x,y-1), x), & \text{otherwise} \end{cases}$$

where $g(x,y,z) = succ(\pi_2^3(x,y,z))$. Thus $add(x,y)$ is obtained by recursion and composition from the initial functions.

We now have

Definition An m-place function h is a *primitive recursive function* if

(i) it is an initial function;
(ii) it is obtained from primitive recursive functions f and $g_1, \ldots, g_m$ by composition;
(iii) it is obtained from a primitive recursive function f by recursion.

Thus, each primitive recursive function is defined by a finite number of applications of composition and recursion with the initial functions. Since the initial functions are total and composition and recursion preserve totality, all primitive recursion functions are total.

Example 7.6.4 Example 7.6.3 has demonstrated that *add* is primitive recursive. In a similar manner we can prove that $mult(x,y) = x \times y$ is also primitive recursive.

We have

$$mult(x,y) = \begin{cases} 0, & \text{if } y = 0 \\ add(x, mult(x,y-1)), & \text{otherwise} \end{cases}$$

which can be rewritten as

$$mult(x,y) = \begin{cases} zero(x), \text{ if } y = 0 \\ add(\pi_2^3(y-1, mult(x,y-1),x), \pi_3^3(y-1, mult(x,y-1),x)), \\ \quad \text{otherwise} \end{cases}$$

Now *zero*, *add*, π_2^3, and π_3^3 are total and primitive recursive, hence *mult* is primitive recursive.

Example 7.6.5 It is straightforward to define a function which tests if its single argument is zero or nonzero. We call this function *test*, and it is defined as

$$test(x) = \begin{cases} 1, \text{ if } x = 0 \\ 0, \text{ if } x \neq 0 \end{cases}$$

It is primitive recursive, since it can be defined by

$$test(x) = \begin{cases} 1, \text{ if } x = 0 \\ zero(\pi_1^2(x-1, test(x-1))), \text{ otherwise} \end{cases}$$

The class of primitive recursive functions do not exhaust the computable functions. This is not only because the class is enumerable (see the Exercises), but also because it is a class of total functions. To see this let $f_0, f_1, \ldots, f_n$ be an enumeration of the primitive recursive functions and let us define a function $g : \mathbb{N}_0 \rightarrow \mathbb{N}_0$ by

For all $i \geq 0$, $g(i) = f_i(i, \ldots, i) + 1$

where f_i takes the value i for each of its arguments. Now g is computable, since given $g(i)$, for some $i \geq 0$, we simply evaluate the ith primitive recursive function f_i with i for each of its arguments. However, if g is primitive recursive, then it must equal f_j, for some $j \geq 0$. But this implies, by diagonalization, that

$$f_j(j) = g(j) = f_j(j) + 1$$

— a contradiction, because the primitive recursive functions are total. Therefore, g is not primitive recursive.

Historically, the Ackermann function, which we now define, is a relatively simple function which is not primitive recursive. The proof of this result is beyond the scope of the present text. It is also defined recursively, but not, of course, by primitive recursion. The Ackermann function $A : \mathbb{N}_0 \times \mathbb{N}_0 \rightarrow \mathbb{N}$ is defined by

$$A(x,y) = \begin{cases} succ(y), & \text{if } x = 0 \\ A(x-1,1), & \text{if } y = 0 \\ A(x-1, A(x,y-1)), & \text{otherwise} \end{cases}$$

For this reason we introduce a new operation on functions, which properly extends the primitive recursive functions, since it produces partial functions.

Definition The *minimalization* of a $(m+1)$-place total function g is the m-place function f such that, for all $x_1, \ldots, x_m$ in $\mathbb{N}_0$

$$f(x_1, \ldots, x_m) = \text{the smallest } k \text{ such that } g(x_1, \ldots, x_m, k) = 0,$$
$$\text{if such a } k \text{ exists.}$$

We say that f *is obtained by minimalization from* g. Clearly, f is a partial function, which need not be total.

Definition A function h is (*partial*) *recursive* if

(i) it is an initial function;
(ii) it is obtained from recursive functions f and $g_1, \ldots, g_m$ by composition;
(iii) it is obtained from a recursive function f by recursion;
(iv) it is obtained from a total recursive function f by minimalization.

Example 7.6.6 Let g be a two-place function defined by

$$g(i,j) = 1, \text{ if } i \text{ and } j \text{ are twin primes}$$

Then g is not total. However, defining g as

$$g(i,j) = \begin{cases} 1, & \text{if } i \text{ and } j \text{ are twin primes} \\ 0, & \text{otherwise} \end{cases}$$

results in a total function.

Example 7.6.7 Let g be defined by

$$g(i,j,k) = \begin{cases} 0, & \text{if either } j = 0 \text{ or } jk \geq i \\ 1, & \text{otherwise} \end{cases}$$

Then g is primitive recursive. We can define $\lceil i/j \rceil$ by using minimalization

$$f(i,j) = \text{the smallest } k \text{ such that } g(i,j,k) = 0$$

It is simple to ignore the case of the divisor $j = 0$ since $f(i,j) = 0$ exactly when $i = 0$ or $j = 0$. Note that f is total.

7.7 ADDITIONAL REMARKS

7.7.1 Summary

We have introduced models for the purpose of defining FUNCTIONS and RELATIONS which, in this context, are also known as TRANSDUCTIONS and TRANSLATIONS. The first four models are extensions of earlier models: FINITE TRANSDUCERS, SYNTAX-DIRECTED TRANSLATION GRAMMARS, PUSHDOWN TRANSDUCERS, and TURING TRANSDUCERS. The fifth and final model is also the earliest model based on computing functions recursively from some given functions and methods of combining functions. It defines the class of RECURSIVE FUNCTIONS which includes as a proper subclass the class of PRIMITIVE RECURSIVE FUNCTIONS.

7.7.2 History

Finite transducers were introduced in Elgot and Mezei (1965). It has assumed a prominent role in the theory of languages; see Berstel (1979), for example. Syntax-directed translations were first discussed in print in Irons (1961) and Barnett and Futrelle (1962), although the formalization treated here is due to Lewis and Stearns (1968). Consult Aho and Ullman (1969a, 1969b, 1971, 1972, 1973) for similar schemes. The development of the compiler-compiler YALL (see Johnson, 1974) is based to some extent on these techniques. Pushdown transducers were first formalized in Evey (1963) and have been studied more recently in Choffrut and Culik II (1983). The Turing transducer is one method of formalizing how a *TM* computes a function, that is, how a function is defined to be computable. This notion was introduced by Turing (1936), the paper in which Turing machines were introduced. Davis (1958) and Hermes (1969), for example, provide a detailed investigation of Turing-computable functions. Finally, the notions of primitive recursive functions and recursive functions are due to Kleene (1936, 1943). The books of Peter (1967), Rogers (1967), and Machtey and Young (1978) are good sources for further study.

7.8 SPRINGBOARD

7.8.1 Lexical Analysis

Partitioning a file of characters, representing some PASCAL program, into sequences of tokens or PASCAL symbols is the first step in the compilation of the program; see Example 7.1.2. Investigate how a lexical analysis generator can be constructed. The work of Johnson et al. (1968) on AED RWORD should be consulted and that of Thompson (1968) on QED. Aho and Ullman (1977) is also a useful source.

7.8.2 Syntax-Directed Translations

Section 7.2 gives a first glimpse of the possibilities in this direction. However, it does not exhaust them by any means. In practice *SDTGs* are too restrictive and because of this a number of alternatives have been suggested. Consult Lewis,

Rosenkrantz, and Stearns (1976) and Aho and Ullman (1973).

7.8.3 File Compaction

Any technique for compacting text can also be applied to the compaction of binary files, for example, the compiled versions of programs. However, in this case compaction must be preceded by an investigation of the structure of such files — should a byte be the "character" on which compaction is based and, if so, what is their probability distribution? Such compaction techniques can enable a machine with a small amount of main memory to run programs, which were originally too large to be held in main memory. Investigate how such a system could be implemented.

7.8.4 Translation Expressions

Regular and context-free expressions can be generalized to denote translations. Rather than symbols appearing by themselves we have symbol pairs (a,b), where a is in the input alphabet and b is in the output alphabet. It is essential to allow either symbol in such a pair to be λ as well. Catenation of two word pairs $(x_1,y_1)(x_2,y_2)$ is performed element by element yielding (x_1x_2,y_1y_2). The operation $\cup$ is unchanged but $*$ is also defined element by element. Investigate regular translation expressions and context-free translation expressions. Are there theorems for the former similar to those in Chapter 3 for regular expressions? Consult Vere (1970) and Gray and Harrison(1966).

7.8.5 String Similarity

Johnson (1983) uses regular relations, that is, translations defined by finite transducers, to provide formal models of string similarity. String similarity measures are important in the real world whenever a search key into a large database can be erroneous. For example, airline reservation systems usually use a traveler's name as a search key. This could have been entered incorrectly initially or subsequently, but in both cases it is necessary to retrieve the correct record. For this reason a number of name compression schemes have come into use, which identify similar sounding names, for example NYSIIS described by Moore et al. (1977). In NYSIIS a name is reduced to, at most, a four letter sequence which captures in most cases, the essence of the name. Investigate how this and similar systems can be modeled by finite transducers. String searching programs can then be extended to provide a search for all variants of a name or string under such a model. Design such a system.

7.8.6 Recursive Function Theory

Section 7.5 merely touches the surface of a rich field — recursive function theory. Investigate this area in more detail, see Davis (1958), Hermes (1969), and Rogers (1967).

EXERCISES

2.1 Construct finite transducers for the following translations.

(i) $\{(a^i b^{2j}, b^{2i} a^j) : i,j \geq 0\}$.
(ii) $\{(a^i b^j, a^i b a^j) : i,j \geq 0\} \cup \{(a^i b^j a^k, a^{i+k}) : i,j,k \geq 0\}$.
(iii) $\{(a^{3i+1}, b^i) : i \geq 0\}$.

2.2 *Lazy finite transducers* are an extension of finite transducers in which $\delta \subseteq Q \times \Sigma^* \times Q \times \Delta^*$, that is, words over the input alphabet are allowed in transitions. Prove that $T \subseteq \Sigma^* \times \Delta^*$ is a finite transduction iff it is a lazy finite transduction.

2.3 *A two-tape finite automaton* M is a sextuple $M = (Q,\Sigma,\Delta,\delta,s,F)$, where Q, s, and F are the same as usual, Σ is the alphabet of the first tape, Δ is the alphabet of the second tape, and $\delta \subseteq Q \times (\Sigma \cup \{\lambda\}) \times (\Delta \cup \{\lambda\}) \times Q$ is a finite transition relation. M has two read heads both placed initially over the first symbol of their input. A configuration $p(x,y)$ is a word in $Q(\Sigma^*,\Delta^*)$. It means that p is the current state, x is the remaining input on tape 1, and y is the remaining input on tape 2. A final configuration has the form $f(,)$, where f is in F.

(i) Define $\vdash$, $\vdash^i$, $\vdash^+$, and $\vdash^*$.
(ii) Define the two-tape *FA* relation, that is, the subset of $\Sigma^* \times \Delta^*$, accepted by M. This is denoted by $T(M)$.
(iii) Prove that $T \subseteq \Sigma^* \times \Delta^*$ is a two-tape *FA* relation iff it is a finite transduction.

3.1 Letting G be $S \rightarrow ST+ \mid T$; $T \rightarrow TF* \mid F$; $F \rightarrow S \mid a$, that is, a postfix expression grammar, construct *SSDTGs* to

(i) translate postfix to fully parenthesized infix;
(ii) translate postfix to prefix;
(iii) translate each postfix expression into its operator sequence.

3.2 Since syntax trees are needed in the compiling process it is interesting to extend translation grammars so that the second component is applied to a tree. For example, we might have

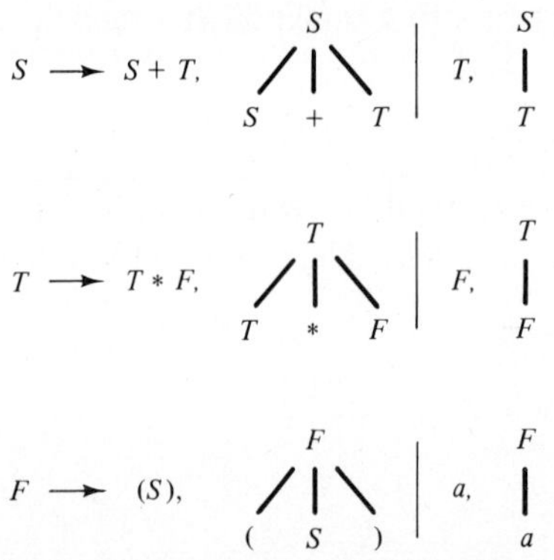

as a tree translation grammar for expressions. It produces the syntax tree corresponding to a derivation, if by leftmost replacement we intend leftmost replacement in the current frontier. With this expression grammar construct a tree translation grammar which produces operators rather than nonterminals at internal nodes. In other words, its output is an expression tree.

4.1 Construct pushdown transducers for the following translations.

(i) $\{(a^i b^i c^j, a^i b^i c^j) : i,j \geq 1\}$.
(ii) $\{(wc, w^R) : w$ is in $\{a,b\}^*\}$.
(iii) $\{(w, w^R) : w$ is in $\{a,b\}^*\}$.

4.2 We can define lazy *PDTs* as we defined lazy *FTs*. Prove that lazy *PDTs* are no more powerful than *PDTs*.

4.3 In analogy with the definition of a two-tape *FA* we can define a two-tape *PDA* and the relation it accepts. Do two-tape *PDAs* characterize pushdown transductions? Justify your answer.

4.4 Prove Theorems 7.3.1 and 7.3.2.

5.1 Design *DTTs* to compute the following functions.

(i) $monus(m,n) = \begin{cases} 0, & \text{if } m < n \\ m-n, & \text{otherwise} \end{cases}$

(ii) $div(m,n) = \begin{cases} 0, & \text{if } n = 0 \\ q, & \text{otherwise, where } m = nq+r \text{ with } 0 \leq r < n \end{cases}$

(iii) $dup(x) = xx$, for all x in $\{a,b\}^*$.
(iv) $half(x) = y$, if $x = yz$ and $|y| = |z|$, for all x in $\{a,b\}^*$.
(v) $rdup(x) = xx^R$, for all x in $\{a,b\}^*$.
(vi) $unary(x) = 1^k$, for all x in $\{0,1\}^*$, where k represents, in unary, the same number represented by x in binary.
(vii) $binary(1^i) =$ the binary representation of i, so $unary(binary(1^i)) = 1^i$.

6.1 Demonstrate that the following functions are primitive recursive.

(i) $div(m,n) = \begin{cases} 0, & \text{if } n = 0 \\ q, & \text{otherwise, where } m = nq+r \text{ with } 0 \leq r < n \end{cases}$

(ii) $exp(m,n) = \begin{cases} 1, & \text{if } n = 0 \\ m^n, & \text{otherwise} \end{cases}$

(iii) $sqrt(m) = \begin{cases} 0, & \text{if } m = 0 \\ n, & \text{otherwise, where } n^2 \leq m < (n+1)^2 \end{cases}$

$$\text{(iv) } gcd(m,n) = \begin{cases} 0, \text{ if } m = 0 \text{ or } n = 0 \\ d, \text{ otherwise, where } m = id, n = jd, \text{ some } i \text{ and } j, \\ \quad \text{and there is no } d' > d \text{ with the same property} \end{cases}$$

6.2 We can define *primitive recursive word functions* over $\{a,b\}$ in a similar manner to primitive recursive functions. We have the initial functions: $empty(x) = \lambda$, for all x in $\{a,b\}^*$, $succ(x,X) = xX$, for X in $\{a,b\}$ and for all x in $\{a,b\}^*$, and if $x \neq \lambda$, then $tail(x) = y$ and $head(x) = X$, where $x = Xy$, for some X in $\{a,b\}$, for all x in $\{a,b\}^+$, otherwise $head(\lambda)$ and $tail(\lambda) =$ ***undef***. *Projection*, *composition*, and *recursion* are defined as before with $\mathbb{N}_0$ replaced by $\{a,b\}^*$ throughout, integers replaced by words, and 0 by λ.

For example, we can define catenation by

$$cat(x,y) = \begin{cases} x, \text{ if } y = \lambda \\ cat(succ(x,head(y)),tail(y)), \text{ otherwise} \end{cases}$$

Prove that the following word functions are primitive recursive.

(i) $dup(x) = xx$, for all x in $\{a,b\}^*$.

$$\text{(ii) } half(x) = \begin{cases} y, \text{ if } x = yz \text{ and } |y| = |z|, \text{ for all } x \text{ in } \{a,b\}^* \\ \textbf{undef}, \text{ otherwise} \end{cases}$$

(iii) $reverse(x) = x^R$, for all x in $\{a,b\}^*$.

(iv) $rdup(x) = xx^R$, for all x in $\{a,b\}^*$.

PROGRAMMING PROJECTS

P2.1 Implement a lexical analyzer for PASCAL or your favorite programming language.

P2.2 Implement a system to automatically produce an encoder program and a decoder program for the encoding scheme discussed in Example 7.1.1.

P2.3 Implement a system to produce lexical analyzers automatically. The key issue is how are they specified, since providing the transition relation is too cumbersome and too low level. Examine LEX, available under UNIX, to see how this exercise might be done. See the references in Section 7.7.1.

P2.4* Implement an un-TROFF system. This should take a TROFF text file and remove all formatting commands, but leave all macro-calls which result in special symbols. TROFF is the typesetting system developed at AT&T Bell Laboratories.

P2.5 Implement a system to produce, when given an alphabet with probabilities, an optimal prefix code, and the associated encoding and decoding programs.

P3.1 Implement a system which given a *CFG* will produce a parsing program, based on the *SSDTG* approach discussed in the text.

P4.1 Implement a program to convert *SSDTGs* to *PDTs*, and vice versa.

P5.1 Implement a program to convert functions computable on a *DTT* into recursive functions.

P6.1 Primitive recursive functions can be used directly as a programming language for functions over the nonnegative integers. Implement an on-line interpreter for such functions.

Part III

PROPERTIES

Chapter 8

Family Relationships

8.1 FINITE, REGULAR, AND CONTEXT-FREE LANGUAGES

8.1.1 Introduction

In Chapters 2-6 we introduced a variety of models for the specification of languages and, hence, families of languages. At the same time we have seen that only four of these can be distinct *in the sense that they specify different families.* The finite automata of Chapter 2 and the regular expressions of Chapter 3 both specify the family of regular languages $\boldsymbol{L}_{REG}$. Similarly, the context-free grammars of Chapter 4 and the nondeterministic pushdown automata of Chapter 5 both specify the family of context-free languages $\boldsymbol{L}_{CF}$. Finally, Chapter 1 introduces $\boldsymbol{L}_{FIN}$, the family of finite sets, while Chapter 6 considers Turing machines and the associated families of recursive languages $\boldsymbol{L}_{REC}$ and *DTM* languages $\boldsymbol{L}_{DTM}$. The *DTM* languages are also known as the *recursively enumerable languages* and the family is, in this case, denoted by $\boldsymbol{L}_{RE}$. We add to these families the family of all languages $\boldsymbol{L}_{ALL}$, that is, $\boldsymbol{L}_{ALL}$ contains for each alphabet Σ, all subsets of Σ^*.

It is the aim of the present chapter to prove that these families[†] form a proper hierarchy, namely,

$$\boldsymbol{L}_{FIN} \subset \boldsymbol{L}_{REG} \subset \boldsymbol{L}_{CF} \subset \boldsymbol{L}_{REC} \subset \boldsymbol{L}_{DTM} \subset \boldsymbol{L}_{ALL}$$

This is carried out in two stages; first, we show that

[†] We actually prove rather more; see Section 8.5.

$$\mathbf{L}_{FIN} \subseteq \mathbf{L}_{REG} \subseteq \mathbf{L}_{CF} \subseteq \mathbf{L}_{REC} \subseteq \mathbf{L}_{DTM} \subseteq \mathbf{L}_{ALL}$$

and, second, we prove that these containments are proper. In this second stage the notion of a nonlanguage is crucial. For example, if we have proved that $\mathbf{L}_{REG} \subseteq \mathbf{L}_{CF}$, then to prove that $\mathbf{L}_{REG} \subset \mathbf{L}_{CF}$, we need to exhibit a language which is context-free but is not in $\mathbf{L}_{REG}$. We say that such a language is a *non-language* with respect to $\mathbf{L}_{REG}$.

Proving that a given language is a nonlanguage with respect to a given family $\mathbf{L}$ is, in general, a nontrivial task. Thus, a number of tools are provided which help us to prove such results. The importance of the proper hierarchy demonstrated in this chapter is that it confirms our suspicions that some of the specification techniques of Chapters 2-6 do have different expressive power.

One final comment: The demonstration of the proper hierarchy, although important, leaves much to be desired. For example, once we know that $\mathbf{L}_{CF} - \mathbf{L}_{REG} \neq \emptyset$ we may also want to know how nonempty this difference is, that is, how many more languages there are in $\mathbf{L}_{CF} - \mathbf{L}_{REG}$ than in $\mathbf{L}_{REG}$. Furthermore, we may also want to know how different such languages are from those in $\mathbf{L}_{REG}$, that is, what features of a nonregular context-free language make it nonregular? As we shall see some partial answers to these questions are provided, in passing, as we prove the hierarchy results. More detailed answers lead us to the forefront of research activity in formal language theory. We begin with the finite, regular, and context-free languages.

The containment $\mathbf{L}_{FIN} \subseteq \mathbf{L}_{REG}$ is most easily seen by way of regular expressions. Let L be an arbitrary finite language over some alphabet Σ. Since L is finite, $\#L = n$, for some $n \geq 0$. If $n = 0$, then $L = \emptyset$ and this is denoted by the regular expression $E = \boldsymbol{\emptyset}$. On the other hand, if $n \geq 1$, then we can enumerate the words in L as $x_1, \ldots, x_n$ say. In this case let $E = x_1 \cup x_2 \cup \cdots \cup x_n$. Clearly, $L(E) = L$ in both cases; therefore, $\mathbf{L}_{FIN} \subseteq \mathbf{L}_{REG}$.

Proper containment is straightforward, since we need only exhibit an *infinite* regular language. For example, $L = \{a\}^*$ is such a language. Clearly, L is infinite and regular since it is denoted by the regular expression a^*.

Thus, we have proved:

Theorem 8.1.1 $\mathbf{L}_{FIN} \subset \mathbf{L}_{REG}$.

As we have pointed out above, this is not the end of the story. For example, for any infinite (regular) language L we can define an infinite sequence of finite languages

$$L_0 \subseteq L_1 \subseteq L_2 \subseteq \cdots$$

where $L_i = \{x : x \text{ is in } L \text{ and } |x| \leq i\}$, $i \geq 0$. Observe that for any $n \geq 0$

$$\bigcup_{i=0}^{n} L_i = L_n$$

but

$$\bigcup_{i=0}^{\infty} L_i = L$$

We say that L_i is the *ith approximation to* L. In practice, as we have discussed in Chapter 1, we deal only with finite, albeit large, languages. However, the importance of infinite regular languages is that their words display a rich structure which enables their specification to be not only finite but also comparatively small.

We can turn the notion of approximation around by defining an infinite regular language $L \subseteq \Sigma^*$ to be an *extension* of a finite language $F \subseteq \Sigma^*$ if $F \subseteq L$. This better reflects our experience, since we usually have finitely many observed words which are words in an infinite language. In this setting we wish to obtain the best extension of F, or at least a good extension. This is called the *inference problem*, which assumes importance in any setting where a language has to be inferred rather than observed or defined.

8.1.2 Regular Languages

Since every regular language can be generated by a regular grammar (see Section 4.3), we have $\boldsymbol{L}_{REG} \subseteq \boldsymbol{L}_{CF}$. However, we prove rather more, namely,

Theorem 8.1.2 $\boldsymbol{L}_{REG} \subseteq \boldsymbol{L}_{DCF}$.

Proof: Let $L \subseteq \Sigma^*$ be a regular language. Then $L = L(M)$, for some *DFA* $M = (Q,\Sigma,\delta,s,f)$. We construct from M a *DPDA* M' such that $L(M') = L(M)$. Essentially, M' ignores its pushdown.

Let $M' = (Q,\Sigma,\{\bot\},\delta',s,F)$, where for all p, q in Q, for all a in Σ, $\delta'(p,a,\bot) = (q,\bot)$ exactly when $\delta(p,a) = q$. Since M' never modifies its pushdown we have $\bot sx \vdash^* \bot f$ in M', for some f in F, exactly when $sx \vdash^* f$ in M. ∎

Proving proper containment is more difficult than for $\boldsymbol{L}_{FIN}$ and $\boldsymbol{L}_{REG}$, although *DCFLs* which are candidates for nonregularity are easy to find. For example, $L_1 = \{a^i b^i : i \geq 1\}$ and $L_2 = \{a^i b^j : 1 \leq i \leq j\}$ are two such examples. It is worthwhile at this stage attempting to construct *DFAs* to accept these languages to gain insight into why they are nonlanguages. However, such insight or intuition is not the same as a rigorous proof, so spend some time trying to prove that there is *no DFA* M_1 with $L_1 = L(M_1)$.

Rather than presenting direct proofs for L_1 and L_2 we present and prove a necessary condition which all *DFAs* must satisfy. This will provide us with a generally useful tool, rather than two specific and limited proofs.

Lemma 8.1.3 The *DFA* Pumping Lemma
Let $M = (Q,\Sigma,\delta,s,F)$ be a DFA and let $p = \#Q$. For all words x in $L(M)$ such that $|x| \geq p$, x can be decomposed into uvw, for some u, v, and w in Σ^* such that

(i) $|uv| \leq p$;
(ii) $|v| \geq 1$; *and*
(iii) *for all* $i \geq 0$, uv^iw *is in* $L(M)$.

The *DFA* pumping lemma provides a partial filter or partial decision maker. On the one hand, if a given language passes through the filter, then it is definitely a nonregular language. On the other hand, if it doesn't pass through the filter, then it may be regular. Diagrammatically we have Figure 8.1.1.

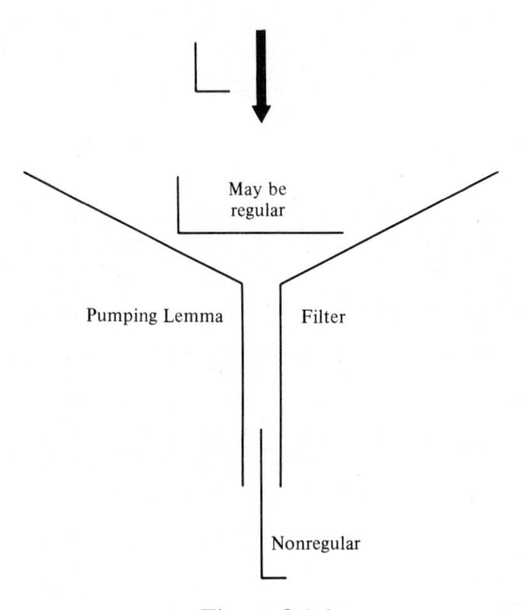

Figure 8.1.1

Before proving the *DFA* pumping lemma, so named because v is "pumped" or iterated, consider an example.

Example 8.1.1 Let M be the *DFA* of Figure 8.1.2. Then

$$L(M) = \{b^{2i} : i \geq 1\} \cup \{ab^{2i} : i \geq 0\}$$

In this case $p = 4$. Now, each word of the form b^{2i}, $i \geq 2$, can be decomposed into uvw, where

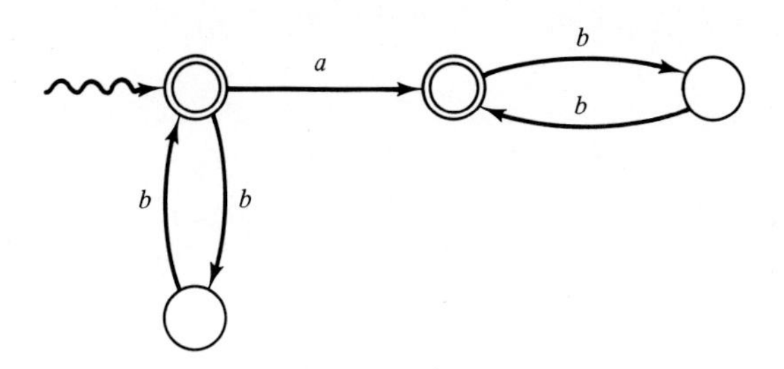

Figure 8.1.2

$$u = bb, \quad v = bb, \quad \text{and} \quad w = b^{2i-4}$$

and $uv^k w$ is in $L(M)$, for all $k \geq 0$, since $bb(bb)^k b^{2i-4} = b^{2(i+k-1)}$ and $i+k-1 \geq 1$.

Similarly, for a word of the form ab^{2i}, $i \geq 2$, we can decompose it into uvw, where

$$u = ab \ , \quad v = bb \ , \quad \text{and} \quad w = b^{2i-3}$$

Once more $uv^k w$ is in $L(M)$, for all $k \geq 0$, since $ab(bb)^k b^{2i-3} = ab^{2(i+k-1)}$ and $i+k-1 \geq 1$.

However, it is important to realize than not all decompositions can be pumped. For example, if $x = b^4$, then $u = b^3$, $v = b$, and $w = \lambda$ is a decomposition, but

$$uv^{2k} w = b^{2k+3} \quad \text{is not in} \quad L(M), \quad \text{for all} \quad k \geq 0$$

We now prove the *DFA* pumping lemma

Proof of *DFA* Pumping Lemma: Let x be in $L(M)$ with $|x| \geq p$. Then M has an accepting configuration sequence

$$q_0 a_1 \cdots a_r \vdash q_1 a_2 \cdots a_r \vdash \cdots \vdash q_{r-1} a_r \vdash q_r$$

for some q_i in Q, $0 \leq i \leq r$, where $q_0 = s$, q_r is in F, and $x = a_1 \cdots a_r$, a_i is in Σ, $1 \leq i \leq r$, and $r \geq p$. Consider the first p transitions. Then $s, q_1, \ldots, q_p$ cannot all be distinct, since there are only p distinct states. This is an application of the pigeonhole principle (see Section 1.5). But this means that there exists i and j, $0 \leq i < j \leq p$ such that $q_i = q_j$. These values determine a decomposition of x, namely,

$$u = a_1 \cdots a_i, \ v = a_{i+1} \cdots a_j \ , \ \text{and} \ \ w = a_{j+1} \cdots a_r$$

since

$$q_0 a_1 \cdots a_i \vdash^* q_i$$

$$q_i a_{i+1} \cdots a_j \vdash^* q_j = q_i$$

and

$$q_j a_{j+1} \cdots a_r \vdash^* q_r$$

Observe that $|uv| \leq p$ and $|v| \geq 1$ by construction.

Since $q_i v \vdash^* q_i$, $q_i v^k \vdash^* q_i$, for all $k \geq 0$. Thus, $q_0 uv^k w \vdash^* q_r$, for all $k \geq 0$. Hence, $uv^k w$ is in $L(M)$, for all $k \geq 0$. ■

Example 8.1.2 Let $L = \{a^i b^i : i \geq 1\}$. Then L is nonregular.

We use the *DFA* pumping lemma in a proof by contradiction.

First, assume L is in $\boldsymbol{L}_{REG}$. Then there is a *DFA* $M = (Q, \{a,b\}, \delta, s, F)$ such that $L = L(M)$. Letting $p = \#Q \geq 1$, observe that there are infinitely many words x in $L(M)$ that satisfy $|x| \geq p$. Consider the word $x = a^p b^p$ in $L(M)$. By the *DFA* pumping lemma x can be decomposed into uvw, for some u, v, and w in $\{a,b\}^*$ such that $|uv| \leq p$, $|v| \geq 1$, and $uv^k w$ is in L, for all $k \geq 0$.

But $|uv| \leq p$ and $|v| \geq 1$ implies that $u = a^r$ and $v = a^s$, for some r and s with $r+s \leq p$ and $s \geq 1$. There are many decompositions of x that satisfy these inequalities. We must show that none of them, under pumping, give words in L.

Now, $x = a^r a^s w$, where $w = a^{p-r-s} b^p$ and the *DFA* pumping lemma implies, in particular, that

$$uv^0 w \text{ is in } L.$$

But $uv^0 w = a^r a^{p-r-s} b^p = a^{p-s} b^p$ is *not in* L, since $s \geq 1$. This contradicts the assumption that L is regular. Therefore L is not regular.

The *DFA* Pumping Lemma is always used to provide the contradiction in proofs by contradiction. Such a proof can be broken down into the following steps.

Assumption Step: Assume the given language L is regular. This implies there is some *DFA*, M, such that $L = L(M)$.

Choice Step: Choose some word z in L with $|z| \geq p$, the number of states of M. Since all words of length greater than p must satisfy the pumping condition, it is sufficient to find one which doesn't.

Decomposition Step: We need to show that *all possible* decompositions of z into uvw that satisfy $|uv| < p$ and $|v| \geq 1$ give rise, under pumping, to words outside L. It is not enough to show that one decomposition leads to words outside the language; all decompositions must do so. See the Exercises.

Theorem 8.1.4 $\boldsymbol{L}_{REG} \subset \boldsymbol{L}_{DCF}$.

Proof: L of Example 8.1.2 is not in $\boldsymbol{L}_{REG}$, but it is in $\boldsymbol{L}_{DCF}$. A *DPDA* that accepts L is easy to construct; see the Exercises of Chapter 5. ■

8.1.3* Unary Regular Languages

Another approach to finding nonlanguages is based on *unary languages*, that is, languages over a single-letter alphabet. For notational convenience we assume the alphabet to be $\{a\}$ for the remainder of this section.

Although regular unary languages appear to be unduly restrictive they are able, via the notion of *length set*, to provide simple proofs for candidate non-languages over arbitrary alphabets.

Definition Let L be an arbitrary language, then the *length set of* L is denoted by $LS(L)$ and is defined by

$$LS(L) = \{ |x| : x \text{ is in } L\}$$

It is simply the set of all word lengths provided by the language. The connection between unary languages, length sets, and regularity is given by

Theorem 8.1.5 *Let $L \subseteq \Sigma^*$ be an arbitrary regular language. Then $L_a = \{a^i : i$ is in $LS(L)\}$ is a regular language.*

Proof: Consider a regular expression E with $L(E) = L$. Replace all appearances of symbols from Σ in E by a to give E_a. Clearly, $L(E_a) = L_a$. ■

Returning to regular unary languages we consider their associated *DFAs*. To ease our discussion we assume the *DFAs* are complete and have no useless states. Since each state in such a *DFA* must have exactly one transition with symbol a, the generic *DFA*, M, must look like Figure 8.1.3. Clearly, $L(M)$ is infinite if some final state occurs in the loop, and finite otherwise.

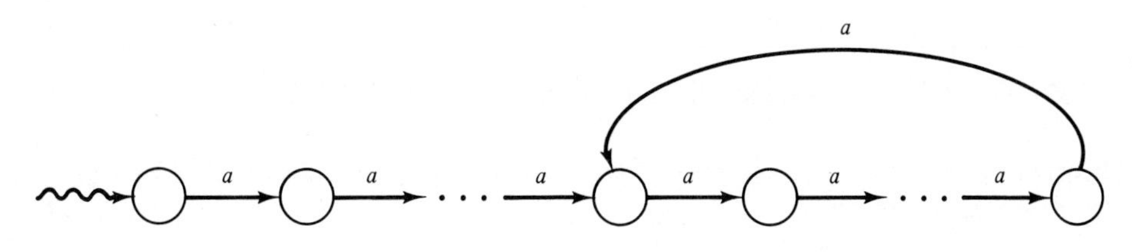

Figure 8.1.3

Example 8.1.3A Let $L = \{a^2, a^3, a^5\}$; then we have the state diagram of Figure 8.1.4, while with $L = \{a^2\} \cup \{a^{5+2i} : i \geq 0\}$ we obtain Figure 8.1.5. Note how the loop defines the length of the period.

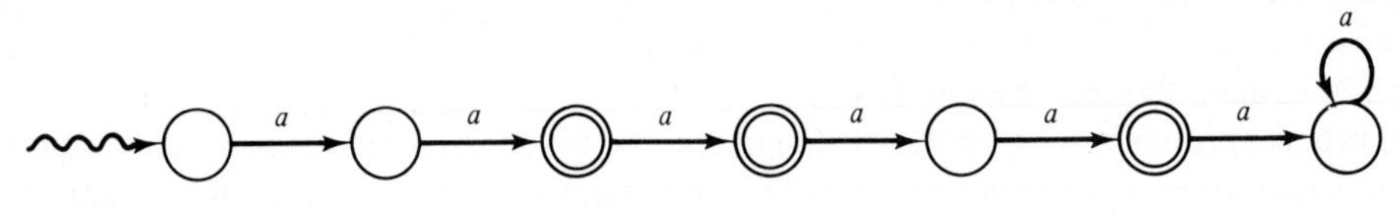

Figure 8.1.4

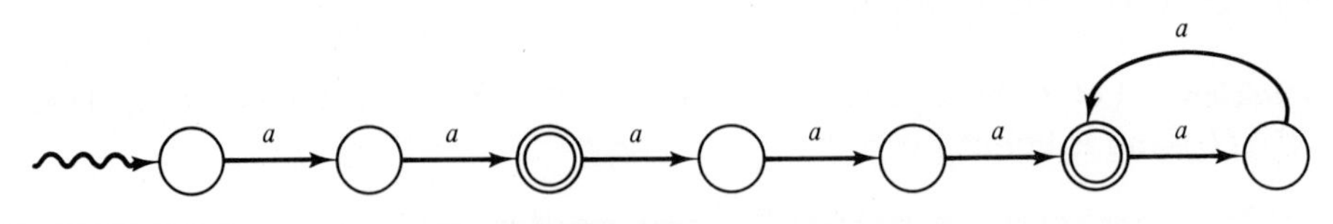

Figure 8.1.5

The unary languages have, as we would expect from their corresponding *DFAs*, a simple periodic structure.

Definition Let N be a set of nonnegative integers. We say N is an *ultimately periodic set* with *threshold* $t \geq 0$ and *period* $p \geq 1$, if for all $n \geq t$, n is in N implies $n+p$ is in N. Observe that every finite set of integers F is an ultimately periodic set, because we can simply choose the threshold to be larger than the largest integer in F, when the condition of periodicity is vacuously satisfied.

Example 8.1.3B From $L = \{a^2,a^3,a^5\}$ we obtain $N = \{2,3,5\} = LS(L)$. In this case choose $t = 6$ and $p = 1$.

Similarly, from $L = \{a^2\} \cup \{a^{5+2i} : i \geq 0\}$ we obtain $N = \{2\} \cup \{5+2i : i \geq 0\}$, which is ultimately periodic with threshold $t = 5$ and period $p = 2$.

We now prove the sought characterization theorem.

Theorem 8.1.6 *Let $L \subseteq \{a\}^*$. Then L is in $\mathbf{L}_{REG}$ if and only if $LS(L)$ is an ultimately periodic set.*

Proof: We prove this theorem in two stages. First, assuming L is a *REGL* we prove $LS(L)$ is ultimately periodic and, second, we prove the converse.

(a) Assume L is a *REGL*. If L is finite, then $LS(L)$ is trivially, ultimately periodic; therefore, assume L is infinite. L is accepted by some complete *DFA*, M, having no useless states. Now, M has the single-loop generic form of Figure 8.1.6, where we have enumerated the states in a suggestive way; t corresponds to the threshold and p to the period. Clearly, $t \geq 0$ and $p \geq 1$. Consider an arbitrary word a^s in L, $s \geq 0$. If $0 \leq s < t$, then q_s is a final state. Furthermore, by construction there is no $i > 0$ such that $q_s a^i \vdash^+ q_s$, that is, q_s is not in the loop. On the other hand, if $s \geq t$, then $q_0 a^s \vdash^* q_r$, for some r, $t \leq r < t+p$, and by construction $q_0 a^{s+p} \vdash^* q_r a^p \vdash^* q_r$; therefore, a^{s+p} is in L also. We have deduced that for $s \geq t$, a^s is in L implies a^{s+p} is in L. But this means that for $s \geq t$, s is in $LS(L)$ implies $s+p$ is in $LS(L)$. Hence, $LS(L)$ is ultimately periodic.

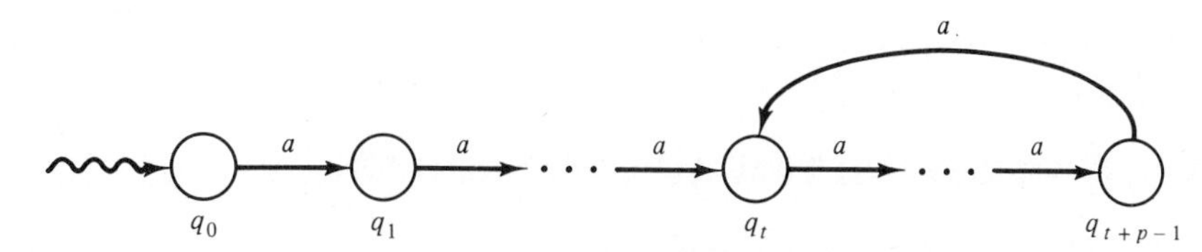

Figure 8.1.6

(b) Assume $LS(L)$ is ultimately periodic. If $LS(L)$ is finite, then L is immediately regular; hence, assume once more that $LS(L)$ is infinite. Since $LS(L)$ is ultimately periodic there is a threshold $t \geq 0$ and a period $p \geq 1$, such that for all n in $LS(L)$ with $n \geq t$,

$$n = s + ip, \text{ for some } s, t \leq s < t+p, \text{ and } i \geq 0$$

Construct a DFA, M, with states $q_0, \ldots, q_{t+p-1}$ satisfying $\delta(q_i, a) = q_{i+1}$, for all i, $0 \leq i < t+p-1$ and $\delta(q_{t+p-1}, a) = q_t$. Further, let q_n be a final state, for all n in $LS(L)$ such that $0 \leq n \leq t+p-1$. We claim that $L(M) = L$. This follows by observing that $LS(L(M)) = LS(L)$. ■

It is tempting to assume Theorem 8.1.6 can be extended to arbitrary regular languages. In other words, $L \subseteq \Sigma^*$ is regular if and only if $LS(L)$ is ultimately periodic. To see that this extension does not hold, consider $L = \{a^i b^i : i \geq 0\}$. L is a nonregular language and $LS(L) = \{2i : i \geq 0\}$ is ultimately periodic.

To see how this theorem is applied consider the following example.

Example 8.1.4 Let $L = \{a^{i^2} : i \geq 0\}$. Then L is not regular. The words of L are

$$\lambda, a^1, a^4, a^9, a^{16}, \ldots$$

If L is regular, then $N = LS(L)$ is ultimately periodic. We argue by contradiction. Assume N *is* ultimately periodic, in which case there is a threshold $t \geq 0$ and a period $p \geq 1$, such that, for all m^2 in N, $m^2 \geq t$

$$m^2 = s + ip, \text{ for some } s, \; t \leq s < t+p, \text{ and some } i \geq 0$$

Because L is infinite there is an m which satisfies both $m > p$ and $m^2 > t$. Consider such an m. We show that there is a word a^k in L, where k is not in

$LS(L)$. By definition

$$m^2 = s + jp, \text{ for some } s, t \leq s < t+p \text{ , and } j \geq 0$$

Now, $(m+1)^2 = m^2 + 2m + 1 > s + jp + 2p$, since $m > p$. Moreover, $s + jp + 2p > m^2$, since $p \geq 1$. Therefore, we have constructed a value $s + (j+2)p$ which is in N, since N is ultimately periodic, but is not in $LS(L)$, by definition. This gives us our contradiction, that is, $LS(L)$ is not ultimately periodic, which in turn implies L is not a *REGL*.

8.2 CONTEXT-FREE AND TRACTABLE LANGUAGES

8.2.1 Deterministic Context-Free Languages

Clearly, $\boldsymbol{L}_{DCF} \subseteq \boldsymbol{L}_{CF}$, and we now prove proper containment.

Theorem 8.2.1 $\boldsymbol{L}_{DCF} \subset \boldsymbol{L}_{CF}$.

Proof: We prove rather less than claimed, namely, there is a *CFL* which is not a *DPDA* language. To prove the theorem we need to prove that this language is not even a λ-*DPDA* language. We leave this task to the Exercises.

Consider the language $L = \{a^i b^i, a^i b^{2i} : i \geq 1\}$. We prove, by contradiction, that it is not a *DPDA* language. Assume that there is some *DPDA*, $M = (Q, \{a, b\}, \Gamma, \delta, s, F)$, such that $L(M) = L$. We claim that there are positive integers i and j, $i < j$, such that

$$\perp s a^i b^{2i} \vdash^{2i} \perp p b^i \vdash^{i} \perp f$$

and

$$\perp s a^j b^{2j} \vdash^{2j} \perp p b^j \vdash^{i} \perp f b^{j-i}$$

For some p in Q and f in F. Observe that this claim implies

$$\perp s a^j b^{i+j} \vdash^{2j} \perp p b^i \vdash^{i} \perp f$$

that is, $a^j b^{i+j}$ is in L — a contradiction.

To establish the claim, note that, for all $k \geq 1$ and for some p in F,

$$\perp s a^k b^k \vdash^{2k} \perp p$$

since $a^k b^k$ is in L. Now, F is finite and there are an infinite number of words of the form $a^k b^k$ in L. Therefore, there are at least two distinct values of k, say $k = i$ and $k = j$, where $i < j$, such that

$$\perp s a^i b^i \vdash^{2i} \perp p$$

and

$$\bot s a^j b^j \vdash^{2j} \bot p$$

for some p in F. Immediately, we have

$$\bot s a^i b^{2i} \vdash^{2i} \bot p b^i$$

$$\bot s a^j b^{2j} \vdash^{2j} \bot p b^j$$

Because $a^i b^{2i}$ is also in L we must also have

$$\bot p b^i \vdash^i \bot f$$

for some f in F, and, therefore,

$$\bot p b^j \vdash^i \bot f b^{j-i}$$

This establishes the claim; thus, $L \neq L(M)$, for any *DPDA* M.

In a similar manner we can prove that $L \neq L(M)$, for any λ-*DPDA*, M; see the Exercises. ∎

8.2.2* Tractable Languages

In order to prove that $\boldsymbol{L}_{CF} \subset \boldsymbol{L}_{DPTIME}$ we must first demonstrate that each *CFL* can be accepted by a polynomial-time-bounded *DTM*. We approach this by showing that an *NPDA* can be simulated by a *DTM* in polynomial time, where the polynomial is, indeed, a polynomial in the input word size. To this end we first introduce the notion of a *surface configuration* of an *NPDA* and show that *realizable pairs* of surface configurations are sufficient to characterize acceptance.

We know, by the results of Chapter 5, that every *CFL* is accepted by some *NPDA* and, more importantly, a highly restricted *NPDA*, the *2SNF-PDA* based on *2-SNF CFGs*. The *2SNF-PDA* accepts an input word if it reads all of the input word, is in a final state, and it leaves the pushdown empty. Moreover at each transition the *2SNF-PDA* performs one of three possible transitions, while reading an input symbol: (i) pop one symbol from the pushdown; (ii) pop one symbol and push one symbol; and (iii) pop one symbol and push two symbols.

We first break down the transitions of an *NPDA* into the more fundamental operations of pop and push. Let $M = (Q,\Sigma,\Gamma,\delta,s,F)$ be an *NPDA*. Given two configurations gqx and $g'q'x'$, we write

$$gqx \underset{\text{push } A}{\vdash} g'q'x' \text{ if } gqx \vdash g'q'x' \text{ and } g' = gA$$

In other words, A is pushed onto the pushdown. Similarly, we write

$$gqx \underset{\text{pop } A}{\vdash} g'q'x' \text{ if } gqx \vdash g'q'x' \text{ and } g = g'A$$

that is, A is popped from the pushdown. Note that we are only concerned with the effect on the pushdown. Whether the reading head moves or remains in the same position is of no interest. It is straightforward to obtain from a *2SNF-PDA*, an equivalent *NPDA* in which only moves of the above two types are permitted. We call such an *PDA* a *restricted NPDA*, or *rNPDA*.

A *surface configuration* of an *NPDA* is a word in $(\Gamma \cup \{\perp\})Q\Sigma^*$. Given a configuration gqy in which $g \neq \lambda$ its corresponding surface configuration is Aqy if A is the rightmost symbol of g. Given an input word x, $|x| = n$, there are at most $\#\Gamma \times \#Q \times (n+1)$ surface configurations appearing during the computation of the *NPDA* on x, that is, $an+b$ surface configurations, where a and b are constants for the *specific NPDA*, M. The importance of surface configurations is that they contain all the information needed for an *NPDA* to perform a transition. We need to define $\underset{\text{push } A}{\vdash}$ and $\underset{\text{pop } A}{\vdash}$ for surface configurations. We write

$$Aqy \underset{\text{push } B}{\vdash} Bq'y'$$

if there exists g such that

$$gAqy \vdash gABq'y'$$

and we write

$$Aqy \underset{\text{pop } A}{\vdash} Bq'y'$$

if there exists gB such that

$$gBAqy \vdash gBq'y'$$

During an accepting computation of M on input x, the pushdown grows and shrinks, until it is empty, M has read all of x, and M has finished in a final state; see Figure 8.2.1. An *initial surface configuration* of M is $\perp sx$, for some x in Σ^*, and a *final surface configuration* is $\perp f$, for some final state f. A word x is accepted by M if and only if $\perp f$ is reachable from $\perp sx$, for some p in F. Such a pair of surface configurations is said to be *realizable*. We now give a formal definition of this notion.

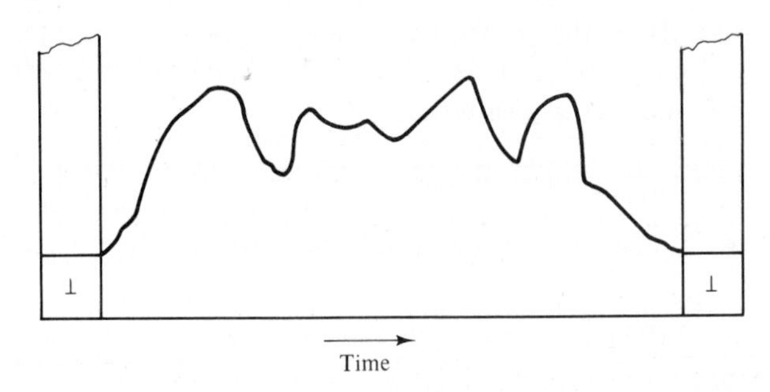

Figure 8.2.1

Definition Let $M = (Q,\Sigma,\Gamma,\delta,s,F)$ be an *rNPDA* and x in Σ^* be an input word. Then two surface configurations C_1 and C_2 (with respect to x) form a *realizable pair* (C_1,C_2) if and only if one of the following holds

(1) $C_1 = C_2$.
(2) There exists a surface configuration C_3 such that (C_1,C_3) and (C_3,C_2) are realizable. Pictorially we have Figure 8.2.2.
(3) There exists an A in Γ and a realizable pair (C_3,C_4) such that $C_1 \underset{\text{push } A}{\vdash} C_3$, $C_4 \underset{\text{pop } A}{\vdash} C_2$, and $top(C_1) = top(C_2)$. Pictorially, we have Figure 8.2.3, where $top(C)$ is the pushdown symbol appearing in C.

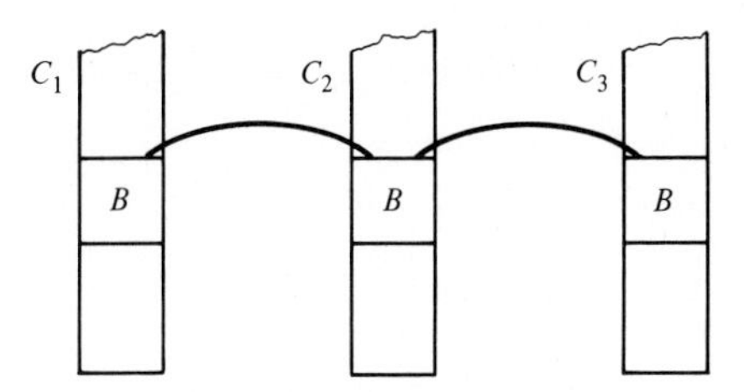

Figure 8.2.2

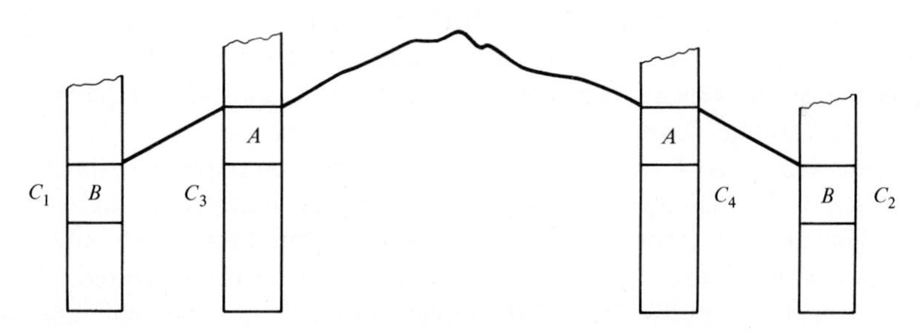

Figure 8.2.3

This inductive definition of realizable pairs of surface configurations guarantees that whenever (C_1,C_2) is a realizable pair the pushdown corresponding to C_1 remains unchanged during the computation of M, taking C_1 to C_2. This is exactly what we require for the pair $(\bot sx, \bot f)$, where f is in F.

We now give a high-level algorithm to compute for a given *rNPDA* and an input word x in Σ^* all realizable pairs of surface configurations for x.

Algorithm *Surface Configuration.*
On entry: A word x in Σ^* and an *rNPDA*, $M = (Q,\Sigma,\Gamma,\delta,s,F)$.
On exit: The set R of realizable pairs of corresponding surface configurations.

begin
Let
$Z = \{Aqy : A \text{ is in } \Gamma \cup \{\perp\},\ q \text{ is in } Q \text{ ,and } x = zy, \text{ for some } z \text{ in } \Sigma^*\}$,
that is, Z is the set of all possible surface configurations for x.
Initialize R to $\{(C,C) : C \text{ is in } Z\}$;
repeat
$T := \varnothing$;
for all C_1 and C_2 **in** Z **do**
if (C_1, C_2) **not in** R **then**
begin Determine if there is
either a C_3 in Z with (C_1, C_3), (C_3, C_2) in R (Rule 2), or there is (C_3, C_4) in R and an A in Γ such that $C_1 \underset{\text{push } A}{\vdash} C_3$, $C_4 \underset{\text{pop } A}{\vdash} C_2$,
and $top(C_1) = top(C_2)$ (Rule 3);
if (C_1, C_2) is realizable **then** $T := T \cup \{(C_1, C_2)\}$
end;
$R := R \cup T$;
until $T = \varnothing$;
end of Algorithm.

Since $\#Z = O(n)$, where $|x| = n$, and $\#R = O(n^2)$, there are $O(n^5)$ steps of the algorithm. This follows by observing that the for loop is executed $O(n^2)$ times and each execution requires $O(n^2)$ steps to determine if (C_1, C_2) is not in R, and if so at most $O(n^3)$ steps to determine whether or not there is a C_3 for Rule 2 to be applied. For Rule 3 $O(n^2)$ steps are required; thus, the for loop requires $O(n^5)$ steps. At worst only one newly found realizable pair is found at each iteration of the repeat loop, and there can be at most $O(n^2)$ pairs added to R. Hence, this implies $O(n^5)$ steps are required at most.

This analysis makes no assumptions about the computer on which *Surface Configuration* is implemented, it only determines the number of high-level steps required. However, an implementation of this algorithm on a *DTM* only increases the exponent of n. To see this we sketch how it can be implemented. It is convenient to consider a multi-tape *DTM* implementation. On the first tape we find x, on the second, third, and fourth, three copies of the elements of Z are given, on the fifth and sixth, two copies of the current elements of R are to be found, and on the seventh are the elements of T. Assuming Z, R, and T have been initialized, the repeat statement scans the second and third tapes simultaneously to run through all possible pairs of surface configurations. For each pair tape five is scanned to see that it is not present. If it is not present tape four (the third copy of Z) is scanned and for each C_3, tape five and six are scanned for (C_1, C_3) and (C_3, C_2) in turn. If both are found, then (C_1, C_2) is added to tape seven. If not, then tape five is scanned for a (C_3, C_4) that satisfies Rule 3. After one iteration of the repeat statement, if tape seven is empty, then the *DTM* terminates; otherwise tape seven is copied to both tapes five and six and it is also erased. The process is then repeated.

From this multitape *DTM* requiring $T(n)$ steps for $|x| = n$, can be constructed a one-tape *TM* which requires at most $T^2(n)$ steps for the same input; see Section 6.6.

This discussion leads to the following

Theorem 8.2.2 *Let $L \subseteq \Sigma^*$ be a CFL. Then there is a DTM, M, and a constant $c \geq 1$ such that for all x in Σ^*, $|x| = n$, M determines whether or not x is in L in time bounded by cn^{10}. Hence, $\boldsymbol{L}_{CF} \subseteq \boldsymbol{L}_{DPTIME}$.*

Proof: See the Exercises. ■

To prove proper containment we need to be able to demonstrate that a given language is a nonlanguage with respect to $\boldsymbol{L}_{CF}$. Such a language is provided in Section 8.2.4

8.2.3 Context-Free Pumping Lemma

In order to prove that some languages are not context-free we provide a pumping lemma for context-free grammars. Although there are languages that it cannot demonstrate to be noncontext-free, it is nevertheless a useful tool.

Before stating and proving the *CFG* pumping lemma we attempt to give some insight into why it holds.

Example 8.2.1 Given the *CFG*, *G*,

$S \rightarrow a \mid (S) \mid F + S \mid S * F$

$F \rightarrow a \mid (S)$

consider the syntax tree for $a * (a + a)$ shown in Figure 8.2.4.

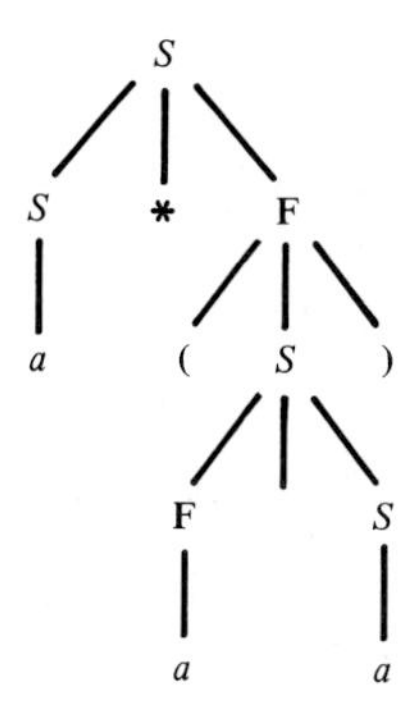

Figure 8.2.4

We can break this syntax tree into three parts at the two appearances of F displayed in boldface in Figure 8.2.4. The result of the breakup is displayed in Figure 8.2.5.

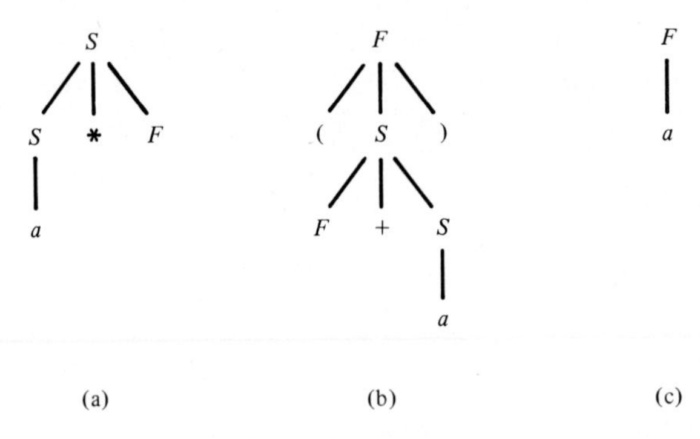

Figure 8.2.5

Schematically we view this break up as shown in Figure 8.2.6. We can treat these three trees as pieces in a jig-saw puzzle, if we assume we have as many copies of type (b) as we need. First observe that we can fit (a) and (c) together without (b) as seen in Figure 8.2.7 (a) and (b). We can also fit two (b)'s between (a) and (c) as shown in Figure 8.2.8 (a) and (b).

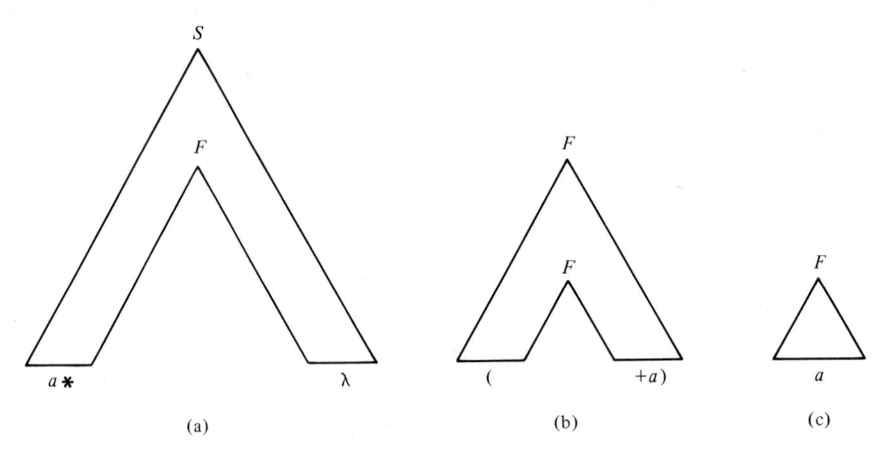

Figure 8.2.6

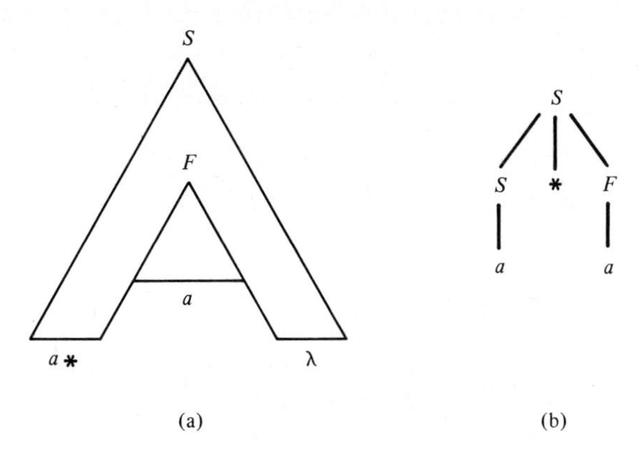

Figure 8.2.7

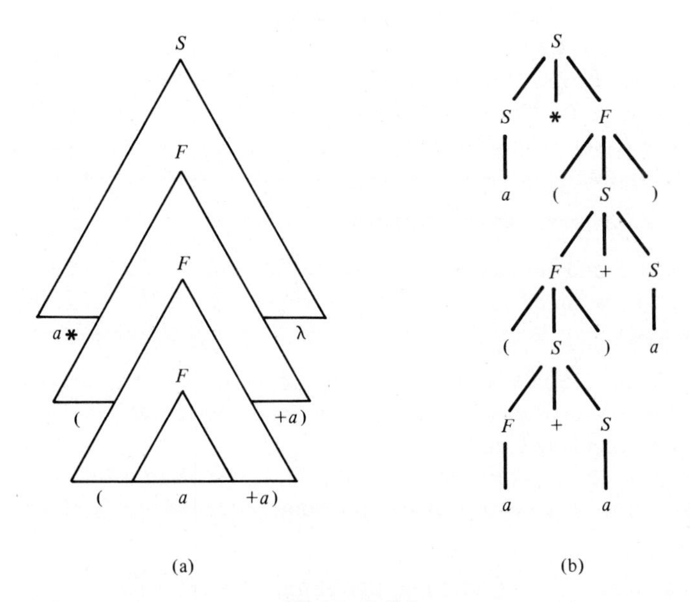

Figure 8.2.8

Indeed we can add any number of (b)'s between the (a) and (c). This implies that each word of the form $a * ({}^{i}a[+a)]^{i}$ is in $L(G)$, for all $i \geq 0$.

Lemma 8.2.3 The *CFG* Pumping Lemma
Let $G = (N, \Sigma, P, S)$ be a λ-free, unit-free CFG, such that $m = \max(\{\,|\alpha| : A \to \alpha$ is in $P\})$, and $p = 1 + m^{\#N+1}$.

Then, for all words z *in* $L(G)$ *such that* $|z| \geq p$, z *has a derivation sequence*

$$S \Rightarrow^* uAv \Rightarrow^+ uxAyv \Rightarrow^+ uxwyv$$

for some A *in* N *and some* u, v, w, x, y *in* Σ^* *such that*

(i) $|xwy| < p$; *and*
(ii) $|xy| \geq 1$.

This implies ux^iwy^iv *is in* $L(G)$, *for all* $i \geq 0$.

Compare the *CFG* and *DFA* pumping lemmas. Before proving the *CFG* pumping lemma we demonstrate how it is used.

Example 8.2.1 Let $L = \{a^ib^ic^i : i \geq 1\}$; this is a classical noncontext-free language. We prove L is noncontext-free as follows. Compare this argument by contradiction with the three steps given in Section 8.1.

Assume L is a *CFL*, that is, $L = L(G)$, for some λ-free, unit-free *CFG* $G = (N, \{a,b,c\}, P, S)$. We use the *CFG* pumping lemma to provide a contradiction. Let m and p be the constants given by G and consider the word $z = a^pb^pc^p$ in $L(G)$. Clearly, $|z| \geq p$.

By the *CFG* pumping lemma we know that there is a derivation

$$S \Rightarrow^* uAv \Rightarrow^+ uxAyv \Rightarrow^+ uxwyv = a^pb^pc^p$$

for some A in N and some u, v, w, x, y in Σ^* with $|xwy| < p$. This implies that either xwy is in $a^* \cup b^* \cup c^*$ or xwy is in $a^+b^+ \cup b^+c^+$. We only examine the cases xwy is in a^* and xwy is in a^+b^+; the others are similar.

(i) xwy is in a^*. Since $|xy| \geq 1$, xy is in a^+. Now, the *CFG* pumping lemma implies that $uwv = ux^0wy^0v$ is in L. But $uwv = a^ib^pc^p$, for some $i < p$, and this is not in L.
(ii) xwy is in a^+b^+. Again, since $|xy| \geq 1$, xy is in $a^*b^+ \cup a^+b^*$. Once more $ux^0wy^0v = uwv = a^ib^jc^p$, for some i and j, $i+j < 2p$, and this is not in L.

In all cases we obtain a contradiction; therefore, L is not a *CFL*.

Proof of the *CFG* Pumping Lemma: There is a strong relationship between the height of a syntax tree and the longest words it can yield. This observation is used to confirm that sufficiently long words have a syntax tree with a root-to-frontier path containing at least two occurrences of the same nonterminal; see Figure 8.2.9. This argument is similar to that used in proving the *DFA* pumping lemma. To complete the proof we only need deduce that there is a nonterminal A, say, having two appearances which provides a decomposition of the terminal word z into $uxwyv$. We can remove the shaded portion of the tree in Figure 8.2.9, attaching the w tree to the upper A to yield a valid syntax tree for uwv; see Figure

8.2.10(a). Similarly, we may iterate the shaded portion of the tree, as in Figure 8.2.10(b), to give ux^iwy^iv, for any $i > 1$.

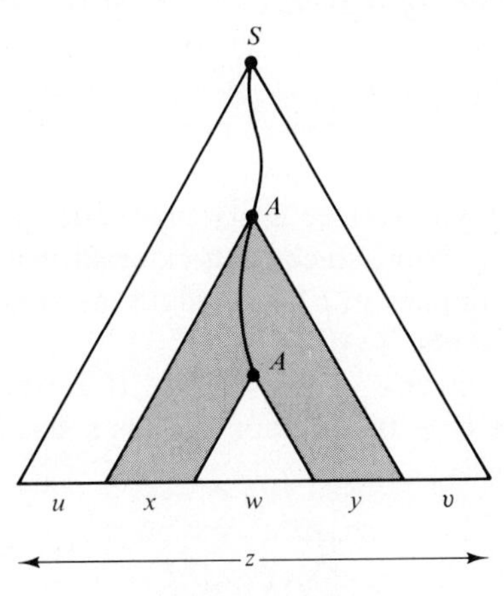

Figure 8.2.9

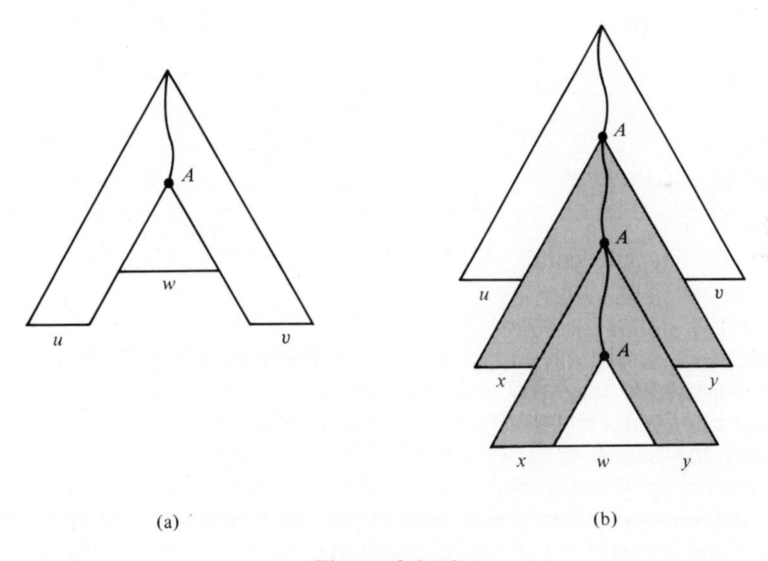

Figure 8.2.10

We now formalize this approach by first bounding the length of the frontier of a syntax tree of a given height h.

Claim: *Let h be the height of a syntax tree T of G. Then its yield $y(T)$ satisfies $y(T) \leq m^h$.*

The bound follows by considering a syntax tree in which every nonterminal node has m-ary branching. See Figure 8.2.11 for a tree of height 3. M-ary branching is the highest possible since every production in G has at most m symbols on its right-hand side. Now, since each external node is labeled with at most one terminal symbol, this implies $y(T)$ has length, at most, equal to the number of external nodes. In other words, $|y(T)| \leq m^h$.

In fact, we use the converse of this claim. If a word z in $L(G)$ has length greater than m^h, for some $h \geq 0$, a syntax tree for z has height greater than h.

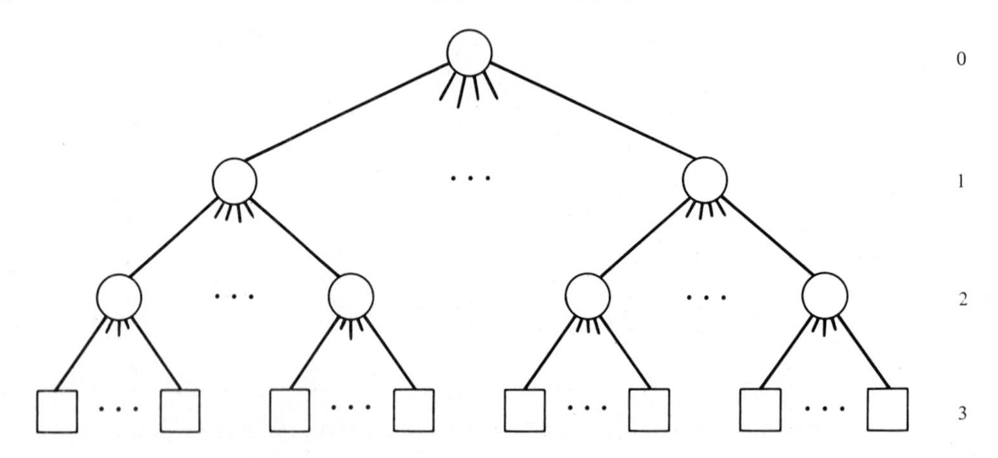

Figure 8.2.11

Second, we assume there are words z in $L(G)$ such that $|z| \geq p$. (For if no such word exists the *CFG* pumping lemma is immediate.) Consider a syntax tree T for z. Since $|z| \geq 1+m^{\#N+1}$, this implies, by the converse of the above claim, that the height of $T \geq \#N+2$. In other words, there is some root-to-frontier path π in T containing at least $\#N+3$ nodes. Now, π contains at least $\#N+2$ nonterminal nodes, which implies, by the pigeonhole principle, that there is at least one nonterminal A in N which appears twice. We have the situation displayed in Figure 8.2.9. The two appearances of A divide T into three syntax trees: the syntax tree T_{finish} having the lower A as root, the syntax tree T_{repeat}, having the upper A as root with T_{finish} removed, and the syntax tree T_{start}, obtained by removing both T_{finish} and T_{repeat} from T; see Figure 8.2.12. Let the frontiers be labeled as in Figure 8.2.12. We can attach T_{finish} to T_{start} to obtain the word $uwv = ux^0wy^0v$. Alternatively, we can attach T_{repeat} to itself i times, for some $i > 0$, and then reattach T_{start} and T_{finish}. This yields the word $ux^i wy^i v$ and the derivation sequence $S \Rightarrow^* uAv \Rightarrow^+ uxAyv \Rightarrow^+ \cdots \Rightarrow^+ ux^i Ay^i v \Rightarrow^+ ux^i wy^i v$. However, condition (i) may not be satisfied.

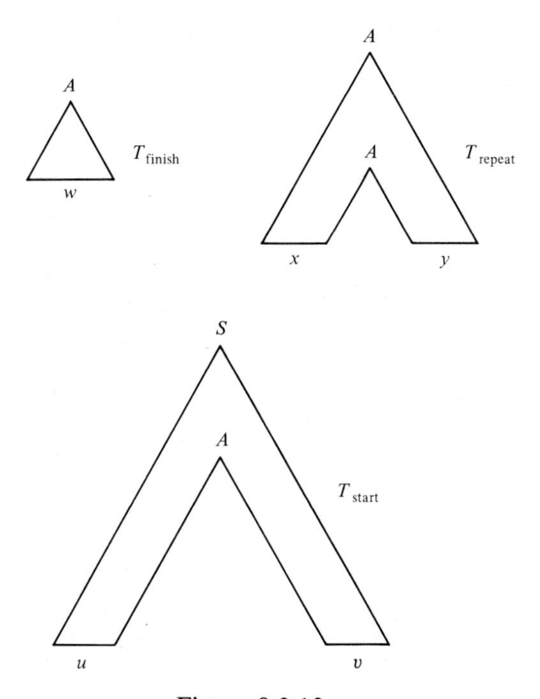

Figure 8.2.12

To satisfy condition (i) take a longest root-to-frontier path π in T, rather than any path which is long enough. Clearly, π contains at least $\#N+2$ nonterminal nodes. Now, truncate the path by removing the root node. This leaves a path containing at least $\#N+1$ nonterminal nodes. By the pigeonhole principle at least one nonterminal must appear two or more times in the truncated portion. Carry out a frontier-to-root scan of the truncated path choosing the first nonterminal to appear twice as the nonterminal A. The number of nodes from the lower A to the end of π is at most $\#N$. Moreover, the height of T_A is this number. This implies that $|y(T_A)| \leq m^{\#N} < p$, and, therefore, condition (i) is now satisfied.

Condition (ii) is satisfied immediately, since G is λ-free and unit-free and, therefore, T_{repeat} contains at least two external nodes. Hence, $|y(T_{repeat})| \geq 1$. ■

We now consider a second example of the application of the *CFG* pumping lemma.

Example 8.2.2 Let $L = \{wcw : w \text{ is in } \{a,b\}^*\} \subseteq \{a,b,c\}^*$. We prove L is not a CFL in an argument by contradiction. Assume L is a CFL. Then there is a λ-free, unit-free CFG $G = (N,\{a,b,c\},P,S)$ such that $L = L(G)$. Let p be the constant of the CFG pumping lemma and consider a word zcz in $L(G)$ with $|z| = p$. The CFG pumping lemma states that zcz has a derivation

$$S \Rightarrow^* uAv \Rightarrow^+ uxAyv \Rightarrow^+ uxwyv = zcz$$

for some A in N, and some u, v, w, x, y in Σ^+. That is zcz has a decomposition into $uxwyv$, where $|xwy| < p$, $|xy| \geq 1$, such that ux^iwy^iv is in $L(G)$, for all $i \geq 0$.

There are three cases to consider

(i) xwy appears in the first z.
(ii) xwy appears in the second z.
(iii) xwy contains c.

Cases (i) and (ii) lead to an immediate contradiction by considering $ux^0wy^0v = uwv$, since either the first or second z has shrunk, while the other z remains the same.

In case (iii) c does not appear in either x or y, since this would imply that ux^2wy^2v contains two c's. Therefore, c is in w. Now, if $x \neq y$, then an immediate contradiction is obtained with uwv. Therefore, $x = y$ and x and y must occur at the same positions in the first and second z, that is, $zcz = z_1xz_2cz_1yz_2$, where $w = z_2cz_1$, $u = z_1$ and $v = z_2$. Otherwise ux^2wy^2v is not in L. But this implies $|xwy| = |xz_2cz_1y| = |z| + 1 + 2|x| > p+1$, since $|z| = p$, and this contradicts the assumption that $|xwy| < p$.

Therefore, in all cases the assumption that L is a CFL has led to a contradiction, hence L is not a CFL.

8.2.4* Parikh's Theorem

We now prove a remarkable result, namely, every CFL is identical to some $REGL$ *if we ignore the order of letters in their words.* This is known as *Parikh's theorem.* We first define this notion formally

Definition Let Σ be an alphabet and denote its symbols by $a_1, a_2, \ldots, a_m$, where $m = \#\Sigma \geq 1$. Given a word x in Σ^* we denote by $\Pi(x)$ the vector

$$(|x|_{a_1}, |x|_{a_2}, \ldots, |x|_{a_m}) \text{ in } \mathbb{N}^m$$

called the *Parikh vector of* x. For $L \subseteq \Sigma^*$, $\Pi(L) = \{\Pi(x) : x \text{ is in } L\}$ is the *Parikh set of* L.

Example 8.2.3 Let $\Sigma = \{a,b,c\}$, where $a_1 = a$, $a_2 = b$, and $a_3 = c$. Then $\Pi(aabbbcabcb) = (3,5,2)$ and $\Pi(\lambda) = (0,0,0)$. Note that $\Sigma_{i=1}^{m} \Pi(x)_i = |x|$, for example, $\Sigma_{i=1}^{3} \Pi(aabbbcabcb)_i = 3+5+2 = 10$.

We first state Parikh's theorem and give one application before proving it.

Theorem 8.2.4 Parikh's Theorem.
Let $\Sigma = \{a_1, \ldots, a_m\}$ *be an arbitrary alphabet and* $L \subseteq \Sigma^*$ *be a CFL. Then there exists a REGL* $R \subseteq \Sigma^*$ *such that* $\Pi(L) = \Pi(R)$.

Theorem 8.2.5 *Let* $L \subseteq \{a\}^*$ *be an arbitrary CFL. Then L is a REGL.*

Proof: Since we have a unary alphabet $\{a\}$, $\Pi(x) = (|x|)$, for all x in $\{a\}^*$, and $\Pi(x) = \Pi(y)$ if and only if $x = y$ in this case. Now, by Parikh's theorem, there exists $R \subseteq \{a\}^*$ with $\Pi(R) = \Pi(L)$ and, therefore, $R = L$. ■

We prove Parikh's theorem by way of an extension of the *CFG* pumping lemma.

Lemma 8.2.6 The Extended *CFG* Pumping Lemma
Let $G = (N, \Sigma, P, S)$ *be a* λ*-free, unit-free CFG,* $m = \max(\{|\alpha| : A \to \alpha \text{ is in } P\})$*. Then, for all* $k \geq 1$, $p_k = 1 + m^{k\#N+1}$*, and for all z in* $L(G)$ *with* $|z| \geq p_k$*, there is a derivation sequence*

$$S \Rightarrow^* uAv \Rightarrow^* ux_1Ay_1v \Rightarrow^+ \cdots$$

$$\Rightarrow^+ ux_1 \cdots x_kAy_k \cdots y_1v \Rightarrow^+ ux_1 \cdots x_kwy_k \cdots y_1v = z$$

in G, for some nonterminal A, for some u, v, w in Σ^**, and for some* x_i, y_i *in* Σ^*, $1 \leq i \leq k$*, such that (i) and (ii) hold.*

(i) $|x_1 \cdots x_kwy_k \cdots y_1| < p_k$.
(ii) *For all i,* $1 \leq i \leq k$, $|x_iy_i| \geq 1$.

Proof: First note that when $k = 1$, we have the *CFG* pumping lemma once again. This is because $p_1 = 1 + m^{\#N+1}$, $z = ux_1wy_1v$, and $S \Rightarrow^* uAv \Rightarrow^* ux_1Ay_1v \Rightarrow^+ ux_1wy_1v$, that is, $ux_1^iwy_1^iv$ is in $L(G)$, for all $i \geq 0$. For this reason we sketch the proof briefly, leaving the details to the Exercises.

A syntax tree for z has height $> k\#N+1$. Therefore, it contains a root-to-frontier path of length at least $k\#N+2$, which by the extended pigeonhole principle implies that there is some nonterminal A which appears at least $k+1$ times. A provides the sought decomposition. By taking a longest path in the syntax tree together with $k+1$ appearances of a nonterminal none of which appears at the root condition (i) is satisfied. ■

Apart from this extension of the *CFG* pumping lemma we also need the following notion.

Definition Let $G = (N, \Sigma, P, S)$ be a *CFG* and $Q \subseteq N$. We say a derivation $S \Rightarrow^+ x$, for x in Σ^*, is a *Q-derivation* if every nonterminal in Q appears in the derivation and every nonterminal in $N - Q$ does not appear. The *Q-language of G* is denoted by $L(G,Q)$ and is defined by

$$L(G,Q) = \{x : x \text{ is in } \Sigma^* \text{ and there is a } Q\text{-derivation of } x \text{ in} G\}$$

If S is not in Q, then $L(G,Q) = \varnothing$. Since G can be ambiguous, it is possible that $L(G,Q_1) \cap L(G,Q_2) \neq \varnothing$, for some Q_1 and Q_2, $Q_1 \neq Q_2$. However, we still obtain a decomposition of $L(G)$ as

$$L(G) = \bigcup_{Q \subseteq N} L(G,Q)$$

In order to prove Parikh's theorem for a *CFL*, $L = L(G)$, for some *CFG*, $G = (N,\Sigma,P,S)$, it is sufficient to establish it for $L_Q = L(G,Q)$, for some arbitrary but fixed Q.

Lemma 8.2.7 *Let* $G = (N,\Sigma,P,S)$ *be a* λ*-free, unit-free CFG and* $Q \subseteq N$, *where* $m = \max(\{\,|\alpha| : A \to \alpha \text{ is in } P\})$, $k = \#Q$, $p = 1+m^{k\#N+1}$, $F = \{w : w \text{ is in } L(G,Q) \text{ and } |w| < p\}$, *and* $H = \{xy : A \Rightarrow^+ xAy, \text{ for some } A \text{ in } Q, \text{ and } 1 \leq |xy| < p\}$.

Then $\Pi(L(G,Q)) = \Pi(FH^*)$.

Proof:

(1) $\Pi(L(G,Q)) \subseteq \Pi(FH^*)$. We need to show that for each z in $L(G,Q)$, there is a word z' in FH^* with $\Pi(z) = \Pi(z')$. We prove this by induction on the length of words in $L(G,Q)$.

Basis: $0 \leq |z| < p$. Then z is in F and therefore in FH^*. The result holds vacuously in this case.

Induction Hypothesis: Assume that for each z in $L(G,Q)$ of length at most n, for some $n \geq p-1$, there is a word z' in FH^* with $\Pi(z') = \Pi(z)$.

Induction Step: Let z be a word in $L(G,Q)$ of length $n+1$. Then $|z| \geq p$ and there is a Q-derivation $S \Rightarrow^* z$ in G.

By the extended *CFG* pumping lemma z has a Q-derivation

$$S \underset{0}{\Rightarrow^*} uAv \underset{1}{\Rightarrow^+} ux_1Ay_1v \underset{2}{\Rightarrow^+} \cdots$$

$$\underset{k}{\Rightarrow^+} ux_1 \cdots x_kAy_k \cdots y_1v \underset{k+1}{\Rightarrow^+} ux_1 \cdots x_kwy_k \cdots y_1v = z$$

for some A in Q, and for some x_iy_i in H, $1 \leq i \leq k$. We have numbered the $k+2$ subderivations as $0, 1, \ldots, k+1$.

Consider the k nonterminals in $Q-\{A\}$. Each of them appears in at least one of the subderivations. For each nonterminal check off a subderivation in which it appears. This leaves at least one unchecked subderivation numbered i, say, where $1 \leq i \leq k$. Remove this subderivation giving

$$S \Rightarrow^* ux_1 \cdots x_{i-1}Ay_{i-1} \cdots y_1v$$
$$\Rightarrow^+ ux_1 \cdots x_{i-1}x_{i+1} \cdots x_k w y_k \cdots y_{i+1}y_{i-1} \cdots y_1 v = \bar{z}$$

Now, $|\bar{z}| < |z|$; therefore, there is a word $\bar{z}'$ in FH^* with $\Pi(\bar{z}') = \Pi(\bar{z})$. But $\bar{z}'x_iy_i$ is also in FH^* and $\Pi(\bar{z}'x_iy_i) = \Pi(z)$, as desired.

(2) $\Pi(FH^*) \subseteq \Pi(L(G,Q))$. We prove by induction that for each z in FH^* there is a word z' in $L(G,Q)$ with $\Pi(z') = \Pi(z)$.

Basis: $0 \leq |z| < p$. Then z is in F and also in $L(G,Q)$.

Induction Hypothesis: Assume that for each z in FH^* of length at most n, for some $n \geq p-1$, there is a word z' in $L(G,Q)$ with $\Pi(z') = \Pi(z)$.

Induction Step: Let z be a word in FH^* with $|z| = n+1$. Since $|z| \geq p$, $z = ft$ for some words f in FH^* and t in H. Since $|f| < |z|$, there is a word f' in $L(G,Q)$ with $\Pi(f') = \Pi(f)$. But $t = x_1y_1$, where $A \Rightarrow^* x_1Ay_1$ is a Q-derivation. But there is a Q-derivation

$$S \Rightarrow^* uAv \Rightarrow^+ uwv = f'$$

in G, for some u and v, since each nonterminal in Q must appear. This implies

$$S \Rightarrow^* uAv \Rightarrow^+ ux_1Ay_1v \Rightarrow^+ ux_1wy_1v$$

is also a Q-derivation in G. Now, $\Pi(ux_1wy_1v) = \Pi(fx_1y_1)$, so we have constructed a word z' in $L(G,Q)$ from z in FH^* with $\Pi(z') = \Pi(z)$ as required.

This completes the proof that $\Pi(FH^*) = \Pi(L(G,Q))$. ■

We now have the final result of this subsection.

Theorem 8.2.8 $\boldsymbol{L}_{CF} \subset \boldsymbol{L}_{DPTIME}$.

Proof: Let $L = \{a^{i^2} : i \geq 0\}$. It has been shown that L is not a *REGL* (see Example 8.1.4), and, therefore, by Theorem 8.2.5, it is not a *CFL* either.

However, L is in $\boldsymbol{L}_{DPTIME}$; see Section 6.2. ■

8.3* TRACTABLE AND RECURSIVE LANGUAGES

In this section we prove that $\boldsymbol{L}_{DPSPACE} = \boldsymbol{L}_{NPSPACE}$, $\boldsymbol{L}_{DPSPACE} \subset \boldsymbol{L}_{REC}$, and $\boldsymbol{L}_{DPTIME} \subseteq \boldsymbol{L}_{NPTIME} \subseteq \boldsymbol{L}_{NPSPACE}$. Neither of the last two inclusions is known to be proper, although at least one of them must be. Indeed, the equality or inequality of $\boldsymbol{L}_{DPTIME}$ and $\boldsymbol{L}_{NPTIME}$ (*P* and *NP* as they are usually known) is a fundamental open problem. *P* represents all problems that have polynomial-time algorithms while *NP* represents those problems that have polynomial-time algorithms to verify that a proposed solution is indeed a solution. It appears that *NP* problems do not have polynomial-time algorithms, and proving $\boldsymbol{L}_{DPTIME} \subset \boldsymbol{L}_{NPTIME}$ would confirm this reasonable conjecture.

Similarly, establishing that $\boldsymbol{L}_{NPTIME} \subset \boldsymbol{L}_{DPSPACE}$ would show that polynomially bounded work space is more powerful than polynomially bounded verification time.

Our first result is:

Theorem 8.3.1 $\boldsymbol{L}_{DPSPACE} = \boldsymbol{L}_{NPSPACE}$.

Proof: Since $\boldsymbol{L}_{DPSPACE} \subseteq \boldsymbol{L}_{NPSPACE}$ by definition, we are only left with proving that $\boldsymbol{L}_{NPSPACE} \subseteq \boldsymbol{L}_{DPSPACE}$.

Consider an arbitrary language $L \subseteq \Sigma^*$, which is accepted by some $S(n)$-space-bounded NTM, $M = (Q,\Sigma,\Gamma,\delta,s,f)$, where $S(n)$ is a polynomial. We wish to prove that L is also accepted by polynomially space-bounded DTM, M'.

First note that the simulation of an NTM by a DTM given in Theorem 6.4.2 is insufficient for this purpose. This is because the simulating DTM runs through all possible control words (a control word specifies the choices to be made at each step of a nondeterministic computation). Now, the number of computation steps is less than $\#QS(n)\#\Gamma^{S(n)}$, since this is the number of distinct configurations. This is surely less than $\#Q(\#\Gamma+1)^{S(n)+1}$, which because Q and Γ are constants with respect to M, becomes $cd^{S(n)}$, where $d = \#\Gamma+1$ and $c = d\#Q$. In other words, a configuration sequence can have length exponential in the input size and, therefore, a control word can have exponential length. This implies that allowable control words do not have polynomially bounded size, which means that the simulating DTM of Theorem 6.4.2 is not polynomially space bounded.

To simulate the given polynomially space-bounded NTM, M, by a polynomially space-bounded DTM, M', we need a different approach. We use divide-and-conquer for this purpose.

Given two configurations C_1 and C_2 of M (initially C_1 is sx and C_2 is y_1fy_2), to determine whether or not $C_1 \vdash^m C_2$, for some m, $1 \leq m \leq cd^{S(n)}$, we ask if there is a configuration C such that

$$C_1 \vdash^{\lfloor m/2 \rfloor} C \text{ and } C \vdash^{\lceil m/2 \rceil} C_2$$

This involves, first, an exhaustive search through the, at most, $cd^{S(n)}$ configurations to determine if such a C exists. Second, the two subproblems are solved recursively using the same technique. Let us illustrate this with an input word x which has an accepting configuration sequence of length $m = 2^k$, for some $k \geq 1$. Then we have

$$C_0 \vdash C_1 \vdash \cdots \vdash C_m$$

in M, where $C_0 = sx$ and $C_m = y_1fy_2$. Now, the procedure outlined above can be pictured as the search tree of Figure 8.3.1 in which the frontier nodes are as shown in Figure 8.3.2. The usual method of simulating M is to attempt to traverse the frontier directly; we, on the other hand, traverse it using a preorder traversal of the tree. This implies that we need keep only the target configurations along one root-to-frontier path. But there are only $\lceil \log_2 m \rceil$ of these, that is, at most

$\lceil \log_2(cd^{S(n)}) \rceil = \lceil S(n)\log_2(cd) \rceil \leq eS(n)$, where $e = \lceil \log_2 cd \rceil$. In other words, only configurations proportional to $S(n)$ need be kept. This means that space proportional to $S^2(n)$ is needed, since each configuration has size bounded by $S(n)$. But the square of a polynomial is also a polynomial so the simulating *DTM* *is* polynomially space bounded.

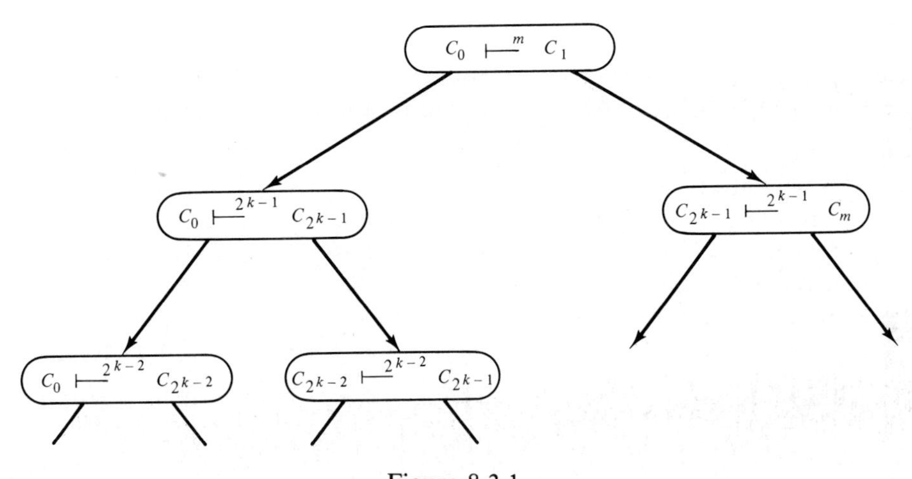

Figure 8.3.1

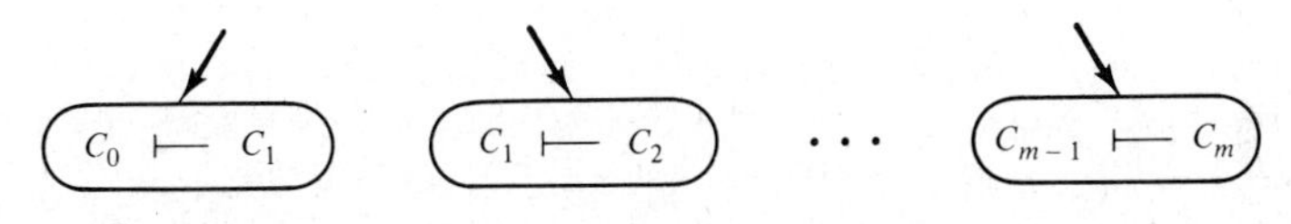

Figure 8.3.2

Of course, there isn't just one such search tree for an input word x, but many. This is because there are at most $d^{S(n)+1}$ possible final configurations and at most $cd^{S(n)}$ intermediate configurations. At each node in the search tree each possible configuration is a candidate target configuration. However, we still need keep only the candidate target configurations along the current search path. Essentially we are programming the *DTM* to deal with recursive subprograms.

The simulation can be described with a PASCAL-like subprogram as follows. The possible configurations of size $S(n)+1$ can be enumerated; hence, we assume the existence of functions to retrieve the ith configuration and ith final configuration.

function *Simulate*(x : *input word*) : **boolean**;
{Returns **true** if x is accepted and **false** otherwise}
begin $n := |x|$; $S := S(n)$;
    $d := \#\Gamma+1$; $c := \#Q * S * d$;

```
    D := c * (d * * (S+1)); {The number of distinct configurations}
    F := c div #Q; {The number of final configurations}
    for i := 1 to F do
        if Check(sx, Final Configuration(i),D) then
                    return true;
    return false
end {of Simulate};
function Check(Source,Target,N) : boolean;
{Return true if M beginning in configuration Source can be reach configuration
Target in at most N steps, otherwise it returns false}

begin
    if N = 1 then
        if Source |— Target in M then return true
                                 else return false
            else {N > 1}
begin
    for i := 1 to D do
    begin Next := Configuration(i);
        if Check(Source,Next,⌊D/2⌋)
            and Check(Next,Target,⌈D/2⌉) then return true;
    end;
    return false
end
end {of Check};
```

The simulating *DTM*, M', must keep a stack of recursive calls of *Check*. For each call the local environment of the calling instance of Check must be kept on the stack. In other words, the local values of *Source*, *Target*, N, i, and *Next*. *Source*, *Target*, and *Next* have length $S(n)+1$ and since i and N are bounded from above by D, they can be represented in binary in $\lceil \log_2 D \rceil$ space. But $\lceil \log_2 D \rceil \equiv eS(n)$, for some constant $e \geq 1$, so i and N also require space proportional to $S(n)$. Note that e is a constant with respect to M, it is independent of the value of x.

The detailed definition of M' is left to the Exercises. It is sufficient to observe that it need keep track of at most $\lceil \log_2 D \rceil$ calls of *Check* on its stack, which implies that the stack requires space proportional to $S^2(n)$. This is the space bound we sought; hence, M' is polynomially space bounded. ■

From the definition of $\boldsymbol{L}_{NPTIME}$ a language L is in $\boldsymbol{L}_{NPTIME}$ if there is an *NTM*, M, and a polynomial $T(n)$ such that $L(M) = L$ and x is in L if and only if x has an accepting configuration sequence of length at most $T(|x|)$. Observe that such an accepting configuration sequence cannot visit more than $T(|x|)$ distinct cells, so L is in $\boldsymbol{L}_{NPSPACE}$. But by Theorem 8.3.1 this implies L is also in $\boldsymbol{L}_{DPSPACE}$, thus $\boldsymbol{L}_{NPTIME} \subseteq \boldsymbol{L}_{DPSPACE}$.

To show that $\boldsymbol{L}_{DPTIME}$, $\boldsymbol{L}_{NPTIME}$, and $\boldsymbol{L}_{DPSPACE}$ are contained in $\boldsymbol{L}_{REC}$ it is sufficient to prove that $\boldsymbol{L}_{DPSPACE}$ is contained in $\boldsymbol{L}_{REC}$.

Theorem 8.3.2 $\boldsymbol{L}_{DPSPACE} \subseteq \boldsymbol{L}_{REC}$.

Proof: Let $L \subseteq \Sigma^*$ be an arbitrary language in $\boldsymbol{L}_{DPSPACE}$. Then L is accepted by an $S(n)$-space-bounded DTM, $M = (Q,\Sigma,\Gamma,\delta,s,f)$, where $S(n)$ is a polynomial. We wish to prove that L is decidable, that is, that L is in $\boldsymbol{L}_{REC}$.

On a given input word x in Σ^*, $|x| = n$, M can respond in four different ways.

(i) M can accept x having visited at most $S(n)$ distinct cells.
(ii) M can visit more than $S(n)$ cells.
(iii) M can hang.
(iv) M can loop.

In order to prove that L is decidable we embed M within a second DTM $M' = (Q',\Sigma,\Gamma',\delta',s',f')$. M' acts as a supervisor or big brother of M (see Figure 8.3.3), in that it allows M to function as usual, except that it continually checks M's progress. If M' sees that M will enter state f as a result of its next move, then M' erases the tape, writes **y** on it, and halts in f'. Informally, if M' observes that (ii), (iii), or (iv) is about to occur, then it erases the tape, writes **n** on it, and halts in state f'. But how is M' able to see that one of (ii), (iii), or (iv) is about to occur? Consider each of them in turn.

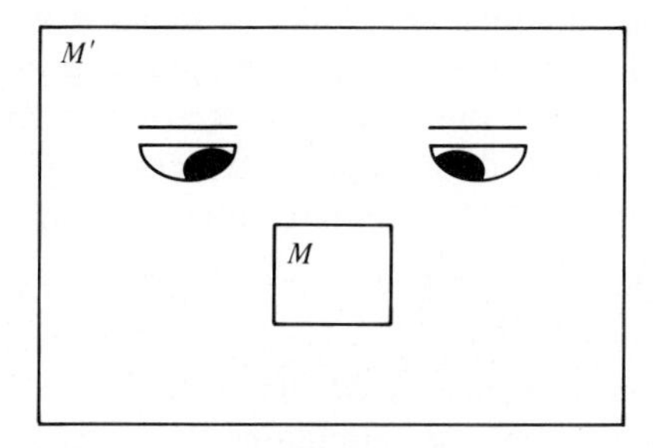

Figure 8.3.3

(ii) *M visits more than $S(n)$ cells.*
Initially, M' computes the value of $S(n)$, where $n = |x|$ and x is the input word. Since S is a given polynomial, $S(n)$ can be evaluated using a fixed sequence (or program) of multiplications, additions, and subtractions. Clearly, this computation terminates for all values of n. Let M' have a second tape which contains $S(n)$ in unary and has its read-write head over the leftmost cell. M' mimics, on the second tape, the head movement of M on its first tape. If at any time the head on the second tape reaches a blank cell, M has visited more than $S(n)$ cells.

(iii) *M hangs.*
This occurs because either M moves left on the leftmost cell or there is no transition for the current state and current symbol.

To detect the first situation M' shifts the input word right one cell, and inserts the symbol $\boldsymbol{l}$ in the first cell of both the first and second tapes. This means that the second tape should contain $S(n)$ from the second cell onward. Both heads are positioned over the second cells of their tapes and M is started. If the first and second heads ever end up over the cell containing $\boldsymbol{l}$, then M' has detected a hang situation of the first kind.

To detect the second situation, add a new state q_{undef} to M'. For all p in Q, for all c in Γ, such that $\delta(p,c) = \boldsymbol{undef}$ in M, add the transition $\delta'(p,c) = (q_{undef},c,\lambda)$ to M'. Whenever M' enters q_{undef} an undefined transition in M has been met.

(iv) *M loops.*
This occurs if a configuration is repeated in a configuration sequence. For an input word x in Σ^*, $|x| = n$, there are $\#Q\#\Gamma^{S(n)}S(n)$ distinct configurations of length at most $S(n)$. (Each cell may contain any one of $\#\Gamma$ symbols; hence, there are $\#\Gamma^{S(n)}$ distinct cell sequences of length at most $S(n)$. The read-write head can be over any one of these cells and there are $\#Q$ distinct states giving $\#Q\#\Gamma^{S(n)}S(n)$ distinct configurations). So M' has a third tape with $\boldsymbol{l}$ in its leftmost cell which contains every configuration met so far during M's computation. The configurations are separated by $\$$. Initially, before beginning the computation of M, M' writes the starting configuration sx, followed by a sufficient number of blanks to make a word of length $S(n)+1$, which is followed in turn by $\$$. Then a single step of M is simulated. M' now checks whether or not the new configuration of M appears on the third tape. If it does, then M has looped. Otherwise the new configuration is added to the third tape and M' simulates the next step of M.

M' is therefore, a three-tape DTM, the first tape is, essentially, M's tape; the second contains the value of $S(|x|)$, where x is the input word; and the third contains the configuration sequence of M on input x. After the appropriate initializations have been carried out, M' simulates the computation of M one step at a time. After each step has been completed M' checks to see if any one of (i)-(iv) have occurred. If none of them have occurred, then the next step of M is simulated. If (i) has occurred, then, M' erases its input tape, writes **y**, and enters f'. If (ii), (iii), or (iv) has occurred, then M' erases its input tape, writes **n**, and enters f'. Finally, we can replace the three tapes of M' by a single tape with six tracks (see Theorem 6.4.4), and thus we have shown that L is decidable. ■

We leave to the next section the proof of

Theorem 8.3.3 $\boldsymbol{L}_{DPSPACE} \subset \boldsymbol{L}_{REC}$.

8.4 RECURSIVE AND *DTM* LANGUAGES

To complete the picture of the various family relationships we prove that $\boldsymbol{L}_{REC} \subset \boldsymbol{L}_{DTM} \subset \boldsymbol{L}_{ALL}$. The proofs use two languages obtained from the encoding of *DTMs* over $\{a,b\}$ given in Section 6.5. Recall that if M is *DTM* with tape alphabet $\{a,b,\boldsymbol{B}\}$, then $\ll M \gg$ is a word over $\{a,b\}$ denoting the encoding of M.

The first language is

$$L_{ACCEPT} = \{\ll M \gg x : M \text{ is a } DTM \text{ with tape alphabet } \{a,b,\boldsymbol{B}\},\ x \text{ is in } \{a,b\}^*, \text{ and } M \text{ accepts } x\}$$

The universal *DTM*, U, is given words of the form $\ll M \gg x$ or equivalently $\ll M,x \gg$ and it simulates the computation of M on input x. Therefore, L_{ACCEPT} consists of exactly those words, which the universal Turing machine U accepts. This implies L_{ACCEPT} is in $\boldsymbol{L}_{DTM}$.

The second language is

$$L_{SELFACCEPT} = \{\ll M \gg : M \text{ is a } DTM \text{ with tape alphabet } \{a,b,\boldsymbol{B}\} \text{and } M \text{ accepts } \ll M \gg\}$$

In other words, $L_{SELFACCEPT}$ consists of those encodings of *DTMs* which are accepted by their corresponding *DTMs*. $L_{SELFACCEPT}$ is well defined since $\ll M \gg$ is a word over $\{a,b\}$, the input alphabet of M. Note that, since encodings are not unique a *DTM*, M, may have some encodings which it accepts and some it doesn't. These two languages play an important role in the undecidability results of Chapter 10. We first argue that $L_{SELFACCEPT}$ is in $\boldsymbol{L}_{DTM}$, before proving that its complement is not in $\boldsymbol{L}_{DTM}$.

We construct a *DTM*, V, with $L(V) = L_{SELFACCEPT}$, which is based on the universal *DTM*, U. V first checks that its input word x is the encoding of some *DTM*, M. This is a straightforward task. If $x = \ll M \gg$, then V instantly calls U with input xx. If U accepts xx, then V accepts x. This implies that $L(V) = L_{SELFACCEPT}$; therefore, $L_{SELFACCEPT}$ is in $\boldsymbol{L}_{DTM}$.

To prove that $L_{SELFACCEPT}$ is not in $\boldsymbol{L}_{REC}$, we prove that $\overline{L}_{SELFACCEPT}$ is not in $\boldsymbol{L}_{DTM}$. This implies that $L_{SELFACCEPT}$ is not in $\boldsymbol{L}_{REC}$ by way of the following theorem.

Theorem 8.4.1 *Let L be an arbitrary language. Then L is in $\boldsymbol{L}_{REC}$ if and only if $\overline{L}$ is in $\boldsymbol{L}_{REC}$.*

Proof: We prove the result in only one direction, since the proof in the other direction is identical.

Assume L is in $\boldsymbol{L}_{REC}$ and $L \subseteq \Sigma^*$. Then $L = Y(M)$, for some *DTM*, $M = (Q,\Sigma,\Gamma,\delta,s,f)$. Now, $\overline{L} = N(M)$; therefore, we modify M, to give $\overline{M} = (Q,\Sigma,\Gamma,\overline{\delta},s,f)$, by interchanging the appearances of the symbols **n** and **y** in δ, giving $\overline{\delta}$. Clearly, $\overline{L} = Y(M)$, that is, $\overline{L}$ is in $\boldsymbol{L}_{REC}$. ■

If $L_{SELFACCEPT}$ is in $\boldsymbol{L}_{REC}$, then $K = \overline{L}_{SELFACCEPT}$ is also in $\boldsymbol{L}_{REC}$ and, hence, in $\boldsymbol{L}_{DTM}$. But as we now prove K is not even in $\boldsymbol{L}_{DTM}$.

Theorem 8.4.2 *$K = \overline{L}_{SELFACCEPT}$ is not in $\boldsymbol{L}_{DTM}$.*

Proof: We use diagonalization in an argument by contradiction. To gain some insight into the diagonalization argument consider an infinite matrix that is indexed by an enumeration of those words in $\{a,b\}^*$ which are encodings of *DTMs* having a tape alphabet $\{a,b,\boldsymbol{B}\}$. By the results of Sections 0.4 and 1.2 such an enumeration exists. Let the enumeration of these words be $x_1, x_2, \ldots, x_i$, and let M_{x_i} denote the *DTM* which has encoding x_i. The i,jth entry of the matrix is 1 if M_{x_i} accepts x_j and is 0 otherwise; see Figure 8.4.1. This implies that $L_{SELFACCEPT}$ corresponds to the 1 entries on the diagonal and K contains the words corresponding to the 0 entries on the diagonal. Assuming K is in $\boldsymbol{L}_{DTM}$ we must have $K = L(M_{x_k})$, for some *DTM*, M_{x_k}, corresponding to some encoding x_k.

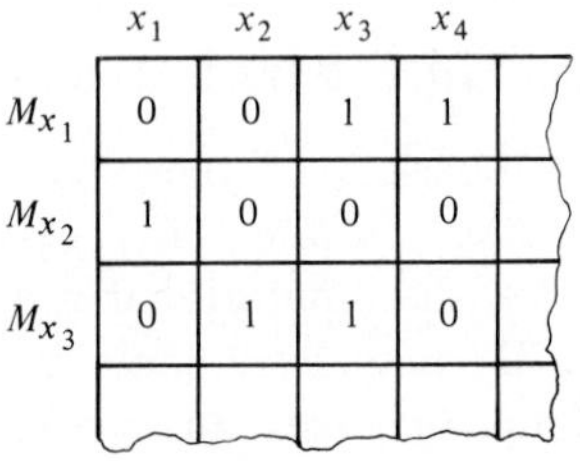

	x_1	x_2	x_3	x_4	
M_{x_1}	0	0	1	1	
M_{x_2}	1	0	0	0	
M_{x_3}	0	1	1	0	

Figure 8.4.1

Consider the kth diagonal entry. If x_k is accepted by M_{x_k}, then the diagonal entry is 1. But x_k is in $L(M_{x_k}) = K$ implies the diagonal entry is 0 — a contradiction.

If x_k is not accepted by M_{x_k}, then the diagonal entry is 0. But x_k is not in $L(M_{x_k}) = K$ implies x_k is in $L_{SELFACCEPT}$, that is, the diagonal entry is 1. Again, a contradiction.

In both cases a contradiction is obtained; thus, K is not in $\boldsymbol{L}_{DTM}$. ■

Corollary 8.4.3 $\boldsymbol{L}_{REC} \subset \boldsymbol{L}_{DTM}$.

Proof: $L_{SELFACCEPT}$ is in $\boldsymbol{L}_{DTM}$ and K is not in $\boldsymbol{L}_{DTM}$. Therefore, $L_{SELFACCEPT}$ is not in $\boldsymbol{L}_{REC}$, by Theorem 8.4.1. ■

Using a similar approach we now prove Theorem 8.3.3. For this purpose we consider a subset of $L_{SELFACCEPT}$ which we call $L_{POLYSELFACCEPT}$, or L_{PSA} for short. L_{PSA} is defined by

$$L_{PSA} = \{x : x \text{ is the encoding of some } DTM\ M \text{ and } M \text{ accepts } x \text{ using at most } |x|^{|x|} \text{ space}\}$$

Although neither L_{ACCEPT} nor $L_{SELFACCEPT}$ are recursive, L_{PSA} is recursive.

Lemma 8.4.4 *L_{PSA} and $\overline{L}_{PSA}$ are in $\boldsymbol{L}_{REC}$.*

Proof: We modify the universal Turing machine U to give a DTM, V, with $Y(V) = L_{PSA}$.

V checks the input word x to determine if it is the encoding of a DTM M. If not, then V returns **n**. Otherwise it calls U on xx and oversees its operation, as in Theorem 8.3.2, to determine if it accepts x within $|x|^{|x|}$ space. If U does accept xx, then V returns **y**, otherwise it returns **n**. Therefore, $Y(V) = L_{PSA}$ and L_{PSA} is in $\boldsymbol{L}_{REC}$. By Theorem 8.4.1 $\overline{L}_{PSA}$ is also in $\boldsymbol{L}_{REC}$. ■

We are now in a position to prove Theorem 8.3.3.

Proof of Theorem 8.3.3: We assume $\overline{L}_{PSA}$ is in $\boldsymbol{L}_{DPSPACE}$ and provide a contradiction based on diagonalization. Assume we have an infinite bit matrix whose rows and columns are indexed by words over $\{a,b\}$ which are encodings of $DTMs$. If x is the encoding of a DTM, then denote the encoded DTM by M_x. Recall that words in L_{PSA} are the encodings of $DTMs$. The i,jth entry in the matrix is 1 if M_{x_i} accepts x_j in space n^n, there $n = |x_j|$ and is zero otherwise. Hence, the 1 entries on the diagonal give L_{PSA}.

As usual in diagonalization arguments we consider the words corresponding to 0 entries on the diagonal; let these be D. Then $D \subseteq \overline{L}_{PSA}$. We wish to prove that $\overline{L}_{PSA}$ and, hence, D are not in $\boldsymbol{L}_{DPSPACE}$.

First, note that for each DTM, M, which has an encoding x, there is a DTM, M', which is equivalent to M such that every encoding x' of M' satisfies $|x'| > |x|$ and, further, M' is $S(n)$-space bounded if M is. Letting $M = (Q,\{a,b\},\{a,b,\boldsymbol{B}\},\delta,s,f)$ we may add a new state q to Q and a new transition $\delta(q,a) = (q,a,\lambda)$ to give M'. Since q is not reachable from any state in Q we have $L(M') = L(M)$ and, by the definition of an encoding, $|x'| > |x|$ for all encodings x' of M'. This observation implies that there exist arbitrarily large encodings of an $S(n)$-space bounded DTM with respect to equivalence.

Assume $\overline{L}_{PSA}$ is accepted by some DTM, $M = (Q,\{a,b\}, \{a,b,\boldsymbol{B}\},\delta,s,f)$, in space bounded by some polynomial $S(n)$. Then there is an encoding x_k of M. Consider the k,kth entry in the matrix. We first argue that $S(n) \leq n^n$, for $n = |x_k|$.

For all $n \geq 1$, $S(n) \leq cn^r$, for some constant $c \geq 1$ and some $r \geq 1$. (For example, $S(n) = a^2x^2 - bx + c \leq (a+b+c)x^2$, for all $a,b,c \geq 0$.) However, for all $n \geq 2$, $c \leq n^t$, for some $t \geq 0$. In particular, $S(n) \leq n^{r+t}$, for $n = |x_k|$ and some $r,t \geq 0$. If $r+t > n$, then using the above observation about encodings of equivalent $DTMs$ we can find an equivalent DTM whose encoding x satisfies $|x| \geq r+t$. Thus, we may assume that $S(n)$ satisfies $S(n) \leq n^n$, for $n = |x_k|$.

If x_k is accepted by $M_{x_k} = M$ in space bounded by $S(n)$, where $n = |x_k|$, then x_k is in $\overline{L}_{PSA}$. But because $S(n) \leq n^n$, for this value of n, x_k is in L_{PSA} and we obtain a contradiction.

On the other hand, if x_k is not accepted by $M_{x_k} = M$ in space bounded by $S(n)$, then x_k is not in $\overline{L}_{PSA}$; that is, x_k is in L_{PSA}. But this implies x_k is accepted by M_{x_k} in space bounded by n^n. Now, by the definition of space-bounded *DTMs*, x_k must be accepted by M_{x_k} in space $S(n)$; hence, x_k is in $\overline{L}_{PSA}$ — a contradiction.

Therefore, $\overline{L}_{PSA}$ is not in $\mathbf{L}_{DPSPACE}$. ■

Finally, we obtain:

Theorem 8.4.4 $\mathbf{L}_{DTM} \subset \mathbf{L}_{ALL}$.

Proof: $K = \overline{L}_{SELFACCEPT}$ is not in $\mathbf{L}_{DTM}$, by Theorem 8.4.2. ■

8.5 ADDITIONAL REMARKS

8.5.1 Summary

We have established the hierarchy displayed in Figure 8.5.1, where a broken line indicates containment which is not necessarily proper and an unbroken line indicates proper containment.

We have used PUMPING LEMMAS, PARIKH'S THEOREM, and DIAGONALIZATION as our main tools to establish these results.

8.5.2 History

The basic hierarchy $\mathbf{L}_{REG} \subset \mathbf{L}_{CF} \subset \mathbf{L}_{CS} \subset \mathbf{L}_{DTM}$, where $\mathbf{L}suig L_{DTM}$, where $\mathbf{L}_{CS}$ is the family of context-sensitive languages is known as the Chomsky hierarchy; see Chomsky (1956, 1959). Révész (1983) and Salomaa (1973) provide good introductions to context-sensitive languages. The two pumping lemmas are due to Bar-Hillel et al. (1961), where they are stated for languages rather than *DFAs* and *CFGs*; see the Exercises. The ultimate periodicity of unary regular languages was first demonstrated in Myhill (1957). Parikh's theorem dates from 1961 and is to be found in Parikh (1966). There have been a number of attempts to give an easier proof; see Goldstine (c. 1980), Greibach (1972), Pilling (1973), and van Leeuwen (1974a). The proof given here which is based on an extension of the *CFG* pumping lemma is due to Goldstine (c. 1980). Ginsburg and Rice (1962) proved that unary *CFLs* are regular.

Surface configurations and their application to the polynomially time-bounded simulation of pushdown automata by *DTMs* is a result due to Cook (1971b). He, in fact, proves that 2-way *NPDAs* can be simulated in this manner.

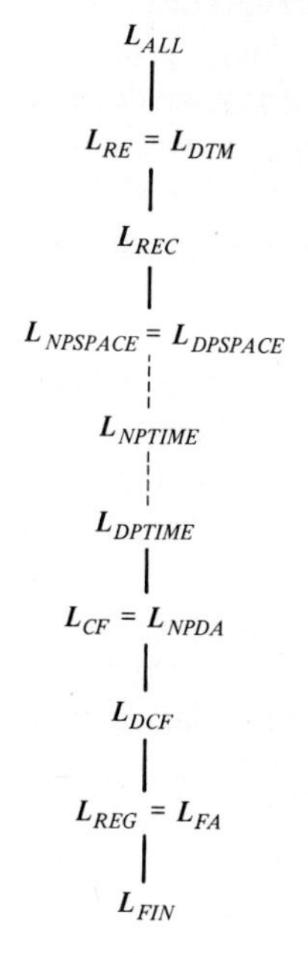

Figure 8.5.1

The proper inclusion of $\boldsymbol{L}_{DSPACE}$ in $\boldsymbol{L}_{REC}$ is a special case of Savitch's theorem (Savitch, 1970), while the proper inclusions $\boldsymbol{L}_{REC} \subset \boldsymbol{L}_{DTM} \subset \boldsymbol{L}_{ALL}$ are due to Turing (1936).

8.6 SPRINGBOARD

8.6.1 Hierarchy Results

The results given here only touch the surface; this hierarchy can be refined substantially. For example, there are hierarchies within $\boldsymbol{L}_{CF}$, $\boldsymbol{L}_{DPSPACE}$, and above $\boldsymbol{L}_{DPSPACE}$. Consult Harrison (1978) and Hopcroft and Ullman (1979).

8.6.2 Pumping Lemmas

There are a number of variations of the *CFG* pumping lemma, the most important being Ogden's lemma (Ogden, 1968), and a characterization of *CFLs* due to Wise (1976). If a language passes through the *DFA* pumping lemma filter, then it may or may not be regular. It is natural to enquire whether or not there exists a non-regular language that passes through the filter. In other words does the *DFA* pumping lemma characterize regular languages? The answer is no via a complex example; see Ehrenfeucht et al. (1981).

8.6.3 Nonlanguages and Closure Properties

In Chapter 9 we consider the closure properties of the regular and context-free languages; for example, given two regular languages their intersection is also regular. There are nonregular languages whose nonregularity is easy to prove if we make use of the closure properties of regular languages. Attempting to use the *DFA* pumping lemma directly may not even give the sought contradiction or, alternatively, it may involve a detailed case analysis. For example, let $L_1 = \{a^i b^j : i \geq j \geq 1\}$. This can be proved to be nonregular using the *DFA* pumping lemma. Based on this we can prove that $L_2 = \{a^i b^j : j > i \geq 1\}$ is also nonregular. This is simply because $L_1 = (\{a,b\}^* - L_2) \cap \{a\}^+\{b\}^+$. Since regular languages are closed under complementation and intersection, if L_2 were regular, then L_1 would be regular. This gives an immediate contradiction, since L_1 is nonregular.

A similar approach can be taken in showing that a language is noncontext-free. That the context-free languages are closed under fewer operations must be taken into account, however. Additional discussion of this topic is to be found in Section 9.5 and the associated Exercises.

EXERCISES

1.1 Try to prove directly that $\{a^i b^j : 1 \leq i \leq j\}$ is not a regular language.

1.2 Prove or disprove that the following languages are *REGLs*.

(i) $\{a^i b^{2i} : i \geq 1\}$.
(ii) $\{a^i b^j : 1 \leq i \leq j\}$.
(iii) $\{ww^R : w \text{ is in } \{a,b\}^*\}$.
(iv) $\{(ab)^i : i \geq 1\}$.
(v) $\{w : w \text{ is in } \{a,b\}^* \text{ and } |w|_a = |w|_b\}$.
(vi) $\{a^{2^m} : m \geq 0\}$.

To prove a language is nonregular use either ultimate periodicity or the *DFA* pumping lemma as appropriate.

1.3 Prove, using diagonalization, that there are languages which are not regular.

1.4 The *DFA* pumping lemma is usually stated for *REGLs* as follows.

Let $L \subseteq \Sigma^*$ be a REGL. Then there exists a constant $p > 0$ such that all words z in L with $|z| \geq p$ can be expressed as $z = uvw$, for some u, v, and w in Σ^* such that (a) $|uv| < p$, (b) $|v| \geq 1$, *and* (c) *for all $i \geq 0$, uv^iw is in L.*

Prove the *REGL* pumping lemma.

1.5 Modify the *DFA* pumping lemma so that condition (iii) reads: *for all $i \geq 1$, uv^iw is in $L(M)$.* Give an example of a nonregular language that passes through this modified filter.

1.6 Consider the following "proof" that $L = \{a^i b^j : i,j \geq 1\}$ is not regular. Point out the fallacious step.

Proof: Assume L is regular. Then $L = L(M)$, for some DFA, M. Letting p be the number of states in M, the DFA pumping lemma tells us that for all z in L, $|z| \geq p$, z can be expressed as $z = uvw$ such that uv^iw is in L. Therefore, consider $z = a^i b^j$ in L with $i \geq 1$, $j \geq 1$, $i+j \geq p$, and $i+1 < p$. We can express z as $a^r a^s b^t b^m$ where $s \geq 1$, $t \geq 1$, $i = r+s$, $j = t+m$, and $r+s+t < p$. This implies $a^r(a^s b^t)^2 b^m$ is in L. Clearly, it is not. Therefore, L is not regular. ■

2.1 Prove directly that the following languages are not $DCFLs$.

(i) $\{a^i b^j : i \geq j \geq 1\}$.
(ii) $\{a^i b^i, a^i b^{2i} : i \geq 1\}$.
(iii) $\{a^i b^i c^i : i \geq 1\}$.

2.2 Prove or disprove that the following languages are $CFLs$.

(i) $\{0^i 10^i 10^i : i \geq 0\}$.
(ii) $\{a^{pq} : p$ and q are primes$\}$.
(iii) $\{ww^R w : w$ is in $\{a,b\}^*\}$.
(iv) $\{x_1 y_1 x_2 y_2 \cdots x_m y_m : x_i$ is in $\{a,b\}^*$, $\{y_i$ is in $\{\bar{a},\bar{b}\}^*$, $m \geq 1$, and $x_1 \cdots x_m$ equals $y_1 \cdots y_m$ if the bars are ignored$\}$.
(v) $\{nHc_1 \cdots c_n : n$ is a positive binary integer and c_i is in $\{a,b\}, 1 \leq i \leq n\} \subseteq \{0,1,a,b,H\}^*$.
(vi) $\{w : w$ is in $\{a,b,c\}^*$ and $|w|_a = |w|_b = |w|_c\}$.

Use either Parikh's theorem or the CFG pumping lemma as appropriate.

2.3 Consider the following restricted classes of $CFGs$. Clearly, each of them defines a subfamily of $\mathbf{L}_{CF}$, but are they proper subfamilies? How are they related to one another?

(i) A CFG, $G = (N,\Sigma,P,S)$, is *meta-linear* if
   (a) for all A in $N - \{S\}$, all A-productions are linear;
   (b) all S-productions have the form $S \rightarrow A_1 \cdots A_k$, for some A_i in $N - \{S\}$, $1 \leq i \leq k$, and some $k \geq 1$; and
   (c) S does not appear on the right-hand side of any production in P.

(ii) A CFG, $G = (N,\Sigma,P,S)$, is said to be *pure* if $\Sigma = \emptyset$ and $L(G) = \{\alpha : S \Rightarrow^* \alpha$ in G and α is in $N^*\}$.

(iii) A CFG, $G = (N,\Sigma,P,S)$, is *unambiguous* if, for all words x in $L(G)$, there is exactly one syntax tree T in G with $y(T) = x$.

(iv) A CFG, $G = (N,\Sigma,P,S)$, is *sequential* if the nonterminals can be numbered from 1 to $\#N = n$ as $A_1, \ldots, A_n$ such that for all i, $1 \leq i \leq n$, $A_i \rightarrow \alpha$ implies α contains no occurrences of nonterminals A_j with $j < i$.

(v) A CFG, $G = (N,\Sigma,P,S)$, is *ultralinear* if there exists an integer $k \geq 1$ such that for all words α in $(N \cup \Sigma)^*$, if $S \Rightarrow^* \alpha$, then α contains at most k occurrences of nonterminals.

(vi) A CFG, $G = (N,\Sigma,P,S)$, is *left ultralinear* if there exists an integer $k \geq 1$ such that for all words α in $(N \cup \Sigma)^*$, if $S \overset{L}{\Rightarrow}{}^* \alpha$, then α contains at most k occurrences of nonterminals.

(vii) A CFG, $G = (N,\Sigma,P,S)$, is *derivation bounded* if there exists an integer $k \geq 1$ such that for all x in $L(G)$, there exists a derivation

$$S = \alpha_0 \Rightarrow \alpha_1 \Rightarrow \cdots \Rightarrow \alpha_m = x$$

in G such that each α_i contains at most k occurrences of nonterminals, $0 \leq i \leq m$.

(viii) A CFG, $G = (N,\Sigma,P,S)$, is *left-derivation bounded* if there exists an integer $k \geq 1$ such that for all x in $L(G)$, there exists a leftmost derivation

$$S = \alpha_0 \overset{L}{\Rightarrow} \alpha_1 \overset{L}{\Rightarrow} \cdots \overset{L}{\Rightarrow} \alpha_m = x$$

in G such that each α_i contains at most k occurrences of nonterminals, $0 \leq i \leq m$.

2.4 Prove, using diagonalization, that there exist noncontext-free languages.

2.5 The CFG pumping lemma is usually stated for $CFLs$ as follows.

Let $L \subseteq \Sigma^*$ be a CFL. Then there exists a constant $p > 0$ such that all words z in L with $|z| \geq p$ can be expressed as $z = uxwyv$, for some u, v, w, x, y in Σ^* such that (a) $|xwy| < p$; (b) $|xy| \geq 1$; *and* (c) for all $i \geq 0$, ux^iwy^iv *is in* L.

Prove the CFL pumping lemma.

2.6 Replace condition (c) in the CFL pumping lemma of Exercise 8.2.5 by: (c) *for all* $i \geq 1$, ux^iwy^iv *is in* L. Give an example of a noncontext-free language that passes through this modified pumping lemma filter.

2.7 The $LINL$ pumping lemma can be stated as follows.

Let $L \subseteq \Sigma^*$ be a LINL. Then there exists a constant $p > 0$ such that for all words z in L with $|z| \geq p$ can be expressed as $z = uxwyv$, for some u, v, w, x, y in Σ^* such that (a) $|uxyv| < p$; (b) $|xy| \geq 1$; *and* (c) *for all* $i \geq 0$, ux^iwy^iv *is in* L.

(i) Prove the $LINL$ pumping lemma.
(ii) Using the $LINL$ pumping lemma prove that the following languages are not linear.

(a) $\{a^i b^i c^j d^j : i,j \geq 1\}$.
(b) $\{x : x$ is in $\{a,b\}^*$ and $|x|_a = |x|_b\}$.
(c) $\{a^i b^i c^i : i \geq 1\}$.

2.8 Let $G = (\{S\},\Sigma,P,S)$ be a single nonterminal *CFG*. Prove Parikh's theorem directly for such *CFGs*.

[*Hint:* Essentially, terminals can be placed anywhere on the right-hand side of productions, since it is only their number that we are interested in. Consider how to transform such a *CFG* into an *ECFG* such that the right-hand sides of productions contain at most one nonterminal in the rightmost position.]

2.9 Prove Parikh's theorem for arbitrary *ECFGs* using induction on the number of nonterminals. Exercise 8.2.8 can be considered to be the basis as well as the method to be used in the induction step.

3.1 A more "natural" proof of Theorem 8.3.3 is to use a diagonalization argument based directly on encodings and polynomials. Let rows be indexed by words over $\{a,b\}$ which are encodings of *DTMs* and columns be indexed by representations of polynomials. First, prove that the class of polynomials of a single variable with integer coefficients is enumerable. Second, prove that $\boldsymbol{L}_{DPSPACE} \subset \boldsymbol{L}_{REC}$ based on this.

Chapter 9

Closure Properties

9.1 BOOLEAN AND REVERSAL OPERATIONS

The study of closure properties of language families is as basic as the study of sets and closure operations in algebra, from which we obtain monoids, groups, rings, and fields, for example,. In each of these cases we have a set of elements and one or more operations defined upon them. That an operation acting upon the elements in a set results in an element of the set is axiomatic in each of these cases. The particular object of study, be it, for example, a group or a ring, is said to be *closed* with respect to its given operations, since its application does not give an element outside the base set.

One application of this concept is that once we know a particular family is closed with respect to some operations we can combine languages in the family, using these operations, to obtain more complex languages. Moreover, this usually implies that we have a corresponding building-block approach for the model defining the family. This is important, both practically and theoretically, for λ-*NFAs* and *CFGs*.

Another application is in the demonstration of nonlanguages. Rather than proving directly that a given language is not in a specific family, we can often prove this indirectly by using the closure properties of the family. This is discussed further in Section 9.5.

In this chapter we concentrate on the families $\boldsymbol{L}_{REG}$, $\boldsymbol{L}_{CF}$, and $\boldsymbol{L}_{DTM}$; other families are treated in the Exercises. The operations we consider range from the classical Boolean operations (union, intersection, and complementation), through a variety of monoid-based operations, to transductions, leading to the abstract notions of cylinders, cones, and full *AFLs*.

We first treat the Boolean operations on $\boldsymbol{L}_{REG}$, $\boldsymbol{L}_{CF}$, and $\boldsymbol{L}_{DTM}$, before treating reversal. We recall, using the Boolean operations, the notion of being *closed* with respect to an operation, and the companion notion of *closure*; see Section 0.3.

Definition Let $\boldsymbol{L}$ be a family of languages. We say $\boldsymbol{L}$ is *closed under union* if for all languages L_1 and L_2 in $\boldsymbol{L}$, $L_1 \cup L_2$ is also a language in $\boldsymbol{L}$. Similarly, $\boldsymbol{L}$ is *closed under intersection* if for all L_1 and L_2 in $\boldsymbol{L}$, $L_1 \cap L_2$ is in $\boldsymbol{L}$, and it is *closed under complementation* if for all L in $\boldsymbol{L}$ and all alphabets Σ with $L \subseteq \Sigma^*$, the complement of L with respect to Σ is in $\boldsymbol{L}$. If $\boldsymbol{L}$ is closed under all three operations we say that it is a *Boolean algebra of languages.*

We say that $\boldsymbol{L}'$ is *the closure of* $\boldsymbol{L}$ *under union* if $\boldsymbol{L}$ satisfies the following conditions

(i) $\boldsymbol{L} \subseteq \boldsymbol{L}'$, that is, all languages in $\boldsymbol{L}$ are also in $\boldsymbol{L}'$;
(ii) $\boldsymbol{L}'$ is closed under union; and
(iii) $\boldsymbol{L}'$ is the smallest family satisfying conditions (i) and (ii), that is, there is no family $\boldsymbol{L}''$ also closed under union and satisfying $\boldsymbol{L} \subseteq \boldsymbol{L}'' \subset \boldsymbol{L}'$.

The existence of such an $\boldsymbol{L}'$ is demonstrated in Section 0.3.

We denote $\boldsymbol{L}'$ by $cl(\boldsymbol{L}, \cup)$.

If $\boldsymbol{L}$ is closed under union $\boldsymbol{L}' = \boldsymbol{L}$, otherwise $\boldsymbol{L} \subset \boldsymbol{L}'$. The closure of $\boldsymbol{L}$ with respect to some operation indicates how far we have to extend it in order to remain inside it whenever the given operation is applied.

Theorem 9.1.1 *$\boldsymbol{L}_{REG}$ is a Boolean algebra of languages.*

Proof: Let L_1 and L_2 be two *REGLs*. Then there are *DFAs*, M_1 and M_2, with $L(M_1) = L_1$ and $L(M_2) = L_2$ such that M_1 and M_2 have disjoint state sets. By Theorem 3.2.1 we can construct a λ-*NFA*, M, from M_1 and M_2 such that $L(M) = L(M_1) \cup L(M_2)$. But this implies, by Theorems 2.3.4 and 2.6.1 that $L_1 \cup L_2$ is a *REGL*. In other words, $\boldsymbol{L}_{REG}$ is closed under union.

We now prove that $\overline{L_1}$ is a *REGL*. We can assume that $M_1 = (Q_1, \Sigma_1, \delta_1, s_1, F_1)$ is complete (see Theorem 2.2.1). Now, make every nonfinal state in M_1 final and every final state nonfinal, giving $M_1' = (Q_1, \Sigma_1, \delta_1, s_1, Q_1 - F_1)$. Since M_1 is complete, for all x in Σ_1^*, $s_1 x \vdash^* q$, for some q in Q_1, hence x is not in L_1 if and only if q is in $Q_1 - F_1$, if and only if x is in $\Sigma_1^* - L_1 = L(M_1') = \overline{L_1}$. Thus, $\overline{L_1}$ is in $\boldsymbol{L}_{REG}$ and $\boldsymbol{L}_{REG}$ is closed under complementation.

Finally, by De Morgan's rules, $\boldsymbol{L}_{REG}$ is closed under intersection. ∎

With $\boldsymbol{L}_{CF}$ and $\boldsymbol{L}_{DTM}$ we are not so fortunate.

Theorem 9.1.2 *$\boldsymbol{L}_{CF}$ is closed under union. $\boldsymbol{L}_{CF}$ is not closed under complementation and intersection.*

Proof: Let L_1 and L_2 be two arbitrary context-free languages. Then there exist $G_1 = (N_1, \Sigma_1, P_1, S_1)$ and $G_2 = (N_2, \Sigma_2, P_2, S_2)$ such that $L_1 = L(G_1)$ and $L_2 = L(G_2)$. We may assume that $N_1 \cap N_2 = \emptyset$. Construct a new context-free grammar $G = (N_1 \cup N_2 \cup \{S\}, \Sigma_1 \cup \Sigma_2, P, S)$, where $P = P_1 \cup P_2 \cup \{S \rightarrow S_1, S \rightarrow S_2\}$. We claim that $L(G) = L(G_1) \cup L(G_2)$. The proof of this claim is left to the Exercises.

Second, we prove that $\boldsymbol{L}_{CF}$ is not closed under intersection. This implies $\boldsymbol{L}_{CF}$ is not closed under complementation also, because of de Morgan's rules.

Consider the two context-free languages $L_1 = \{a^i b^i c^j : i,j \geq 1\}$ and $L_2 = \{a^i b^j c^j : i,j \geq 1\}$. Now, $L_1 \cap L_2 = \{a^i b^i c^i : i \geq 1\}$ and this is not a *CFL*; see Example 8.2.1. ■

Although $\boldsymbol{L}_{CF}$ is not closed under intersection it is closed under a restricted intersection operation.

Definition Let $\boldsymbol{L}$ be a family of languages. Then $\boldsymbol{L}$ is *closed under regular intersection* if for all L in $\boldsymbol{L}$ and all R in $\boldsymbol{L}_{REG}$, $L \cap R$ is in $\boldsymbol{L}$.

This restricted intersection operation turns out to be as fundamental in language theory as the usual intersection operation is in set theory. We now have the following.

Theorem 9.1.3 *$\boldsymbol{L}_{CF}$ is closed under regular intersection.*

Proof: Consider an arbitrary *CFL* $L \subseteq \Sigma^*$, then $L - \{\lambda\} = L'$ is generated by a *CFG*, $G = (N, \Sigma, P, S)$, in *2-SF*, that is, its productions only have the forms $A \rightarrow a$, $A \rightarrow aB$, and $A \rightarrow aBC$, for a terminal symbol a, and not necessarily distinct nonterminals A, B, and C.

Second, consider an arbitrary *REGL*, $R \subseteq \Delta^*$. R is accepted by some *DFA*, $M = (Q, \Delta, \delta, s, F)$, that is, $R = L(M)$.

We now construct a new *CFG*, H, from G and M such that $L(H) = L(G) \cap L(M) = L' \cap R$. If $\Sigma \cap \Delta = \emptyset$, then $L' \cap R = \emptyset$ and $L \cap R$ is either $\emptyset$ or $\{\lambda\}$. In both cases $L \cap R$ is in $\boldsymbol{L}_{CF}$. From now on we assume that $\Sigma \cap \Delta \neq \emptyset$. Let

$$H = (U, \Sigma \cap \Delta, T, Z)$$

where Z is a new nonterminal,

$$U = \{[p,A,q] : p,q \text{ are in } Q \text{ and } A \text{ is in } N\} \cup \{Z\}$$

and

$$T = \{[p,A,q] \rightarrow a : \text{if } \delta(p,a) = q \text{ and } A \rightarrow a \text{ is in } P\}$$

$$\cup \{[p,A,q] \rightarrow a[r,B,q] : \text{if } \delta(p,a) = r,$$

$$A \to aB \text{ is in } P, \text{ and } q \text{ is in } Q\}$$
$$\cup \{[p,A,q] \to a[r,B,t][t,C,q] : \text{if } \delta(p,a) = r,$$
$$A \to aBC \text{ is in } P, \text{ and } q,t \text{ are in } Q\}$$
$$\cup \{Z \to [s,S,f] : f \text{ is in } F\}.$$

The basic idea behind the construction is that a new nonterminal $[p,A,q]$ generates the terminal words obtained from A in G which also cause M to move from state p to state q. Thus, $[s,S,f]$, for f in F, generates words in $L(G) \cap R$. This can be proved formally by means of the following claim.

Claim: Let x be in $(\Sigma \cap \Delta)^*$ and k be a positive integer. Then for all A in N, for all p and q in Q, $[p,A,q] \Rightarrow^k x$ in H if and only if $A \Rightarrow^k x$ in G and $px \vdash^k q$ in M.

Proof of Claim: We can prove this claim by induction on k.

Basis: $k = 1$. $[p,A,q] \Rightarrow x$ in H iff $[p,A,q] \to x$ is in T iff $A \to x$ is in P and $\delta(p,a) = q$, by definition, iff $A \Rightarrow x$ in G and $pa \vdash q$ in M.

Induction Hypothesis: Assume the claim holds for all k with $1 \leq k \leq m$, for some $m \geq 1$.

Induction Step: Consider a word x in $(\Sigma \cap \Delta)^*$ having an $(m+1)$-step derivation $[p,A,q] \Rightarrow^{m+1} x$ in H.

Since $m+1 \geq 2$, $|x| \geq 2$ and $x = ay$, for some a in $\Sigma \cap \Delta$ and some y. Moreover, the first derivation step is either $[p,A,q] \Rightarrow a[r,B,q]$, where $\delta(p,a) = r$ and $A \to aB$ is in P, or $[p,A,q] \to a[r,B,t][t,C,q]$, for some t in Q, where $\delta(p,a) = r$ and $A \to aBC$ is in P. We have either $[r,B,q] \Rightarrow^m y$ or $[r,B,t] \Rightarrow^u y_1$ and $[t,C,q] \Rightarrow^v y_2$, respectively, where $y = y_1y_2$ and $u+v = m$. Therefore, by the induction hypothesis we obtain either $B \Rightarrow^m y$ in G and $ry \vdash^m q$ in M, or $B \Rightarrow^u y_1$, $C \Rightarrow^v y_2$ in G and $ry_1 \vdash^u t$, $ty_2 \vdash^v q$ in M, respectively.

Putting these together we obtain in both cases $A \Rightarrow^{m+1} x$ in G and $px \vdash^{m+1} q$ in M, as desired.

To complete the proof we must argue that the converse also holds. But this is a similar argument to the one above, which we leave the reader to complete. Thus, by the principle of mathematical induction the claim has been proved. ■

Now, $Z \Rightarrow^* x$ in $(\Sigma \cap \Delta)^*$ if and only if $[s,S,f] \Rightarrow^* x$, for some f in F, if and only if x is in $L(G)$ and x is in $L(M)$ if and only if x is in $L(G) \cap L(M) = L' \cap R$, that is, $L(H) = L' \cap R$. But this means that $L' \cap R$ is in $\mathbf{L}_{CF}$.

To complete the proof observe that if λ is in L, then

$$L \cap R = (L' \cup \{\lambda\}) \cap R = (L' \cap R) \cup (\{\lambda\} \cap R).$$

Now, $\{\lambda\} \cap R$ is either $\emptyset$ or $\{\lambda\}$, and in either case $\{\lambda\} \cap R$ is in $\mathbf{L}_{CF}$; hence, $(L' \cap R) \cup (\{\lambda\} \cap R)$ is the union of two *CFLs*, which is a *CFL* by Theorem 9.1.2. ■

Example 9.1.1 Let G be the *CFG*

$S \to a \mid aCS \mid (SBH \mid (SD$

$B \to)$

$C \to + \mid *$

$D \to)CS$

This is a variant of an expression grammar. Let M have the state diagram of Figure 9.1.1. Then the new *CFG*, $H = (U, \{+, *, a, (,)\}, T, Z)$, where $N = \{S, B, C, D\}$ and $U = \{[i,A,j] : 0 \leq i,j \leq 1 \text{ and } A \text{ is in } N\} \cup \{Z\}$.

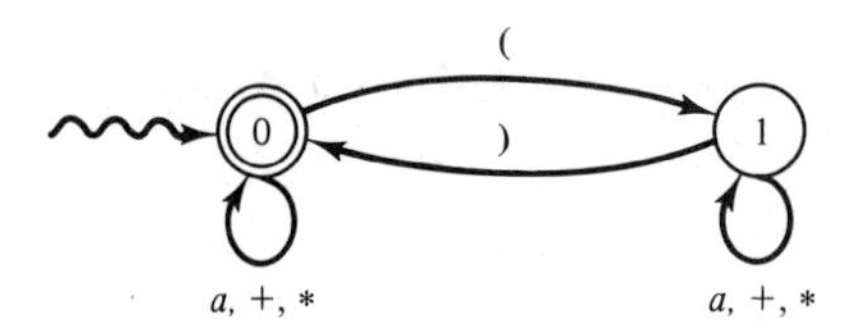

Figure 9.1.1

Now, we have

$Z \to [0,S,0]$

as the only Z-production, since 0 is the only final state. Thus,

$[0,S,0] \to a$

since $\delta(0,a) = 0$.

$[0,S,0] \to a[0,C,0][0,S,0] \mid ([1,S,1][1,B,0] \mid ([1,S,1][1,D,0]$

since $\delta(0,() = 1$, and $\delta(1,)) = 0$.

$[1,S,1] \to a \mid a[1,C,1][1,S,0]$

since $\delta(1,+) = 1$ and $\delta(1,()$ is undefined.

$[1,S,0] \to a[1,C,1][1,S,0]$

is the only $[1,S,0]$-production, and thus is nonterminating, so

$[1,S,1] \to a;$

is the only terminating production for $[1,S,1]$. Finally, H consists of

$[0,C,0] \rightarrow + \mid *$
$[1,B,0] \rightarrow)$
$[1,D,0] \rightarrow)[0,C,0][0,S,0]$
$[1,C,1] \rightarrow + \mid *$

together with

$Z \rightarrow [0,S,0]$
$[0,S,0] \rightarrow a \mid a[0,C,0][0,S,0] \mid ([1,S,1][1,B,0] \mid ([1,S,1][1,D,0]$
$[1,S,1] \rightarrow a$

that is, H generates all "expressions" of G which do not have nested parentheses.

The above theorem also provides us with a direct proof that $\boldsymbol{L}_{REG}$ is closed under intersection.

Corollary 9.1.4 *$\boldsymbol{L}_{REG}$ is closed under intersection.*

Proof: Every regular language, L, has a regular grammar $G = (N,\Sigma,P,S)$ with $L(G) = L - \{\lambda\}$. Recall that a regular grammar only has productions of the forms $A \rightarrow a$ and $A \rightarrow aB$. Now, the construction of Theorem 9.1.3 preserves the form of the productions, so for an arbitrary *REGL*, R, we obtain a regular grammar H with $L(H) = L(G) \cap R$, that is, $L(H)$ is a *REGL*. The empty word is dealt with as in the proof of Theorem 9.1.3. Thus, $\boldsymbol{L}_{REG}$ is closed under intersection. ■

Given two *REGLs*, L_1 and L_2, the determination of their intersection is usually based on Corollary 9.1.4. It is called the *cross-product construction.* Observe that because of Theorem 4.3.1 we may assume both L_1 and L_2 are given by *DFAs* (or both by regular grammars) and perform the construction directly. See the Exercises.

Theorem 9.1.5 *$\boldsymbol{L}_{DTM}$ is closed under union and intersection. $\boldsymbol{L}_{DTM}$ is not closed under complementation.*

Proof: Let L_1 and L_2 be two arbitrary *DTM* languages. Then there are *DTMs*, $M_1 = (Q_1,\Sigma_1,\Gamma_1,\delta_1,s_1,f_1)$ and $M_2 = (Q_2,\Sigma_2,\Gamma_2,\delta_2,s_2,f_2)$, with $L(M_1) = L_1$ and $L(M_2) = L_2$.

I. *$L_1 \cup L_2$ is in $\boldsymbol{L}_{DTM}$.*

Assuming $Q_1 \cap Q_2 = \varnothing$ and $\Gamma_1 \cap \Gamma_2 = \{\boldsymbol{B}\}$, as in the proof of Theorem 9.1.1, we construct an *NTM*, $M = (Q_1 \cup Q_2 \cup Q,\ \Sigma_1 \cup \Sigma_2,\ \Gamma_1 \cup \Gamma_2,\ \delta, s, f)$ which is the merger of M_1 and M_2, where s and f are in Q, a set of new states. Pictorially, we have Figure 9.1.2. Indeed the construction appears to be completely analogous to that of Theorem 9.1.1 until we observe that *DTMs* are not required to scan their input completely before accepting. Because of this when M starts in state s it first scans the input word to check its purity, that is, whether or not it is a word in $\Sigma_1^* \cup \Sigma_2^*$. This is the reason for the additional set of states Q. Finally, we add all transitions of the form $\delta(f_i,a) = (f,a,\lambda))$, for all a in Σ_i, for $i = 1,2$. The details are left to the Exercises.

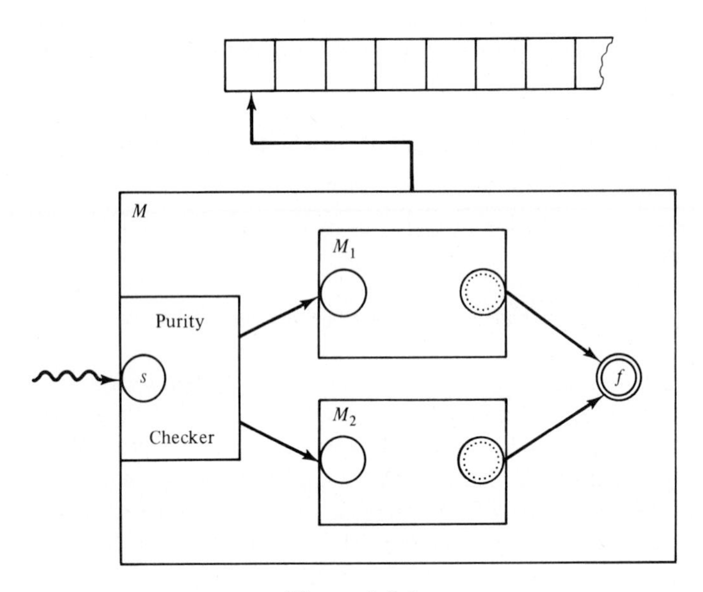

Figure 9.1.2

II. *$L_1 \cap L_2$ is in $\boldsymbol{L}_{DTM}$.*

We again assume $Q_1 \cap Q_2 = \varnothing$ and $\Gamma_1 \cap \Gamma_2 = \{\boldsymbol{B}\}$. We construct a *DTM*, M, which on input x first simulates M_1 and, if successful, simulates M_2 on x, and if it is also successful, then x is accepted. Thus, we identify the final state of M_1 with the start state of M_2 and the final state of M_2 becomes the final state of M. However, there is one remaining difficulty, namely, M_1 can be expected to rewrite its input tape. Hence, M needs to keep a copy of the input word. The simplest way to incorporate this feature is to assume M is an off-line *DTM*. Then M need only reset the input and work tapes after an accepting simulation of M_1. Pictorially, we have Figure 9.1.3. The details of this proof are also also left to the Exercises.

III. *$\boldsymbol{L}_{DTM}$ is not closed under complementation.*

This follows from Theorem 8.4.2, since this provides an example of a language in $\boldsymbol{L}_{DTM}$ whose complement is not in $\boldsymbol{L}_{DTM}$. ■

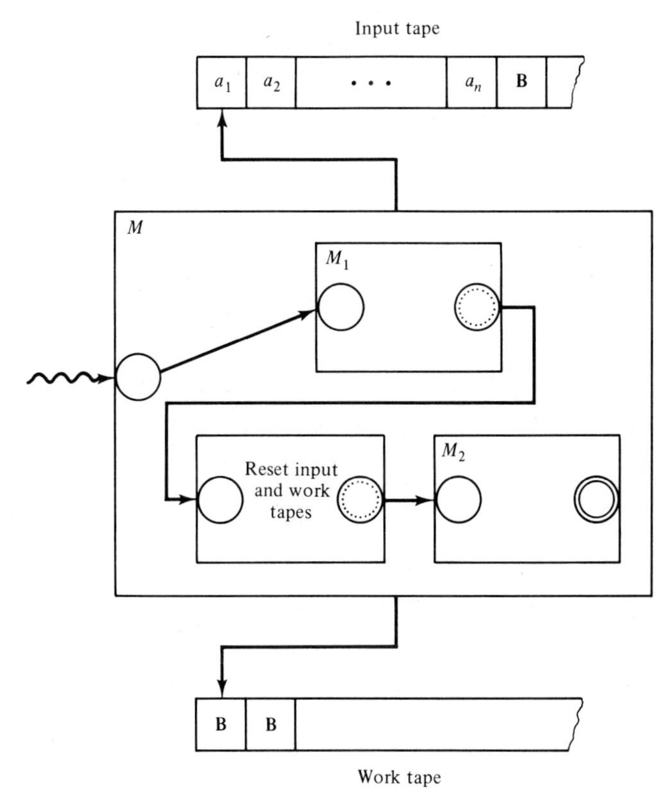

Figure 9.1.3

To close this section we turn our attention to reversal. Since *DFAs* and *PDAs* read their input from left to right it is somewhat surprising, from a machine viewpoint, that $\boldsymbol{L}_{REG}$ and $\boldsymbol{L}_{CF}$ are closed under reversal. However, it should come as no surprise that $\boldsymbol{L}_{DTM}$ is closed under reversal.

Theorem 9.1.6 *$\boldsymbol{L}_{REG}$, $\boldsymbol{L}_{CF}$, and $\boldsymbol{L}_{DTM}$ are closed under reversal.*

Proof:

I. $\boldsymbol{L}_{REG}$.

Let $L \subseteq \Sigma^*$ be a regular language. Then $L = L(M)$ for some *DFA*, $M = (Q,\Sigma,\delta,s,f)$. Consider the λ-*NFA*, $M^R = (Q^R,\Sigma,\delta^R,s^R,\{s\})$, where s^R is a new state, $Q^R = Q \cup \{s^R\}$, and δ^R consists of the following transitions.

(i) (q,a,p) is in δ^R if $\delta(p,a) = q$, that is, we reverse the direction of all original transitions.

(ii) (s^R,λ,f) is in δ^R, for all f in F, that is, the final states in M become, essentially, start states in M^R.

Whenever

$$sa_1 \cdots a_n \vdash q_1a_2 \cdots a_n \vdash \cdots \vdash q_{n-1}a_n \vdash q_n$$

in M we have

$$q_na_n \cdots a_1 \vdash q_{n-1}a_{n-1} \cdots a_1 \vdash \cdots \vdash q_1a_1 \vdash s$$

in M^R. The converse also holds if only states in Q are considered. Thus, $L(M^R) = L(M)^R = L^R$ and L^R is a *REGL*.

II. $\boldsymbol{L}_{CF}$.

Let $L \subseteq \Sigma^*$ be a context-free language. Then there exists a context-free grammar, $G = (N,\Sigma,P,S)$ with $L(G) = L$. Construct a *CFG*, $G^R = (N,\Sigma,P^R,S)$, where $P^R = \{A \to \alpha^R : A \to \alpha \text{ is in } P\}$. The proof that $L(G^R) = L^R$ is left to the Exercises.

III. $\boldsymbol{L}_{DTM}$.

This case we leave completely to the Exercises. ■

9.2 MAPPINGS

9.2.1 Morphism and Substitution

We begin with the following.

Definition Let Σ and Δ be alphabets. Then a mapping $h : \Sigma^* \to \Delta^*$ is said to be a (monoid) *morphism* if $h(\lambda) = \lambda$ and $h(a_1 \cdots a_n) = h(a_1)h(a_2) \cdots h(a_n)$, for all $a_1, \ldots, a_n$ in Σ.

This means that we need only specify $h(a)$ for each a in Σ to be able to specify h in totality. A morphism is also known as a *homomorphism*. Such mappings are well known in coding and cryptography.

Example 9.2.1 Consider the table

input →	$A\ B\ C\ D\ E\ F\ G\ H\ I\ J\ K\ L\ M$	$\cdots$	$W\ X\ Y\ Z$	
	$Z\ A\ B\ C\ D\ E\ F\ G\ H\ I\ J\ K\ L$	$\cdots$	$V\ W\ X\ Y$	→ output

which is an example of a Caesar cipher. Given an input word, for example,

$$IBM$$

we look up each letter on the input line and replace it by the letter below it on the output line. So in this case we obtain

$$HAL$$

Letting $\Gamma = \{A,B,C,\ldots,X,Y,Z\}$, then we can specify this cipher by a morphism $h_1 : \Gamma^* \rightarrow \Gamma^*$ in which $h_1(A) = Z$, $h_1(B) = A, \ldots, h_1(Z) = Y$. Then $h_1(IBM) = h_1(I)h_1(B)h_1(M) = HAL$ as desired.

Rotating the alphabet i places rather than 1 place we obtain h_i, $1 \leq i \leq 26$. That is, there are 26 distinct Caesar ciphers. Each Caesar cipher is a *renaming*, since the image of a letter is also a letter and is unique. Clearly, any permutation of the alphabet gives a renaming.

Example 9.2.2 Morse code is another example of a morphism, in this case $h : \Gamma^* \rightarrow \{\cdot, -\}^*$ is given by Table 9.2.1. Now, $h(IBM) = \cdot\cdot - \cdot\cdot\cdot - -$. Note that in this case the *images* of letters have different lengths.

Table 9.2.1

$h(A)$	$= \cdot -$	$h(N)$	$= - \cdot$
$h(B)$	$= - \cdot\cdot\cdot$	$h(O)$	$= - - -$
$h(C)$	$= - \cdot -$	$h(P)$	$= \cdot - - \cdot$
$h(D)$	$= - \cdot\cdot$	$h(Q)$	$= - - \cdot -$
$h(E)$	$= \cdot$	$h(R)$	$= \cdot - \cdot$
$h(F)$	$= \cdot\cdot - \cdot$	$h(S)$	$= \cdot\cdot\cdot$
$h(G)$	$= - - \cdot$	$h(T)$	$= -$
$h(H)$	$= \cdot\cdot\cdot\cdot$	$h(U)$	$= \cdot\cdot -$
$h(I)$	$= \cdot\cdot$	$h(V)$	$= \cdot\cdot\cdot -$
$h(J)$	$= \cdot - - -$	$h(W)$	$= \cdot - -$
$h(K)$	$= - \cdot -$	$h(X)$	$= - \cdot\cdot -$
$h(L)$	$= \cdot - \cdot\cdot$	$h(Y)$	$= - \cdot - -$
$h(M)$	$= - -$	$h(Z)$	$= - - \cdot\cdot$

The coding view of life can be displayed pictorially as in Figure 9.2.1. Since h and $h(message)$ are known, the receiver needs to discover the original, uncoded, message by considering the "inverse" of h.

Definition Let Σ, Δ be alphabets, and $h : \Sigma^* \rightarrow \Delta^*$ be a morphism. Then the *inverse morphism* $h^{-1} : \Delta^* \rightarrow 2^{\Sigma^*}$ is defined by

For all y in Δ^*, $h^{-1}(y) = \{x : x \text{ is in } \Sigma^* \text{ and } h(x) = y\}$

In our code setting the receiver needs to compute $h^{-1}(h(message))$. The receiver can do this by considering all possible messages M which satisfy $|h(M)| \leq |h(message)|$, and for each choice of M checking whether or not $h(M) = h(message)$. There are, however, two problems, at least, with this approach.

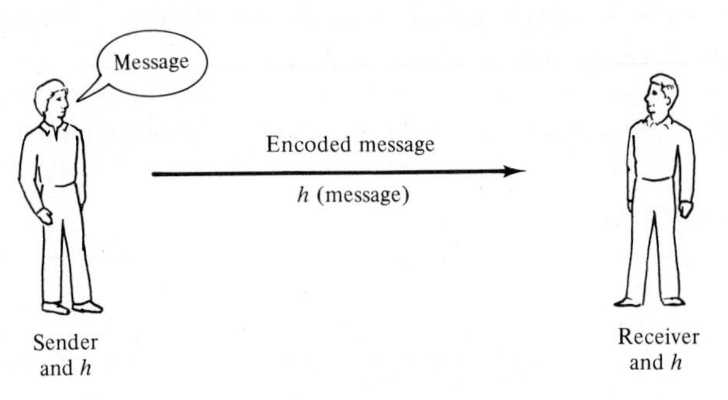

Figure 9.2.1

First, there may be two different messages M_1 and M_2 that satisfy $h(M_1) = h(M_2)$. Fortunately, in the first example we have given this is not the case. In fact, it can be proved that $h^{-1}(h(x)) = x$, for all x in Γ^*, in this case. We say that such a morphism is *uniquely decodable*, and it is a necessary condition for a morphism to be a code in the technical and intuitive sense. Morse code, as given, does not satisfy this requirement. For example, $h(A) = h(E)h(T)$. In practice we have a third symbol "pause" which terminates the Morse code for each letter. Once this is assumed Morse code can be proved to be uniquely decodable.

Example 9.2.3A Let $\Sigma = \{a,b,c\}$, $\Delta = \{0,1\}$, and $h : \Sigma^* \to \Delta^*$ be defined by

$$h(a) = 0$$

$$h(b) = 11$$

$$h(c) = 011$$

Now, h is not uniquely decodable, since

$$h(c) = 011$$

and

$$h(ab) = 011$$

that is, $h^{-1}(011) = \{c,ab\}$; see the Exercises.

Second, the method of decoding we have suggested is impractical, except for very short messages. This is simply because there are exponentially many messages M satisfying $|h(M)| \leq |h(message)|$, that is, exponentially many with respect to $|h(message)|$. Fortunately, decoding can be accomplished much more easily for Caesar ciphers, Morse code, and other codes; see the Exercises. This is because Caesar ciphers and Morse code are *bounded delay codes*. A code is a bounded delay code if there is a constant $b \geq 0$, such that during decoding at most b symbols beyond the current one need be examined to determine how the decoding should proceed. That unique decodability does not necessarily imply the bounded delay property is seen by the following

---

Example 9.2.4 Let $\Sigma = \{a,b,c\}$, $\Delta = \{0,1\}$ and $h : \Sigma^* \to \Delta^*$ be defined by

$$h(a) = 00$$
$$h(b) = 10$$
$$h(c) = 1$$

Then $h(ca^i) = 10^{2i}$ agrees with $h(ba^i) = 10^{2i+1} = 10^{2i}0$ up to and including the first $2i+1$ symbols, for all $i \geq 0$. Therefore, h is not a bounded delay code, although it is uniquely decodable. Note that 10^i is the encoding of ca^m if $i = 2m$, for some m, that is, i is even and it is the encoding of ba^m if $i = 2m+1$, for some m, that is, i is odd.

This brief excursion into codes and ciphers has demonstrated the need for not only considering morphisms but also their inverses. Although the morphic image of a given word is unique, the converse need not hold as Example 9.2.3A shows. Moreover, the inverse morphic image may, indeed, consist of no words at all.

Example 9.2.3B

$$h^{-1}(1) = h^{-1}(1^{2i+1}) = \varnothing, \qquad i \geq 0$$
$$h^{-1}(0^i 1) = \varnothing, \qquad i \geq 0$$

A natural generalization of the replacement of symbols by words as found in morphisms is the replacement of symbols by sets of words, that is, *substitution*. This notion is a useful tool in a number of proofs, as we shall see.

Definition Let $\boldsymbol{L}$ be a family of languages and Σ and Δ be two alphabets. Then a mapping $\sigma : \Sigma^* \to 2^{\Delta^*}$ is said to be a *substitution* if $\sigma(\lambda) = \{\lambda\}$ and $\sigma(a_1 \cdots a_n) = \sigma(a_1) \cdots \sigma(a_n)$, for all $a_1, \ldots, a_n$ in Σ.

Again this means that we need specify only $\sigma(a)$, for each a in Σ, to be able to specify σ in totality. If, additionally, for all a in Σ, $\sigma(a)$ is in $\boldsymbol{L}$, then we say that σ is an $\boldsymbol{L}$ *substitution*.

Example 9.2.5 Let $\Sigma = \{a,b,c\}$, $\Delta = \{0,1\}$, and $\sigma : \Sigma^* \to 2^{\Delta^*}$ be defined by

$$\sigma(a) = \{01,001\}$$

$$\sigma(b) = \{11,111\}$$
$$\sigma(c) = \{1\}$$

then

$$\sigma(abac) = \{0111011,01110011,01111011,011110011,\\ 00111011,001110011,001111011,0011110011\}$$

Since $\sigma(a)$, $\sigma(b)$, $\sigma(c)$ are in $\boldsymbol{L}_{FIN}$, that is, are finite languages, σ is a *finite substitution.*

Both morphism and substitution are "context-free" in their application, in the sense that $h(a)$ and $\sigma(a)$ are the same, for the given h and σ, respectively, whatever position the symbol a occupies in the given word. But in Chapter 7 we introduced a number of mappings, the so-called transductions, which, in general, do not have this property. One kind of transduction has become fundamental, namely, the *finite transduction.* However, when we introduced finite transductions we assumed that all words over the input alphabet were eligible input words. We now restrict the input words to belong to some given language.

Definition Let L be a language over some alphabet Σ_1, and let $M = (Q,\Sigma_2,\Delta,\delta,s,F)$ be a finite transducer. *The finite transduction of L with respect to M* is denoted by $M(L)$ and is defined by

$$M(L) = \bigcup_{x \text{ is in } L} M(x)$$

If $\Sigma_1 \cap \Sigma_2 = \varnothing$, then immediately $M(L) = \varnothing$.

Example 9.2.6 Let M be given by the state diagram of Figure 9.2.2. Then $M(011011) = \{abab,abc,cab,cc\}$, that is, $M(x) = h^{-1}(x)$ for h of Example 9.2.3. Hence, $M(\{0,1\}^*) = \{a,b,c\}^*$.

We now prove that $\boldsymbol{L}_{REG}$, $\boldsymbol{L}_{CF}$, and $\boldsymbol{L}_{DTM}$ are closed under regular, context-free, and *DTM* substitution, respectively. These results enable us to prove that these three families are closed under morphism, inverse morphism, and finite transduction, respectively.

Theorem 9.2.1 *$\boldsymbol{L}_{REG}$ is closed under regular substitution and, hence, under both morphism and finite substitution.*

Proof: The results for morphisms and finite substitutions are special cases of regular substitutions. Let $L \subseteq \Sigma^*$ be an arbitrary *REGL*, for some alphabet Σ^*. Now, given an arbitrary regular substitution $\sigma : \Sigma^* \rightarrow 2^{\Delta^*}$, for some alphabet Δ, we wish to prove that $\sigma(L)$ is in $\boldsymbol{L}_{REG}$.

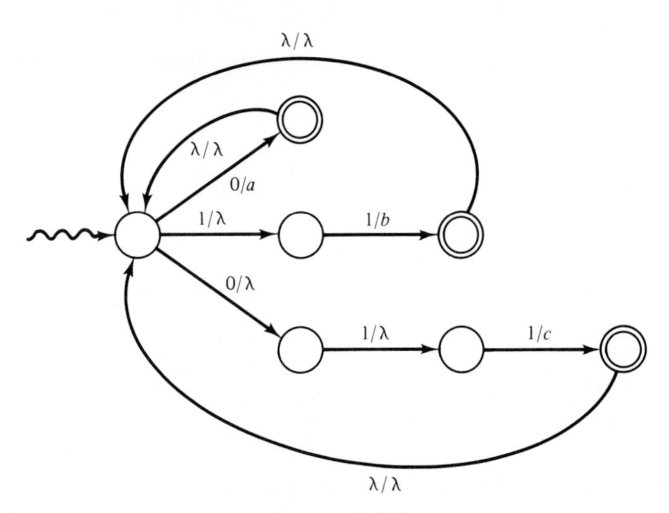

Figure 9.2.2

But there is a regular expression E, for L, and there are regular expressions E_a for $\sigma(a)$, for each a in Σ^*. Consider the regular expression E_σ obtained from E by replacing, simultaneously, every appearance of a symbol a by its corresponding regular expression E_a. We claim that $L(E_\sigma) = \sigma(L)$, and therefore $\sigma(L)$ is a *REGL* and $\boldsymbol{L}_{REG}$ is closed under regular substitution.

We do not prove the claim rigorously, since it appears to be, almost, self-evident. Observe that any word $a_1 \cdots a_n$ in L, where a_i is in Σ^*, $1 \leq i \leq n$, for some $n \geq 1$, is denoted by one of the clauses of E, which implies that in the corresponding clause in E_σ we obtain any word in $\sigma(a_1) \cdots \sigma(a_n)$. Conversely, any word $b_1 \cdots b_n$ in $L(E_\sigma)$ can be broken down into $\beta_1 \cdots \beta_m$ where β_i is in $\sigma(a_i)$, for some a_i in Σ^*, $1 \leq i \leq m$, and hence $b_1 \cdots b_n$ is $\sigma(L)$.

These are proof sketches that $\sigma(L) \subseteq L(E_\sigma)$ and $L(E_\sigma) \subseteq \sigma(L)$. The detailed proofs are left to the Exercises. ■

There are a number of other proofs of this theorem, which are investigated in the Exercises.

Example 9.2.7 Let $E = [a \cup b\,c]^*\, b\,c\,[a^*\, b\,c]^*$ over $\{a,b,c\}^*$ denote the given regular language L, and let

$$E_a = \varnothing$$

$$E_b = [001 \cup 10]^* 1$$

$$E_c = \lambda$$

define a regular substitution $\sigma : \{a,b,c\}^* \rightarrow 2^{\{0,1\}^*}$. Then $\sigma(L)$ is denoted by

$$E_\sigma = [\varnothing \cup E_b \lambda]^* \; E_b \lambda [\varnothing \; E_b \lambda]^*$$

where we have delayed the actual substitution of E_b, because of its length. Simplifying E_σ we see that

$$E_\sigma = E_b^* \; E_b$$

since $\varnothing \; E \equiv \varnothing$, $\varnothing^* \equiv \lambda$, and $\varnothing \cup E \equiv E$. Finally,

$$E_\sigma = [[001 \cup 10]^* \; 1]^* \; [[001 \cup 10]^* \, 1]$$

Analogously, we have the following.

Theorem 9.2.2 *$\boldsymbol{L}_{CF}$ is closed under context-free substitution and, hence, under morphism and finite and regular substitution.*

Proof: Again we only prove the first statement; the others follow as special cases.

Let L be an arbitrary context-free language with $L \subseteq \Sigma^*$, and let $\sigma : \Sigma^* \rightarrow 2^{\Delta^*}$ be an arbitrary context-free substitution. We show that $\sigma(L)$ is a *CFL* by construction of a *CFG* for $\sigma(L)$.

Let the *CFG*, $G = (N,\Sigma,P,S)$, satisfy $L = L(G)$ and, further, let $G_a = (N_a,\Delta,P_a,S_a)$ satisfy $L(G_a) = \sigma(a)$, for all a in Σ^*. We may assume that $N \cap N_a = \varnothing$, for all a in Σ^*, and $N_a \cap N_b = \varnothing$, for all a, b in Σ^*, $a \neq b$, since nonterminals can always be uniformly renamed without affecting the generated language.

First, we modify G to give $G' = (N,\Sigma',P',S)$ where $\Sigma' = \{S_a : a \text{ in } \Sigma\}$, the set of sentence symbols of the G_a, and P' is the same as P except that every appearance of a terminal a in Σ is replaced by S_a. This uniform and unique renaming of terminals implies

$$a_1 \cdots a_n \text{ is in } L(G) \text{ iff } S_{a_1} \cdots S_{a_n} \text{ is in } L(G')$$

Second, we combine these grammars to give $G_\sigma = (N_\sigma,\Delta,P_\sigma,S)$, where

$$N_\sigma = N \cup \bigcup_{a \text{ in } \Sigma} N_a, \text{ and } P_\sigma = P' \cup \bigcup_{a \text{ in } \Sigma} P_a$$

We claim that $L(G_\sigma) = \sigma(L)$. To see this consider an arbitrary word $a_1 \cdots a_n$ in Σ^*, for some $n \geq 1$.

Now, $a_1 \cdots a_n$ is in L iff

$S \Rightarrow^* a_1 \cdots a_n$ in G iff

$S \Rightarrow^* S_{a_1} \cdots S_{a_n}$ in G' iff

$S \Rightarrow^* S_{a_1} \cdots S_{a_n} \Rightarrow^* x_1 S_{a_2} \cdots S_{a_n}$

$\Rightarrow^* \cdots \Rightarrow^* x_1 \cdots x_n$ in G_σ, for all $x_1, \ldots, x_n$ in Δ^*,

where $S_{a_i} \Rightarrow^* x_i$ in G_{a_i}, $1 \leq i \leq n$, iff $\sigma(a_1 \cdots a_n) \subseteq L(G_\sigma)$. ■

Example 9.2.8 Let L be the CFL defined by the CFG, G,

$$S \rightarrow S + T \mid T$$
$$T \rightarrow a \mid (S)$$

and $\sigma : \{a, +, (,)\}^* \rightarrow 2^{\{a,b,',(,)\}^*}$ be defined by

$$\sigma(a) = \{a,b\}^+ \qquad \sigma(+) = \{'\}\{a,b\}^+\{'\}$$
$$\sigma(() = \{(\} \qquad \sigma()) = \{)\}$$

Then G' is

$$S \rightarrow SS_+T \mid T$$
$$T \rightarrow S_a \mid S_(SS_)$$

G_a is

$$S_a \rightarrow a \mid b \mid aS_a \mid bS_b$$

G_+ is

$$S_+ \rightarrow A_+$$
$$A_+ \rightarrow a \mid b \mid aA_+ \mid bA_+$$

$G_($ is

$$S_(\rightarrow ($$

and $G_)$ is

$$S_) \rightarrow)$$

These yield the final CFG, G_σ, the "union" of G', G_a, G_+, $G_($, and $G_)$ with sentence symbol S and $\sigma(L) = L(G_\sigma)$.

We sketch the proof of the third result.

Theorem 9.2.3 *$\mathbf{L}_{DTM}$ is closed under DTM substitution and, hence, under morphism and finite, regular, and context-free substitution.*

Proof: Closure under DTM substitution implies closure under the other four operations, so we deal only with DTM substitution.

For simplicity of exposition assume we have a DTM language $L \subseteq \{a,b\}^*$, a DTM substitution $\sigma : \{a,b\}^* \rightarrow 2^{\Delta^*}$, for some alphabet Δ, and DTM languages L_a and L_b, where $\sigma(a) = L_a$ and $\sigma(b) = L_b$.

Since L, L_a, and L_b are in $\boldsymbol{L}_{DTM}$, there are *DTMs* M, M_a, and M_b with $L = L(M)$, $L_a = L(M_a)$, and $L_a = L(M_b)$. We construct an *NTM*, M_σ, such that $L(M_\sigma) = \sigma(L)$. M_σ has three tapes. The first tape, which contains an input word $c_1 \cdots c_m$ in Δ^*, is read-only. The other two tapes are initially blank and all three heads are over the leftmost cells of their tape. M_σ begins by marking the first cells of tapes 2 and 3 with $\boldsymbol{l}$, a new symbol. Then M_σ copies some nondeterministically chosen prefix $c_1 \cdots c_i$ of the input word to tape 2. This may be the empty word. It then repositions the second head over the second cell and nondeterministically chooses to simulate either M_a or M_b on $c_1 \cdots c_i$. If the chosen *DTM* accepts $c_1 \cdots c_i$, then either a and b is written to tape 3, depending on whether M_a or M_b was chosen.

Now, M_σ erases tape 2 apart from the $\boldsymbol{l}$ and it copies a nondeterministically chosen prefix $c_{i+1} \cdots c_j$ of the remaining input $c_{i+1} \cdots c_m$, repeating the above process.

M_σ continues this nondeterministic copy/simulate/write/symbol process until the head on tape 1 is over the blank cell immediately after c_m. Because $\sigma(a)$ and/or $\sigma(b)$ may contain the empty word, M_σ nondeterministically chooses to either continue the above process with the empty word on tape 2 or to move into the final stage.

When M_σ enters the final stage it resets the head on tape 3 to be over the second cell and simulates M with tape 3 as its input tape. If M accepts, then M_σ also accepts.

The idea behind this construction is that if $c_1 \cdots c_m$ in $\sigma(L)$, then there must be some $d_1 \cdots d_n$ in L such that $c_1 \cdots c_m$ is in $\sigma(d_1 \cdots d_n)$, for some d_i in $\{a,b\}$, $1 \leq i \leq n$, for some $n \geq 0$. This implies there is some partition of $c_1 \cdots c_m$ into

$$c_1 \cdots c_{i_1}, c_{i_1+1} \ . \ . \ . \ c_{i_2+1}, \ . \ . \ . \ , c_{i_{n-1}+1} \ . \ . \ . \ c_{i_n}$$

such that $c_1 \cdots c_{i_1}$ is in $\sigma(d_1)$, and $c_{i_j+1} \cdots c_{i_{j+1}}$ is in $\sigma(d_{j+1})$, $1 \leq j < n$. M_σ attempts to find such a partition. Since n may be much larger than m many elements of the partition can be the empty word.

A formal proof that $L(M_\sigma) = \sigma(L)$ is left to the Exercises. ■

An interesting application of the above three theorems is the following

Corollary 9.2.4 *$\boldsymbol{L}_{REG}$, $\boldsymbol{L}_{CF}$, and $\boldsymbol{L}_{DTM}$ are closed under union.*

Proof: We give the proof for $\boldsymbol{L}_{REG}$. The proofs for $\boldsymbol{L}_{CF}$ and $\boldsymbol{L}_{DTM}$ are similar. Consider two arbitrary languages L_1 and L_2 in $\boldsymbol{L}_{REG}$. Now, the language $L = \{a,b\}$ is in $\boldsymbol{L}_{REG}$. Define a substitution σ on L by $\sigma(a) = L_1$ and $\sigma(b) = L_2$. Then $\sigma(L) = L_1 \cup L_2$ and since σ is a regular substitution, by Theorem 9.2.1 $\sigma(L)$ is a *REGL*; hence, $L_1 \cup L_2$ is in $\boldsymbol{L}_{REG}$ as desired. ■

9.2.2* Inverse Morphism and Finite Transduction

We are now in a position to prove closure under inverse morphism. Inverse morphism is perhaps the most difficult of the operations to visualize; therefore, our proof proceeds by characterizing inverse morphisms in terms of some of the operations already introduced. Unlike our previous proofs this characterization is, to a large extent, independent of the particular language family involved.

Construction 9.2.1 Let $L \subseteq \Sigma^*$ be an arbitrary language and let $h : \Delta^* \rightarrow \Sigma^*$ be a morphism. The construction requires five steps.

(1) Let $\bar{\Delta} = \{\bar{a} : a \text{ in } \Delta\}$ be a set of completely new symbols and $\Delta_\lambda = \{a : a \text{ is in } \Delta \text{ and } h(a) = \lambda\}$ be the set of those symbols whose image under h is the empty word.
(2) Let y_a denote $h(a)$ and $F = \{\bar{a}y_a : a \text{ in } \Delta\}$.
(3) Let $\sigma : \Sigma^* \rightarrow 2^{(\Sigma \cup \bar{\Delta})^*}$ be a regular substitution defined by $\sigma(a) = \bar{\Delta}^* a \bar{\Delta}^*$, for all a in Σ^*.
(4) Let $g : (\Sigma \cup \bar{\Delta})^* \rightarrow \Delta^*$ be a morphism defined by $g(a) = \lambda$, for all a in Σ, and $g(\bar{a}) = a$, for all $\bar{a}$ in Δ.
(5) Consider $g(\sigma(L - \{\lambda\}) \cap F^+) \cup \Delta_\lambda^*$ which we claim equals $h^{-1}(L)$. ■

To illustrate this construction and gain some insight into why it works consider

---

Example 9.2.9 Let $\Delta = \{a,b,c,d\}$, $\Sigma = \{0,1\}$ and h be defined by

$$h(a) = \lambda \qquad h(b) = 0$$
$$h(c) = 10 \qquad h(d) = 1$$

Consider the word $x = 1010$, since $h(c) = 10$

$$cc \text{ is in } h^{-1}(x)$$

as are

$$dbc,\ cdb,\ dbdb,\ adbc,\ dabc,\ acaca$$

Indeed, $h^{-1}(1010)$ is an infinite set. From Construction 9.2.1 $\bar{\Delta} = \{\bar{a},\bar{b},\bar{c},\bar{d}\}$, $\Delta_\lambda = \{a\}$, and $F = \{\bar{a},\bar{b}0,\bar{c}10,\bar{d}1\}$. Immediately, we obtain $\sigma(0) = \bar{\Delta}^*0\bar{\Delta}^*$ and $\sigma(1) = \bar{\Delta}^*1\bar{\Delta}^*$; and $g(0) = g(1) = \lambda$, $g(\bar{a}) = a$, $g(\bar{b}) = b$, $g(\bar{c}) = c$, and $g(\bar{d}) = d$.

Now, $\sigma(x) = \bar{\Delta}^*1\bar{\Delta}^*0\bar{\Delta}^*1\bar{\Delta}^*0\bar{\Delta}^*$ and $\sigma(x) \cap F^+$ contains only words of the forms

(1) $\bar{a}^*\bar{d}1\bar{a}^*\bar{b}0\bar{a}^*\bar{d}1\bar{a}^*\bar{b}0\bar{a}^*$.
(2) $\bar{a}^*\bar{c}10\bar{a}^*\bar{d}1\bar{a}^*\bar{b}0\bar{a}^*$.
(3) $\bar{a}^*\bar{d}1\bar{a}^*\bar{b}0\bar{a}^*\bar{c}10\bar{a}^*$.
(4) $\bar{a}^*\bar{c}10\bar{a}^*\bar{c}10a^*$.

which under g, that is, $g(\sigma(x) \cap F^+)$, yield

(1) $a^*da^*ba^*da^*ba^*$.
(2) $a^*ca^*da^*ba^*$.
(3) $a^*da^*ba^*ca^*$.
(4) $a^*ca^*ca^*$.

The morphism g erases Σ-symbols. Therefore, define a morphism g' to erase $\overline{\Delta}$-symbols, that is,

$$g'(\bar{a}) = g'(\bar{b}) = g'(\bar{c}) = g'(\bar{d}) = \lambda$$

and

$$g'(0) = 0 \text{ and } g'(1) = 1$$

Then $g'(\sigma(x) \cap F^+)$ gives 1010 in all cases — the original word x.

Thus, every word in $\sigma(x) \cap F^+$ contains x and also contains the barred version of a word y in Δ^* satisfying $h(y) = x$. The trick in the construction is to pad out x with barred letters in all possible positions with σ, and then use F^+ to restrict this padding to exactly those words that have x as their image.

Theorem 9.2.5 $\boldsymbol{L}_{REG}$, $\boldsymbol{L}_{CF}$, *and* $\boldsymbol{L}_{DTM}$ *are closed under inverse morphism.*

Proof: We prove the result only for $\boldsymbol{L}_{CF}$. The other two proofs are almost identical. Given a language $L \subseteq \Sigma^*$ in $\boldsymbol{L}_{CF}$ and a morphism $h : \Delta^* \to \Sigma^*$ we apply Construction 9.2.1 to give

$$K = g(\sigma(L - \{\lambda\}) \cap F^+) \cup \Delta_\lambda^*.$$

First observe that K is indeed in $\boldsymbol{L}_{CF}$ if L is in $\boldsymbol{L}_{CF}$. This follows because $\boldsymbol{L}_{CF}$ is closed under morphism, regular intersection, regular substitution, and union, because F^+ and Δ_λ^* are both *REGLs*, and because $L - \{\lambda\} = L \cap \Sigma^+$ is a *CFL* if L is a *CFL*.

Thus, it only remains to prove that $h^{-1}(L) = K$. We proceed in two stages. First, assume λ is not in L, in which case $L - \{\lambda\} = L$. It is convenient to let $D = \sigma(L) \cap F^+$, that is, $K = g(D)$. We wish to prove that

$$g(D) = h^{-1}(L) \tag{9.2.1}$$

when λ is not in L.

Observe that both sides of (9.2.1) do not contain the empty word. The left-hand side doesn't contain the empty word, because $D \subseteq F^+$; each word in F^+ contains at least one barred symbol; and g changes barred symbols to unbarred symbols. The right-hand side doesn't contain the empty word, because L doesn't and λ is in $h^{-1}(x)$ if and only if $h(\lambda) = x$, that is, $x = \lambda$.

Now, by the definition of inverse morphism, for all $m \geq 1$ and for all a_i in Δ, $1 \leq i \leq m$,

$$a_1 \cdots a_m \text{ is in } h^{-1}(L) \text{ iff } h(a_1 \cdots a_m) = y_1 \cdots y_m \text{ is in } L$$

for some y_i in Σ^*, $1 \leq i \leq m$, where $h(a_i) = y_i$, $1 \leq i \leq m$. But, by the definition of D,

$$y_1 \cdots y_m \text{ is in } L \text{ iff } \bar{a}_1 y_1 \cdots \bar{a}_m y_m \text{ is in } D$$

where $\bar{a}_i y_i$ is in F, $1 \leq i \leq m$, and

$$\bar{a}_1 y_1 \cdots \bar{a}_m y_m \text{ is in } D \text{ iff } g(\bar{a}_1 y_1 \cdots \bar{a}_m y_m) = a_1 \cdots a_m \text{ is in } g(D)$$

Therefore, we have established that $h^{-1}(L) = g(D)$ when λ is not in L.

Second, assume λ is in L and let $L' = L - \{\lambda\}$, that is, $L = L' \cup \{\lambda\}$. Then $h^{-1}(L) = h^{-1}(L' \cup \{\lambda\}) = h^{-1}(L') \cup h^{-1}(\{\lambda\})$, by definition of inverse morphism. If there is at least one symbol in Δ whose image under h is the empty word, then $h^{-1}(\{\lambda\}) = \Delta_\lambda^*$. Otherwise $h^{-1}(\{\lambda\}) = \{\lambda\} = \varnothing^*$. In both cases $h^{-1}(\{\lambda\}) = \Delta_\lambda^*$.

Since λ is not in L', $h^{-1}(L') = g(\sigma(L') \cap F^+)$ by the first part. Taken together this implies that

$$\begin{aligned} h^{-1}(L) &= h^{-1}(L') \cup h^{-1}(\{\lambda\}) \\ &= g(\sigma(L - \{\lambda\}) \cap F^+) \cup \Delta_\lambda^* \end{aligned}$$

as desired. ■

To close this section we consider finite transductions. As for inverse morphisms we first give a characterization of finite transductions in terms of simpler operations.

Construction 9.2.2 Let $M = (Q,\Sigma,\Delta,\delta,s,F)$ be a finite transducer.

1. Let $\Gamma = \{[p,a,q] : p, q \text{ in } Q,\ a \text{ in } \Sigma \cup \{\lambda\}\}$ be a completely new alphabet.
2. Let $g : \Gamma^* \rightarrow \Sigma^*$ be defined by $g([p,a,q]) = a$, for all $[p,a,q]$ in Γ.
3. Let $R = \{[p_0,a_1,p_1][p_1,a_2,p_2] \cdots [p_{r-1},a_r,p_r] : p_0, \ldots, p_r$ are in Q, $p_0 = s$, p_r is in F, and (p_{i-1},a_i,p_i,y_i) is in δ, for some y_i in Δ^*, for some a_i in $\Sigma \cup \{\lambda\}$, $1 \leq i \leq r\}$.
4. Let $h : \Gamma^* \rightarrow \Delta^*$ be defined by $h([p,a,q]) = y$, where (q,a,q,y) is in δ.
5. Then $M(x) = h(g^{-1}(x) \cap R)$, for all x in Σ^*. ■

To gain some insight into this construction consider

Example 9.2.10 Let M be the finite transducer of Figure 9.2.3, which accepts words over $\Sigma = \{a,b\}$ containing an odd number of letters. It replaces the first a, third a, ..., $(2n+1)$st a, by 01, the second a, fourth a, ..., $2n$th a, by 0111 and, similarly for the b's with 011 and 01111.

Then $\Gamma = \{[i,c,j] : 2 \leq i,j \leq 5$, and $c = a$ or $b\}$, since there are no λ-transitions. Now, R consists of all accepting configuration sequences, for example,

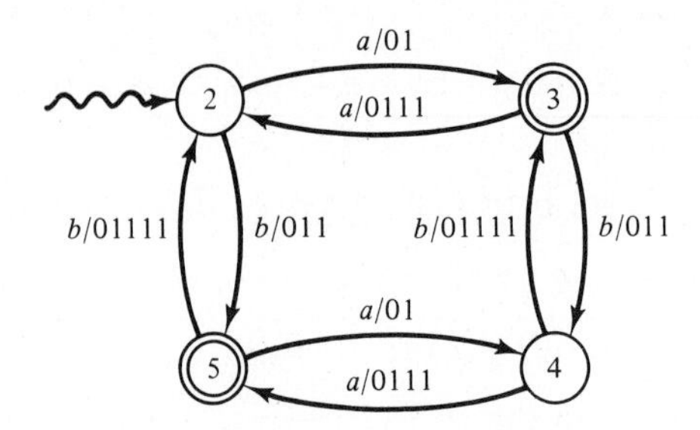

Figure 9.2.3

$$[2,a,3][3,b,5][5,a,4]$$

is in R since

$$2aba \vdash 013ba \vdash 010115a \vdash 0101101114$$

in M.

Now, $g([i,c,j]) = c$, for all $[i,c,j]$ in Γ, hence,

$$g^{-1}(aba) = \{[i,a,j][k,b,l][m,a,n] : 2 \leq i,j,k,l,m,n \leq 5\}$$

and, therefore,

$$g^{-1}(aba) \cap R = \{[2,a,3][3,b,5][5,a,4]\}$$

since this is the only accepting configuration sequence for aba. Finally, since $h([2,a,3]) = 01$, $h([3,b,5]) = 011$ and $h([5,a,4]) = 0111$, we find that $h(g^{-1}(aba) \cap R) = 010110111$, which is indeed $M(aba)$.

We now prove the following.

Theorem 9.2.6 *$\boldsymbol{L}_{REG}$, $\boldsymbol{L}_{CF}$, and $\boldsymbol{L}_{DTM}$ are closed under finite transduction.*

Proof: We again only prove the result for $\boldsymbol{L}_{CF}$. Let $L \subseteq \Sigma^*$ be an arbitrary language in $\boldsymbol{L}_{CF}$ and $M = (Q,\Sigma,\Delta,\delta,s,F)$ be an arbitrary finite transducer (without loss of generality we may always assume M's input alphabet and L's alphabet are the same; see the Exercises.

Let g, h and R be as defined in Construction 9.2.2. We claim that $M(L) = g(h^{-1}(L) \cap R)$. We only prove $M(L) \subseteq h(g^{-1}(L) \cap R)$, since the converse follows by a similar argument; see the Exercises. Indeed, we need only prove that $M(x) \subseteq h(g^{-1}(x) \cap R)$, for all x in Σ^*.

Consider some word y in $M(x)$. If no such word exists then $M(x) = \varnothing$ and the result follows immediately. Otherwise y is determined by some accepting configuration sequence for x in M

$$p_0x_0 \vdash y_1p_1x_1 \vdash y_1y_2p_2x_2 \vdash \cdots \vdash y_1 \cdots y_rp_r$$

where $p_0 = s$, $x_0 = x$, $y = y_1 \cdots y_r$, $x_i = a_{i+1}x_{i+1}$, a_{i+1} is in $\Sigma \cup \{\lambda\}$, and $(p_i, a_{i+1}, p_{i+1}, y_{i+1})$ is in $\delta, 0 \leq i < r$. But by construction

$$w = [p_0, a_1, p_1][p_1, a_2, p_2] \cdots [p_{r-1}, a_r, p_r]$$

is in R, is, clearly, in $g^{-1}(x)$, and, hence, in $g^{-1}(x) \cap R$. But $h(w)$, again by construction, is

$$y_1 \cdots y_r = y$$

in other words, y is in $h(g^{-1}(x) \cap R)$, which implies that we have shown $M(x) \subseteq h(g^{-1}(x) \cap R)$ as required.

We now argue $h(g^{-1}(L) \cap R)$ is a *CFL* if L is a *CFL*. First, by Theorem 9.2.4 $g^{-1}(L)$ is a *CFL* if L is a *CFL*. Further, if R is a *REGL*, then $g^{-1}(L) \cap R$ is a *CFL*, by Theorem 9.1.3, and hence $h(g^{-1}(L) \cap R)$ is a *CFL*, by Theorem 9.2.2. But is R a *REGL*? We prove this is the case in the following way.

R was obtained from the finite transducer $M = (Q, \Sigma, \Delta, \delta, s, F)$, which we modify to produce a λ-*NFA* $M' = (Q, \Sigma, \delta', s, F)$. For all p and q in Q and a in $\Sigma \cup \{\lambda\}$, the transition (p, a, q) is in σ' exactly when there is a transition (p, a, q, y) in δ, for some y in Δ^*. Now, $Domain(M) = L(M')$, since M' is the same as M except that the output is ignored. Second, we modify M' to give $M'' = (Q, \Gamma, \delta', s, F)$. For all $[p, a, q]$ in Γ, we have a transition $(p, [p, a, q], q)$ in δ' exactly when there is a transition (p, a, q) in δ'.

Immediately,

$$s[s, a_1, p_1][p_1, a_2, p_2] \cdots [p_{r-1}, a_r, p_r] \vdash p_1[p_1, a_2, p_2] \cdots [p_{r-1}, a_r, p_r]$$
$$\vdash \cdots \vdash p_{r-1}[p_{r-1}, a_r, p_r] \vdash p_r$$

in M'', exactly when

$$sa_1 \cdots a_r \vdash p_1a_2 \cdots a_r \vdash \cdots \vdash p_{r-1}a_r \vdash p_r$$

in M'. Thus, $L(M'') = R$. We leave to the Exercises the inductive proof of this equality, which implies R is a *REGL*.

Thus, $h(g^{-1}(L) \cap R)$ is a *CFL* if L is a *CFL*. Moreover, $M(L) = h(g^{-1}(L) \cap R)$ we have proved that $\boldsymbol{L}_{CF}$ is closed under finite transduction. ■

In the above proof we have really shown that a language family $\boldsymbol{L}$ is closed under finite transduction if it is closed under morphism, inverse morphism, and regular intersection, that is, it is a *cone* or *full trio*; see Section 9.5. The converse also holds (see the Exercises), yielding the following.

Theorem 9.2.7 *Let* $\boldsymbol{L}$ *be a family of languages. Then* $\boldsymbol{L}$ *is closed under finite transduction if and only if it is closed under morphism, inverse morphism, and regular intersection.*

9.3 CATENATION, QUOTIENT, AND STAR

We consider three operations based on the catenation of words. The operations of the catenation of two languages and the star of a language have been introduced already, so we define only the third.

Definition Let $L_1 \subseteq \Sigma_1^*$ and $L_2 \subseteq \Sigma_2^*$ be two languages.

The *left quotient* of L_1 with respect to L_2, denoted by $L_2\backslash L_1$, is defined by

$$L_2\backslash L_1 = \{y : x_2y \text{ is in } L_1, \text{ for some } x_2 \text{ in } L_2 \text{ and } y \text{ in } \Sigma_1^*\}$$

and the *right quotient* of L_1 with respect to L_2, denoted by L_1/L_2, is defined as

$$L_1/L_2 = \{y : yx_2 \text{ is in } L_1, \text{ for some } x_2 \text{ in } L_2 \text{ and } y \text{ in } \Sigma_1^*\}$$

If L_2 is a *REGL*, then we have *regular left quotient* and *regular right quotient*.

Theorem 9.3.1 *$\mathbf{L}_{REG}$, $\mathbf{L}_{CF}$, and $\mathbf{L}_{DTM}$ are closed under catenation and star.*

Proof: Let L_1 and L_2 be two arbitrary languages, $K_1 = \{ab\}$, and $K_2 = \{a\}^*$. Then K_1 and K_2 are in $\mathbf{L}_{REG}$, $\mathbf{L}_{CF}$, and $\mathbf{L}_{DTM}$. Define substitutions σ_1 and σ_2 by $\sigma_1(a) = L_1$, $\sigma_1(b) = L_2$, and $\sigma_2(a) = L_1$. Immediately, $\sigma_1(K_1) = L_1L_2$ and $\sigma_2(K_2) = L_1^*$. But σ_1 and σ_2 are regular substitutions if L_1 and L_2 are *REGLs*, in which case L_1L_2 and L_1^* are also *REGLs*. Similarly, σ_1 and σ_2 are context-free substitutions if L_1 and L_2 are *CFLs* and, hence, L_1L_2 and L_1^* are *CFLs* in this case. Finally, σ_1 and σ_2 are *DTM* substitutions if L_1 and L_2 are *DTM* languages, so L_1L_2 and L_1^* are both *DTM* languages. In all cases we have the desired result. ■

In Section 3.2 direct constructions of λ-*NFAs*, from *DFAs*, are given which accept the catenation product of two regular languages and the star of a regular language.

We also have the following.

Theorem 9.3.2 *$\mathbf{L}_{REG}$, $\mathbf{L}_{CF}$, and $\mathbf{L}_{DTM}$ are closed under regular left and right quotients.*

Proof: We only treat the case of $\mathbf{L}_{CF}$ directly. The results for $\mathbf{L}_{REG}$ and $\mathbf{L}_{DTM}$ follow similarly.

Let $L \subseteq \Sigma^*$ be an arbitrary *CFL* and $R \subseteq \Sigma^*$ be an arbitrary *REGL*. We first consider regular right quotient, that is, L/R. There is a morphism g, a finite substitution σ, and an alphabet $\overline{\Sigma} = \{\overline{a} : a \text{ is in } \Sigma\}$ such that

$$L/R = g(\sigma(L) \cap \overline{\Sigma}^*R).$$

Let $g : (\Sigma \cup \overline{\Sigma})^* \rightarrow \Sigma^*$ be defined by $g(a) = \lambda$, for all a in Σ, and $g(\overline{a}) = a$,

for all $\overline{a}$ in $\overline{\Sigma}$, and let $\sigma : \Sigma^* \rightarrow 2^{(\Sigma \cup \overline{\Sigma})^*}$ be defined by $\sigma(a) = \{a, \overline{a}\}$, for all a in Σ^*. Then, immediately, for all $x = a_1 \cdots a_n$, for some a_i in Σ and some $n \geq 1$, we have

$$\sigma(x) = \{a_1, \overline{a}_1\} \cdots \{a_n, \overline{a}_n\}$$

Therefore, $a_j \cdots a_n$ is in R, for some j, $1 \leq j \leq n$, if and only if

$$\overline{a}_1 \cdots \overline{a}_{j-1} a_j \cdots a_n \text{ is in } \sigma(x) \cap \overline{\Sigma}^*R.sp-0.5v$$

if and only if

$$a_1 \cdots a_{j-1} \text{ is in } g(\sigma(L) \cap \overline{\Sigma}^*R)$$

if and only if

$$a_1 \cdots a_{j-1} \text{ is in } L/R$$

The proof for left quotient is symmetric to the proof for right quotient, based on

$$R\backslash L = g(\sigma(L) \cap R\overline{\Sigma}^*)$$

■

9.4 COMPOSITION

Since transductions are relations it is natural to consider their relational composition, or series composition, where the output of the first transduction becomes the input for the second. In UNIX this construction is called a *pipe*.

Definition Let $T_1 \subseteq \Sigma_1^* \times \Sigma_2^*$ and $T_2 \subseteq \Sigma_2^* \times \Sigma_3^*$; then the *composition* of T_1 and T_2 is denoted by $T_1 \circ T_2$, (see Section 0.3), and is defined by

$$T_1 \circ T_2 = \{(x_1, x_3) : (x_1, x_2) \text{ is in } T_1 \text{ and } (x_2, x_3) \text{ is in } T_2 \text{ for some } x_2 \text{ in } \Sigma_2^*\}$$

We consider the composition of transductions for finite, pushdown, and Turing transducers. We begin with finite transducers.

Theorem 9.4.1 *Let $T_1 \subseteq \Sigma_1^* \times \Sigma_2^*$ and $T_2 \subseteq \Sigma_2^* \times \Sigma_3^*$ be two finite transductions. Then $T = T_1 \circ T_2$ is also a finite transduction and, therefore, finite transductions are closed under composition.*

Proof: The proof is by construction. Indeed, the proof demonstrates how a pipe could be implemented. We are given $M_1 = (Q_1,\Sigma_1,\Sigma_2,\delta_1,s_1,F_1)$ and $M_2 = (Q_2,\Sigma_2,\Sigma_3,\delta_2,s_2,F_2)$, two finite transducers, with $T_1 = T(M_1)$ and $T_2 = T(M_2)$.

First, we may assume that the output of each transition in a finite transducer is either a symbol or the empty word. Although this normal form is not strictly necessary it simplifies the construction somewhat.

Second, assume, there is a buffer B, initially empty, which never contains more than one symbol. We construct a finite transducer M, so that whenever B is empty it simulates M_1. M_1 reads input and changes state as usual, except that any output is placed in B. As soon as B contains an output symbol, M_1 is interrupted and M_2 takes over using B as the current cell of its input. As soon as B has been read, M_2 is interrupted and M_1 is reactivated. When an interrupt of M_1 occurs the current state and the position of the read head is preserved, while for M_2 the current state and the position of write head is preserved. Pictorially we have Figure 9.4.1. (In UNIX we would write the pipe $M = M_1 \parallel M_2$.) More formally M has a state set

$$Q = \{[q_1,q_2,B] : q_i \text{ is in } Q_i\ , i = 1,2, \text{ and } B \text{ is in } \Sigma_2 \cup \{\lambda\}\}$$

a start state $[s_1,s_2,\lambda]$, and final states $[f_1,f_2,\lambda]$, for all f_i in F_i, $i = 1,2$, where the triples $[p,q,B]$ are new symbols. Define δ, the transition function of M, by

(1) For all $[q_1,q_2,\lambda]$ in Q, and all a in $\Sigma_1 \cup \{\lambda\}$,
$([q_1,q_2,\lambda],a,[r_1,q_2,B],\lambda)$ is in δ, if (q_1,a,r_1,B) is in δ_1.

(2) For all $[q_1,q_2,a]$ in Q, where a is in Σ_2,
$([q_1,q_2,a],\lambda,[q_1,r_2,a],b)$ is in δ, if (q_2,λ,r_2,b) is in δ_2, and
$([q_1,q_2,a],\lambda,[q_1,r_2,\lambda],b)$ is in δ, if (q_2,a,r_2,b) is in δ_2.

We leave to the Exercises the proof of the equality of $T(M)$ and $T_1 \circ T_2$. It is sufficient to prove that

$$[q_1,q_2,B]x \vdash^+ z[f_1,f_2,\lambda] \text{ in } M$$

for f_1 in F_1 and f_2 in F_2, exactly when

$$q_1x \vdash^+ yf_1 \text{ in } M_1$$

and

$$q_2y \vdash^+ zf_2 \text{ in } M_2$$

for some y in Σ_2^*. ■

For pushdown transductions we obtain a negative result.

Theorem 9.4.2 *Given two pushdown transductions $T_1 \subseteq \Sigma_1^* \times \Sigma_2^*$ and $T_2 \subseteq \Sigma_2^* \times \Sigma_3^*$. Then $T = T_1 \circ T_2$ is not necessarily a pushdown transduction.*

Proof: Let M_1 and M_2 be two pushdown transducers, whose input and output alphabets equal $\{a,b,c\}$. Let

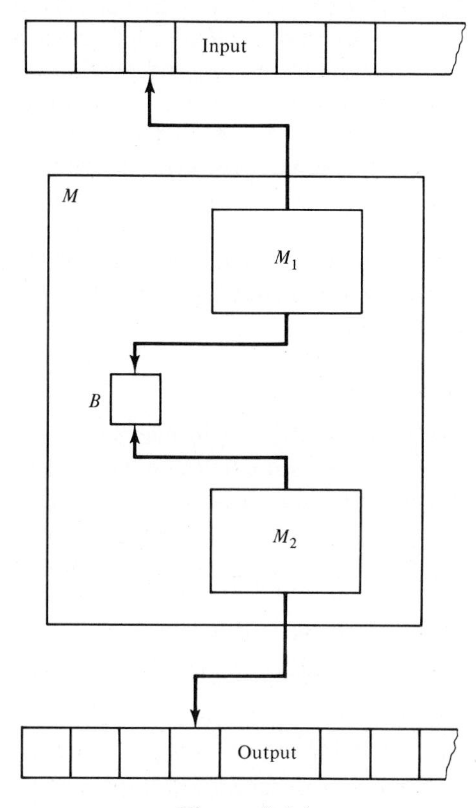

Figure 9.4.1

$$T(M_1) = \{(x,x) : x = a^i b^i c^j, \text{ for some } i,j \geq 1\}$$

and

$$T(M_2) = \{(x,x) : x = a^i b^j c^j, \text{ for some } i,j \geq 1\}.$$

Immediately, $T = T(M_1) \circ T(M_2) = \{(x,x) : x = a^i b^i c^i, \text{ for some } i \geq 1\}$.

We demonstrated in Section 8.2 that $Domain(T) = \{a^i b^i c^i : i \geq 1\}$ is not a *CFL*, and we claimed in Section 7.3 that, for each *PDT*, T, $Domain(T)$ is a *CFL*. Therefore, we conclude that T cannot be a pushdown transduction. ■

Finally, we sketch a proof that Turing transductions are closed under composition. On the one hand, although we introduced *DTTs* in Section 7.4 we only used them to compute functions. However, it is straightforward to define $T(M)$, for a *DTT*, $M = (Q,\Sigma,\Delta,\Gamma,\delta,s,f)$, as those pairs of words (x,y) in $\Sigma^* \times \Delta^*$ such that $sx \vdash^* y_1 f y_2$, for some y_1 and y_2 with $y = y_1 y_2$. On the other hand, the proof of the following theorem can be modified to prove that Turing-computable functions are closed under functional composition; see the Exercises.

Theorem 9.4.3 *Let $T_1 \subseteq \Sigma_1^* \times \Sigma_2^*$ and $T_2 = \Sigma_2^* \times \Sigma_3^*$ be two deterministic Turing transductions. Then $T = T_1 \circ T_2$ is a deterministic Turing transduction.*

Proof: Let M_1 and M_2 be two *DTTs* with $T_1 = T(M_1)$ and $T_2 = T(M_2)$. Construct a *DTT*, M, with $T(M) = T$. M first shifts its input word right by one cell, leaving a new symbol $\boldsymbol{l}$ in the leftmost cell. Second, it simulates M_1 on the input word, except that it begins at the second cell. Third, if M_1 halts in its final state, then M resets the read-write head to the second cell and simulates M_2. Fourth, and, finally, if M_2 halts in its final state, then M shifts the resulting output left one cell and halts in M's final state. ■

9.5 CYLINDERS, CONES, AND AFLs

When examining the research literature in formal language theory from the mid-sixties to the early seventies one is struck by the number of different language families introduced. There are the programmed languages, the indexed languages, the macro languages, the nonterminal-bounded languages, the *ET0L* languages, and many others. Whenever such a new family is introduced, or rather a new means of specifying languages is introduced, one first step is to determine

How powerful is the new method?
What is the relationship of its language family with previously introduced families?
What closure properties does it have?

As far as this latter question is concerned, time and time again, it has been found that the new family is closed under many of the operations discussed in Sections 9.1-9.3. This observation confirmed suspicions that these operations are basic ones for language families. This in turn has led to the classification of language families with respect to these operations. The first steps in the theory selected six of these operations as basic, giving *full abstract families of languages*, but, in time, it became clear that finite transductions were more central.

In this section we introduce the classification of language families and, in particular, of $\boldsymbol{L}_{REG}$ and $\boldsymbol{L}_{CF}$. Other families are discussed in the Exercises.

Definition Let $\boldsymbol{L}$ be a family of languages. Then

(i) $\boldsymbol{L}$ is a *cylinder* if it closed under inverse morphism and union.
(ii) $\boldsymbol{L}$ is a *cone* or *full trio* if it is closed under finite transduction.
(iii) $\boldsymbol{L}$ is a *full semi-AFL* (abstract family of languages) if it is a cone and also closed under union.
(iv) $\boldsymbol{L}$ is a *full AFL* if it is a cone and also closed under union, catenation, and star.

By the results in Sections 9.1-9.3 immediately we have

Theorem 9.5.1 *$\boldsymbol{L}_{REG}$, $\boldsymbol{L}_{CF}$, and $\boldsymbol{L}_{DTM}$ are cylinders, cones, full semi-AFLs, and full AFLs.*

An additional, and attractive, method of classifying language families is based on the notion of principality, that is, one language is, in some sense, the key to the whole family.

Definition Let $\boldsymbol{L}$ be a family of languages and X a set of operations. Then $\boldsymbol{L}$ is said to be *principal*, with respect to X, if $\boldsymbol{L}$ is closed under the operations in X, and there exists a language L in $\boldsymbol{L}$ such that the smallest language family containing L and closed under X, denoted by $cl(L,X)$, is equal to $\boldsymbol{L}$. L is said to be an *X-generator* of $\boldsymbol{L}$.

Specifically, it can be proved that $\boldsymbol{L}_{CF}$ is, for example, a principal cone or a principal cylinder. The importance of this latter result comes from the following scenario. Let $L \subseteq \Sigma^*$ be a *CFL* which is a cylindrical generator of $\boldsymbol{L}_{CF}$, and assume we have a black box for L which given a word in Σ^* determines whether or not x is in L (see Sections 4.1 and 10.1.2); see Figure 9.5.1. Every *CFL*, $L' \subseteq \Delta^*$, can be written as $h^{-1}(L)$, for some morphism h. Therefore, to decide whether or not a given word y in Δ^* is in L', we only need test if $h(y)$ is in L using the L box. We don't have to construct an L' box; the L box operates as an L' box. Moreover, if we can obtain a highly efficient L box, then immediately we have, essentially, highly efficient L' boxes. For this reason L is said to be a *hardest CFL*.

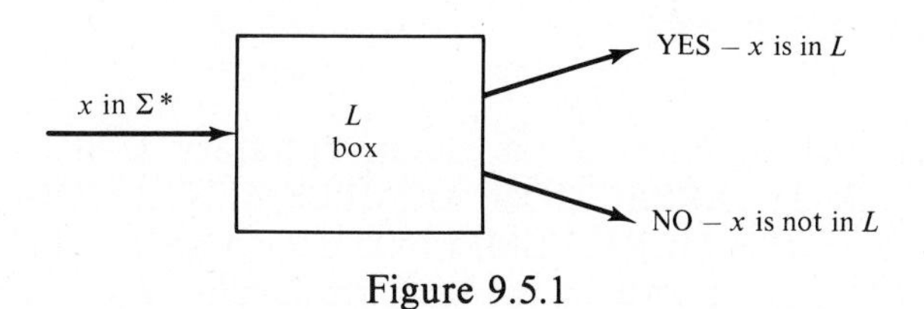

Figure 9.5.1

The cone of *CFLs* leads to many difficult and intriguing problems. For example, it is known that there are infinitely many subfamilies of $\boldsymbol{L}_{CF}$ which are also cones; in particular, $\boldsymbol{L}_{REG}$ is such a family. However, no subfamily of $\boldsymbol{L}_{REG}$ is a cone—it is the *smallest* cone containing all the finite sets, that is, $cl(\boldsymbol{L}_{FIN}$, finite transduction$) = \boldsymbol{L}_{REG}$; see the Exercises. But is there a cone $\boldsymbol{L}$ satisfying

(i) $\boldsymbol{L}_{REG} \subset \boldsymbol{L} \subset \boldsymbol{L}_{CF}$; and

(ii) there is no cone $\boldsymbol{L}'$ with $\boldsymbol{L}_{REG} \subset \boldsymbol{L}' \subset \boldsymbol{L}$.

In other words, is there a smallest cone properly containing $\boldsymbol{L}_{REG}$? This remains open.

Finally, we consider how to prove that a given language is not in a given family. In Chapter 8 we have provided pumping lemmas and Parikh's theorem as tools in such proofs. However, we can often shorten such proofs by using the closure properties of the given family — assuming it has some!

Example 9.5.1 Let $L = \{0^i 1 0^i : i \geq 0\}$. Prove that L is not in $\boldsymbol{L}_{REG}$.

We already know that $K = \{a^i b^i : i \geq 1\}$ is not in $\boldsymbol{L}_{REG}$. Therefore, we try to obtain K from L using the operations that $\boldsymbol{L}_{REG}$ is closed under.

Define a morphism $h : \{a,b,c\}^* \to \{0,1\}^*$ by $h(a) = h(b) = 0$ and $h(c) = 1$. Then $h^{-1}(L) = \{xcy : x,y \text{ are in } \{a,b\}^* \text{ and } |x| = |y|\}$. Let this be L_1 and let $L_2 = L_1 \cap a^+cb^+$. Then $L_2 = \{a^i cb^i : i \geq 1\}$. Finally, define a morphism $g : \{a,b,c\}^* \to \{a,b\}^*$ by $g(a) = a$, $g(b) = b$, and $g(c) = \lambda$. Immediately, $g(L_2) = K$.

Summarizing, we have shown that

$$g(h^{-1}(L) \cap a^+cb^+) = K.$$

Because $\boldsymbol{L}_{REG}$ is closed under morphism, inverse morphism, and regular intersection if L is in $\boldsymbol{L}_{REG}$, then K is in $\boldsymbol{L}_{REG}$. But K is not in $\boldsymbol{L}_{REG}$; therefore, L is not in $\boldsymbol{L}_{REG}$.

We explore this approach further in the Exercises.

9.6 ADDITIONAL REMARKS

9.6.1 Summary

We have investigated the effect of the closure operations UNION, INTERSECTION, REGULAR INTERSECTION, COMPLEMENTATION, REVERSAL, MORPHISM, INVERSE MORPHISM, FINITE TRANSDUCTION, CATENATION, QUOTIENT, and STAR on the language families $\boldsymbol{L}_{REG}$, $\boldsymbol{L}_{CF}$, and $\boldsymbol{L}_{DTM}$. We have also examined the COMPOSITION of finite, pushdown, and Turing transductions. Language families closed under the same operations have come to be of interest since they share many common properties. We briefly introduced the abstract families known as CYLINDERS, CONES, SEMI-AFLs, and AFLs.

9.6.2 History

The closure of $\boldsymbol{L}_{CF}$ under regular intersection was first proved in Bar-Hillel et al. (1961). The nonclosure of $\boldsymbol{L}_{REG}$ under intersection and complementation is due to Scheinberg (1960).

The closure of $\boldsymbol{L}_{REG}$ and $\boldsymbol{L}_{CF}$ under regular and context-free substitution, respectively, is due to Bar-Hillel et al. (1961) and closure under inverse morphism is proved in Ginsburg and Rose (1963b). The results on right quotient are to be found in Ginsburg and Spanier (1963).

The notion of *AFLs* is due to Ginsburg and Greibach (1969); semi-*AFLs* were introduced by Greibach (1970); cylinders by Greibach (1973); and cones by Berstel and Boasson (1974). Programmed grammars were introduced by Rosenkrantz (1969); indexed grammars by Aho (1968); macro grammars by Fischer (1968); nonterminal-bounded grammars by Altman (1964) and Banerji (1963); and *ET0L* grammars by Rozenberg (1973).

9.7 SPRINGBOARD

9.7.1 Closure Properties

Investigate the closure properties of the other families discussed in Chapter 8.

9.7.2 Cones

The notion of a cone is very basic. Carry out an in-depth study beginning with Berstel (1979).

EXERCISES

1.1 Prove or disprove the following claims.

- (i) $\boldsymbol{L}_{DCF}$ is closed under complementation.
- (ii) $\boldsymbol{L}_{DCF}$ is closed under union.
- (iii) $\boldsymbol{L}_{DCF}$ is closed under regular intersection.
- (iv) $\boldsymbol{L}_{DCF}$ is closed under reversal.
- (v) $\boldsymbol{L}_{LIN}$ is closed under union.
- (vi) $\boldsymbol{L}_{LIN}$ is closed under intersection.
- (vii) $\boldsymbol{L}_{LIN}$ is closed under reversal.
- (viii) $\boldsymbol{L}_{DPTIME}$ and $\boldsymbol{L}_{NPTIME}$ are closed under the Boolean operations.
- (ix) $\boldsymbol{L}_{REC}$ is closed under reversal.

1.2 Consider the restricted classes of *CFGs* which are defined in Exercise 8.2.3. Which Boolean operations are these subfamilies closed under, if any. Justify your answers in all cases.

1.3 Prove the first part of Theorem 9.1.2.

1.4 Prove Theorem 9.1.3 using *NPDAs* and *DFAs* rather than *CFGs* and *DFAs*.

1.5 Prove Corollary 9.1.4 when the two regular languages are both specified by *DFAs*, rather than one by a *DFA* and one by a regular grammar.

1.6 Complete the details of the proof of Theorem 9.1.5.

1.7 Prove Theorem 9.1.6.

2.1 Are the context-free subfamilies introduced in Exercise 8.2.3 closed under (i) morphism, (ii) inverse morphism, and (iii) finite transduction? Justify your answers in each case.

2.2 Consider the morphism h defined in Example 9.2.3.

(i) What are $h(abbbca)$ and $h(bab)$?
(ii) What are $h^{-1}(\lambda)$, $h^{-1}(1)$, $h^{-1}(0111)$, and $h^{-1}(00110110)$?

2.3 Give a decoding algorithm for a Caesar cipher.

2.4 Suggest a decoding algorithm for Morse code (assume each encoded letter is followed by a "pause").

2.5 You are given the morphism $h : \{a,b,c\}^* \rightarrow \{a,b\}^*$ defined by $h(a) = a^5$, $h(b) = aba$, and $h(c) = b^3$.

(i) Is it uniquely decodable?
(ii) Is it bounded delay?
(iii) Is $h^{-1}(x) \neq \varnothing$, for all x in $\{a,b\}^*$?

2.6 For the substitution of Example 9.2.5 what are $\sigma(cbc)$ and $\sigma(cab)$?

2.7 Complete the proof of Theorem 9.2.1.

2.8 Prove Theorem 9.2.1 using λ-*NFAs* rather than regular expressions.

2.9 Prove Theorem 9.2.1 using right linear grammars rather than regular expressions.

2.10 Complete the proof of Theorem 9.2.3.

2.11 Prove Theorem 9.2.5 for the cases of $\boldsymbol{L}_{REG}$ and $\boldsymbol{L}_{DTM}$.

2.12 In the proof of Theorem 9.2.6 we require that L's alphabet and M's input alphabet be the same. Prove that this can always be assumed.

2.13 Prove Theorem 9.2.7.

3.1 Are the families introduced in Exercise 8.2.3 closed under (i) catenation, (ii) star, and (iii) regular left and right quotient? Justify your answers.

3.2 Letting $L \subseteq \Sigma^*$, define *half*(L) by

$$half(L) = \{x : xy \text{ is in } L \text{ and } |x| = |y|\}.$$

Are $\boldsymbol{L}_{REG}$, $\boldsymbol{L}_{CF}$, and $\boldsymbol{L}_{DTM}$ closed under halving?

3.3 Generalize Exercise 3.2 as follows. Let $L \subseteq \Sigma^*$, and p and q be integers satisfying $1 \leq p < q$. Then $p/q(L)$ is defined as

$$p/q(L) = \{x : x \text{ is in } L \text{ and } q\,|x| = p\,|y|\,\}$$

Are $\boldsymbol{L}_{REG}$, $\boldsymbol{L}_{CF}$, and $\boldsymbol{L}_{DTM}$ closed under these operations?

3.4 Are $\boldsymbol{L}_{REG}$, $\boldsymbol{L}_{CF}$, and $\boldsymbol{L}_{DTM}$ closed under context-free left quotient?

3.5 Let L_1 and L_2 be two languages. Then the quotient of L_1 with L_2 is denoted by

$$\frac{L_1}{L_2}$$

and is defined by

$$\frac{L_1}{L_2} = \{xz : xyz \text{ is in } L_1 \text{ and } y \text{ is in } L_2\}$$

Are $\boldsymbol{L}_{REG}$, $\boldsymbol{L}_{CF}$, and $\boldsymbol{L}_{DTM}$ closed under quotient?

3.6 Let L be an arbitrary language and R a regular language. Prove that $L \backslash R$ is regular.

3.7 For a given language $L \subseteq \Sigma^*$, define the following:

$$Pre(L) = \{x : xy \text{ is in } L, \text{ for some } y \text{ in } \Sigma^*\},$$

$$In(L) = \{y : xyz \text{ is in } L, \text{ for some } x \text{ and } z \text{ in } \Sigma^*\},$$

$$Suf(L) = \{y : xy \text{ is in } L, \text{ for some } x \text{ in } \Sigma^*\}.$$

Are $\boldsymbol{L}_{REG}$, $\boldsymbol{L}_{CF}$, and $\boldsymbol{L}_{DTM}$ closed under these operations?

3.8 Let $L_1 \subseteq \Sigma_1^*$ and $L_2 \subseteq \Sigma_2^*$ be languages. Then

$$shuffle(L_1, L_2) = \{x_1 y_1 \cdots x_m y_m : x_i \text{ is in } \Sigma_1^*,$$
$$y_i \text{ is in } \Sigma_2^*,\ 1 \leq i \leq m,\ x_1 \cdots x_m \text{ is in } L_1,$$
$$\text{and } y_1 \cdots y_m \text{ is in } L_2\}.$$

Are $\boldsymbol{L}_{REG}$, $\boldsymbol{L}_{CF}$, and $\boldsymbol{L}_{DTM}$ closed under shuffle?

4.1 An n-ary relation $R \subseteq \Sigma_1^* \times \cdots \times \Sigma_n^*$ is a *regular relation* if it is accepted by an n-tape finite automata. If $(x_1, \ldots, x_n)$ is an input n-tuple, then x_i is placed on the ith tape, $1 \leq i \leq n$. Such an automaton is similar to a multitape Turing machine, except that the tapes cannot be rewritten and the n read heads cannot move left. At each move each of the read heads either moves one cell right or stays where it is. Prove that regular relations are closed under composition.

4.2 Pushdown and Turing relations can be defined in a similar manner to regular relations. Generalize Theorems 9.4.2 and 9.4.3 for n-ary relations?

4.3 Complete the proof of Theorem 9.4.1.

4.4 Prove Theorem 9.4.3.

5.1 Given the various families defined in Exercise 8.2.3 which of them are cylinders, cones, semi-*AFLs*, and *AFLs*?

5.2 Prove that $\mathbf{L}_{REG} = cl(\mathbf{L}_{FIN}, \text{finite transduction})$.

5.3 For $n \geq 1$, let $R \subseteq \Sigma_1^* \times \cdots \times \Sigma_n^*$ be an n-ary relation. We define three component operations (a)-(c). Let k be an integer, $1 \leq k \leq n$.

(a) *Regular intersection.* Let L be a regular language. Then

$$R \cap_k L = \{(x_1, \ldots, x_n) : (x_1, \ldots, x_n) \text{ is in } R \text{ and } x_k \text{ is in } L\}$$

(b) *Morphism.* Let $h : \Sigma^* \to \Delta^*$ be a morphism. Then

$$h(R)_k = \{(x_1, \ldots, h(x_k), \ldots, x_n) : (x_1, \ldots, x_n) \text{ is in } R\}$$

(c) *Inverse morphism.* Let $h : \Delta^* \to \Sigma^*$ be a morphism. Then

$$h^{-1}(R)_k = \{(x_1, \ldots, y_k, \ldots, x_n) : (x_1, \ldots, x_n) \text{ is in } R \text{ and } y_k \text{ is in } h^{-1}(x_k)\}$$

Now, we define the generalization of product and star.

(d) For two n-ary relations R_1 and R_2 define their *product* R_1R_2 as

$$R_1R_2 = \{(x_1y_1, \ldots, x_ny_n) : (x_1, \ldots, x_n) \text{ is in } R_1 \text{ and } (y_1, \ldots, y_n) \text{ is in } R_2\}$$

(e) For an n-ary relation R define its *star* R^* as

$$R^* = \bigcup_{i=0}^{\infty} R^i$$

where $R^0 = \{(\lambda, \ldots, \lambda)\}$ and $R^{i+1} = RR^i$, for all $i \geq 0$.

A *full abstract family of n-ary relations* or *full n-AFR* is defined as a family of n-ary relations closed under union, product, star, component regular intersection, component morphism, and component inverse morphism.

(i) Prove that the finite transductions are full 2-*AFR*.
(ii) Prove that a fuli 1-*AFR* is a full *AFL*.
(iii) Is the family of *DTTs* a 2-*AFR*? Justify your answer.
(iv) Is the family of *PDTs* a 2-*AFR*? Justify your answer.

5.4 Using the closure properties of $\boldsymbol{L}_{REG}$ show that the following languages are not regular.

(i) $\{x : |x|_a = |x|_b \text{ and } x \text{ is in } \{a,b\}^*\}$.
(ii) $\{a^i b a^i : i \geq 1\}$.
(iii) $\{a^i b^j c^i : i \geq j \geq 1\}$.
(iv) $\{xx^R : x \text{ is in } \{a,b\}^*\}$.
(v) $\{a^i b a^j : i,j \geq 1 \text{ and } i \neq j\}$.

5.5 Using the closure properties of $\boldsymbol{L}_{CF}$ show that the following languages are not context-free.

(i) $\{x : |x|_a = 2|x|_b, |x|_c = 3|x|_b, \text{ and } x \text{ is in } \{a,b,c\}^*\}$.
(ii) $\{0^i 1 0^i 1 0^i 1 0^i : i \geq 1\}$.
(iii) $\{xcy : |x| = 2^{|y|} \text{ and } x,y \text{ are in } \{a,b\}^*\}$.
(iv) $\{xcx^R cx : x \text{ is in } \{a,b\}^*\}$.
(v) $\{a^i b^j c^k : i,j,k \geq 1, i \neq j \text{ or } j \neq k \text{ or } i \neq k\}$.

Chapter 10

Decision Problems

10.1 DECIDABILITY AND MEMBERSHIP

10.1.1 Introduction

The subject matter of this chapter is both fundamental and important. It is concerned with the existence of decision makers for various decision problems. Recall that a decision problem is *decidable* if there is a decision maker for it. We demonstrate that a problem is decidable by providing an informal description of a decision-making program.

A decision problem is *undecidable* if there is no decision maker for it. A negative result of this kind requires us to have available a formal model of decision makers. For this purpose we use decision-making Turing machines, which by Church's thesis are one of the most general models.

We show not only that undecidable problems exist, but also that they abound. Most problems of interest for *CFGs* and *DTMs* are undecidable, while the same problems for *FAs* are decidable. The most famous undecidable problem is

HALTING
INSTANCE: A *DTM*, $M = (Q,\Sigma,\Gamma,\delta,s,f)$, and a word x in Σ^*.
QUESTION: Does M halt (in state f) when given input x, that is, is x in $L(M)$?

which we study in this section. (Note that a *DTM* can stop in one of two ways. It can halt in the final state or it can hang in some other state.)

Since the class of functions computable by *DTMs* does not exhaust all functions, there are functions which cannot be computed. We say a function is *uncomputable* if there is no *DTM* which computes it. Undecidable problems provide examples of uncomputable Boolean functions, but there are uncomputable functions which are not Boolean. Such functions correspond to programs which *cannot be written.*

Hand in hand with decidability and computability we have tractability. If a problem is decidable or computable, then is there an efficient program to solve it? We consider this issue in Section 11.1. In the remainder of this chapter we examine the basic decision problems for *FAs*, *CFGs*, and *DTMs*.

Testing whether or not an arbitrarily given word is accepted or generated by a class of models, is, perhaps, the most fundamental decision problem. We prove that it is decidable for *FAs* and *CFGs*, but for *DTMs* we show it to be undecidable. The latter result is demonstrated by proving that HALTING is undecidable. The undecidability of HALTING is not an isolated result; it provides a basis for proving the undecidability of many other decision problems.

10.1.2 Finite Automata

The problem can be stated as

DFA MEMBERSHIP
INSTANCE: A *DFA*, $M = (Q,\Sigma,\delta,s,F)$, and a word x in Σ^*.
QUESTION: Is x in $L(M)$?

A decision algorithm for this problem simply runs M on input x. Either x is accepted by M in which case it returns **true**, otherwise x is rejected and the algorithm returns **false**.

Variants of this problem are discussed in the Exercises.

10.1.3 Context-Free Grammars

We have

CFG MEMBERSHIP
INSTANCE: A *CFG*, $G = (N,\Sigma,P,S)$, and a word x in Σ^*.
QUESTION: Is x in $L(G)$?

A decision algorithm for this problem first converts G into an equivalent unit-free, λ-free *CFG*, $G' = (N,\Sigma,P',S)$. Second, if $x = \lambda$, then whether or not λ is in $L(G)$ can be tested using *λ-Nonterminals*; see Section 4.2.2. Thirdly, and, finally, if $x \neq \lambda$, then the algorithm enumerates all leftmost derivations in G' with length $\leq 2|x|-1$. If x is generated by G', then there must be a leftmost derivation for x with length $\leq 2|x|-1$, because of the following claim.

Claim: Let $G = (N,\Sigma,P,S)$ be a unit-free, λ-free *CFG*. Then, for all A in N and all x in Σ^* such that $A \overset{L}{\Rightarrow}{}^k x$ in G, for some $k \geq 1$,

$$k \leq 2|x|-1.$$

Proof of Claim: By induction on the length of x.

Basis: $|x| = 1$. $A \rightarrow x$ is the only possibility since G is unit-free and λ-free. Hence, $k \leq 2.1-1 = 1$ as desired.

Induction Hypothesis: Assume the claim holds for all words x with $|x| \leq n$, for some $n \geq 1$.

Induction Step: Consider a nonterminal A and a word x with $|x| = n+1$ and $A \Rightarrow^k x$, for some $k \geq 1$.

Either $A \rightarrow x$ is in P, in which case $k = 1$ and $k \leq 2n-1$ is satisfied since $n \geq 1$, or $A \rightarrow x$ is not in P. In the latter case the derivation $A \Rightarrow^k x$ satisfies $k \geq 2$ and $A \overset{L}{\Rightarrow} \alpha \overset{L}{\Rightarrow}^{k-1} x$, for some α in $(N \cup \Sigma)^*$. Now, $|\alpha| \geq 2$ and α can be decomposed into $\alpha = x_0B_1x_1 \cdots B_rx_r$, for some $r \geq 1$, where $x_0x_1 \cdots x_r$ is in Σ^* and $B_1, \ldots, B_r$ are in N. This implies

$$A \overset{L}{\Rightarrow} x_0B_1x_1 \cdots B_rx_r \overset{L}{\Rightarrow}^{k_1} x_0y_1x_1 \overset{L}{\Rightarrow}^{k_2}$$

$$\cdots \overset{L}{\Rightarrow}^{k_r} x_0y_1x_1 \cdots y_rx_r = x$$

where $k_1+k_2+ \cdots +k_r = k-1$. Applying the induction hypothesis we have $k_i \leq 2|y_i|-1$, $1 \leq i \leq r$, that is, $k-1 \leq 2\Sigma_{i=1}^{r} |y_i| - r$. There are two cases to consider. If $x_0x_1 \cdots x_r = \lambda$, then $r \geq 2$,

$$k-1 \leq 2\sum_{i=1}^{r} |y_i| - 2$$

and, hence,

$$k \leq 2|x| - 1$$

Otherwise $x_0x_1 \cdots x_r \neq \lambda$, in which case $r \geq 1$ and $|y_1 \cdots y_r| < |x|$, yielding

$$k-1 \leq 2(|x| - 1) - 1$$

that is,

$$k \leq 2|x| - 1$$

This completes the induction step and the proof of the claim. ■

The enumeration of all leftmost derivations from S of length at most $2|x|-1$ is straightforward; therefore, *CFG* MEMBERSHIP is decidable.

10.1.4 Deterministic Turing Machines

The halting problem is

HALTING
INSTANCE: A *DTM*, $M = (Q,\Sigma,\Gamma,\delta,s,f)$, and a word x in Σ^*.
QUESTION: Does M halt (in state f) when given input x, that is, does M accept x?

HALTING defined in this way could also be called *DTM* MEMBERSHIP. In the Exercises HALTING-HANGING is also considered.

A decision algorithm for this problem is given a representation of a *DTM* and a word over the *DTMs* input alphabet. This format is similar to words in the language L_{ACCEPT} of Section 8.4. Recall that

$$L_{ACCEPT} = \{\ll M \gg x : M = (Q,\{a,b\},\{a,b,\mathbf{B}\},\delta,s,f) \text{ is a } DTM,$$
$$x \text{ is a word over } \{a,b\}, \text{ and } x \text{ is accepted by } M\}$$

However, in HALTING we have arbitrary input and tape alphabets. This leads us to consider

RESTRICTED HALTING
INSTANCE: A *DTM*, $M = (Q,\{a,b\},\{a,b,\mathbf{B}\},\delta,s,f)$, and a word x in $\{a,b\}^*$.
QUESTION: Does M halt when given input x?

Now, if HALTING is decidable, then, clearly, RESTRICTED HALTING is also decidable. Furthermore, if RESTRICTED HALTING is decidable then L_{ACCEPT} is decided by some *DTM*. Having established the connection between the two decision problems and L_{ACCEPT} we can now prove the following.

Theorem 10.1.1 *HALTING is undecidable.*

Proof: Recall that $L_{SELFACCEPT}$, defined as

$$L_{SELFACCEPT} = \{x : x \text{ is the } \{a,b\} \text{ encoding of a } DTM,$$
$$M = (Q,\{a,b\},\{a,b,\mathbf{B}\},\delta,s,f) \text{ and } M \text{ accepts } x\},$$

was shown not to be in $\boldsymbol{L}_{REC}$ by Theorems 8.4.1 and 8.4.2. In other words, there is no *DTM* that decides $L_{SELFACCEPT}$. We conclude via the following claim that L_{ACCEPT} is also not in $\boldsymbol{L}_{REC}$.

Claim: If L_{ACCEPT} is in $\boldsymbol{L}_{REC}$, then $L_{SELFACCEPT}$ is in $\boldsymbol{L}_{REC}$.

Proof of Claim: We assume that L_{ACCEPT} is decided by some *DTM*, $M = (Q,\Sigma,\Gamma,\delta,s,f)$, that is, $L_{ACCEPT} = Y(M)$. We modify M to obtain an M' with $Y(M') = L_{SELFACCEPT}$.

M' checks its input to determine that it has the form x, for some x in $\{a,b\}^*$ which is the encoding of some *DTM* $\overline{M} = (\overline{Q},\{a,b\}, \{a,b,\mathbf{B}\},\overline{\delta},\overline{s},\overline{f})$. This check is laborious, but straightforward. If the input word fails to pass this test, then M' returns **n** and halts. Otherwise M' calls M with input xx, returning whatever M returns when M halts. Hence, $Y(M') = L_{SELFACCEPT}$ as desired and $L_{SELFACCEPT}$ is in $\boldsymbol{L}_{REC}$. ■

By Theorem 8.4.2 $L_{SELFACCEPT}$ is not in $\boldsymbol{L}_{REC}$; therefore, L_{ACCEPT} is not in $\boldsymbol{L}_{REC}$. But this implies that RESTRICTED HALTING is undecidable and, therefore, that HALTING is also undecidable. ■

HALTING is a fundamental undecidability result as we shall see.

10.2 EMPTINESS AND FINITENESS

A weaker question than MEMBERSHIP is: Does this model accept or generate any words? For *DFAs* and *CFGs* this is a decidable problem, while, yet again for *DTMs* it is undecidable. We also consider a variant of this question: Does this model accept or generate a finite number of words. These problems are called EMPTINESS and FINITENESS, respectively.

10.2.1 Finite Automata

DFA EMPTINESS
INSTANCE: A *DFA*, $M = (Q,\Sigma,\delta,s,F)$.
QUESTION: Is $L(M) = \varnothing$?

If M accepts words of length at least $\#Q$, then the *DFA* pumping lemma tells us that M accepts an infinite number of words. This suggests that a necessary and sufficient condition for $L(M) = \varnothing$ is: M accepts no words of length less than $\#Q$. We prove that this is indeed the case.

Lemma 10.2.1 *Let $M = (Q,\Sigma,\delta,s,F)$ be a DFA. Then $L(M) = \varnothing$ if and only if for all x in Σ^*, $|x| < \#Q$, x is not accepted by M.*

Proof:

if: Assume that M accepts no words of length less than $\#Q$. If $L(M) \neq \varnothing$, then M accepts some word x with $|x| = m \geq \#Q$. But this implies by the *DFA* pumping lemma that $x = uvw$, for some u, v, and w in Σ^* such that $|w| \geq 1$, $|uw| < \#Q$, and uv^iw is in $L(M)$, for all $i \geq 0$. In particular $uw = x'$ is in $L(M)$. Now, this argument is repeated until we obtain a word x' in $L(M)$ such that $|x'| < \#Q$. But this contradicts the assumption that M accepts no words of this form; therefore, the assumption that $L(M) \neq \varnothing$ is false. In other words, $L(M) = \varnothing$.

only if: If $L(M) = \varnothing$, then M certainly accepts no words of length less than $\#Q$. ■

The above lemma is the basis of a decision algorithm for EMPTINESS. It simply checks all words over Σ of length less than $\#Q$ to see if any of them are accepted.

A related problem is

DFA UNIVERSALITY
INSTANCE: A *DFA*, $M = (Q,\Sigma,\delta,s,F)$.
QUESTION: Is $L(M) = \Sigma^*$?

without loss of generality we may assume M is complete, in which case we construct $\overline{M} = (Q,\Sigma,\delta,s,Q-F)$. Now, $L(M) = \Sigma^*$ if and only if $L(\overline{M}) = \varnothing$. Hence, UNIVERSALITY is decidable.

The final problem of this subsection is

DFA FINITENESS
INSTANCE: A *DFA*, $M = (Q,\Sigma,\delta,s,F)$.
QUESTION: Is $L(M)$ finite?

Again the *DFA* pumping lemma tells us that if M accepts any words of length at least $\#Q$, then $L(M)$ is infinite. To devise a decision algorithm for *DFA* FINITENESS we need to be able to bound the search for such words, as we did for *DFA* EMPTINESS. We restrict the search to configuration sequences with a single repeated state. This is because configuration sequences with multiple repetitions of states can be replaced by shorter configuration sequences with a single repetition. We have

Lemma 10.2.2 *Let $M = (Q,\Sigma,\delta,s,F)$ be a DFA. Then $L(M)$ is finite if and only if M accepts no words x that satisfy $\#Q \leq |x| < 2\#Q$.*

Proof:

if: Assume M accepts no words x that satisfy $\#Q \leq |x| < 2\#Q$. If $L(M)$ is infinite, then M accepts some word x with $|x| \geq 2\#Q$. Apply the *DFA* pumping lemma as in the proof of Lemma 10.2.1 obtaining a new word x' in $L(M)$ with $|x'| < |x|$. Either $|x'| < 2\#Q$ or we can repeat the above reduction until such an x' is found. It is crucial to this reduction that at a single step we cannot reduce x by too much. Otherwise we may obtain an x' with $|x'| < \#Q$. However, the *DFA* pumping lemma tells us that the pumped subword has length less than $\#Q$. This implies that when $|x| \geq 2\#Q$, the smallest x' that can be obtained after one reduction step has length a it least $\#Q+1$.

only if: This is immediate. ■

A decision algorithm for *DFA* FINITENESS simply checks all words that satisfy the bounds of Lemma 10.2.2 to see if any are accepted by M.

10.2.2 Context-Free Grammars

We solve

CFG EMPTINESS
INSTANCE: A *CFG*, $G = (N,\Sigma,P,S)$.
QUESTION: Is $L(G) = \varnothing$?

and

CFG FINITENESS
INSTANCE: A *CFG*, $G = (N,\Sigma,P,S)$.
QUESTION: Is $L(G)$ finite?

by techniques similar to those for *DFAs*. We characterize those *CFGs* for which $L(G) = \emptyset$ and $L(G)$ is finite based on the *CFG* pumping lemma.

Lemma 10.2.3 *Let $G = (N,\Sigma,P,S)$ be a unit-free, λ-free CFG. Then*

(i) *$L(G) = \emptyset$ if and only if there is no word z in $L(G)$ such that $|z| < p$.*
(ii) *$L(G)$ is finite if and only if there is no word z in $L(G)$ such that $p \leq |z| < 2p$.*

Proof: In both cases the "only if" direction is trivial, so we only deal with the "if" direction.

(i) Assume there is no word z in $L(G)$, $|z| < p$, but $L(G) \neq \emptyset$. Then there must be a word z in $L(G)$ with $|z| \geq p$. By the *CFG* pumping lemma there is a decomposition of $z = uxwyv$ such that $|xy| \geq 1$, $|xwy| < p$, and ux^iwy^iv is in $L(G)$, for all $i \geq 0$. This implies, in particular, that $z' = uwv$ is in $L(G)$ and $|z'| < |z|$. If $|z'| \geq p$, repeat this process with z equal to z'. Otherwise $|z'| < p$ and we have found a word z' in $L(G)$, $|z'| < p$, which contradicts one of the assumptions. Therefore, the other assumption is false and $L(G) = \emptyset$ as desired.
(ii) Assume there is no word z in $L(G)$, $p \leq |z| < 2p$, but $L(G)$ is infinite. Then there must be a word z in $L(G)$ with $|z| \geq 2p$. By the *CFG* pumping lemma there is a decomposition of $z = uxwyv$ such that $|xy| \geq 1$, $|xwy| < p$, and ux^iwy^iv is in $L(G)$, for all $i \geq 0$. This implies, in particular, that $z' = uwv$ is in $L(G)$, where $|z'| < |z|$. Furthermore, $|z'| > |z| - p > p$, since $|xwy| < p$ and $|z| \geq 2p$.

    If $|z'| \geq 2p$, then repeat this process with z equal to z'. Otherwise we have found a z' in $L(G)$, satisfying $p \leq |z'| < 2p$, a contradiction. Therefore, $L(G)$ is finite. ■

Since *CFLs* are not closed under complementation, testing whether or not $L(G) = \Sigma^*$ is not reducible to *CFG* EMPTINESS as it is for *DFAs*. As we shall see later *CFG* UNIVERSALITY is undecidable.

10.2.3 Deterministic Turing Machines

Not surprisingly both

DTM EMPTINESS
INSTANCE: A *DTM*, $M = (Q,\Sigma,\Gamma,\delta,s,f)$.
QUESTION: Is $L(M) = \emptyset$?

and

DTM FINITENESS
INSTANCE: A *DTM*, $M = (Q,\Sigma,\Gamma,\delta,s,f)$.
QUESTION: Is $L(M)$ finite?

are undecidable. To prove these undecidability results we first prove an undecidability result of interest in its own right.

> *DTM* EMPTY WORD MEMBERSHIP
> INSTANCE: A *DTM*, $M = (Q, \Sigma, \Gamma, \delta, s, f)$.
> QUESTION: Is λ in $L(M)$?

Lemma 10.2.4 *DTM* EMPTY WORD MEMBERSHIP is undecidable.

Proof: This is equivalent to proving that

$$L_{EMPTYWORD} = \{<<M>> : <<M>> \text{ is the encoding of a } DTM,$$
$$M = (Q, \{a,b\}, \{a,b,\boldsymbol{B}\}, \delta, s, f) \text{ which accepts } \lambda\}$$

is not recursive. Observe that $L_{EMPTYWORD} \subseteq L_{ACCEPT}$. Assume, for some *DTM*, V, $L_{EMPTYWORD} = Y(V)$. We show that this implies there is a *DTM*, V', with $Y(V') = L_{ACCEPT}$ — a contradiction, so V doesn't exist.

V' operates as follows. (1) If the input to V' is not of the form $<<M>>x$, then V' immediately returns no. Otherwise the input is of the form $<<M>>x$ and V' modifies it to give $<<M(x)>>$. $M(x)$ is the same as M except that when given the empty word as input it first replaces it by x, before calling M. (2) V' calls V on input $<<M(x)>>$.

First, observe that the transformation of $<<M>>x$ into $<<M(x)>>$ can be accomplished by a completely general-purpose subprogram, that is, a *DTM*. Second, observe that M halts on input x if and only if $M(x)$ halts on input λ. Thirdly, and, finally, V' decides $<<M>>x$ if and only if V decides $<<M(x)>>$. In other words, $Y(V') = L_{ACCEPT}$ if and only if $Y(V) = L_{EMPTYWORD}$. But $L_{ACCEPT} \neq Y(T)$, for any *DTM*, T, by Theorem 10.1.1; therefore, $L_{EMPTYWORD} \neq Y(T)$, for any *DTM*, T. That is, $L_{EMPTYWORD}$ is not in $\boldsymbol{L}_{REC}$. ■

We can now prove the following.

Theorem 10.2.5 *DTM EMPTINESS is undecidable.*

Proof: Assume *DTM* EMPTINESS is decidable for *DTMs* with a tape alphabet $\{a,b,\boldsymbol{B}\}$, that is,

$$L_{EMPTY} = \{<<M>> : M = (Q, \{a,b\}, \{a,b,\boldsymbol{B}\}, \delta, s, f)$$
$$\text{is a } DTM \text{ and } L(M) = \varnothing\}$$

is decided by some *DTM*, V. In other words, $L_{EMPTY} = Y(V)$. We construct a *DTM*, V' from V, with $Y(V') = L_{EMPTYWORD}$ — a contradiction.

V' operates as follows.

(1) It checks its input word x to see if $x = \ll M \gg$ for some encoding of some DTM, $M = (Q, \{a,b\}, \{a,b,\mathbf{B}\}, \delta, s, f)$. If $x \neq \ll M \gg$, for some DTM, M, then V' immediately returns no. Otherwise V' modifies $\ll M \gg$ to obtain $\ll \overline{M} \gg$, where $\overline{M}$ is the same as M except that it first replaces its input word with λ before entering M.

(2) V' enters V with input $\ll \overline{M} \gg$. V' returns no if V returns yes, and it returns yes if V returns no. Note that $L(\overline{M}) = \{a,b\}^*$ if and only if M accepts λ. Therefore, $L(\overline{M}) = \varnothing$ if and only if M does not accept λ. This implies $Y(V') = L_{EMPTYWORD}$, since, by assumption, $Y(V) = L_{EMPTY}$. But this contradicts Lemma 10.2.4. Therefore, DTM EMPTINESS is undecidable. ∎

Using a similar technique we prove the following.

Theorem 10.2.6 *DTM FINITENESS is undecidable.*

Proof: We only give a proof sketch. The details are left to the Exercises. As before consider the language

$$L_{FINITE} = \{\ll M \gg : M = (Q, \{a,b\}, \{a,b,\mathbf{B}\}, \delta, s, f) \text{ is a } DTM \text{ and } L(M) \text{ is finite}\}$$

Assume there is a DTM, V with $Y(V') = L_{EMPTYWORD}$ as in Theorem 10.2.5 noting that $L(\overline{M})$ is finite if and only if $L(\overline{M}) = \varnothing$ if and only if M does not accept λ. ∎

We leave to the Exercises a number of other proofs.

10.3 CONTAINMENT, EQUIVALENCE, AND INTERSECTION

In this section we wish to compare two generators or two recognizers to see if their languages are the same, if the language of one is contained in the language of the other, or if they have no words in common. We only consider *DFAs* and *CFGs* since these problems are decidable for *DFAs*, but already undecidable for *CFGs*.

10.3.1 Finite Automata

The problems are

DFA CONTAINMENT
INSTANCE: Two *DFAs*, $M_1 = (Q_1, \Sigma_1, \delta_1, s_1, F_1)$ and $M_2 = (Q_2, \Sigma_2, \delta_2, s_2, F_2)$.
QUESTION: Is $L(M_1) \subseteq L(M_2)$?

DFA EQUIVALENCE
INSTANCE: Two *DFAs*, $M_1 = (Q_1, \Sigma_1, \delta_1, s_1, F_1)$
and $M_2 = (Q_2, \Sigma_2, \delta_2, s_2, F_2)$.
QUESTION: Is $L(M_1) = L(M_2)$?

and

DFA INTERSECTION
INSTANCE: Two *DFAs*, $M_1 = (Q_1, \Sigma_1, \delta_1, s_1, F_1)$
and $M_2 = (Q_2, \Sigma_2, \delta_2, s_2, F_2)$.
QUESTION: Is $L(M_1) \cap L(M_2) = \varnothing$?

Since $L_1 \subseteq L_2$ if and only if $L_1 \cap \overline{L_2} = \varnothing$, $L_1 = L_2$ if and only if $(L_1 \cap \overline{L_2}) \cup (\overline{L_1} \cap L_2) = \varnothing$, and $\boldsymbol{L}_{REG}$ is closed under the Boolean operations, the first two problems can be solved if the third one can be solved.

But given two *DFAs*, M_1 and M_2, we can construct a *DFA*, M, with $L(M) = L(M_1) \cap L(M_2)$, by way of Corollary 9.1.4. Now, this implies $L(M_1) \cap L(M_2) = \varnothing$ iff $L(M) = \varnothing$ and this latter problem is decidable as demonstrated in Section 10.2.1.

Hence, all three problems are decidable.

10.3.2 Context-Free Grammars

We will prove that the three decision problems

CFG CONTAINMENT
INSTANCE: Two *CFGs*, $G_1 = (N_1, \Sigma_1, P_1, S_1)$
and $G_2 = (N_2, \Sigma_2, P_2, S_2)$.
QUESTION: Is $L(G_1) \subseteq L(G_2)$?

CFG EQUIVALENCE
INSTANCE: Two *CFGs*, $G_1 = (N_1, \Sigma_1, P_1, S_1)$
and $G_2 = (N_2, \Sigma_2, P_2, S_2)$.
QUESTION: Is $L(G_1) = L(G_2)$?

and

CFG INTERSECTION
INSTANCE: Two *CFGs*, $G_1 = (N_1, \Sigma_1, P_1, S_1)$
and $G_2 = (N_2, \Sigma_2, P_2, S_2)$.
QUESTION: Is $L(G_1) \cap L(G_2) = \varnothing$?

are undecidable. We do this by constructing two languages from a given *DTM*, *M*, whose intersection consists of the accepting configuration sequences of *M*. Now, the intersection is empty if and only if $L(M) = \varnothing$. But we have already proved that *DTM* EMPTINESS is undecidable in Section 10.2.3; therefore, *DTM* INTERSECTION is also undecidable. But this proves even more since the two languages are both context-free; therefore, *CFG* INTERSECTION is undecidable. Further, the complement of this intersection is also a *CFL* which enables us to deduce that *CFG* CONTAINMENT and *CFG* EQUIVALENCE are also undecidable.

Let $M = (Q,\Sigma,\Gamma,\delta,s,f)$ be a given *DTM*, $\Sigma = \{a,b\}$, and $\Gamma = \{a,b,\boldsymbol{B}\}$. Without loss of generality we can assume that if $sx = C_0 \vdash C_1 \vdash \cdots \vdash C_m$ is an accepting configuration sequence for some word x in Σ^*, then m is even. In other words, an accepting configuration sequence always has an even number of steps, possibly zero. In general this doesn't hold, but we can always transform a *DTM*, M, as follows. Associate to every transition in M a new unique state. Now, replace each transition by two transitions. The first maintains the status quo apart from changing state to its associated new state and the second carries out the original actions except that it begins in the new state. For example, given $\delta(p,a) = (q,b,R)$ and new state r we obtain the transitions $\delta(p,a) = (r,a,\lambda)$ and $\delta(r,a) = (q,b,R)$. The only effect of this transformation is to double the length of each accepting configuration sequence, that is, they contain an even number of steps.

We now construct the two context-free languages whose intersection gives accepting configuration sequences. Given an accepting configuration sequence

$$C_0 \vdash C_1 \vdash \cdots \vdash C_{2m}$$

We encode it as the word

$$C_0\$C_1^R\$C_2\$ \cdots \$C_{2m-1}^R\$C_{2m}$$

We reverse each odd indexed configuration and we separate configurations with \$, a new symbol. The first language L_M^1 consists of words of the form

$$x_0\$x_1\$ \cdots \$x_{2m-1}\$x_{2m}$$

where x_i is in $\Gamma^*Q\Gamma^*$, $0 \leq i \leq 2m$, for some $m \geq 0$. However, this is not all. We also require that $x_0 \vdash x_1^R$, $x_2 \vdash x_3^R, \ldots, x_{2m-2} \vdash x_{2m-1}^R$ in M and that x_{2m} is a final configuration. The second language L_M^2 also consists of words of the form

$$y_0\$y_1\$ \cdots \$y_{2n-1}\$y_{2n}$$

where y_i is $\Gamma^*Q\Gamma^*$, $0 \leq i \leq 2n$, for some $n \geq 0$. In this case we require that $y_1^R \vdash y_2$, $y_3^R \vdash y_4, \ldots, y_{2n-1}^R \vdash y_{2n}$ in M and that y_0 is a starting configuration.

It should be clear that $L_M^1 \cap L_M^2$ consists of words of the form

$$z_0\$z_1\$ \cdots \$z_{2k-1}\$z_{2k}$$

where $z_0 \vdash z_1^R \vdash z_2 \vdash \cdots \vdash z_{2k}$ in M, where z_0 is a starting configuration and z_{2k} is a final configuration. But this implies it is an accepting configuration sequence!

Lemma 10.3.1 *L_M^1 and L_M^2 are CFLs.*

Proof: First, we prove that L_M^1 is a *CFL*. Now, $L_M^1 = K_1^*K_F$, where

$$K_1 = \{x\$y\$: x,y \text{ are in } \Gamma^*Q\Gamma^* \text{ and } x \vdash y^R \text{ in } M\}$$

and

$$K_F = \{xfy : xy \text{ is in } \Gamma^*\}$$

Since $\mathbf{L}_{CF}$ is closed under star and catenation we need prove only that K_1 and K_F are *CFLs*.

Let G_F be

$S_F \rightarrow AS_F \mid S_FA \mid f$

for all A in $\{a,b,\mathbf{B}\}$. Then $L(G_F) = K_F$ and K_F is a *CFL*.

The *CFG*, G_1, for K_1 is somewhat more complex. The idea is that for each word $x\$y\$$ in K_1 y^R is *almost* the same x. The exceptions are restricted to at most two adjacent symbols in x and y at corresponding positions. They are the result of M's computation on x. We can deal with this minor perturbation quite easily using a linear grammar.

Let G_1 be

$S_1 \rightarrow T_1\$$	
$T_1 \rightarrow aT_1a \mid bT_1b \mid Q_1$	
$Q_1 \rightarrow pcR_1dq,$	$\delta(p,c) = (q,d,\lambda), c,d$ in Γ
$Q_1 \rightarrow pcR_1qd$	$\delta(p,c) = (q,d,R), c,d$ in Γ
$Q_1 \rightarrow apcR_1daq \mid bpcR_1dbq$	$\delta(p,c) = (q,d,L), c,d$ in Γ
$Q_1 \rightarrow p\$dq$	$\delta(p,\mathbf{B}) = (q,d,\lambda), d$ in Σ
$Q_1 \rightarrow p\$q$	$\delta(p,\mathbf{B}) = (q,\mathbf{B},\lambda)$
$Q_1 \rightarrow p\$qd$	$\delta(p,\mathbf{B}) = (q,\mathbf{B},R), d$ in Γ
$Q_1 \rightarrow ap\$daq \mid bp\$dbq$	$\delta(p,\mathbf{B}) = (q,d,L), d$ in Σ
$Q_1 \rightarrow ap\$aq \mid bp\$bq$	$\delta(p,\mathbf{B}) = (q,\mathbf{B},L)$
$R_1 \rightarrow aR_1a \mid bR_1b \mid \$$	

The formal proof that $L(G_1) = K_1$ is left to the Exercises. We note only that the Q_1-productions deal with the effects of a transition. The number of Q_1-productions is indeed a reflection of the number of possibilities at each transition as the reader should verify. Thus, K_1 and K_F are context-free and, hence, L_M^1 is context-free.

The proof that L_M^2 is a *CFL* is similar. We can decompose L_M^2 as $K_IK_2^*$, where

$$K_I = \{sx : x \text{ is in } \Sigma^*\}$$

and

$$K_2 = \{\$x\$y : x,y \text{ are in } \Gamma^*Q\Gamma^* \text{ and } x^R \vdash y \text{ in } M\}$$

We leave the details to the Exercises. ■

Theorem 10.3.2 *CFG INTERSECTION is undecidable.*

Proof: Assume *CFG* INTERSECTION is decidable. Then we can decide *DTM* EMPTINESS as follows. Given an instance $M = (Q, \{a,b\}, \{a,b,\mathbf{B}\},\delta,s,f)$ of *DTM* EMPTINESS, we construct *CFGs*, H_1 and H_2 with $L(H_1) = L_M^1$ and $L(H_2) = L_M^2$. This step can be carried out straightforwardly by way of the proof of Lemma 10.3.1. Now, using the decision algorithm for *CFG* INTERSECTION we determine whether or not $L(H_1) \cap L(H_2) = \emptyset$. But $L(H_1) \cap L(H_2) = \emptyset$ if and only if $L(M) = \emptyset$; therefore, we have solved *DTM* EMPTINESS. This is a contradiction; thus, *CFG* INTERSECTION is undecidable. ■

For the remaining undecidability results we deal with the complement of $L_M^1 \cap L_M^2$. Let $L_{ACS}^M = L_M^1 \cap L_M^2$ be the language of accepting configuration sequences. Then $\overline{L}_{ACS}^M = (\{a,b,\mathbf{B},\$\} \cup Q)^* - L_{ACS}^M$. Although L_{ACS}^M is not a *CFL*, we show that $\overline{L}_{ACS}^M$ is.

Lemma 10.3.3 *$\overline{L}_{ACS}^M$ is a CFL.*

Proof: $\overline{L}_{ACS}^M$ consists of two kinds of words, those which look like configuration sequences and those which don't.

The words which look like configuration sequences contain an even number of $\$$ symbols and the subwords separated by the $\$$ symbols contain one state. These words may not be configuration sequences because either they begin or end wrongly or they contain two "configurations" C_i and C_{i+1} such that M does not yield C_{i+1} in one step when given C_i. This latter possibility is called a bad computation.

Words which do not look like configuration sequences contain an odd number of $\$$ symbols, no $\$$ symbols, or a "configuration" with either no state or at least two states.

Words of each kind can be generated by linear *CFGs*. Indeed, most possibilities can be generated by regular *CFGs*! We only require a linear grammar to generate bad computations. We first decompose $\overline{L}_{ACS}^M$ as

$$\overline{L}_{ACS}^M = L_{nostate} \cup L_{\geq 2\ states} \cup L_{blank} \cup L_{no\$}$$
$$\cup\ L_{odd\$} \cup L_{badstart} \cup L_{badend} \cup L_{badcomp}$$

Apart from $L_{badcomp}$ we can specify each of them directly as a regular language.

$$L_{nostate} = (\Gamma \cup \{\$\} \cup Q)^*\{\$\}\Gamma^*\{\$\}(\Gamma \cup \{\$\} \cup Q)^*$$

$$L_{\geq 2\ states} = (\Gamma \cup \{\$\} \cup Q)^*\{\$\}(\Gamma \cup Q)^*$$
$$\Gamma^* Q(\Gamma \cup Q)^*\{\$\}(\Gamma \cup \{\$\} \cup Q)^*$$

$$L_{blank} = (\Gamma \cup \{\$\} \cup Q)^*\{\$\}\Gamma^* Q\Gamma^*\{\mathbf{B}\,\$\}(\Gamma \cup \{\$\} \cup Q)^*$$

$$L_{no\$} = (\Gamma \cup Q)^*$$

$$L_{odd\$} = (\Gamma \cup Q)^*\{\$\}((\Gamma \cup Q)^*\{\$\}(\Gamma \cup Q)^*\{\$\})^*(\Gamma \cup Q)^*$$

$$L_{badstart} = [\Gamma^*\{\$\} \cup (\Gamma \cup Q)^*Q(\Gamma \cup Q)^*Q(\Gamma \cup Q)^*\{\$\} \cup (\Gamma \cup (Q-\{s\})) \cup (\Gamma \cup Q)^*\{\boldsymbol{B}\}\{\$\}](\Gamma \cup \{\$\} \cup Q)^*$$

$$L_{badend} = (\Gamma \cup \{\$\} \cup Q)^*\{\$\}[\Gamma^* \cup (\Gamma \cup Q)^*Q(\Gamma \cup Q)^*Q(\Gamma \cup Q)^* \cup (\Gamma \cup Q)^*(Q-\{f\})(\Gamma \cup Q)^* \cup (\Gamma \cup Q)^*\{\boldsymbol{B}\}]$$

$$L_{badcomp} = ((\Gamma \cup Q)^*\{\$\}(\Gamma \cup Q)^*\{\$\})^*N_1\{\$\} ((\Gamma \cup Q)^*\{\$\}(\Gamma \cup Q)^*\{\$\}(\Gamma \cup Q)^*\{\$\})^* \cup (\Gamma \cup Q)^*(\{\$\}(\Gamma \cup Q)^*\{\$\}(\Gamma \cup Q)^*)^* \{\$\}N_2(\{\$\}(\Gamma \cup Q)^*\{\$\}(\Gamma \cup Q)^*)^*\{\$\}(\Gamma \cup Q)^*$$

To complete the decomposition of $\overline{L}_{ACS}^M$ we need to specify N_1 and N_2, where

$$N_1 = \{x\$y : x,y \text{ are in } \Gamma^*Q\Gamma^* \text{ and } x \not\vdash y^R \text{ in } M\}$$

and

$$N_2 = \{x\$y : x,y \text{ are in } \Gamma^*Q\Gamma^* \text{ and } x^R \not\vdash y \text{ in } M\}.$$

We construct a linear grammar G_1, for N_1, which closely follows the linear grammar for K_1 in Lemma 10.3.1. A linear grammar G_2, for N_2, can be constructed in a similar fashion. From these it is straightforward to construct a linear grammar $\overline{H}$ such that $L(\overline{H}) = \overline{L}_{ACS}^M$. To simplify the presentation of G_1 we decompose N_1 further into L_{left}, L_{right}, $L_{leftlength}$, $L_{rightlength}$, and $L_{transition}$. If $u \vdash v$ in M, then we can decompose u into u_1cpdu_2 and v into u_1wu_2, for some u_1, u_2 in Γ^*, some c, d in Γ, some p in Q, and some w. Essentially, if $\delta(p,d) = (r,e,\lambda)$, then $w = cre$; if $\delta(p,d) = (r,e,L)$, then $w = rce$; and if $\delta(p,d) = (r,e,R)$, then $w = cer$. If $u_2 = \lambda$, then we have to take account of the cases when $d = \boldsymbol{B}$, $e = \boldsymbol{B}$, and $c = \boldsymbol{B}$. The important issue is that $|cpd| = |w|$ if $u_2 \neq \lambda$, the numbers of symbols to their left are equal, and also to their right. We refer to cpd and w as the *active parts* of the configurations u and v, respectively. $L_{leftlength}$ contains all words $x\$y$ in which the numbers of symbols to the left of the active parts in x and y^R differ by at least one. $L_{rightlength}$ is the corresponding set for differing lengths to the right of the active parts. L_{left} contains all words $x\$y$ in which there are different symbols at the same positions in x and y^R to the left of the read-write head. L_{right} is contains those words which are different to the right of the read-write head. Finally, $L_{transition}$ contains all words $x\$y$ in which either the implied transition is not in M or its context is incorrect.

We give linear grammars for $L_{leftlength}$, L_{left}, and $L_{transition}$. The linear grammars for $L_{rightlength}$ and L_{right} are constructed in a similar manner. We form G_1 by taking the "union" of these five linear grammars (see the proof that $\boldsymbol{L}_{CF}$ is closed under union in Section 9.1). In each case we give the grammar in an abbreviated form.

The linear grammar for $L_{leftlength}$ has the following ten production types.

$1 : S_1 \rightarrow cS_1d$	$6 : B_1 \rightarrow B_1c$
$2 : S_1 \rightarrow cA_1$	$7 : B_1 \rightarrow cpC_1qe \mid cpC_1q \mid cpC_1qef$
$3 : S_1 \rightarrow B_1c$	$8 : C_1 \rightarrow cC_1$
$4 : A_1 \rightarrow cA_1$	$9 : C_1 \rightarrow C_1c$
$5 : A_1 \rightarrow cpC_1qe \mid cpC_1q \mid cpC_1qef$	$10 : C_1 \rightarrow \$$

We establish the convention that appearances of c and d correspond to any symbols in $\{a,b,\boldsymbol{B}\}$. Thus, the type 4 production denotes the three productions $A_1 \rightarrow aA_1$, $A_1 \rightarrow bA_1$, and $A_1 \rightarrow \boldsymbol{B}A_1$. Similarly, appearances of p and q correspond to any states in Q. So the type 5 production denotes the $\#Q^2$ productions $\{A_1 \rightarrow pC_1q : p,q \text{ in } Q\}$.

Type 1 productions produce equal length prefixes of x and y^R, where types 2 and 3 ensure they differ in length by at least one. Type 4 increases the length of x and type 6 the length of y. The remaining types deposit the states and further symbols terminating with production 10.

The linear grammar for L_{left} is quite similar.

$1 : S_2 \rightarrow cS_2c$	$5 : A_2 \rightarrow pB_2q$
$2 : S_2 \rightarrow cA_2\bar{c}$	$6 : B_2 \rightarrow cB_2$
$3 : A_2 \rightarrow cA_2c$	$7 : B_2 \rightarrow B_2c$
$4 : A_2 \rightarrow A_2c$	$8 : B_2 \rightarrow \$$

However, in this case type 1 productions produce equal prefixes of x and y. In the type 2 production the notation $\bar{c}$ means a symbol from $\{a,b,\boldsymbol{B}\} - \{c\}$. So we have $S_2 \rightarrow aA_2b$, $S_2 \rightarrow aA_2\boldsymbol{B}$, but not $S_2 \rightarrow aA_2a$. These productions introduce symbols that differ to the left of the read-write head.

The linear grammar for $L_{transition}$ requires more productions than the grammars for L_{left} and $L_{leftlength}$, but conceptually it is no more difficult. First of all we have production types

$1 : S_3 \rightarrow cS_3c$

$2 : A_3 \rightarrow cA_3c$

$3 : A_3 \rightarrow \$$

Type 1 produces equal prefixes and types 2 and 3 equal suffixes. We now add additional S_3-productions to produce $L_{transition}$.

If $x\$y$ is in $L_{transition}$, then either the implied transition is not a transition in M or the implied transition is a transition in M, but the active parts do not agree. First, assume that $x = x_1cpdx_2$, $y^R = x_1wx_2$, and $w = eqf$. Then the implied transition is $\delta(p,d) = (q,f,\lambda)$. For $x \vdash y^R$ in M we require that this is a transition of M and that $c = e$. Therefore, we add productions

$4 : S_3 \rightarrow cpdA_3fqe,$	$\delta(p,d) \neq (q,f,\lambda)$ in M
$5 : S_3 \rightarrow cp\overline{\boldsymbol{B}}\$qe,$	$\delta(p,d) \neq (q,\boldsymbol{B},\lambda)$, $d \neq \boldsymbol{B}$

$6 : S_3 \rightarrow cp\$\overline{\boldsymbol{B}}qe,$	$\delta(p,\boldsymbol{B}) \neq (q,f,\lambda), f \neq \boldsymbol{B}$
$7 : S_3 \rightarrow cp\$qe,$	$\delta(p,\boldsymbol{B}) \neq (q,\boldsymbol{B},\lambda)$
$8 : S_3 \rightarrow cpdA_3fq\bar{c},$	$\delta(p,d) = (q,f,\lambda)$ in M
$9 : S_3 \rightarrow cpd\$q\bar{c},$	$\delta(p,d) = (q,\boldsymbol{B},\lambda)$ in M, $d \neq \boldsymbol{B}$
$10 : S_3 \rightarrow cp\$fq\bar{c},$	$\delta(p,d) = (q,f,\lambda)$ in M, $f \neq \boldsymbol{B}$
$11 : S_3 \rightarrow cp\$q\bar{c},$	$\delta(p,\boldsymbol{B}) = (q,\boldsymbol{B},\lambda)$ in M

Type 4 introduces an incorrect implied transition, while types 5-7 detail the cases in which $x_2 = \lambda$. Type 8 introduces a correct implied transition but with an incorrect context. Again, types 9-11 detail the case $x_2 = \lambda$.

Second, assume $w = efq$, that is, the implied transition is $\delta(p,d) = (q,f,R)$. In a similar manner we obtain productions

$12 : S_3 \rightarrow cpdA_3qfe,$	$\delta(p,d) \neq (q,f,R)$ in M
$13 : S_3 \rightarrow cp\$qfe,$	$\delta(p,\boldsymbol{B}) \neq (q,f,R)$ in M
$14 : S_3 \rightarrow cpdA_3qf\bar{c},$	$\delta(p,d) = (q,f,R)$ in M
$15 : S_3 \rightarrow cp\$qf\bar{c},$	$\delta(p,\boldsymbol{B}) = (q,f,R)$ in M

Third, and, finally, assume $W = qef$ and the implied transition is $\delta(p,d) = (q,f,L)$. We leave the details of this case to the reader.

We leave to the Exercises the detailed proof that $N_1 = L(G_1)$ and $\overline{L}^M_{ACS}$ is indeed equal to the union of the given languages. ■

Theorem 10.3.4 *CFG CONTAINMENT and CFG EQUIVALENCE are undecidable.*

Proof: That *CFG* CONTAINMENT is undecidable follows from the undecidability of *CFG* EQUIVALENCE.

Assume *CFG* EQUIVALENCE is decidable. Then we demonstrate that *DTM* EMPTINESS is decidable. Let $M = (Q,\{a,b\},\{a,b,\boldsymbol{B}\},\delta,s,f)$ be an instance of *DTM* EMPTINESS. From M construct $\overline{H}$ of Lemma 10.3.3. Now, $L(\overline{H}) = \overline{L}^M_{ACS}$ and $\overline{L}^M_{ACS} = (\{a,b,\boldsymbol{B},\$\} \cup Q)^*$ if and only if $L(M) = \emptyset$. But using the decision algorithm for *CFG* EQUIVALENCE we can decide whether or not $L(\overline{H}) = L(G)$, where G is a regular grammar for $(\{a,b,\boldsymbol{B},\$\} \cup Q)^*$. That is, we can decide whether or not $L(M) = \emptyset$. This is a contradiction; therefore, *CFG* EQUIVALENCE is undecidable.

If *CFG* CONTAINMENT were decidable we would be able to decide *CFG* EQUIVALENCE by two successive calls of its decision algorithm. Therefore, *CFG* CONTAINMENT is also undecidable. ■

The above proof implies that we have the following promised theorem.

Theorem 10.3.5 *CFG UNIVERSALITY is undecidable.*

Proof: $\overline{L}^{M}_{ACS} = (\{a,b,\boldsymbol{B},\$\} \cup Q)^*$ if and only if $L(M) = \varnothing$. ■

10.4 CONTEXT-FREE AMBIGUITY

In this brief section we pose and prove undecidable the following problem

CFG AMBIGUITY
INSTANCE: A *CFG*, $G = (N,\Sigma,P,S)$
QUESTION: Is G ambiguous?

Theorem 10.4.1 *CFG AMBIGUITY is undecidable.*

Proof: Assume *CFG* AMBIGUITY is decidable. Then for an instance $M = (Q,\{a,b\},\{a,b,\boldsymbol{B}\},\delta,s,f)$ of *DTM* EMPTINESS construct H_1 and H_2 such that $L(H_1) = L^1_M$ and $L(H_2) = L^2_M$. Let Z_1 and Z_2 be the sentence symbols of H_1 and H_2, respectively, and assume their nonterminal sets are disjoint. Letting S be a new nonterminal form the "union" H of H_1 and H_2, that is, H contains all the productions of H_1 and H_2 together with $S \rightarrow Z_1 \mid Z_2$, where S is the sentence symbol of H. Then $L(H) = L(H_1) \cup L(H_2)$. Now, H is ambiguous if and only if $L(H_1) \cap L(H_2) \neq \varnothing$ if and only if $L(M) \neq \varnothing$.

Therefore, using the assumed decision algorithm for *CFG* AMBIGUITY we can decide whether or not $L(M) \neq \varnothing$. In other words, this implies *DTM* EMPTINESS is decidable. This is a contradiction; therefore, *CFG* AMBIGUITY is undecidable. ■

10.5 ADDITIONAL REMARKS

10.5.1 Summary

We have considered a number of basic decision problems for *DFAs*, *CFGs*, and *DTMs*. These include HALTING, MEMBERSHIP, EMPTINESS, FINITENESS, CONTAINMENT, and EQUIVALENCE. These problems are all decidable for *DFAs* and all undecidable for *DTMs*. For *CFGs* the first four problems are decidable, while the last two are undecidable.

10.5.2 History

Hoare and Allison (1972) provide a very readable introduction to undecidability and Davis (1965) reprints the basic papers in this area. HALTING was shown to be undecidable by Turing (1936). It was the first decision problem to be proved undecidable. Moore (1956) proved the decidability of *DFA* EMPTINESS, *DFA* FINITENESS, *DFA* CONTAINMENT, and *DFA* EQUIVALENCE. Rabin and Scott (1959) investigated a number of decision problems for finite automata. *CFG* FINITENESS was proved decidable in Bar-Hillel, Perles, and Shamir (1961). The undecidability of *CFG* CONTAINMENT, *CFG* EQUIVALENCE, and *CFG* INTERSECTION was demonstrated in Bar-Hillel et al. (1961) and Ginsburg and

Rose (1963a). The undecidability of *CFG* AMBIGUITY was proved by Cantor (1962), Floyd (1962a), and Chomsky and Schutzenberger (1963). The approach we have taken to prove undecidability results for *CFGs* is due to Floyd (1964) and Hartmanis (1967).

Other famous undecidable problems are the domino problem (see Berger, 1966); the tiling problem (see Robinson, 1971); and Hilbert's tenth problem (see Davis, 1973).

10.6 SPRINGBOARD

10.6.1 Post's Correspondence Problem

Post (1946) introduced a problem that is usually taken as the basis for proving undecidability properties. It can be stated as follows.

PCP
INSTANCE: Two alphabets Σ and Δ and two λ-free morphisms $g,h : \Sigma^* \rightarrow \Delta^*$.
QUESTION: Is there a word x in Σ^+ such that $g(x) = h(x)$?

Note that we require x to be nonempty and g and h to be λ-free to avoid trivial solutions.

Example 10.6.1 Let $\Sigma = \{a,b,c\}$, $\Delta = \{a,b\}$, $h(a) = aab$, $h(b) = abb$, $h(c) = a$, $g(a) = a$, $g(b) = abab$, and $g(c) = ba$. Then $h(abc) = g(abc) = aababba$. So this instance of the *PCP* has a positive answer.

It can be proved that *PCP* is undecidable by reducing HALTING to it; see the Exercises. Once the *PCP* has been proved to be undecidable we can use it to prove the undecidability results for *CFGs*; see the Exercises once more.

In recent years the *PCP* has been the subject of intensive study. One reason for this is that we can parameterize *PCP* in terms of the size of Σ as follows.

PCP(n)
INSTANCE: Two alphabets Σ and Δ, $\#\Sigma = n$, and two λ-free morphisms $g,h : \rightarrow \Delta^*$.
QUESTION: Is there a word x in Σ^+ such that $g(x) = h(x)$?

PCP(1) is decidable; see the Exercises. It is not so obvious that *PCP*(2) is decidable; see Ehrenfeucht et al. (1982) and Pavlenko (1981). In contrast to these results it has been proved that *PCP*(9) is undecidable; see Pansiot (1981). This leaves the status of *PCP*(i), $3 \leq i \leq 8$, completely open. Is *PCP*(3) undecidable? *PCP*(4)? or . . . ?

EXERCISES

1.1 Prove that the following problems are decidable.

(i) *DFA* SUFFIX MEMBERSHIP
INSTANCE: A *DFA*, $M = (Q,\Sigma,\delta,s,F)$ and a word x in Σ^*.
QUESTION: Is there a word ux in $L(M)$, for some u in Σ^*?

(ii) *DFA* SUBWORD MEMBERSHIP
INSTANCE: A *DFA*, $M = (Q,\Sigma,\delta,s,F)$ and a word x in Σ^*.
QUESTION: Is there a word uxv in $L(M)$, for some u and v in Σ^*?

(iii) *DFA* ENCODING MEMBERSHIP
INSTANCE: A *DFA*, $M = (Q,\Sigma,\delta,s,F)$.
QUESTION: Is $\ll M \gg$ in $L(M)$, where $\ll M \gg$ is defined in a similar manner to the $\ll M \gg$ for *DTMs*?

(iv) *EFA* MEMBERSHIP
INSTANCE: An *EFA*, $M = (Q,\Sigma,\delta,s,F)$ and a word x in Σ^*.
QUESTION: Is x in $L(M)$?

1.2 Can you prove directly that *CFG* MEMBERSHIP is decidable even when the given *CFG* is not unit-free and λ-free? In other words, you should not transform the *CFG* into an equivalent unit-free, λ-free *CFG*.

1.3 Are the following problems decidable? Justify your answers.

(i) *CFG* PREFIX MEMBERSHIP.
(ii) *CFG* SUFFIX MEMBERSHIP.
(iii) *CFG* SUBWORD MEMBERSHIP.
(iv) *ECFG* MEMBERSHIP.
(v) *NPDA* MEMBERSHIP.
(vi) *CFG* PARIKH MEMBERSHIP
INSTANCE: A *CFG*, $G = (N,\Sigma,P,S)$, an ordering of $\Sigma = \{a_1, \ldots, a_m\}$, and a Parikh vector $(n_1, \ldots, n_m)$.
QUESTION: Is there a word x in $L(G)$ with $\Pi(x) = (n_1, \ldots, n_m)$?

1.4 Is following problem decidable or undecidable? Justify your answer.

BOUNDED HALTING
INSTANCE: A *DTM*, $M = (Q,\Sigma,\Gamma,\delta,s,f)$, a bound $B \geq 1$, and a word x in Σ^*.
QUESTION: Does M halt after at most B steps when given x as input?

1.5 Is the following problem decidable or undecidable? Justify your answer.

HALTING-HANGING
INSTANCE: A *DTM*, $M = (Q,\Sigma,\Gamma,\delta,s,f)$ and a word x in Σ^*.
QUESTION: Does M halt or hang when given input x?

2.1 Are the following problems decidable or undecidable? Justify your answers.

(i) *DFA* COFINITENESS
INSTANCE: A *DFA*, $M = (Q,\Sigma,\delta,s,F)$.
QUESTION: Is $\Sigma^* - L(M)$ finite?

(ii) *DFA* ALL-STATE EMPTINESS
INSTANCE: A *DFA*, $M = (Q,\Sigma,\delta,s,F)$.
QUESTION: Are there no words x in Σ^+ such that $qx \vdash^+ f$, for some q in Q and f in F.

(iii) *EFA* EMPTINESS.
(iv) *EFA* FINITENESS.

2.2 Are the following problems decidable or undecidable? Justify your answers.

(i) *CFG* COFINITENESS.
(ii) *ECFG* EMPTINESS.
(iii) *ECFG* FINITENESS.

2.3 Complete the proof of Theorem 10.2.6.

2.4 Are the following problems decidable or undecidable? Justify your answers.

(i) *DTM* STATE REACHABILITY
INSTANCE: A *DTM*, $M = (Q,\Sigma,\Gamma,\delta,s,f)$, a state q in Q, and a word x in Σ^*.
QUESTION: Does M ever enter state q when started on input x?

(ii) *DTM* MULTIPLE CELL SCAN
INSTANCE: A *DTM*, $M = (Q,\Sigma,\Gamma,\delta,s,f)$ and an integer $n \geq 1$.
QUESTION: When M is started on a blank tape does it scan any cell more than n times?

(iii) *DTM* WRITE SYMBOL
INSTANCE: A *DTM*, $M = (Q,\Sigma,\Gamma,\delta,s,f)$ and a symbol a in Γ.
QUESTION: Does M ever write a when started with a blank tape?

(iv) *DTM* WRITING
INSTANCE: A *DTM*, $M = (Q,\Sigma,\Gamma,\delta,s,f)$ and a word x in Σ^+.
QUESTION: Does there exist some input word w such that when M is started on w it writes the symbols of x into consecutive cells during the computation.

(v) *DTM* NONBLANK
INSTANCE: A *DTM*, $M = (Q,\Sigma,\Gamma,\delta,s,f)$.
QUESTION: Does M ever write a nonblank symbol when started with a blank tape.

(vi) *DTM* NO MOVE

INSTANCE: A *DTM*, $M = (Q,\Sigma,\Gamma,\delta,s,f)$ and a word x in Σ^*.

QUESTION: Does M ever execute a no-move transition when started with input x?

(vii) *DTM* TAPE-BOUNDED

INSTANCE: A *DTM*, $M = (Q,\Sigma,\Gamma,\delta,s,f)$, a bound $B \geq 1$, and a word x in Σ^*.

QUESTION: Does M ever visit the $(B+1)$st cell when started on input x?

(viii) *DTM* CELL VISIT

INSTANCE: A *DTM*, $M = (Q,\Sigma,\Gamma,\delta,s,f)$ and a bound $B \geq 1$.

QUESTION: Is there an input word x in Σ^*, such that when M is started on x, M visits the Bth cell during the computation?

3.1 Prove that *DTM* CONTAINMENT, *DTM* EQUIVALENCE, and *DTM* INTERSECTION are undecidable.

3.2 Which of the following problems are decidable or undecidable? Justify your answers.

(i) *CFG* REGULAR CONTAINMENT

INSTANCE: A *CFG*, $G = (N,\Sigma,P,S)$ and a *DFA* $M = (Q,\Delta,\delta,s,F)$.

QUESTION: Is $L(G) \subseteq L(M)$?

(ii) *DFA* CONTEXT-FREE CONTAINMENT

INSTANCE: A *DFA*, $M = (Q,\Sigma,\delta,s,F)$ and a *CFG* $G = (N,\Delta,P,S)$.

QUESTION: Is $L(M) \subseteq L(G)$?

(iii) FINITE *DTM* CONTAINMENT

INSTANCE: A finite set F and a *DTM*, $M = (Q,\Sigma,\Gamma,\delta,s,f)$.

QUESTION: Is $F \subseteq L(M)$?

(iv) *CFG* SENTENTIAL FORM EQUIVALENCE

INSTANCE: Two *CFGs*, $G_1 = (N_1,\Sigma_1,P_1,S_1)$ and $G_2 = (N_2,\Sigma_2,P_2,S_2)$.

QUESTION: Letting $SF(G_i) = \{\alpha : S_i \Rightarrow^* \alpha \text{ in } G_i\}$, is $SF(G_1) = SF(G_2)$?

5.1 Use *PCP* to prove the following problems for *CFGs* are undecidable.

(i) $L(G_1) \cap L(G_2) = \emptyset$.

(ii) G is ambiguous.

(iii) $L(G_1) \subseteq L(G_2)$.
(iv) $L(G)$ is regular.
(v) $L(G) = R$, a given regular language.
(vi) $L(G)$ is cofinite.

5.2* The proof that *PCP* is undecidable is usually done in two steps via the MODIFIED *PCP*, which is defined as follows.

MODIFIED *PCP*
INSTANCE: Two alphabets Σ and Δ, two λ-free morphisms g and $h : \Sigma^* \to \Delta^*$, and a symbol a in Σ.
QUESTION: Is there a word ax in Σ^+ such that $g(ax) = h(ax)$?

(i) Prove that *PCP* is decidable if and only if MODIFIED *PCP* is decidable.
(ii) Prove that MODIFIED *PCP* is undecidable by way of HALTING.

5.3 Prove that $PCP(1)$ is decidable.

5.4 Consider the following variant of the *PCP*.

UNIFORM *PCP*
INSTANCE: Two alphabets Σ and Δ, an integer $n \geq 1$, and two λ-free morphisms $g,h : \Sigma^* \to \Delta^*$ so $|g(a)| = |h(a)| = n$, for all a in Σ.
QUESTION: Is there a word x in Σ^+ such that $g(x) = h(x)$?

Prove that UNIFORM *PCP* is decidable.

5.5 Define the following restricted version of *PCP*.

nPCP
INSTANCE: Two alphabets Σ and Δ and two λ-free morphisms $g,h : \Sigma^* \to \Delta^*$ such that $|g(a)|, |h(a)| \leq n$, for all a in Σ.
QUESTION: Is there a word x in Σ^+ such that $g(x) = h(x)$?

Prove that *nPCP* is undecidable, for all $n \geq 2$.

Part IV

ONWARD

Chapter 11

Further Topics

We have dealt with three problem areas in Chapters 8, 9, and 10: family relationships, closure properties, and decision problems. These are central problems but they by no means exhaust important areas of investigation. In this, our final chapter, we treat three other areas. In the first and lengthiest investigation, we re-examine decision problems from the viewpoint of time-bounded computations. Our interest does not lie with the decidability status of these problems, but rather with the time taken to decide them. We concentrate on the classes of deterministic and nondeterministic tractable problems, referred to as $\boldsymbol{P}$ and $\boldsymbol{NP}$, respectively. The major open problem in this area is whether $\boldsymbol{P} = \boldsymbol{NP}$. Because of this we focus our attention on those decisions problems in $\boldsymbol{NP}$ which are the hardest problems. These problems are said to be NP-complete and to show $\boldsymbol{P} \neq \boldsymbol{NP}$. It is sufficient to prove that one of them is not accepted by a DTM in polynomial time. We prove that TRAVELING SALESMAN CHECKING and RECTANGLE INTERSECTION are NP-complete. To arrive at these results we prove a number of other problems are NP-complete. The most fundamental of these is SATISFIABILITY.

Second, we provide additional evidence in favor of Church's thesis. For this purpose we introduce phrase structure grammars and interactive Lindenmayer systems. In both cases we show that they have the same power as $DTMs$.

Third, and, finally, we reconsider CFG MEMBERSHIP. We give a general and efficient algorithm for testing membership, the Cocke-Kasami-Younger algorithm. This is followed by two approaches for very efficient parsing, as it often called: a top-down and a bottom-up method. Both techniques depend on bounded look-ahead during the parsing process. The top-down, or LL, method works for subclasses of $\boldsymbol{L}_{DCF}$ and the bottom-up, or LR, method works for $\boldsymbol{L}_{DCF}$.

11.1 *NP*-COMPLETE PROBLEMS

In Section 6.6 we introduced polynomially time-bounded *DTMs* as a model for decision problems which can be solved in polynomial time. Indeed, polynomial time-bounded *DTMs* provide a model for functions which are computable in polynomial time. In other words, polynomial time-bounded *DTMs* can be used to characterize those problems that can be solved in deterministic polynomial time.

If a language is accepted by a polynomial-time-bounded *NTM*, then a candidate accepting configuration sequence for a given word can be checked for correctness in deterministic polynomial time. Thus, polynomial time-bounded *NTMs* characterize those problems whose solutions can be verified in deterministic polynomial time.

Let ***P*** denote those decision problems that can be solved in deterministic polynomial time, and ***NP*** those that can be solved in nondeterministic polynomial time. All deterministic simulations of polynomial-time-bounded *NTMs* that have been considered require exponential time. However, no one has managed to prove that there is an *NTM* all of whose equivalent *DTMs* are non-polynomially time-bounded. This, apparently technical, result would imply that many decision problems cannot be solved in deterministic polynomial time. Conversely, no one has managed to prove that each such *NTM* can be simulated by a *DTM* in polynomial time. This result would imply that many decision problems can be solved in deterministic polynomial time. The problem of whether or not ***P*** = ***NP*** is such a fundamental one that it is known as the ***P*** = ***NP*** *problem.*

The ***P*** = ***NP*** problem is a difficult one. In one attempt to simplify it an ordering has been defined between problems in ***NP***, according to their relative difficulty. The equivalence classes defined by this ordering consist of equally difficult problems. If ***P*** = ***NP***, then this ordering collapses. For this reason results are based the assumption that ***P*** ≠ ***NP*** — which most people believe. It turns out that there is one maximal equivalence class, the class of hardest problems in ***NP***. This is the class of *NP-complete problems.*

In this section we define an ordering between problems and demonstrate that there is an *NP*-complete problem. The remainder of the section is devoted to proving two specific problems are *NP*-complete. The first one is TRAVELING SALESMAN CHECKING (*TSC*) and the second is RECTANGLE INTERSECTION. We have already met *TSC* in Section 6.6. RECTANGLE INTERSECTION is a special membership problem for a class of grammars similar to those used in describing VLSI layouts, for example, Caltech Intermediate Form. Along the way we also prove that a number of other interesting problems are *NP*-complete.

11.1.1 The Ordering $\propto$

Given two languages $L_1 \subseteq \Sigma_1^*$ and $L_2 \subseteq \Sigma_2^*$ we say that L_1 can be *transformed* into L_2 if there is a function $t : \Sigma_1^* \rightarrow \Sigma_2^*$ such that

For all x in Σ_1^*, x is in L_1 if and only if $t(x)$ is in L_2

We call t a *transformation*. The definition implies that $t(L_1) \subseteq L_2$ and $t(\bar{L}_1) \subseteq \bar{L}_2$; see Figure 11.1.1. There may be words in Σ_2^* which are not images of any word in Σ_1^*. Given such a transformation t we can "decide" if a word x in Σ_1^* is in L_1 by first computing $t(x)$ and testing to see if $t(x)$ is in L_2. This general approach should be compared with the one for context-free languages and inverse morphisms in Section 9.5, where t is a morphism.

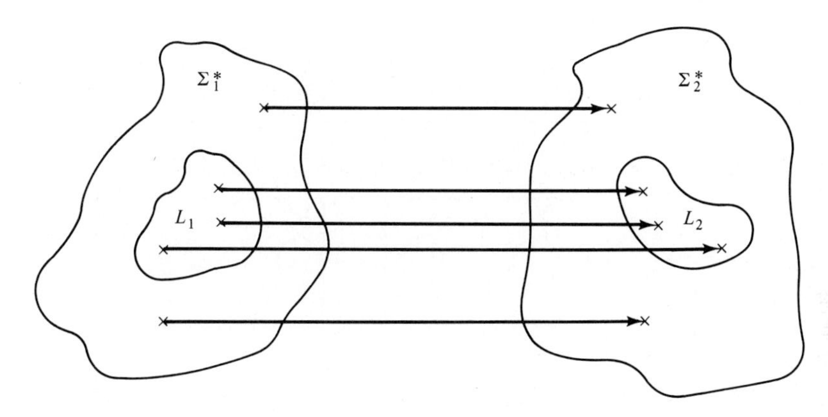

Figure 11.1.1

Our interest lies in a more restricted transformation since we wish to compare languages in $\boldsymbol{L}_{NPTIME}$ (and problems in ***NP***). Since we wish to transform the membership problem for one language into a membership problem for a second language, the transformation should be computable in deterministic polynomial time.

We say that a transformation $t : \Sigma_1^* \to \Sigma_2^*$ is a *polynomial transformation* if t is computable by some $T(n)$-time-bounded DTM, where T is some polynomial.

If L_1 can be *polynomially transformed* into L_2 we write $L_1 \propto L_2$. This terminology means that L_1 is no more difficult to compute than L_2. It may, of course, be a lot easier! The following lemma formalizes this intuition.

Lemma 11.1.1 *Let L_1 and L_2 be two languages such that $L_1 \propto L_2$.*

(i) *If L_2 is in $\boldsymbol{L}_{DPTIME}$, then L_1 is in $\boldsymbol{L}_{DPTIME}$. Equivalently, if L_1 is not in $\boldsymbol{L}_{DPTIME}$, then L_2 is not in $\boldsymbol{L}_{DPTIME}$.*
(ii) *If L_2 is in $\boldsymbol{L}_{NPTIME}$, then $\boldsymbol{L}_1$ is in $\boldsymbol{L}_{NPTIME}$.*

Proof:

(i) If L_2 is in $\boldsymbol{L}_{DPTIME}$, then it is accepted by a $T_2(n)$-time-bounded DTM, M_2, where $T_2(n)$ is some polynomial. Since $L_1 \propto L_2$ there is a polynomial transformation t that transforms L_1 into L_2. This implies that there is a $T(n)$-time-bounded DTM, M, which computes t, for some polynomial $T(n)$.

Construct a polynomially time-bounded DTM, M_1, that accepts L_1, as follows.

M_1 first calls M to compute $t(x)$ from its input word x. On completion of this computation M_1 calls M_2 with input $t(x)$; see Figure 11.1.2. M_1 accepts x if M_2 accepts $t(x)$ and M_1 rejects x otherwise. Therefore, $L(M_1) = L_1$. Moreover, it requires at most $T(|x|)$ steps for the call of M and at most $T_2(T(|x|))$ steps for the second step. The latter bound follows by observing that $|t(x)| \leq T(|x|)$. This implies M_1 requires at most $T(|x|) + T_2(T(|x|))$ steps. Since T and T_2 are polynomials this composition and sum is also polynomial; therefore, L_1 is in $\boldsymbol{L}_{DPTIME}$.

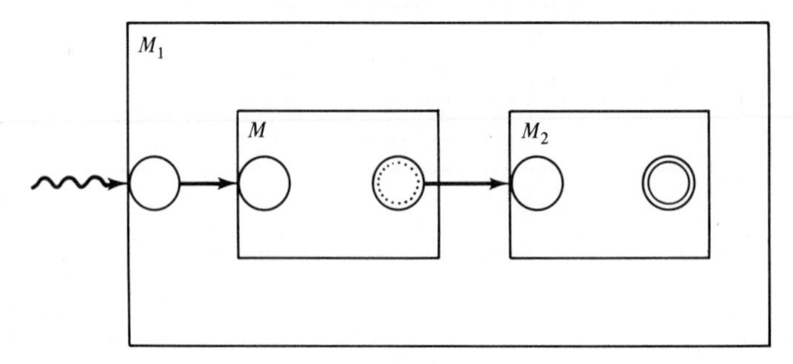

Figure 11.1.2

(ii) This proof follows by a similar argument. ■

The relation $\propto$ is transitive simply because the composition and sum of two polynomials is again a polynomial. The details of the following lemma are left to the Exercises.

Lemma 11.1.2 *$\propto$ is transitive, therefore $\propto$ is a pre-order.*

Definition Whenever we have two languages L_1 and L_2 such that $L_1 \propto L_2$ and $L_2 \propto L_1$, we say that L_1 and L_2 are *(polynomially) equivalent.* This implies that neither L_1 nor L_2 is the more difficult language with respect to membership.

We are particularly interested in the existence of languages which are the most difficult or hardest in $\boldsymbol{L}_{NPTIME}$. We say that such a language is *NP-hard.*

Definition A language L is *NP-hard* if for every language L' in $\boldsymbol{L}_{NPTIME}$, $L' \propto L$. If, further, L is itself in $\boldsymbol{L}_{NPTIME}$ we say that L is *NP-complete.* It is a hardest language within $\boldsymbol{L}_{NPTIME}$.

If we have two NP-complete languages L_1 and L_2, then L_1 and L_2 are equivalent. Therefore, if NP-complete languages exist, they form a single equivalence class, $\boldsymbol{L}_{NP-complete}$. Pictorially this refinement of $\boldsymbol{L}_{NPTIME}$ is shown in

Figure 11.1.3. It is sufficient to prove that one language in $\boldsymbol{L}_{NP-complete}$ is either in or not in $\boldsymbol{L}_{DPTIME}$ to prove either equality or inequality of $\boldsymbol{L}_{NPTIME}$ and $\boldsymbol{L}_{DPTIME}$. Recall that we have built the picture of Figure 11.1.3 on "sand," since we have assumed that $\boldsymbol{L}_{NPTIME} \neq \boldsymbol{L}_{DPTIME}$. Moreover, we have yet to prove that *NP*-complete languages exist! We will prove this by reverting to decision problems and encodings of yes instances in the remainder of this section.

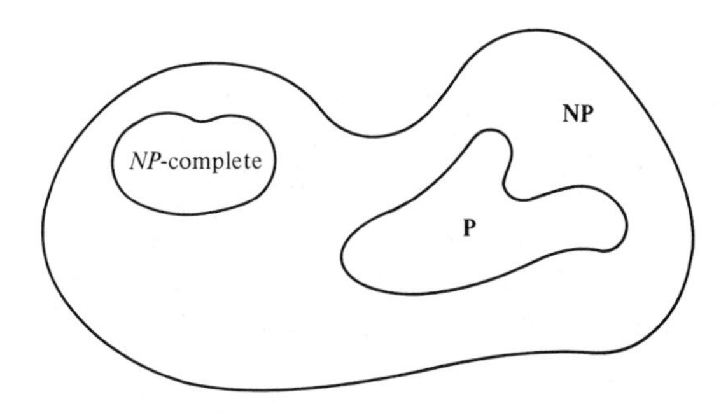

Figure 11.1.3

In Chapters 1 and 6 we introduced the notion of an encoding e of a decision problem Π and the language $L(\Pi, e)$ consisting of encodings of yes instances of Π. For example, we considered encodings of *DTMs* in Section 6.3. We will, henceforth, not specify a particular encoding for a decision problem; we will assume a reasonable encoding exists. For this reason we will write about polynomial transformations between problems rather than between the encodings of their yes instances *under some encoding.*

Given two problems Π_1 and Π_2 we write $\Pi_1 \propto \Pi_2$ if for all "reasonable" encodings e_1 of Π_1 and e_2 of Π_2, $L(\Pi_1, e_1) \propto L(\Pi_2, e_2)$.

11.1.2 An *NP*-Complete Problem

We demonstrate that *NP*-complete problems exist. For simplicity we restrict ourselves to languages over the alphabet $\{a, b\}$.

The problem we consider is a variation of *NTM* HALTING.

NTM TIME-BOUNDED HALTING (TBH)
INSTANCE: An *NTM*, $M = (Q, \{a,b\}, \{a,b,\boldsymbol{B}\}, \delta, s, f)$, a word x in $\{a,b\}^*$, and a time bound $B \geq 1$. B is represented in unary.
QUESTION: Is x accepted by M in at most B steps.

Theorem 11.1.3 *TBH is NP-complete for* $\{L : L \subseteq \{a,b\}^*$ *and* L *is in* $\boldsymbol{L}_{NPTIME}\}$.

Proof: We prove this in two steps. First, we prove that *TBH* is *NP*-hard and, second, we prove it is in ***NP***.

Let

$$L(TBH,e) = \{<<M>>a^B bx : M,x, \text{ and } B \text{ determine an instance of } TBH\}$$

where e is based on the encoding of *DTMs* introduced in Section 6.3.

We must prove that $L' \propto L(TBH,e)$, for all L' in $\boldsymbol{L}_{NPTIME}$. We consider an arbitrary language $L' \subseteq \{a,b\}^*$ in $\boldsymbol{L}_{NPTIME}$. Therefore, there is an *NTM*, $M' = (Q,\{a,b\},\{a,b,\boldsymbol{B}\},\delta,s',f')$ and a polynomial $T(n)$ such that $L(M') = L'$ and each x in L' is accepted by M' in at most $T(|x|)$ steps. Define a function $f_{L'} : \{a,b\}^* \rightarrow \{a,b\}^*$ by

$$f_{L'}(x) = <<M'>>a^{T(|x|)}bx$$

If x is in L', then x is accepted by M' in at most $T(|x|)$ steps. This implies $<<M'>>a^{T(|x|)}bx$ is in $L(TBH,e)$. If x is not in L', then x is not accepted by M' in at most $T(|x|)$ steps. This implies $<<M'>>a^{T(|x|)}bx$ is not in $L(TBH,e)$.

We have shown that $f_{L'}$ is a transformation; we need to show it is polynomial.

We can construct a *DTM*, $M_{L'}$ such that on input x it produces output $<<M'>>a^{T(|x|)}bx$. First, $M_{L'}$ computes $a^{T(|x|)}$ in at most $p(|x|)$ steps, for some polynomial p. It then assembles on its tape the word $<<M'>>a^{T(|x|)}bx$. Clearly, this takes at most $q(|x|)$ steps, for some polynomial q; hence, $M_{L'}$ requires a total of at most $p(|x|)+q(|x|)$ steps — again a polynomial bound. Hence, $f_{L'}$ is a polynomial transformation and $L' \propto L(TBH,e)$.

We claim $L(TBH,e)$ is in $\boldsymbol{L}_{NPTIME}$. Consider an *NTM*, M, which given an input word w in $\{a,b\}^*$ first checks whether or not it equals $<<M'>>a^B bx$, for some *NTM*, M'. This requires at most $p(|w|)$ steps, for some polynomial p.

Second, M chooses a sequence of transitions from M' of length at most B. If each choice is made during a scan of $<<M'>>$, there are at most B scans. M' requires at most $q(|x|)$ steps, for some polynomial q, for this phase. Thirdly, M applies these transitions (if possible) to x. If this results in x being accepted, then M accepts w. This takes, at most, a number of steps linear in $|w|$.

Overall, M' requires a number of steps bounded by some polynomial $T(|w|)$ to determine if w is in $L(TBH,e)$. Hence, $L(TBH,e)$ is in $\boldsymbol{L}_{NPTIME}$, $L(TBH,e)$ is *NP*-complete, and *TBH* is *NP*-complete. ∎

It is crucial that B is represented in unary in the above theorem. *NP*-hardness of *TBH* can still be proved even when B is represented in binary. However, we can no longer prove that $L(TBH,e)$ is in $\boldsymbol{L}_{NPTIME}$; see the Exercises. Having demonstrated that there is an *NP*-complete problem we turn to SATISFIABILITY, the foundational *NP*-complete problem.

11.1.3 Satisfiability

Given a Boolean expression such as

$$E(a,b,c) = (a \textbf{ or } b) \textbf{ and } (b \textbf{ or not } a \textbf{ or } c)$$

we may ask: Are there values for the Boolean variables, a, b, and c such that $E(a,b,c)$ has the value **true**? In this case the answer is: yes, there are. Choose b to be **true** and a and c to have arbitrary truth values. Immediately, $E(a,\textbf{true},c) = (a \textbf{ or true}) \textbf{ and } (\textbf{true or not } a \textbf{ or } c) = \textbf{true}$. We say $E(a,b,c)$ is *satisfiable*. Alternatively, $E(\textbf{true},b,\textbf{true}) = \textbf{true}$ also. Solving this question for $E(a,b,c)$ is particularly simple. We can do it by inspection. However, if E is a function of a hundred variables, rather than three, this is just not possible. In the general case we can arrive at a solution by trying all combinations of truth values. For n variables we have 2^n combinations; therefore, it is unclear whether an efficient, that is, a polynomial-time, solution exists. We will show that no efficient algorithm exists — if $\boldsymbol{NP} \neq \boldsymbol{P}$ — by demonstrating that it is *NP*-complete. But first we define SATISFIABILITY more carefully.

Let $U = \{u_1, \ldots, u_m\}$ be a set of Boolean variables. If u is a variable, then u and $\bar{u}$ are said to be *literals* over U, where $\bar{u}$ denotes $\textbf{not}\, u$. A *clause* over U is a set of literals over U, for example, $\{\bar{a},b,c\}$. A clause represents a disjunction of its literals, for example, $\{\bar{a},b,c\}$ represents $\textbf{not}\, a \textbf{ or } b \textbf{ or } c$. A collection C of clauses over U represents the conjunction of its clauses. For example, $C = \{\{a,b\},\{\bar{a},b,c\}\}$ represents $(a \textbf{ or } b) \textbf{ and } (\textbf{not}\, a \textbf{ or } b \textbf{ or } c)$. We say that a collection C of clauses over U is *satisfiable* if there are values of the variables in U for which C has the value true. Such an assignment of values to the variables in U is said to be a *satisfying truth assignment*.

SATISFIABILITY or, simply *SAT*, can be stated as

SATISFIABILITY (*SAT*)
INSTANCE: A set U of variables and a collection C of clauses over U.
QUESTION: Is there a satisfying truth assignment for C?

We can encode a literal u_i by ab^ia and $\bar{u}_i$ by aab^ia. A clause can be encoded by encoding its literals and surrounding them by aa, while a collection can be encoded by encoding its clauses and surrounding them with $aaaaa$. Let e be this encoding. $L(SAT,e)$ is in $\boldsymbol{L}_{NPTIME}$. $L(SAT,e)$ is in $\boldsymbol{L}_{NPTIME}$ for other reasonable encodings. For this reason we simply refer to $L(SAT)$; a reasonable encoding is understood. In general we write $L(\Pi)$ with the same meaning.

We construct an *NTM* which chooses a truth assignment for the variables and evaluates the collection for the chosen assignment. Both steps are linear in the size of the instance; hence, $L(SAT)$ is in $\boldsymbol{L}_{NPTIME}$.

We now prove that $L \propto L(SAT)$ for every language L in $\boldsymbol{L}_{NPTIME}$, thus establishing the *NP*-completeness of $L(SAT)$ and, hence, of *SAT* itself. We use a generic proof technique. That is, given an arbitrary language $L \subseteq \Sigma^*$ in $\boldsymbol{L}_{NPTIME}$, we construct, in a generic manner, a polynomial transformation $f_L : \Sigma^* \rightarrow \{a,b\}^*$ of L into $L(SAT)$. Rather than basing f_L solely on L, which has little structure, instead we base it on a $T(n)$-time-bounded *NTM*, $M = (Q,\Sigma,\Gamma,\delta,s,f)$, which accepts L, where $T(n)$ is a polynomial. More precisely, we associate with each x in L an accepting configuration sequence

$S : sx \vdash^k yfz$ in M, where $k \leq T(|x|)$. This is, in turn, associated with a yes instance of SAT.

How do we associate a configuration sequence with an instance of SAT? With a Boolean variable we can represent one local property of an NTM, for example, the read-write head is over cell 25, cell 3 contains c, or the state is q. A single configuration is represented as a collection of clauses formed from such variables. Such a collection must have a satisfying truth assignment exactly when it represents a valid configuration. A configuration sequence is then given by a union of such collections together with clauses that capture the computation steps.

The NTM, M, on input x, where $n = |x|$, cannot visit any cells beyond cell $T(n)+1$, since this would require more than $T(n)$ steps. For ease of exposition we assume every configuration sequence has exactly $T(n)$ steps. If a particular configuration sequence has fewer steps, then we pad it out by repeating the terminating configuration. We need to be able to index the elements of Q, Σ, and Γ. Let $Q = \{q_1, \ldots, q_m\}$, where we assume $q_1 = s$ and $q_2 = f$; $\Sigma = \{a_1, \ldots, a_p\}$, $\Gamma = \{a_1, \ldots, a_r\}$, where we assume $a_r = \boldsymbol{B}$, $x = a_{i_1} \cdots a_{i_n}$ and $1 \leq p < r$. Letting $t = T(n)$; a configuration sequence of M has $t+1$ steps, step $0, \ldots,$ step t.

We introduce *state* variables Q_{ij}. If Q_{ij} is **true**, then M is in state q_j at step i. Similarly, we have *head* variables H_{ik}. If H_{ik} is **true**, then the read-write head of M is over cell k at step i. Finally, we have *symbol* variables S_{ikl}. If S_{ikl} is **true**, then cell k contains symbol a_l at step i.

At each step we must ensure that M is in exactly one state, the read-write head is over exactly one square, and there is exactly one symbol in each cell of the tape. We build a collection of clauses that enforce these requirements.

(1) $\{Q_{ij} : 1 \leq j \leq m\}$ — M is in at least one state at step i, $0 \leq i \leq t$.

(2) $\{\overline{Q}_{ij}, \overline{Q}_{ij'}\}$, $1 \leq j < j' \leq m$ — M is in at most one state at step i, $0 \leq i \leq t$.

(3) $\{H_{ik} : 1 \leq k \leq t+1\}$ — the read-write head is over at least one cell at step i, $0 \leq i \leq t$.

(4) $\{\overline{H}_{ik}, \overline{H}_{ik'}\}$, $1 \leq k < k' \leq t+1$ — the read-write head is over at most one cell at step i, $0 \leq i \leq t$.

(5) $\{S_{ikl} : 1 \leq k \leq t+1, 1 \leq l \leq r\}$ — there is at least one symbol in each cell at step i, $0 \leq i \leq t$.

(6) $\{\overline{S}_{ikl}, \overline{S}_{ikl'}\}$, $1 \leq l < l' \leq r$ — there is at most one symbol in cell k at step i, $1 \leq k \leq t+1$, $0 \leq i \leq t$.

Observe that clauses of types (1), (3), and (5) are **true** when at least one literal in each of them is **true**. In contrast clauses of types (2), (4), and (6) are **false** exactly when both their variables are **true**. In other words, when M, at step i, is in two states, has two read-write head positions, or has a cell containing two symbols.

Next, we enforce a correct initial configuration:

(7) $\{Q_{01}\}$, $\{H_{01}\}$, $\{S_{0ki_k} : 1 \leq k \leq n\}$, $\{S_{0kr} : n+1 \leq k \leq t+1\}$;

and an accepting terminal configuration

(8) $\{Q_{t2}\}$.

Finally, we need to ensure for each pair of adjacent configurations that the second is, indeed, obtained from the first by a transition of M. We construct three types of clauses; one for the states, one for the read-write head, and one for the symbols. For all i, j, k, l, $0 \leq i \leq t$, $1 \leq j \leq m$, $1 \leq k \leq t+1$, $1 \leq l \leq r$, they are

(9) $\{\overline{Q_{ji}}, \overline{H_{ik}}, \overline{S_{ikl}}, \overline{Q_{i+1,j'}}\}$;

(10) $\{\overline{Q_{ij}}, \overline{H_{ik}}, \overline{S_{ikl}}, H_{i+1,k'}\}$; and

(11) $\{\overline{Q}_{ij}, \overline{H}_{ik}, \overline{S}_{ikl}, S_{i+1,kl'}\}$;

where, if $j \neq 2$ then $\delta(q_j, a_l) = (q_j, a_l, D)$ and $k' = k-1$, k, or $k+1$ as $D = L, \lambda$, or R, respectively. If $j = 2$, then $j = j'$, $k = k'$, and $l = l'$.

Now, clauses (9), (10), and (11) are satisfied when one of their four literals is **true**. If M, at step i, is in state q_j, over cell k which contains a_l, then (9), (10), and (11) can only be satisfied if M, at step $i+1$, is in state $q_{j'}$ and is over cell k' which contains $a_{l'}$.

By construction x is in L if and only if $f_L(x)$ is a yes instance of *SAT*. Therefore, f_L is a transformation. We have to show that it is a polynomial transformation.

The total number of clauses of types (1)-(11) is easily computed as at most $6m^2r(t+1)^3$. Now, m and r are constants for the chosen language L and *NTM*, M and $t = T(n)$. Moreover, $f_L(x)$ is the encoding of an instance of *SAT*. We need to show that $|f_L(x)|$ is bounded above by a polynomial in n. It contains $m(t+1)+(t+1)^2+r(t+1)^2 \leq (m+r+1)(t+1)^2$ variables. Encoding each variable as a word ab^ia, for some i, we see that $|f_L(x)| \leq c(t+1)^6$, for some constant c. This is indeed a polynomial in $n = |x|$, therefore $|f_L(x)|$ is polynomially bounded. That $f_L(x)$ can be computed by a *DTM* in polynomial time follows immediately. This is because the construction is a mechanical production of clauses of types (1)-(11) above.

We have shown f_L is a polynomial transformation and $L \propto SAT$. This holds for any L in $\boldsymbol{L}_{NPTIME}$. Therefore, *SAT* is in ***NP*** and is *NP*-complete, that is, we have established

Theorem 11.1.4 *SATISFIABILITY is NP-complete.*

We close this section by proving that a restricted version of *SAT* is *NP*-complete. This version is more useful in *NP*-completeness proofs than *SAT* itself.

3-SATISFIABILITY (3-*SAT*)
INSTANCE: A set U of variables and a collection C of clauses over U, where each clause contains exactly three literals.
QUESTION: Is there a satisfying truth assignment for C?

3-*SAT* is in ***NP*** because *SAT* is in ***NP***.

To show 3-*SAT* is *NP*-complete we need only show that *SAT* $\propto$ 3-*SAT*. By transitivity of $\propto$, this implies $L \propto$ 3-*SAT*, for all L in $\boldsymbol{L}_{NPTIME}$. This is the approach that we take with each decision problem Π in the following sections. First, we prove that Π is in ***NP*** — often, but not always, a simple step. Secondly, we prove that $\Pi' \propto \Pi$, for some *NP*-complete problem Π.

Theorem 11.1.5 *3-SAT is NP-complete.*

Proof: We prove *SAT* $\propto$ 3-*SAT* by considering a generic instance of *SAT*. Let $U = \{u_1, \ldots, u_m\}$ be a set of variables and C be a collection of clauses over U. We define a function f which maps the instance (U,C) of *SAT* into an instance (U',C') of 3-*SAT*. (We are writing loosely here; we should write: the instance *determined by* (U,C). But we will continue to sacrifice precision for brevity, in this regard.) With each clause C_i in C we associate a group of clauses in C'. There are four types of clauses in C; those with one, two, three, or more than three literals. We deal with each of them in turn.

(1) $C_i = \{x_j\}$, where $x_j = u_j$ or $\bar{u}_j$, $1 \le j \le m$.
Associate four clauses $\{x_j, v_i, w_i\}$, $\{x_j, \bar{v}_i, w_i\}$, $\{x_j, v_i, \bar{w}_i\}$, and $\{x_j, \bar{v}_i, \bar{w}_i\}$ with C_i, where v_i and w_i are new variables. These four clauses can be satisfied simultaneously only if x_j is **true**, that is, if C_i can be satisfied.

(2) $C_i = \{x_j, x_k\}$, $1 \le j < k \le m$.
Associate two clauses $\{x_j, x_k, v_i\}$, $\{x_j, x_k, \bar{v}_i\}$ with C_i, where v_i is a new variable. These two clauses can be satisfied simultaneously only if C_i can be satisfied.

(3) $C_i = \{x_j, x_k, x_l\}$, $1 \le j < k < l \le m$.
Associate C_i with itself.

(4) $C_i = \{x_{j_1}, \ldots, x_{j_r}\}$, $1 \le j_1 < \cdots < j_r \le m$, $3 < r \le m$.
Associate $r-2$ clauses $\{x_{j_1}, x_{j_2}, v_{i,1}\}$, $\{x_{j_3}, \bar{v}_{i,1}, v_{i,2}\}$, $\ldots$, $\{x_{j_{r-2}}, \bar{v}_{i,r-4}, v_{i,r-3}\}$, $\{x_{j_{r-1}}, x_{j_r}, \bar{v}_{i,r-3}\}$ with C_i, where the variables $v_{i,k}$, $1 \le k \le r-3$, are new. These $r-2$ clauses can be satisfied simultaneously only if C_i can be satisfied. Why is this? If we attempt to satisfy these clauses using only the new variables then we must assign **true** to $v_{i,1}$. But this implies $v_{i,2}$ must be assigned **true**, and so on. Therefore, $v_{i,r-3}$ must be assigned **true**, which implies that the final clause is **false**. However, if one of the literals from C_i is **true**, then this chain of events is broken. The $r-1$ clauses without this literal can be satisfied using only the new variables.

Now, $f(U,C) = (U',C')$, where $U' = U$ together with the new variables and C' is the collection of clauses associated with the clauses in C. By the above remarks, (U,C) is satisfiable if and only if $f(U,C)$ is satisfiable. Hence, f is a transformation.

We need to demonstrate that f is a polynomial transformation to complete the proof.

Note that the total number of literals in the clauses associated with a clause $\{x_{j_1}, \ldots, x_{j_r}\}$ in C is at most $2r+10$, for $r \geq 1$. Therefore, the size of (U',C') is a polynomial in the size of (U,C). (U',C') is easily computed from (U,C) by a *DTM* — a linear scan of the collection is sufficient. Hence, f is a polynomial transformation and $SAT \propto 3\text{-}SAT$. ■

11.1.4 Traveling Salesman

In this section we will prove that TRAVELING SALESMAN CHECKING (*TSC*) is *NP*-complete. For this result we consider two intermediate problems VERTEX COVER (*VC*) and HAMILTONIAN CIRCUIT (*HC*). We show that

$$3\text{-}SAT \propto VC \propto HC \propto TSC$$

to reach our goal. This indirect approach is simpler than proving

$$3\text{-}SAT \propto TSC$$

directly.

We begin by recalling the definition of *TSC* and defining *HC* and *VC*.

TRAVELING SALESMAN CHECKING (*TSC*)
INSTANCE: An integer $n \geq 1$, n distinct cities $C_1, \ldots, C_n$, positive integral distances between every pair C_i, C_j, denoted by $d(C_i, C_j)$, and a positive integral bound B.
QUESTION: Is there a tour $C_{i_1}, \ldots, C_{i_n}$ of the cities such that

$$\sum_{j=1}^{n-1} d(C_{i_j}, C_{i_{j+1}}) + d(C_{i_n}, C_{i_1}) = COST(i_1, \ldots, i_n) \leq B\,?$$

HAMILTONIAN CIRCUIT (*HC*)
INSTANCE: A graph $G = (V,E)$.
QUESTION: Does G contain a Hamiltonian circuit, that is, an ordering $v_1, \ldots, v_n$ of the vertices, where $n = \#V$, such that $\{v_i, v_{i+1}\}$ is in E, $1 \leq i < n$ and $\{v_n, v_1\}$ is in E?

VERTEX COVER (*VC*)
INSTANCE: A graph $G = (V,E)$ and a positive integer $K \leq \#V$.
QUESTION: Is there a subset $V' \subseteq V$ such that $\#V' \leq K$ and, for each edge $\{u,v\}$ in E, $\{u,v\} \cap V' \neq \emptyset$?

We establish the results in order of difficulty. We begin with $HC \propto TSC$, continue with $3\text{-}SAT \propto VC$, and conclude with $VC \propto HC$.

Lemma 11.1.6 $HC \propto TSC$.

Proof: We must show that the instances of HC can be transformed into instances of TSC.

Let $G = (V,E)$ be a graph that determines an instance of HC. Derive an instance of TSC by letting the vertices of G correspond to cities, that is, $V = \{C_1, \ldots, C_n\}$, where $n = \#V$. Now, define the distance function as

$$d(C_i,C_j) = \begin{cases} 1, & \text{if } \{C_i,C_j\} \text{ is in } E. \\ 2, & \text{if } \{C_i,C_j\} \text{ is not in } E. \end{cases}$$

and the bound B as n. This instance of TSC is denoted by $f(G)$.

We claim that G has a Hamiltonian circuit if and only if $f(G)$ has a solution. If G has a Hamiltonian circuit $C_{i_1}, \ldots, C_{i_n}$, then the cost of this tour in $f(G)$ is exactly B. This is because each edge in E has a cost of 1 unit in $f(G)$. Therefore, $f(G)$ has a solution.

Conversely, assume $f(G)$ has a tour $C_{i_1}, \ldots, C_{i_n}$ with cost at most B. Since there are n edges in the tour and $B = n$, each edge must correspond to one unit of distance. But this implies $C_{i_1}, \ldots, C_{i_n}$ is a Hamiltonian circuit in G. The function f is, therefore, a transformation. That it is a polynomial transformation follows straightforwardly; see the Exercises. ■

Let $U = \{u_1, \ldots, u_m\}$ and $C = \{C_1, \ldots, C_n\}$ determine an instance of 3-SAT, where the u_i are the variables and the C_i are the clauses. Since this is an instance of 3-SAT we know that $\#C_i = 3$, $1 \leq i \leq n$.

We define a transformation f such that $f(U,C)$ is an instance of VC. The proof technique is different from the previous ones and it is not as straightforward.

Let the bound K of $f(U,C)$ be $m+2n$. The graph $G = (V,E)$ given by $f(U,C)$ can be viewed as having two types of subgraphs together with connecting edges.

With each variable u_i associate the subgraph shown in Figure 11.1.4(a) — a *truth-setting component* — and with each clause C_i associate the subgraph of Figure 11.1.4(b) — a *satisfaction-testing component*. The set V of vertices of G is defined by

$$V = \{u_i,\overline{u}_i : 1 \leq i \leq m\} \cup \{x_i,y_i,z_i : 1 \leq i \leq n\}$$

hence, $\#V = 2m+3n$.

To complete the specification of G we need to add edges between truth-setting and satisfaction-testing components — *communication edges*. For this purpose we require that the three literals appearing in clause C_i have a fixed correspondence with x_i, y_i, and z_i. For example, if $C_i = \{u_2,\overline{u}_7,u_9\}$, we take (u_2,x_i), $(\overline{u}_7,y_i)$, and (u_9,z_i) as the corresponding pairs.

Continuing with this example we add edges

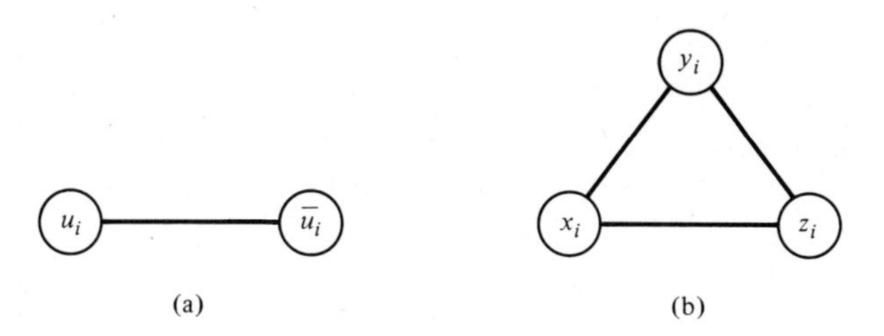

Figure 11.1.4

$$\{u_2, x_i\}, \{\bar{u}_7, y_i\}, \{u_9, z_i\}$$

to G. In general, for each clause $C_i = \{w_{i_1}, w_{i_2}, w_{i_3}\}$ we add edges

$$\{w_{i_1}, x_i\}, \{w_{i_2}, y_i\}, \{w_{i_3}, z_i\}$$

to E.

Example 11.1.1 Given

$$U = \{u_1, u_2, u_3, u_4, u_5, u_6\}$$

and

$$C = \{\{u_2, \bar{u}_4, u_6\}, \{\bar{u}_3, \bar{u}_5, u_6\}, \{u_1, u_4, u_5\}\}$$

we obtain $K = 6+6 = 12$ and the graph G is as shown in Figure 11.1.5. Any vertex cover V' of G must include at least one of u_i and $\bar{u}_i$, for each i, $1 \le i \le 6$, and at least two of x_i, y_i, and z_i, for each i, $1 \le i \le 3$. Therefore, a vertex cover V' of size at most $6 + 2 \times 3 = 12$ must contain exactly these vertices, so $\# V' = 12$.

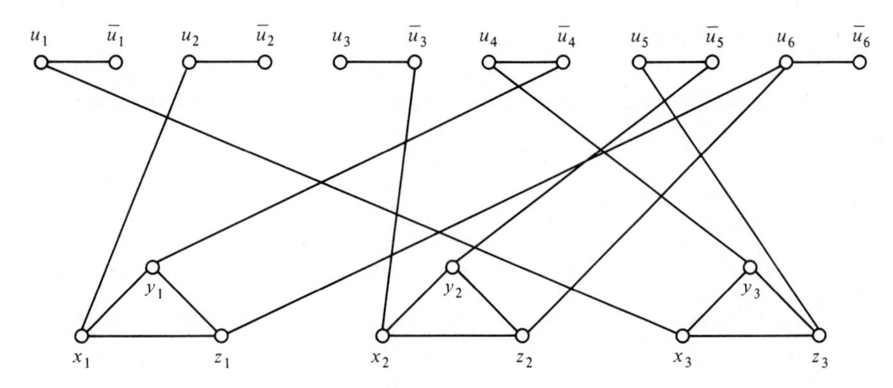

Figure 11.1.5

Choose

$$V' = \{u_1, u_2, \bar{u}_3, u_4, u_5, u_6\} \cup \{y_1, z_1, y_2, z_2, y_3, z_3\}$$

This is, indeed, a vertex cover of the desired size. Setting $u_1 = u_2 = \bar{u}_3 = u_4 = u_5 = u_6 =$ **true**, we see that C is satisfied.

To complete the proof we must demonstrate that f is, indeed, a polynomial transformation. Assume C is satisfiable. Let w_i, $1 \le i \le m$, be literals corresponding to the variables u_i, $1 \le i \le m$, such that $w_i =$ **true**, $1 \le i \le m$ satisfies C. Choose the truth-setting vertices w_i of G to be in V'. Now, each clause C_i contains at least one literal that has the value **true**. Choose the two other vertices in the satisfaction-testing component corresponding to C_i to be the remaining two vertices in V'. We claim that V' is a vertex cover of G. We need only show that every communication edge associated with a satisfaction-testing component has a vertex in V'. If the edge corresponds to the chosen literal in the clause, then the truth-setting vertex is in V'. If the edge corresponds to some other literal its satisfaction-testing vertex is in V'. Therefore, all edges have a vertex in V', and V' is a vertex cover.

Conversely, assume G has a vertex cover V' with $\#V' \le K$. Then, as we argued in Example 11.1.1, $\#V' = K$. Let w_i, $1 \le i \le m$ be the truth-setting vertices in V' and assign the value **true** to the corresponding literals w_i. Each of the three communication edges corresponding to a satisfaction-testing component must have at least one of their vertices in V'. But one of these three vertices must be a truth-setting vertex w_i, for some i, $1 \le i \le m$. This implies the clause corresponding to the satisfaction-testing component contains a literal with the value **true**. This holds for all clauses in C, so C has a satisfying truth assignment. We leave to the Exercises the simple proof that f is polynomial.

To summarize, we have shown the following.

Theorem 11.1.7 $3\text{-}SAT \propto VC$.

We now complete the chain of reasoning by proving that $VC \propto HC$, which implies VC, HC and TSC are NP-complete, by Lemma 11.1.1.

We are given an instance (G,K) of VC determined by a graph $G = (V,E)$ and an integer-bound $K \ge 1$. We must provide a transformation f such that $f(G,K)$ determines an instance $H = (U,F)$ of HC. There are similarities to the previous proof in that we have two types of subgraphs which are connected with additional edges.

The first type of subgraph is the single vertex of Figure 11.1.6(a), for $1 \le i \le K$, called *selector vertices.* To each edge $e = \{u,v\}$ in E there corresponds a second type of subgraph. This subgraph has 12 vertices, six corresponding to u and 6 corresponding to v; see Figure 11.1.6(b). Any additional edges only involve the end vertices $[u,e,1]$, $[v,e,1]$, $[u,e,6]$, and $[v,e,6]$ of this subgraph, which is called a *cover-testing component*. As we will see below it has been cunningly chosen to restrict the ways a Hamiltonian circuit goes through it.

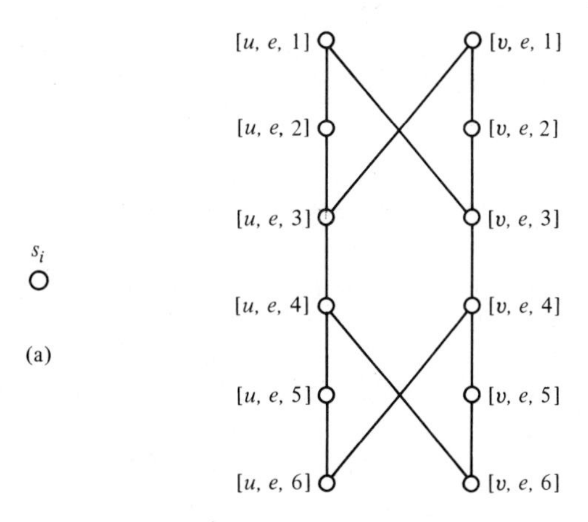

Figure 11.1.6

We add connecting edges between components and between components and selectors as follows.

In order to add the connecting edges assume that the vertices V of G are ordered as $V = \{v_1, \ldots, v_n\}$. At each vertex v_i its edges can be ordered as $e_1, \ldots, e_{d_i}$, where d_i is the degree of v_i, such that for all $e_j = \{v_i, v_p\}$ and $e_k = \{v_i, v_q\}$, $j < k$ if and only if $p < q$. We connect the v_i sides of the cover-testing components for $e_1, \ldots, e_{d_i}$ by adding edges $\{[i, e_j, 6], [i, e_{j+1}, 1]\}$, $1 \leq j \leq d_i$.

To complete the graph $H = (U, F)$ we add edges $\{s_j, [i, e_1, 1]\}$ and $\{s_j, [i, e_{d_i}, 6]\}$, for all j, $1 \leq j \leq k$. This can be seen pictorially in Figure 11.1.7. If G has a vertex cover V' with $\# V' \leq K$, then we need to show that H has a Hamiltonian circuit, and vice versa.

We can assume that $\# V' = K$, since if $\# V' < K$, we add $K - \# V'$ vertices from $V - V'$ to V'. These additional vertices do not affect V''s status as a vertex covering set. Let $V' = \{v_{i(1)}, \ldots, v_{i(K)}\}$, where $1 \leq i(1) < \cdots < i(K) \leq n$.

Beginning at selector vertex s_1 we construct a Hamiltonian circuit. From s_1 we "walk" to $[i(1), e_1, 1]$, entering the first cover-testing component of $v_{i(1)}$. There are two paths through this component which leave at $[i(1), e_1, 6]$ and do not visit a vertex twice. These are shown in Figure 11.1.8. The (a) path is taken if vertex v_i is in V', while the (b) path is taken if v_i is not in V'. Eventually we reach vertex $[i(1), e_{d_{i(1)}}, 6]$, the last vertex of the cover-testing components for $v_{i(1)}$. Therefore, we walk to s_2 along edge $\{[i(1), e_{d_{i(1)}}, 6], s_2\}$.

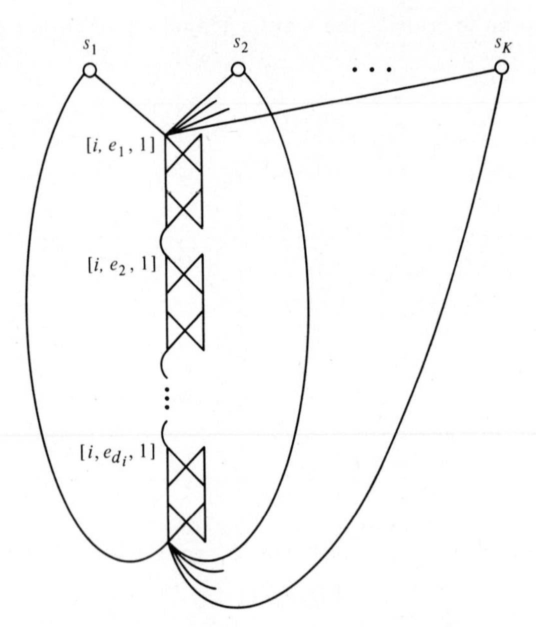

Figure 11.1.7

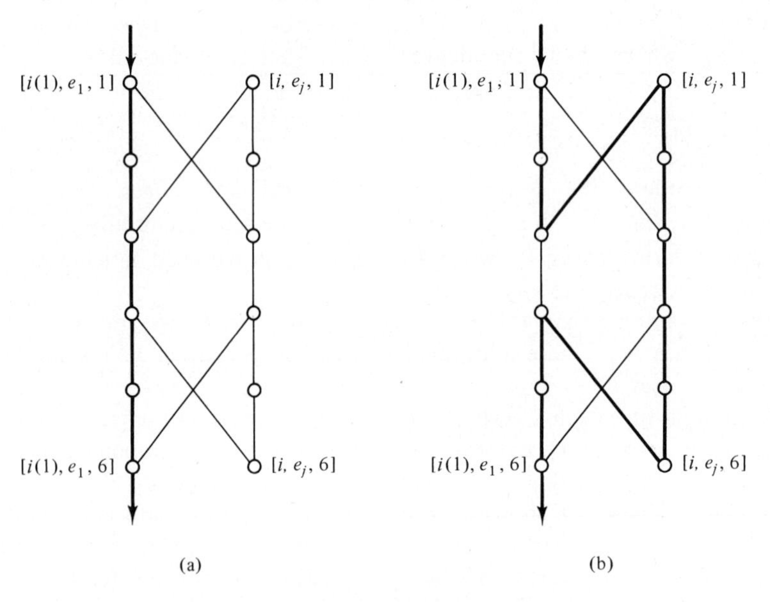

Figure 11.1.8

We carry out a similar walk for $v_{i(2)}$, which leads to s_3. This is repeated for $v_{i(3)}, \ldots, v_{i(K)}$, except that the final edge for $v_{i(K)}$ is $\{[i(K), e_{d_{i(K)}}, 6], s_1\}$ which completes the cycle.

We claim that the cycle we have obtained is a Hamiltonian circuit. First, the cycle does not visit any vertex of U twice. This obviously holds for selector vertices and the vertices $[i(j), e_k, l]$ by construction. But what about vertices $[i, e_k, l]$, where v_i is not in V'? If $[i, e_k, l]$ is visited during the cycle, then it is because the other vertex in e_k is in V'. This means the cycle contains a path of the form shown in Figure 11.1.8(b). But this implies all vertices $[i, e_k, l]$, $1 \leq l \leq 6$, are visited. The only other edges attached to them are the communication edges for v_i, which link this component to other v_i components or selector vertices. By the choice of cycle, this implies none of the vertices $[i, e_k, l]$, $1 \leq l \leq 6$ are visited a second time.

We argue that the cycle visits every vertex in U and is, therefore, a Hamiltonian circuit. We are concerned only about vertices in U obtained from vertices in $V - V'$. Consider the vertex $[i, e_k, l]$, for some l, $1 \leq l \leq 6$, where v_i is not in V'. Immediately, the other vertex v_j in e_k must be in V'. This implies that $[j, e_p, 1]$, where $e_p = e_k$, is visited by the cycle and, therefore, the cycle visits $[i, e_k, l]$, $1 \leq l \leq 6$, using the path of Figure 11.1.8(b).

We have demonstrated that H has a Hamiltonian circuit, if G has a vertex cover of size at most K. We now tackle the converse. Assume H has a Hamiltonian circuit. We partition the circuit into K paths. Each path begins at a selector vertex, ends at selector vertex, and does not pass through a selector vertex. Consider the path beginning at s_1. The second vertex on the path is either $[i, e_1, 1]$, for some i, or $[j, e_{d_j}, 6]$, for some j. We assume that it is $[i, e_1, 1]$; the other case is symmetric. This is the point in the argument where we see that the choice of cover-testing components is, indeed, cunning! In Figure 11.1.9 we display choices for a path other than the two shown in Figure 11.1.8. Each of them causes a middle vertex in the component to be missed. For, as we argued above, a middle vertex in a component has no edges in F other than the ones in the component. But the path is part of a Hamiltonian circuit; therefore, the middle vertices on the entry side must be visited. The only paths that do this are the two in Figure 11.1.8. So the path enters at $[i, e_1, 1]$ and leaves at $[i, e_1, 6]$. There is only one edge from $[i, e_1, 6]$ which leads to either $[i, e_2, 1]$ or to a selector vertex. In the latter case we have completed the path. In the former case a similar argument shows that the path leaves the second component at vertex $[i, e_2, 6]$. Eventually the path leaves the d_ith component of v_i at vertex $[i, e_{d_i}, 6]$ for some selector vertex.

What have we shown? Consider a path, in the partition of the circuit, from a selector vertex to a selector vertex which does not pass through a selector vertex. We have shown it must pass through all the components corresponding to some vertex v_i in V. Moreover, the path through each component has one of two forms shown in Figure 11.1.8.

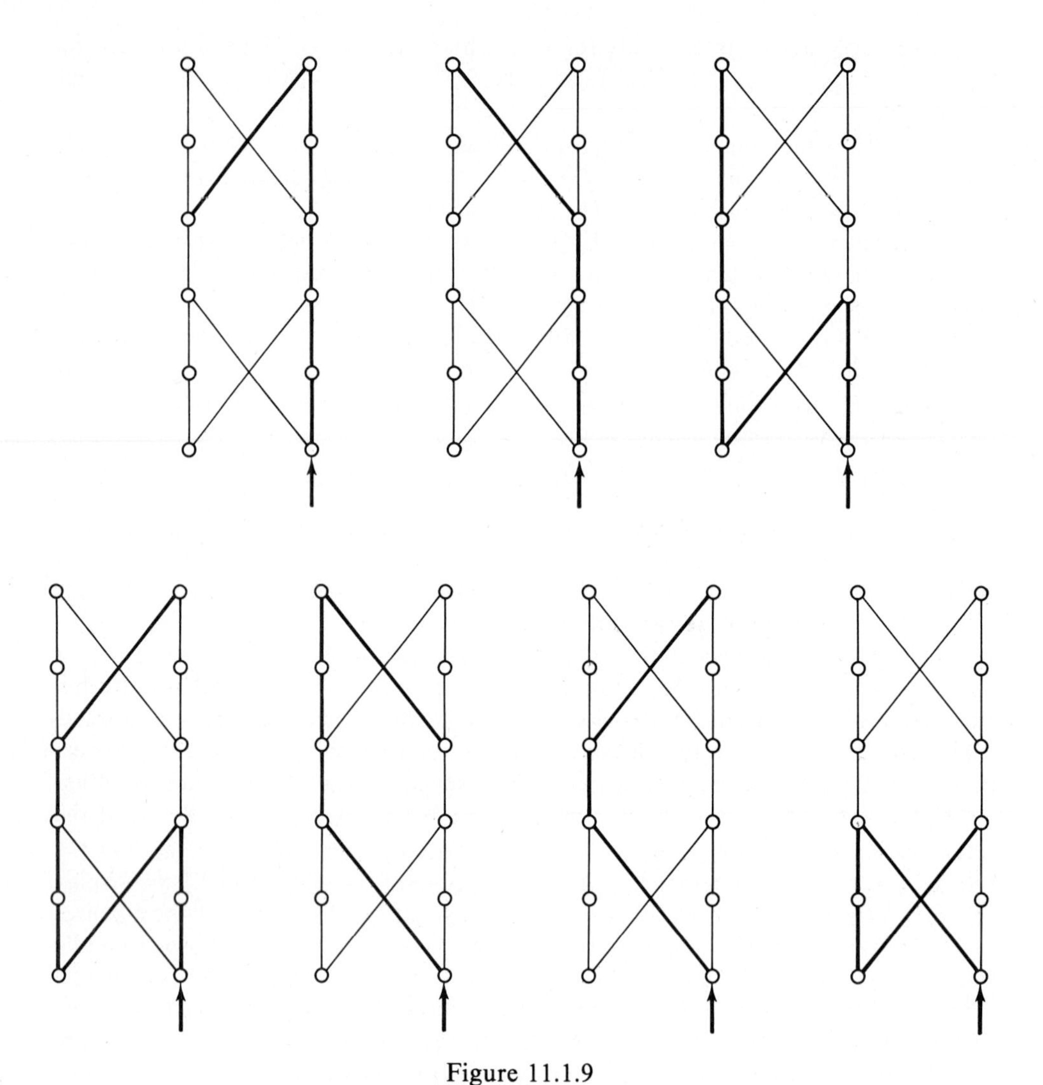

Figure 11.1.9

The K vertices from V obtained in this way form a vertex cover of G, by similar arguments to those used in the first part of the proof. Hence, f is a transformation. That it is a polynomial transformation is left to the Exercises.

We have proved the following.

Theorem 11.1.8 *$VC \propto HC$; therefore HC and TSC are NP-complete.*

11.1.5 Subset Sum

The second major problem we wish to attack is one that occurs in grammars describing VLSI layouts. To prepare the way for this result we need to show the following problem is *NP*-complete.

SUBSET SUM (*SS*)
INSTANCE: A finite set A, a size function $\sigma : A \to \mathbb{N}$, and a positive integer bound B.
QUESTION: Is there a subset $A' \subseteq A$ whose size is B, that is,

$$\sum_{a \text{ in } A'} \sigma(a) = B?$$

Once more we do not prove this result directly. We make use of one other problem. This is

3-DIMENSIONAL MATCHING (3-*DM*)
INSTANCE: Three sets X, Y, Z with $\#X = \#Y = \#Z = n$, say, and a ternary relation $M \subseteq X \times Y \times Z$.
QUESTION: Is there a subset $M' \subseteq M$ such that $\#M' = n$ and every element of $X \cup Y \cup Z$ appears in exactly one triple in M'?

We prove that

$$3\text{-}SAT \propto 3\text{-}DM \propto SS$$

SS is easily seen to be in ***NP***. Simply guess a subset $A' \subseteq A$ and test whether its size is exactly B. Therefore, by transitivity we will have shown that 3-*DM* is ***NP*** also. First, we consider the transformation 3-*DM* $\propto$ *SS*. With each instance (X,Y,Z,M) of 3-*DM* we associate an instance (A,σ,B) of *SS*. Letting $\#X = n$, we associate an n-digit integer $(d_{n-1}, \ldots, d_0)$, to some base b, with each (x,y,z) in M. This implies that we choose $A = M$ and $\sigma((x,y,z))$ to be the n-digit integer. Finally, B is chosen so that (A,σ,B) has a solution if and only if (X,Y,Z,M) has a solution.

In more detail let $X = \{x_i : 0 \le i \le n-1\}$, $Y = \{y_i : 0 \le i \le n-1\}$, and $Z = \{z_i : 0 \le i \le n-1\}$. Letting $\#M = m$, define b to be $(m+1)^3$. The elements x_r, y_r, and z_r are associated with three zones of digit d_r, $0 \le r \le n-1$. The *lower zone* corresponds to z_r and consists of the values $0..m$, the *middle zone* corresponds to y_r and the values $m+1..(m+1)^2-1$, and the *upper zone* to x_r and $(m+1)^2..(m+1)^3-1$; see Figure 11.1.10. Given d_r we use the notation d_r^l, d_r^m, and d_r^u to refer to these.

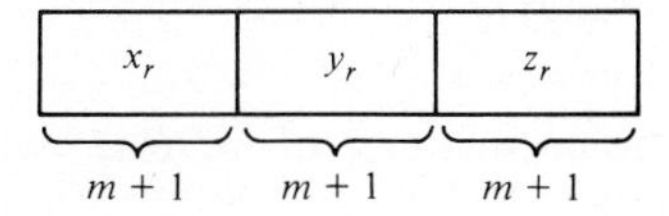

Figure 11.1.10

With each triple (x_i,y_j,z_k) in M associate an integer $(d_{n-1},\ldots,d_0)_b$, where $d_i^u = (m+1)^2$, $d_j^m = m+1$, $d_k^l = 1$, and all other zones and digits are 0. This defines the size function σ. Note that if we add the m associated integers, no zone can have a value greater than m. This implies that overflow from one zone to another cannot take place.

In a solution M' of (X,Y,Z,M) each x_i, y_i, and z_i occurs exactly once. Therefore, we choose B to be the n-digit integer $(B_{n-1},\ldots,B_0)_b$ where $B_i = (m+1)^2+(m+1)+1$, $1 \leq i \leq n$.

Assume (X,Y,Z,M) has a solution M'. Then M' is a solution of (M,σ,B) since $\sigma(M') = B$. Conversely if $M' \subseteq M$ is a solution of (M,σ,B), then $\sigma(M') = B$. Because overflow cannot occur between zones, this implies M' is a solution of (X,Y,Z,M).

The transformation is easily seen to be polynomial, so to summarize the simplest result so far, we have the following.

Lemma 11.1.9 *3-DM* $\propto$ *SS*.

To complete the chain of reasoning we must prove that 3-*SAT* $\propto$ 3-*DM*. This implies that 3-*SAT* $\propto$ 3-*DM* $\propto$ *SS*. Now, 3-*DM* is in ***NP*** since *SS* is and by transitivity we conclude

Theorem 11.1.10 3-*SAT* $\propto$ 3-*DM*. *Therefore, 3-DM and SS are NP-complete.*

We prove 3-*SAT* $\propto$ 3-*DM* in two steps. We introduce a relaxed version of 3-*DM*, 3-*DRM* and prove that 3-*DRM* $\propto$ 3-*DM*. Then we prove 3-*SAT* $\propto$ 3-*DRM* to complete the chain.

The relaxed version of 3-*DM* is defined as

3-DIMENSIONAL RELAXED MATCHING (3-*DRM*)
INSTANCE: Three sets X, Y, and Z with $\#X \geq \#Y = \#Z = n$ and a ternary relation $M \subseteq X \times Y \times Z$.
QUESTION: Is there a *relaxed matching* $M' \subseteq M$? That is $\#M' = n$, each element of X appears at most once, and each element of Y and Z appears exactly once.

To prove 3-*DRM* $\propto$ 3-*DM* is straightforward.

Lemma 11.1.11 3-*DRM* $\propto$ 3-*DM*.

Proof: Let (X,Y,Z,M) be an instance of 3-*DRM*, where $\#Y = \#Z = n$ and $\#X = n+r$, for some $r \geq 0$.

We transform it into an instance (U,V,W,N) of 3-*DM* as follows. Introduce $2r$ new elements and add r to each of Y and Z. Hence, $U = X$, $V = Y \cup \{a_i : 1 \leq i \leq r\}$, and $W = Z \cup \{b_i : 1 \leq i \leq r\}$, where the a_i and b_i are new elements. This ensures that $\#U = \#V = \#W$. Now, add new triples to M to give N. More precisely let

$$N = M \cup \{(x,a_i,b_j) : x \text{ is in } X,\ 1 \le i \le r,\ 1 \le j \le r\}$$

If (X,Y,Z,M) has a relaxed matching M', then exactly r elements of X do not appear in M'. Let these be the elements $x_1, \ldots, x_r$ and let $N' = M' \cup \{(x_i,a_i,b_i) : 1 \le i \le r\}$. N' is clearly a matching for (U,V,W,N).

Conversely, let N' be a matching for (U,V,W,N). There cannot be any triple of the form (x,a_i,z) or (x,y,b_i) in N' since there are no triples of this form in N. Therefore, there are r elements of $X = U$ which are paired with elements from $V-Y$ and $W-Z$. Let these be $x_1, \ldots, x_r$. The elements $x_{r+1}, \ldots, x_{r+n}$ in X must appear in triples containing only elements of Y and Z. But this means that

$$M' = \{(x_i,y_j,z_k) : (x_i,y_j,z_k) \text{ is in } N' \text{ and } r+1 \le i \le r+n\}$$

is a relaxed matching for (X,Y,Z,M). Clearly, the transformation is polynomial; therefore, 3-*DRM* $\propto$ 3-*DM*. ■

To complete the proof of Theorem 11.1.10 we now prove the following.

Lemma 11.1.12 3-*SAT* $\propto$ 3-*DRM*.

Proof: Let $U = \{u_1, \ldots, u_m\}$ be a set of variables and $C = \{c_1, \ldots, c_n\}$ be a set of clauses of an instance of 3-*SAT*. To construct an instance (X,Y,Z,M) of 3-*DRM* we associate two sets of triples with each instance of 3-*SAT*; *truth-setting* and *satisfaction-testing* triples.

With each variable u_i and each clause c_j we associate two truth-setting triples. The first is $(\bar{u}_{ij},a_{ij},b_{ij})$ and the second is either $(u_{ij},a_{i,j+1},b_{ij})$, if $j \ne n$, or (u_{in},a_{i1},b_{in}) otherwise. The elements u_{ij} and $\bar{u}_{ij}$ are in X, the a_{ij} are in Y, and the b_{ij} are in Z, $1 \le i \le m$, $1 \le j \le n$.

Each a_{ij} and b_{ij} occur in exactly two triples. For a given variable u_i the truth-setting triples T_i associated with u_i are best understood pictorially; see Figure 11.1.11. In a solution of (X,Y,Z,M) exactly n triples must be chosen from T_i so that the a_{ij} and b_{ij}, $1 \le j \le n$ are chosen. The definition of these triples, however, ensures that either all u_{ij} triples or all $\bar{u}_{ij}$ triples, $1 \le j \le n$, are chosen.

For example, if (u_{i1},a_{i2},b_{i1}) is chosen, then the triple $(\bar{u}_{i1},a_{i1},b_{i1})$ cannot be chosen, since this would repeat b_{i1}. Similarly, $(\bar{u}_{i2},a_{i2},b_{i2})$ cannot be chosen since this would repeat a_{i2}. On the other hand, if (u_{i2},a_{i3},b_{i2}) is not chosen, then b_{i2} will not appear in any triple. This demonstrates the "domino effect"; for each j, $1 \le j \le n$, either all $\bar{u}_{ij}$ triples or all u_{ij} triples must be chosen; mixtures do not yield solutions.

The second set S of triples, the satisfaction-testing triples, connect together truth-setting triples. With each clause c_j we associate new elements s_j and t_j of Y and Z, respectively. If literal u_i appears in c_j, then add the triple (u_{ij},s_j,t_j) to M and if literal $\bar{u}_i$ appears in c_j add the triple $(\bar{u}_{ij},s_j,t_j)$. Since c_j contains exactly three literals s_j and t_j appear in exactly three triples.

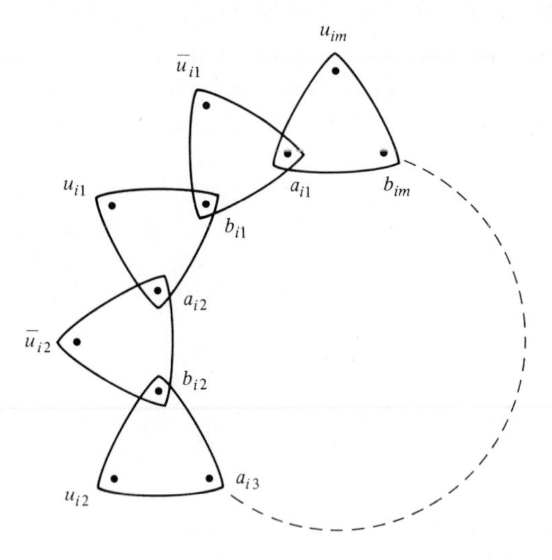

Figure 11.1.11

A simple computation shows that $\#X = 2mn$ and $\#Y = \#Z = mn + n$; therefore, we have obtained an instance of 3-*DRM*.

To see how the triples function, assume (U, C) is satisfiable. In each clause c_j there is a literal w_i (either u_i or $\overline{u}_i$) that has the value **true**. To construct a relaxed matching take the triple (w_{ij}, s_j, t_j) and all triples from T_j which do not contain w_{ij}. Note that this choice of triples for w_i, given c_j, is consistent with all appearances of w_i and $\overline{w}_i$ in all other clauses. This is, simply, because $\overline{w}_i$ is never **true** if w_i is.

Conversely, assume we have a relaxed matching. We set u_i to **true** if (u_{ij}, s_j, t_j) is in M' for some j, $1 \leq j \leq n$, we set u_i to **false** if $(\overline{u}_{ij}, s_j, t_j)$ is in M' for some j, $1 \leq j \leq n$; and otherwise we set u_i to **true**. As we have already observed at most one of these two triples is in M', for any j, $1 \leq j \leq n$. We must prove that this truth assignment is well defined, that is, we do not assign both values to any u_i, $1 \leq i \leq n$.

Assume there is a u_i which is assigned both values. This occurs because u_i occurs in some clause c_j and $\overline{u}_i$ occurs in some clause c_k, $j \neq k$, that is (u_{ij}, s_j, t_j) and $(\overline{u}_{ik}, s_k, t_k)$ are in M'. But we have already seen that either all u_{ir} truth-setting triples or all $\overline{u}_{ir}$ truth-setting triples are in M', $1 \leq r \leq n$. Assume all u_{ir} truth-setting triples are in M'. Then (u_{ij}, s_j, t_j) cannot be in M'. Therefore, u_i cannot be assigned **true**.

To complete the proof, observe that we have defined a polynomial transformation; see the Exercises. ■

11.1.6 Hierarchical Grammars

VLSI designs are usually described hierarchically rather than enumeratively. The reason for this is, simply, description size. A typical VLSI design consists of about one million rectangles, each defined by a quadruple (x_l, x_r, y_b, y_t): see Figure 11.1.12. Their placement can be described individually, that is, as a file of one million quadruples. Alternatively, using the regularity of layouts, their placement may be defined by only one hundred hierarchical productions with a grammar (or layout language) such as the Caltech Intermediate Form (*CIF*).

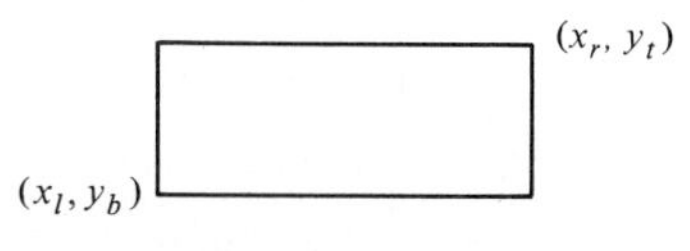

Figure 11.1.12

For example, given the productions

$$A_0 \rightarrow (0,1,0,1)$$
$$A_1 \rightarrow (0,0)A_0(1,1)A_0$$
$$A_2 \rightarrow (0,0)A_1(2,2)A_1$$

A_2 defines the rectangles of Figure 11.1.13. A_0, a terminal production, produces a 1×1 square at the origin. A_1 forms the union of A_0 offset by $(0,0)$ and A_0 offset by $(1,1)$. Hence, A_1 produces a 1×1 square at the origin and one at position $(1,1)$. Now, A_2 forms the union of A_1 offset by $(0,0)$ and A_1 offset by $(2,2)$, that is, the four squares of Figure 11.1.13.

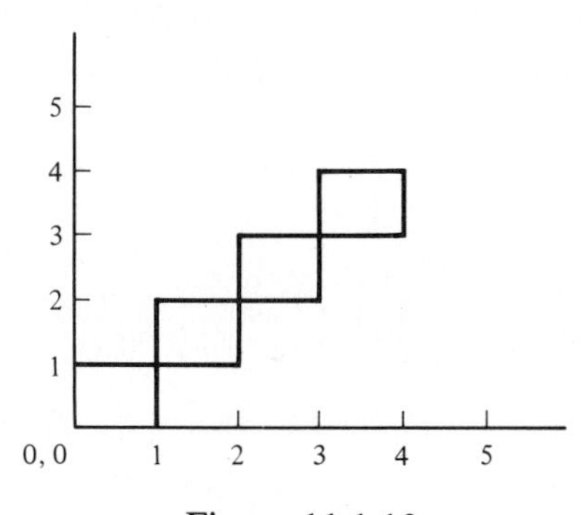

Figure 11.1.13

A_0, a terminal production, produces a 1×1 square at position (0,0).

It is easy to generalize the above example to produce 2^n squares on the diagonal using nonterminals $A_0, \ldots, A_n$, where

$$A_0 \rightarrow (0,1,0,1)$$
$$A_1 \rightarrow (0,0)A_{i-1}(2^{i-1}, 2^{i-1})A_{i-1},\ 1 \leq i \leq n$$

These grammars, called hierarchical *CFGs* or *HCFGs*, are, of course, an abstraction and simplification of actual layout languages. Layout languages contain other primitives besides rectangles and they also contain other operations. However, these additions do not add any power to the model, so we do not consider them.

A *hierarchical context-free grammar* (*HCFG*) G is given by a triple (N,P,S), where

N is a *nonterminal alphabet*;
$P \subseteq (N \times \mathbb{N}_0^4) \cup (N \times (\mathbb{N}_0^2 N)^*)$ is a finite set of *productions*; and
S in N is a *sentence symbol.*

A production in $N \times \mathbb{N}_0^4$ is called a *terminal production*; it is usually written in the form $A \rightarrow (x_1,x_2,y_1,y_2)$. It defines one rectangle. A production in $N \times (\mathbb{N}_0^2 N)^*$ is called a *nonterminal production*; it is usually written in the form $A \rightarrow (x_1,y_1)A_1 \cdots (x_m,y_m)A_m$. It defines the union of m layouts $A_1, \ldots, A_m$, where A_i is offset by (x_i,y_i).

A *derivation step* and a *derivation* are defined as for *CFGs* with one additional rule. Assume $\alpha = \alpha_1(x,y)A\alpha_2$ and $A \rightarrow (x_1,y_1)A_1 \cdots (x_m,y_m)A_m$ is in P. First we obtain

$$\alpha_1(x,y)((x_1,y_1)A_1 \cdots (x_m,y_m)A_m)\alpha_2$$

which we simplify by distributing (x,y). This yields

$$\beta = \alpha_1(x+x_1,y+y_1)A_1 \cdots (x+x_m,y+y_m)A_m\alpha_2$$

which is the result of applying $A \rightarrow (x_1,y_1)A_1 \cdots (x_m,y_m)A_m$ to α. This we denote by

$$\alpha \Rightarrow \beta$$

Note that the production $A \rightarrow \lambda$ leads to

$$\beta = \alpha_1\alpha_2$$

Similarly, if $A \rightarrow (x_1,x_2,y_1,y_2)$ is in P, we obtain

$$\beta = \alpha_1(x+x_1,x_2+x,y+y_1,y+y_2)\alpha_2$$

We obtain $\Rightarrow^i$, $\Rightarrow^+$, and $\Rightarrow^*$ in the usual way. A *terminating derivation* is a derivation

$$S \Rightarrow^+ \alpha$$

where $\alpha = (x_1^1,x_2^1,y_1^1,y_2^1) \cdots (x_1^k,x_2^k,y_1^k,y_2^k)$. It defines the set of rectangles

$$\{(x_1^i,x_2^i,y_1^i,y_2^i) : 1 \leq i \leq k\}$$

which we denote by $R(\alpha)$.

An *HCFG* $G = (N,P,S)$ defines a set of rectangles

$$\bigcup R(\alpha)$$

where the union is taken over all α, such that $S \Rightarrow^+ \alpha$ is a terminating derivation in G. We denote this set by $R(G)$. Rather than studying *HCFGs* with this generality, we prefer to restrict them further. We require that there be *exactly one terminating derivation from* S. From hereon in we assume *HCFGs* satisfy this requirement.

A fundamental question during VLSI design rule checking is whether there are intersecting pairs of rectangles in the layout. Two rectangles intersect if they have a pair of intersecting edges or one rectangle is contained in the other. Alternatively, considering a rectangle to specify a closed set of pairs of reals, two rectangles intersect if they contain a common value. We define the problem more formally as

RECTANGLE INTERSECTION (*RI*)
INSTANCE: An *HCFG*, $G = (N,P,S)$ with exactly one terminating derivation.
QUESTION: Are there rectangles R_1 and R_2 in $R(G)$ such that R_1 and R_2 intersect?

The following argument shows that *RI* is in ***NP***. S can only be replaced in one way — if we require termination — yielding α_1, say. Choose one appearance of a nonterminal in α_1 to be rewritten as α_2, say. This yields $\alpha_1'\alpha_2\alpha_1''$, where we assume the offset of the rewritten nonterminal has been distributed throughout α_2. Again choose some appearance of a nonterminal in α_2 to be rewritten as α_3, say. This yields $\alpha_1'\alpha_2'\alpha_3\alpha_2''\alpha_1''$. Continue this process until a sentential form $\alpha_1' \cdots \alpha'_{m-1}\alpha_m\alpha''_{m-1} \cdots \alpha_1''$ is obtained where α_m is a rectangle R_1. R_1 is one of the rectangles in $R(G)$. We have, however, carried out only the rewriting necessary to obtain R_1. In other words, $m \leq \#N$. In a similar manner we can obtain a rectangle R_2 from G. These are compared to see whether they intersect. If they do this is a yes instance of *RI*.

The above algorithm is a typical guess-and-verify algorithm. We need to show that it requires for each pair of chosen rectangles a number of steps polynomial in the size of G. But how do we measure the size of a grammar? In an encoding we need to provide the sentence symbol, the productions, and, perhaps, the nonterminals. Representing nonterminals in the form ab^ia and integers in binary, the size of an encoding of a grammar $G = (N,P,S)$, which we call *size*(G), is proportional to $\#N^2 + \#N\Sigma_{A\to\alpha \text{ is in } P} |A\alpha|$. A derivation requires a single scan of the productions at each step. This can be performed by an *NTM* within *size*(G) steps. In other words, a single rectangle can be generated in a polynomial number of steps. Testing two rectangles for intersection requires a constant number of steps, so *RI* is, indeed, in ***NP***.

To show *RI* is *NP*-complete we transform SUBSET SUM to *RI*. Let (A,σ,B) be an instance of SUBSET SUM. We construct a instance G of *RI* such that G generates rectangles representing possible sums and rectangles corresponding to B. More precisely, G generates columns of squares, where each column represents a subset of A. The squares in a column are offset vertically from each other by the sizes of the corresponding elements in A; see Figure 11.1.14. This implies that the topmost square in a column corresponding to a subset $A' \subseteq A$, has a vertical offset corresponding to $\sigma(A')$. Therefore, G also generates squares, one for each column, at vertical offset B. By construction the only possible rectangle intersections are between the subset squares and the B squares and, moreover, there is such an intersection if and only there is a subset $A' \subseteq A$ with $\sigma(A') = B$.

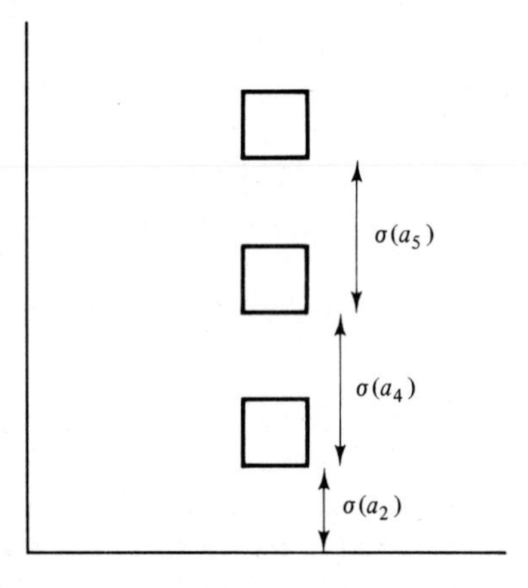

Figure 11.1.14

Having shown the geometrical insight that is, needed to understand the transformation we now define the *HCFG*, G, formally.

Letting $A = \{a_1, \ldots, a_m\}$ we have the set P of productions

$$S \rightarrow (0,0)A_m(0,0)B_m$$
$$A_0 \rightarrow (0,2,0,2)$$
$$A_i \rightarrow (0,0)A_{i-1}(2^2, 2\sigma(a_i))A_{i-1},\ 1 \leq i \leq m$$
$$B_0 \rightarrow (0,1,2,2B,2B+1)$$
$$B_i \rightarrow (0,0)B_{i-1}(0,2^i)B_{i-1},\ 1 \leq i \leq m$$

Thus, $N = \{A_i, B_i : 0 \leq i \leq m\} \cup \{S\}$ and the *HCFG*, $G = (N,P,S)$. A_m generates columns of 2×2 squares. Each square is offset vertically from the one below it by twice the size of the corresponding element of A. B_m generates a row of 1×1 squares at vertical offset $2B$. By the preceding discussion (A,σ,B) has a solution if and only if $R(G)$ contains a rectangle intersection.

Observe that *size*(G) is polynomial in m, $\sigma(A)$, and B. Since the size of *SS* can be considered to be $\#A + \sigma(A) + B$, *size*(G) is polynomial in the size of (A,σ,B). Computing G from (A,σ,B) is straightforward. Therefore, $SS \propto RI$.

To summarize the results of this subsection, we have proved

Theorem 11.1.13 *RECTANGLE INTERSECTION is NP-complete.*

11.2 REWRITING SYSTEMS AND CHURCH'S THESIS

Although we have discussed context-free grammars in detail, there are many other kinds of grammars. Noam Chomsky identified two other natural classes of grammars — phrase structure grammars and context-sensitive grammars. More recently, Aristid Lindenmayer introduced *L* grammars to model biological development. Macro grammars were introduced by Fischer as a model for macros in programming languages, and programmed grammars were introduced by Rosenkrantz as an abstraction of SNOBOL. In this section we consider phrase structure grammars and interactive Lindenmayer grammars. We sketch the proofs that both of them are equivalent in power to *DTMs*; more evidence for the Church(-Turing) thesis.

11.2.1 Phrase Structure Grammars

A *phrase structure grammar*, or *PSG*, G is a quadruple (N,Σ,P,S), where

N is a *nonterminal alphabet*;
Σ is a *terminal alphabet*, $N \cap \Sigma = \varnothing$;
$P \subseteq (N \cup \Sigma)^+ \times (N \cap \Sigma)^*$ is a finite set of *productions*; and
S in N is a *sentence symbol*.

A production (α,β) in P is usually written as $\alpha \rightarrow \beta$. Observe that every *CFG* is a *PSG*. In a *PSG* a nonempty subword is replaced at each rewriting step rather than a single symbol. More formally for α in $(N \cup \Sigma)^+$ and β in $(N \cup \Sigma)^*$ we write

$$\alpha \Rightarrow \beta$$

in G, if $\alpha = \alpha_1\alpha_2\alpha_3$, $\beta = \alpha_1\beta_2\alpha_3$, for some α_1 and α_3 in $(N \cup \Sigma)^*$ and some $\alpha_2 \rightarrow \beta_2$ in P.

We extend $\Rightarrow$ to $\Rightarrow^i$, $\Rightarrow^+$, and $\Rightarrow^*$ in the usual way and we define *the language generated by G*, which is denoted by $L(G)$, as

$$L(G) = \{x : x \text{ is in } \Sigma^* \text{ and } S \Rightarrow^* x \text{ in } G\}$$

The *family of phrase structure languages* is denoted by $\boldsymbol{L}_{PSG}$ and is defined by

$$\boldsymbol{L}_{PSG} = \{L : L = L(G), \text{ for some } PSG,\ G\}$$

Example 11.2.1 We give a *PSG* to generate $\{a^i b^i c^i : i \geq 1\}$, a non-context-free language.

The *PSG* has context-free productions which generate $a^i L(bc)^i R$, $i \geq 1$, where L and R are nonterminals. It has other productions to ripple all the c's to the right of all the b's and, finally productions to move L to the right over b^*c^* until it meets R, when they are both erased.

$$\left.\begin{aligned} S &\to aAbcR \\ A &\to aAbc \mid L \end{aligned}\right\} \quad \text{generates } a^i L(bc)^i R$$

$$cb \to bc \qquad \text{interchange } b\text{'s and } c\text{'s}$$

$$\left.\begin{aligned} Lb &\to bL \\ Lc &\to c\overline{L} \\ \overline{L}c &\to c\overline{L} \\ \overline{L}R &\to \lambda \end{aligned}\right\} \quad \text{checks that all } b\text{'s precede all } c\text{'s.}$$

The production $cb \to bc$ can be applied during the generation of some $a^i L(bc)^i R$ and during the application of the checking productions. But this has no effect on the generated word. However, if a b appears after a c and this is met by $\overline{L}$, then the derivation doesn't terminate. For example,

$$S \Rightarrow aAbcR \Rightarrow aaAbcbcR \Rightarrow aaLbcbcR \Rightarrow aabLcbcR \Rightarrow aabc\overline{L}bcR$$

and no production is applicable. However, we have

$$\begin{aligned} S &\Rightarrow^+ aabLcbcR \Rightarrow aabLbccR \Rightarrow aabbLccR \Rightarrow aabbc\overline{L}cR \\ &\Rightarrow aabbcc\overline{L}R \Rightarrow aabbcc \end{aligned}$$

A formal proof that G generates $\{a^i b^i c^i : i \geq 1\}$ is left to the Exercises.

We sketch the proofs that every *DTM* can be simulated by some *PSG*, and vice versa.

Lemma 11.2.1 *Let $M = (Q,\Sigma,\Gamma,\delta,s,f)$ be an arbitrary DTM. Then a PSG $G = (N,\Sigma,P,S)$ with $L(G) = L(M)$ can be constructed from M.*

Proof sketch: The *PSG* first generates words of the form $x\$x^R s ¢$, for x in Σ^* and for nonterminals s, $\$$, and $¢$. Second, the *PSG* simulates the computation of M on x using $\$x^R s ¢$ (sx is the initial configuration of M). If M accepts x, then the *PSG* derives a word of the form $x\$x_1 f x_2 ¢$. Third, the *PSG* erases $\$x_1 f x_2 ¢$ leaving x.

We require that $Q \subseteq N$ and $\Gamma - \Sigma \subseteq N$. We define P in three groups corresponding to these three stages.

(1) Generating $x\$x^R s¢$.

$$S \to As¢ \quad A \to aAa, \text{ for all } a \text{ in } \Sigma \quad A \to \$$$

(2) Simulating the computation of M.
For all a, c in Γ, for all p in $Q-\{f\}$,

$apc \to bqc$ if $\delta(p,a) = (q,b,\lambda)$

$apc \to qbc$ if $\delta(p,a) = (q,b,R)$

$apc \to bcq$ if $\delta(p,a) = (q,b,L)$

For all c in Γ, for all p in $Q-\{f\}$,

$\$pc \to \$aqc$ if $\delta(p,\boldsymbol{B}) = (q,a,\lambda)$

$\$pc \to \$qac$ if $\delta(p,\boldsymbol{B}) = (q,a,R)$

$\$pc \to \$acq$ if $\delta(p,\boldsymbol{B}) = (q,a,L)$

Since we are simulating M on the reverse of its tape we move right in the *PSG* when a move left occurs in M and analogously for a move right in M.

(3) Erasing a final configuration.

$$\left.\begin{array}{l} fa \to f \\ af \to f \end{array}\right\} \text{ for all } a \text{ in } \Gamma$$

$$\$f¢ \to \lambda$$

It should be clear that $s \Rightarrow^* x$ in Σ^* if and only if x is in $L(M)$. However, because group (2) productions can be applied before group (1) productions are finished a formal proof is nontrivial. We can avoid this difficulty by forcing group (1) productions to be applied before any group (2) production is used. This requires that we modify group (1) productions as follows.

(1′) $S \to A¢$

$A \to aAa$, for all a in Σ

$A \to \$\overline{\$}$

$\overline{\$}a \to a\overline{\$}$, for all a in Σ

$\overline{\$}¢ \to s¢$

The start state of M is only introduced after all group (1) productions have been applied. ■

Theorem 11.2.2 $\boldsymbol{L}_{DTM} = \boldsymbol{L}_{PSG}$.

Proof sketch: Lemma 11.2.1 establishes $\boldsymbol{L}_{DTM} \subseteq \boldsymbol{L}_{PSG}$. The reverse containment requires that we simulate derivations in a *PSG* by a *DTM*. Let $G = (N,\Sigma,P,S)$ be a *PSG*. Construct an off-line *NTM*, $M = (Q, \Sigma, \Gamma,\delta,s,f)$, such that $L(M) = L(G)$ as follows.

M is given some word x on its input tape. M writes $\$S¢$ on its work tape and nondeterministically chooses a production from P with which to rewrite S. After this initial rewrite step, M nondeterministically chooses a production $\alpha \rightarrow \beta$ from P and nondeterministically chooses an occurrence of α on the work tape which is to be rewritten (there may be none). After each rewriting step M checks to see if the work tape contains $\$y¢$ with y in Σ^*. If it does not, then M repeats the rewriting step. If it does, then M tests to see if $x = y$. If $x = y$, then M accepts x, otherwise M hangs.

Clearly, if $S \Rightarrow^* x$ in G, then M has an accepting configuration sequence for x. By the results of Chapter 6, $L(M)$ is in $\mathbf{L}_{DTM}$. ■

11.2.2 Interactive Lindenmayer Grammars

For $m,n \geq 0$, an (m,n) *Lindenmayer grammar*, or $(m,n)LG$, G is a quadruple (N,Σ,P,S), where

- N is a *nonterminal alphabet*;
- Σ is a *terminal alphabet*, $N \cap \Sigma = \varnothing$;
- $P \subseteq (N \cup \Sigma)^{\leq m} \times (N \cup \Sigma) \times (N \cup \Sigma)^{\leq n} \times (N \cup \Sigma)^*$ is a finite set of *productions*, where $A^{\leq j} = \bigcup_{i=0}^{j} A^i$; and
- S in N is a *sentence symbol*.

A production (β,A,γ,α) in P is usually written as $\beta,A,\gamma \rightarrow \alpha$ and is read as: A is rewritten as α if it occurs with left context β and right context γ, where $|\beta| \leq m$ and $|\gamma| \leq n$. If $m = n = 0$ we have a context-free Lindenmayer grammar. An *interactive Lindenmayer grammar*, or *ILG*, is an $(m,n)LG$, for some $m,n \geq 0$.

In an interactive Lindenmayer grammar all symbols in a sentential form must be rewritten simultaneously according to their context. More formally for $\alpha = A_1 \cdots A_k$, where $k \geq 1$ and A_i is in $N \cup \Sigma$, $1 \leq i \leq k$, we write

$$\alpha \Rightarrow \beta$$

in G if

$$A_{i-m} \cdots A_{i-1}, A_i, A_{i+1} \cdots A_{i+n} \rightarrow \alpha_i$$

is in P, $1 \leq i \leq k$, $\beta = \alpha_1 \cdots \alpha_k$, and $A_{1-m} = \cdots = A_0 = A_{k+1} = \cdots = A_{k+m} = \lambda$, by convention.

$\Rightarrow^i$, $\Rightarrow^+$, and $\Rightarrow^*$ are defined as usual, as is the language generated by G which is denoted by $L(G)$ and defined by

$$L(G) = \{x : x \text{ is in } \Sigma^* \text{ and } S \Rightarrow^* x \text{ in } G\}.$$

The *family of interactive Lindenmayer languages* is denoted by $\mathbf{L}_{ILG}$ and is defined as

$$\mathbf{L}_{ILG} = \{L : L = L(G) \text{ for some } ILG\ G\}$$

Example 11.2.2 We show how $\{a^i b^i c^i d : i \geq 1\}$ can be generated by a $(1,0)LG$. It has productions

$$
\begin{array}{lll}
\lambda, S \rightarrow aAbBCc & & \\
a, A \rightarrow aA \mid \lambda & a, a \rightarrow a & \lambda, a \rightarrow a \\
b, B \rightarrow bB \mid \lambda & b, b \rightarrow b & A, b \rightarrow b \\
B, C \rightarrow Cc \mid \lambda & c, c \rightarrow c & C, c \rightarrow c
\end{array}
$$

Now,

$$S \Rightarrow aAbBCc \Rightarrow aaAbbBCcc$$

using $\lambda, a \rightarrow a$; $a, A \rightarrow aA$; $A, b \rightarrow b$; $B \rightarrow bB$; $B, C \rightarrow Cc$; $C, c \rightarrow c$. This in turn yields

$$aabbccd$$

using the erasing productions for A, B, and C. But $a^2b^2c^2d$ cannot be further rewritten since there are no productions for a,b; b,c; or c,d. This absence forces A, B, and C to give either aA, bB, and Cc, or λ in one rewriting step. If we rewrite one of A, B, or C using a terminating production and one of them with a nonterminating production we obtain one of the pairs a,b; b,c; b,c; or B,c; which prevents further rewriting.

Lemma 11.2.1 and Theorem 11.2.2 can be modified without too much difficulty for *ILGs* rather than *PSGs*.

Lemma 11.2.3 *Let* $M = (Q, \Sigma, \Gamma, \delta, s, f)$ *be an arbitrary DTM. Then a* $(1,1)LG$ $G = (N, \Sigma, P, S)$ *with* $L(G) = L(M)$ *can be constructed from M.*

Proof: We leave to the reader the modification of the construction in Lemma 11.2.1. Note that a *PSG* production

$$apc \rightarrow qbc$$

can be simulated with $(1,1)LG$ productions

$$
\begin{array}{ll}
a, p, c \rightarrow b & \\
d, a, p \rightarrow q, & \text{for all } d \text{ in } \Gamma \cup \{\$\} \\
p, c, e \rightarrow c, & \text{for all } e \text{ in } \Gamma \cup \{¢\}
\end{array}
$$

while

$$apc \rightarrow bcq$$

requires

$$a, p, c \rightarrow \bar{q}$$

$d,a,p \to b,$ for all d in $\Gamma \cup \{\$\}$
$p,c,e \to c,$ for all e in $\Gamma \cup \{¢\}$

and

$b,\bar{q},c \to c$
$\bar{q},c,e \to q,$ for all e in $\Gamma \cup \{¢\}$
$d,b,\bar{q} \to b,$ for all d in $\Gamma \cup \{\$\}$

where $\bar{q}$ is a new nonterminal which is uniquely chosen for the production $apc \to bcq$ in the *PSG*. This simulation requires two steps when M moves left.

To produce a faithful simulation of the *DTM* we ensure that each move of the *DTM* requires two steps in the equivalent $(1,1)LG$. The details are left to the Exercises. ■

However, we can strengthen this lemma by simulating each *DTM* with a (1,0)LG. For this purpose we need the following definition and result.

Definition An *ILG*, $G = (N,\Sigma,P,S)$, is said to be *synchronized* if $P \subseteq N^* \times N \times N^* \times (N^* \cup \Sigma^*)$, that is, only nonterminals can be rewritten, they can be rewritten only as a terminal word or a nonterminal word, and they can only be rewritten in a nonterminal context.

Lemma 11.2.4 *Let $G = (N,\Sigma,P,S)$ be an ILG. Then there exists a synchronized ILG, $G' = (N',\Sigma,P',S)$ with $L(G') = L(G)$.*

Proof: Let $\bar{\Sigma} = \{\bar{a} : a \text{ is in } \Sigma\}$ and $N' = N \cup \bar{\Sigma}$. P' contains

(i) $A \to \bar{\alpha}$, if $A \to \alpha$ is in P, where A is in N and $\bar{\alpha}$ in N^* is α with every terminal replaced by its barred version;
(ii) $\bar{a} \to \bar{\alpha}$, if $a \to \alpha$ is in P, where a is in Σ and $\bar{\alpha}$ in N^* is α with every terminal replaced by its barred version; and
(iii) $\bar{a} \to a,$ for all a in Σ.

Now, if

$$S \Rightarrow \alpha_1 \Rightarrow \alpha_2 \Rightarrow \cdots \Rightarrow \alpha_m$$

in G with α_m in Σ^*, then

$$S \Rightarrow \bar{\alpha}_1 \Rightarrow \bar{\alpha}_2 \Rightarrow \cdots \Rightarrow \bar{\alpha}_m$$

in G', and using productions of type (iii)

$$\bar{\alpha}_m \Rightarrow \alpha_m$$

Conversely, if

$$S \Rightarrow \alpha_1 \Rightarrow \cdots \Rightarrow \alpha_m$$

in G' with α_m in Σ^*, then

$$S \Rightarrow \alpha'_1 \Rightarrow \cdots \Rightarrow \alpha'_{m-1}$$

in G, where $\alpha'_{m-1} = \alpha_m$, and in general α'_i is the unbarred version of α_i. Thus, $L(G) = L(G')$. ■

We now strengthen Lemma 11.2.3 to require only left context.

Lemma 11.2.5 *Let $G = (N,\Sigma,P,S)$ be an arbitrary $(1,1)LG$. Then a $(1,0)LG$ $G' = (N',\Sigma,P',S')$ with $L(G') = L(G)$ can be constructed from G.*

Proof sketch: The essential idea is that one rewriting step in G becomes two steps in G' in which the first one accumulates the left context in each symbol. Let $\$$ be a new nonterminal which is not in N. If we have

$$A_1 \cdots A_k \Rightarrow \alpha_1 \cdots \alpha_k$$

where $S \Rightarrow^* A_1 \cdots A_k$ in G, for some A_i in N, $1 \le i \le k$, for some $k \ge 1$, then in G' we have

$$A_1 \cdots A_k\$ \Rightarrow [A_1A_2][A_2A_3] \cdots [A_{k-1}A_k][A_k\$]$$

using the $(1,0)$ productions $\lambda,A_1 \to \lambda$; $A_{i-1},A_i \to [A_{i-1}A_i]$, $1 < i \le k$; and $A_k,\$ \to [A_k,\$]$. At the second step we have $[A_1A_2] \cdots [A_k\$] \Rightarrow \alpha_1 \cdots \alpha_k$ using the productions $\lambda,[A_1A_2] \to \alpha_1$; $[A_{i-1}A_i],[A_iA_{i+1}] \to \alpha_i$, $1 < i < k$; and $[A_{k-1}A_k],[A_k\$] \to \alpha_k\$$.

For this simulation to work correctly we require that G be synchronized (by Lemma 11.2.4 we may assume this property to hold). Furthermore, if $\alpha_1 \cdots \alpha_k$ is in Σ^* we must ensure that we obtain $\alpha_1 \cdots \alpha_k$ in G'. For this purpose we assume that we have productions

$$\lambda,a \to \lambda;\ a,b \to [ab];\ a,\$ \to [a\$]$$

in P', for all a, b in Σ. These enforce the following derivation step:

$$\alpha_1 \cdots \alpha_k \Rightarrow [a_1a_2] \cdots [a_m\$]$$

where $a_1 \cdots a_m = \alpha_1 \cdots \alpha_k$. Finally, we also include productions $[ab] \to a$, for all a, b in Σ, and $[a\$] \to a$, for all a in Σ. These ensure that

$$[a_1a_2] \cdots [a_m\$] \Rightarrow a_1 \cdots a_m$$

as desired. ■

Finally, simulating an *ILG* with an *NTM* is similar and simpler than the simulation of a *PSG*. The proof of the following theorem is left to the reader.

Theorem 11.2.6 $\boldsymbol{L}_{ILG} = \boldsymbol{L}_{DTM}$.

11.3 PARSING

Parsing is such an important step in the compilation of programming languages that it has been a major research area. There are two principal research directions. First, there have been many attempts to define subclasses of context-free grammars which yield efficient parsers. These subclasses should include most of the specific grammars for current programming languages. Second, the production of optimal or nearly optimal parsers with respect to space and time is of great practical interest. In Section 11.3.1 we consider an efficient parser for context-free grammars in general. In Section 11.3.2 we examine deterministic top-down parsing and $LL(k)$ grammars and in Section 11.3.3 we examine deterministic bottom-up parsing and $LR(k)$ grammars.

We use the word "parsing" somewhat loosely to mean an algorithm to solve *CFG* MEMBERSHIP. Strictly speaking, a parsing algorithm also produces a syntax tree of the given word, if it is in the language. However, the three methods we consider can be easily modified to produce a syntax tree.

11.3.1 General *CFG* Parsing

Given a *CFG*, $G = (N,\Sigma,P,S)$ and a word x in Σ^* we can determine if x is generated by G by the simple technique given in Section 10.1. Assuming G is a unit-free, λ-free *CFG* we examine all leftmost sentential derivations of length at most $2|x|-1$. If each nonterminal in G has at most p productions, then there are at most $p^{2|x|-1}$ derivations. Since $p \geq 2$, in general, this implies there are $O(2^{|x|})$ derivations, that is, an exponential number! In order to demonstrate that x is not in $L(G)$ we need to examine all of these! Clearly, this is an absurdly inefficient algorithm, which raises the question of the existence of an efficient algorithm.

In Section 8.2 an $O(|x|^{10})$ time algorithm is given for deciding whether or not a word x is accepted by an *rNPDA*. It is the purpose of this section to sketch an algorithm for *CFGs* which requires $O(|x|^3)$ time.

For simplicity we assume G is in Chomsky normal form, that is, it only has productions of the form $A \rightarrow a$ and $A \rightarrow BC$, where A, B, C are nonterminals and a is a terminal.

Consider $x = a_1 \cdots a_n$, a_i is in Σ, $1 \leq i \leq n$, $n \geq 1$. The algorithm is a bottom-up algorithm and is an example of a dynamic programming algorithm. Beginning with x it attempts to build a "syntax tree" for x with root S. Pictorially we have Figure 11.3.1. Since terminals can only be produced via productions of the form $A \rightarrow a$, we erect a first layer as shown in Figure 11.3.2, where each $S_{i,i}$ is the set of all nonterminals that produce a_i. From hereon in only productions of the form $A \rightarrow BC$ can be applied. This implies that nonterminals in adjacent sets can be merged into one set at the next layer as in Figure 11.3.3, where $S_{i,i+1}$ is the set of all nonterminals A which have a production $A \rightarrow BC$ with B in S_i and C in S_{i+1}, $1 \leq i \leq n-1$. In other words, A is in $S_{i,i+1}$ if and only if there is a derivation $A \Rightarrow^+ a_i a_{i+1}$. At the third layer we construct sets $S_{i,i+2}$, $1 \leq i \leq i-2$, and, in general, at the jth layer we construct sets $S_{i,i+j-1}$,

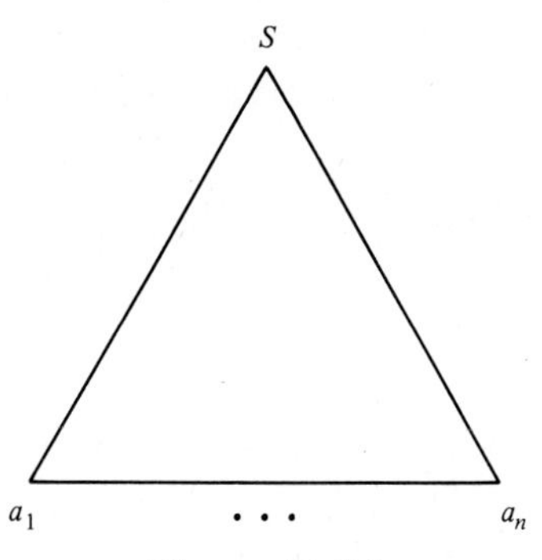

Figure 11.3.1

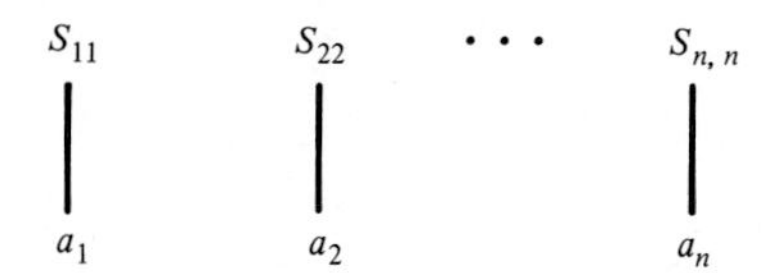

Figure 11.3.2

$1 \leq i \leq n-j+1$. Finally, x is in $L(G)$ if and only if $S_{1,n}$, the only set in the nth layer, is nonempty, since A is in $S_{i,j}$ if and only if $A \Rightarrow^+ a_i \cdots a_j$.

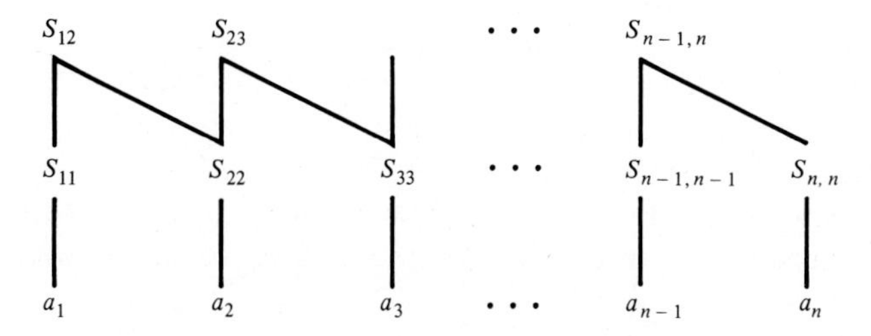

Figure 11.3.3

But how do we compute $S_{i,i+j-1}$? By definition A is in $S_{i,i+j-1}$ if and only if $A \Rightarrow^+ a_i \cdots a_{i+j-1}$ in G. For $j > 1$, this implies there is a production $A \rightarrow BC$ in P such that $B \Rightarrow^+ a_i \cdots a_{i+k-1}$ and $C \Rightarrow^+ a_{i+k} \cdots a_{i+j-1}$, for some k, $1 \leq k \leq j-1$. Equivalently there is a production $A \rightarrow BC$ in P such that B is in $S_{i,i+k-1}$ and C is in $S_{i+k,i+j-1}$, for some k, $1 \leq k < j-1$.

Observe that for each possible value of k the sets $S_{i,i+k-1}$ and $S_{i+k,i+j-1}$ are already available on levels k and $j-k$, respectively. So for each pair (B,C) in $S_{i,i+k-1} \times S_{i+k,i+j-1}$ if there is a production $A \rightarrow BC$ in P, for some nonterminal A, we add A to $S_{i,i+j-1}$. This process is carried out for all values of k,

$1 \le k \le j-1$.

Before analyzing this solution to the membership problem we consider an example.

Example 11.3.1 Let G be given by the productions

$S \to a \mid LB \mid FA \mid SC$

$T \to a \mid LB \mid FA$

$F \to A \mid LB$

$A \to IT;\ B \to SR;\ C \to JT;$

$I \to *;\ J \to +;\ L \to (;\ R \to)$

It is the expression *CFG* of Example 4.2.6 with barred symbols replaced by uppercase letters.

Letting $x = a + a * a$ we obtain the first level; see Figure 11.3.4. At the second level we have Figure 11.3.5; at the third level, Figure 11.3.6; at the fourth level, Figure 11.3.7; and, finally, at the fifth level, Figure 11.3.8. Hence, $S \Rightarrow^+ a + a * a$.

Figure 11.3.4

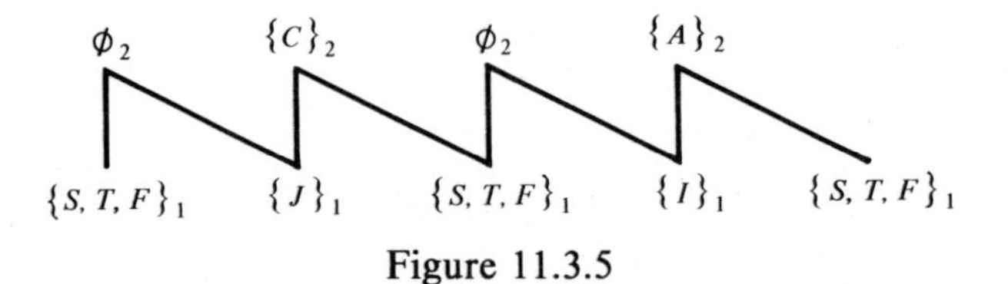

Figure 11.3.5

The size of the nonterminal sets is bounded by $m = \#N$, which is a constant *with respect to the given grammar* G. To compute an $S_{i,i+j-1}$ we consider at most m^2 pairs of nonterminals for each value of k, $1 \le k \le j-1$. Clearly, $k \le n$; therefore, the computation of each $S_{i,i+j-1}$ involves at most $m^2 n$ pairs. For each pair (B,C) we find all productions $A \to BC$ in P. Now, P has at most m^3 productions, so this involves m^3 steps. So the computation of each $S_{i,i+j-1}$ takes $O(m^5 n)$ time. Finally, there are at most n^2 such sets to be computed; therefore, they require $O(m^5 n^3)$ time overall. Since m is a constant with respect to G

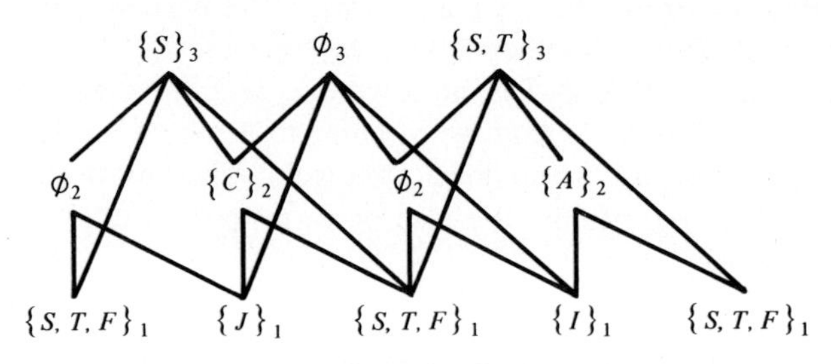

Figure 11.3.6

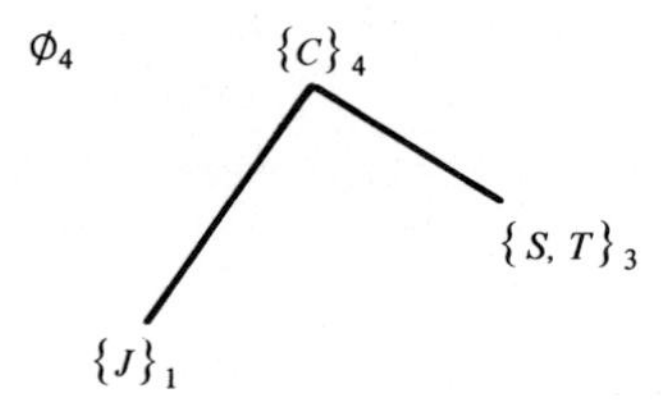

Figure 11.3.7

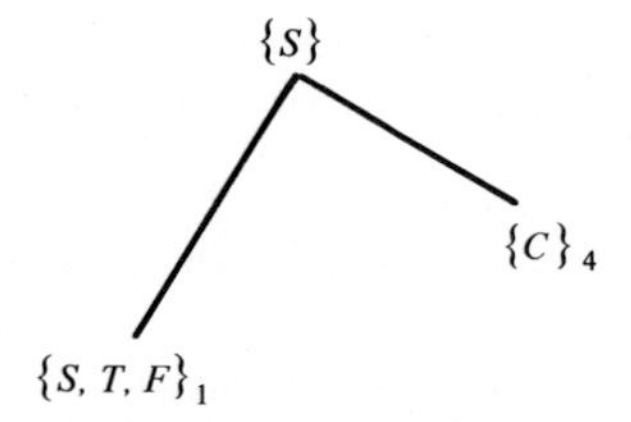

Figure 11.3.8

this is an $O(n^3)$ time algorithm.

It is worth remarking that by some straightforward preprocessing we can avoid the search through P. We set up an $m \times m$ matrix which is indexed by $N \times N$. The entry corresponding to row B and column C is the set of all nonterminals A having a production $A \rightarrow BC$ in P. This reduces the computation of each $S_{i,i+j-1}$ to $O(m^2n)$ time.

11.3.2 Deterministic Top-Down Parsing

The ability to parse a word of length n with unit-free, λ-free *CFGs* in $O(n^3)$ time is an important result. However, as far as compilers are concerned the method of Section 11.3.1 is too inefficient. Therefore, we consider alternative techniques in this and the following subsection. Our aim is solve the membership problem for unambiguous *CFGs* in $O(n)$ time. Clearly, this is the best we can hope for. We arrive at such an algorithm by further restricting the class of *CFGs*.

Given a *CFG*, $G = (N, \Sigma, P, S)$ and a word x in Σ^*, we wish to determine if x is in $L(G)$. The basic idea is that we attempt to generate a leftmost derivation for x from S, that is, we use a top-down approach. This, by itself, is not enough to produce an efficient algorithm. We also perform look-ahead. To illustrate this we consider an example.

Example 11.3.2A Let G have productions

$$S \rightarrow F + S \mid F * S \mid F$$
$$F \rightarrow a$$

and the word $x = a + a * a$. Beginning with S we have three choices of production, $S \rightarrow F + S$, $S \rightarrow F * S$, and $S \rightarrow F$. Looking at the first two symbols of x we can see that neither $S \overset{L}{\Rightarrow} F * S$ nor $S \overset{L}{\Rightarrow} F$ will yield a+, but $S \overset{L}{\Rightarrow} F + S$ will. So we have

$$S \overset{L}{\Rightarrow} F + S$$

Now, $F \rightarrow a$ enables $a+$ to be obtained. So we have

$$F + S \overset{L}{\Rightarrow} a + S$$

and the first two symbols of x have been obtained. At this stage neither $S \rightarrow F + S$ nor $S \rightarrow F$ will yield $a\ *$, the third and fourth symbols. However, $S \rightarrow F * S$ does, so

$$a + S \overset{L}{\Rightarrow} a + F * S \overset{L}{\Rightarrow} a + a * S$$

and the first four symbols of x have been generated. Now, only $S \rightarrow F$ will yield a *and nothing else*, so we obtain finally

$$S \overset{L}{\Rightarrow} F + S \overset{L}{\Rightarrow} a + S \overset{L}{\Rightarrow} a + F * S$$

$$\overset{L}{\Rightarrow} a + a * S \overset{L}{\Rightarrow} a + a * F \overset{L}{\Rightarrow} a + a * a$$

In the above example, we were able, at each step, to choose the unique production that led to a match. This choice was based on the leftmost nonterminal and the next two unmatched symbols in the input word. In general, we are given a nonterminal A, A-productions

$$A \rightarrow \alpha_1 \mid \alpha_2 \mid \cdots \mid \alpha_m$$

and the next two symbols ab of the input. We choose the production $A \rightarrow \alpha_i$ if either $\alpha_i \overset{L}{\Rightarrow}{}^+ ab\beta$, for some β, or $S \overset{L}{\Rightarrow}{}^+ uA\beta \overset{L}{\Rightarrow} u\alpha_i\beta \overset{L}{\Rightarrow}{}^+ uab\gamma$ for some β

and γ. The second case occurs if $\alpha_i \overset{L}{\Rightarrow} a$ and b is supplied by its context. This situation occurred above in

$$F+S \overset{L}{\Rightarrow} a+S$$

since $F \overset{L}{\Rightarrow} a$, not $a+$.

The *CFG* of Example 11.3.2 is said to be an $LL(2)$ grammar.

A parsing algorithm for an $LL(2)$ grammar $G = (N,\Sigma,P,S)$ uses a table indexed by N and $\Sigma^{\leq 2} = \{\lambda\} \cup \Sigma \cup \Sigma^2$. For each nonterminal A and each terminal word z in $\Sigma^{\leq 2}$, the corresponding entry in the table contains either an A-production or an error entry. This is because by definition there can be at most one A-production which, in some context, yields z.

Example 11.3.2B We have Table 11.3.1, where the productions have been numbered as

$1 : S \rightarrow F+S$

$2 : S \rightarrow F*S$

$3 : S \rightarrow F$

$4 : F \rightarrow a$

and E denotes error. We give an algorithm, below, to construct this table for $LL(1)$ grammars.

Table 11.3.1

	λ	a	$+$	$*$	aa	$a+$	$a*$
S	E	3	E	E	E	1	2
F	E	4	E	E	E	4	4

	$+a$	$++$	$+*$	$*a$	$*+$	$**$
S	E	E	E	E	E	E
F	E	E	E	E	E	E

The parsing algorithm keeps track of the leftmost nonterminal and the look-ahead at each step using a stack (in fact a modified pushdown transducer). A simple table look-up yields either an error or a production to be applied. If there is an error, then the input word is not in the language. We illustrate the algorithm by way of an example.

Example 11.3.2C Consider the input word $x = a+a*a$ once more. We can

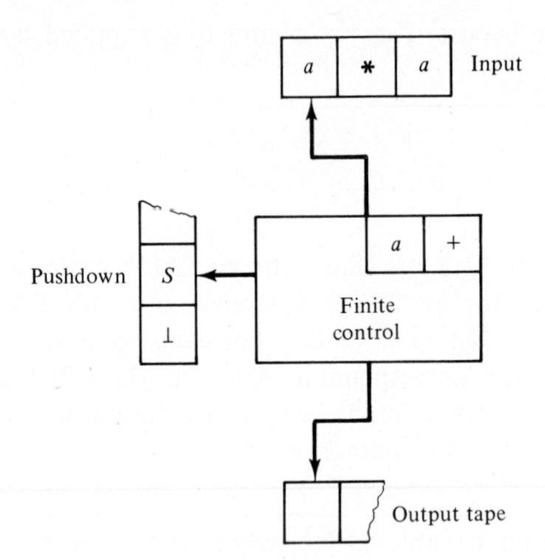

Figure 11.3.9

picture a pushdown transducer M initialized as shown in Figure 11.3.9. The finite control includes the look-ahead and Table 11.3.1, the output tape is blank, and the pushdown contains $\perp S$. M determines from Table 11.3.1 the action to be taken with S, the leftmost nonterminal, and $a+$, the look-ahead. Here it is rewrite S with production $1 : S \to F + S$ and output 1. This yields Figure 11.3.10.

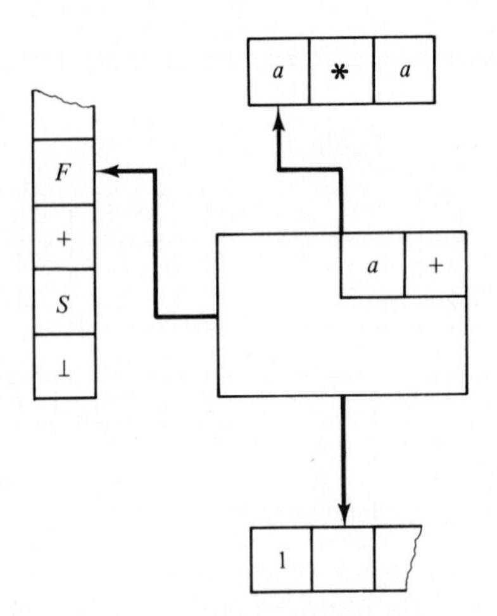

Figure 11.3.10

F with $a+$ results in applying $4 : F \rightarrow a$, giving Figure 11.3.11. Since the top symbol of the pushdown is a terminal we match, pop, and advance the look-ahead. This occurs twice resulting in Figure 11.3.12.

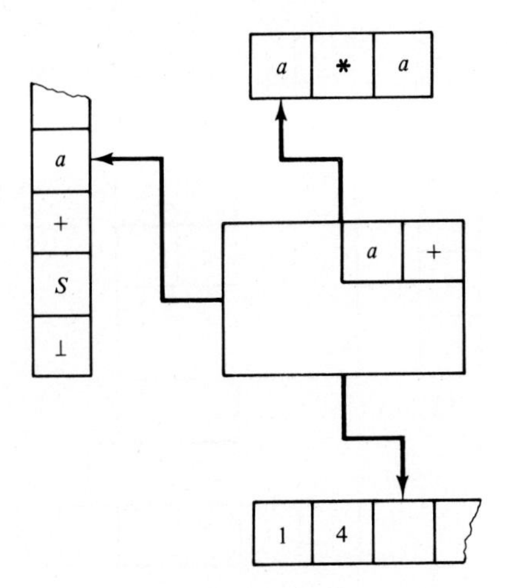

Figure 11.3.11

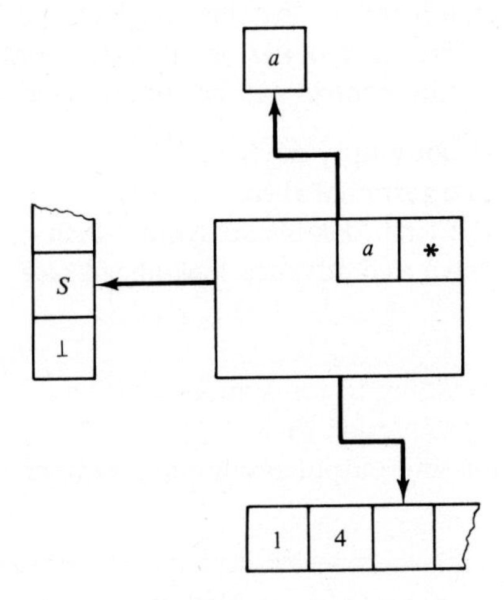

Figure 11.3.12

Now, S with $a*$ results in replacing S with $F*S$ and outputting 2. Then F with $a*$ replaces F with a and outputs 4 giving the situation of Figure 11.3.13 after matching of the input has taken place.

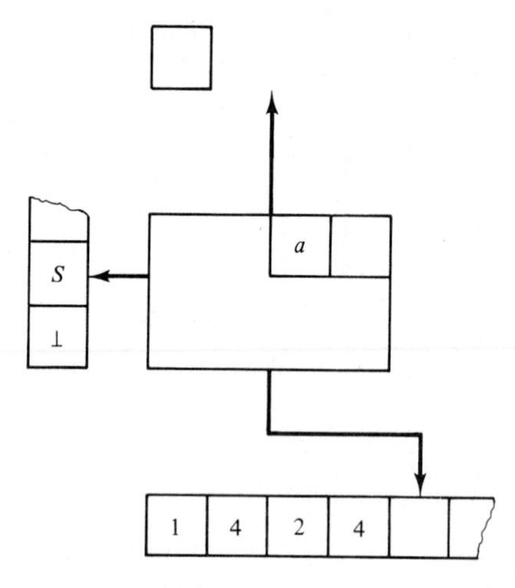

Figure 11.3.13

The final three steps are: (i) Replace S by F and output 3; (ii) replace F by a and output 4; (iii) and match input. At this stage the pushdown is empty and the input is exhausted; therefore, $a+a*a$ is in $L(G)$ and 142434 is its leftmost parse. The action of the finite control can be summarized as follows

while pushdown and input not empty **do**
    if top of pushdown is a terminal **then**
    begin if it matches the first look-ahead symbol **then**
            pop pushdown and advance look-ahead **else**
            error
    end else
    if table entry corresponding to top symbol and
        look-ahead is a production **then**
        apply production and output production number **else**
            error;

Since the table is constant with respect to the given grammar, the parsing algorithm requires $O(|x|)$ time for an input word x, as desired.

We now define $LL(k)$ grammars formally.

Definition Let k be a nonnegative integer and $G = (N,\Sigma,P,S)$ be a CFG. Then $FIRST_k(\alpha)$, where α is in $(N \cup \Sigma)^*$, is defined by

$$FIRST_k(\alpha) = \{x : x \text{ is in } \Sigma^* \text{ and either } |x| < k \text{ and } \alpha \Rightarrow^* x \text{ or}$$
$$|x| = k \text{ and } \alpha \Rightarrow^* xy, \text{ for some } y \text{ in } \Sigma^*\}$$

Now, G is an *$LL(k)$ grammar* if, whenever

$$S \overset{L}{\Rightarrow}{}^+ uAB \overset{L}{\Rightarrow} u\alpha_1 B \overset{L}{\Rightarrow}{}^+ uv$$

$$S \overset{L}{\Rightarrow}{}^+ uAB \Rightarrow u\alpha_2 B \overset{L}{\Rightarrow}{}^+ uw$$

and

$$FIRST_k(v) = FIRST_k(w)$$

then $\alpha_1 = \alpha_2$.

In analogy with the example above we require that the same input portion u has been matched in both cases.

Given a CFG, G, and an integer k we can determine if G is $LL(k)$. For this purpose we need the following characterization theorem.

Theorem 11.3.1 *Let $G = (N,\Sigma,P,S)$ be a CFG. Then G is $LL(k)$ if and only if*

Whenever $A \to \alpha_1$ and $A \to \alpha_2$ are distinct productions in P, then $FIRST_k(\alpha_1\beta) \cap FIRST_k(\alpha_2\beta) = \varnothing$, for all $uA\beta$ such that $S \overset{L}{\Rightarrow}{}^* uA\beta$.

Proof: This is left to the Exercises. ■

Note that $FIRST_k(\alpha_1\beta) = FIRST_k(FIRST_k(\alpha_1)FIRST_k(\beta))$ if we extend $FIRST_k$ to sets of words. Also for a given CFG G there are only finitely many distinct $FIRST_k$ sets.

To prove that a given CFG is $LL(k)$, for a given $k \geq 1$, we demonstrate that the condition $FIRST_k(\alpha_1\beta) \cap FIRST_k(\alpha_2\beta) = \varnothing$ does indeed hold for all distinct productions $A \to \alpha_1$ and $A \to \alpha_2$, and for all $uA\beta$ such that $S \overset{L}{\Rightarrow}{}^* uA\beta$. This test can be carried out, since there are only a finite number of distinct $FIRST_k(\beta)$, for each A, and there are only a finite number of pairs of distinct A-productions.

This sketch leaves one loose end; how are all distinct $FIRST_k(\beta)$ computed for a given nonterminal A? Essentially, we examine all 1-step leftmost derivations from S that yield $\gamma A\beta$, all 2-step derivations, and so on until no new $FIRST_k(\beta)$ is introduced. See the Exercises for the details.

On the other hand, it is undecidable whether or not a CFG is $LL(k)$, for some k. Once again the proof of this result is left to the Exercises.

To round out this discussion of $LL(k)$ grammars we demonstrate how to decide if a λ-free CFG is $LL(1)$. If a CFG is $LL(k)$, for some $k \geq 1$, then it is non-left-recursive. The proof of this fact is left to the Exercises. Moreover, for $LL(1)$ grammars we obtain the following corollary to Theorem 11.3.1.

Corollary 11.3.2 *Let $G = (N,\Sigma,P,S)$ be a λ-free, reduced CFG. Then G is LL(1) if and only if for all A in N.*

If $A \rightarrow X_1\alpha_1$ and $A \rightarrow X_2\alpha_2$ are distinct productions in P, where X_1, X_2 are in $N \cup \Sigma$, then $FIRST_1(A_1) \cap FIRST_1(A_2) = \varnothing$.

Proof: This is left to the Exercises. ■

This implies we only need compute $FIRST_1(X)$, for each production $A \rightarrow X\alpha$ in P. If X is terminal, then $FIRST_1(X) = \{X\}$. Otherwise we can order the nonterminals using *int* as stated in Lemma 4.2.15. We modify the definition of *int* slightly, that is, for each A in N, $int(A)$ is the largest integer k such that $A \overset{k-1}{\underset{L}{\Rightarrow}} B\alpha \underset{L}{\Rightarrow} a\beta$, for some a in Σ, B in N, and α, β in $(N \cup \Sigma)^*$. Allowing α and β to contain terminals is the only modification we require. Fortunately, it requires no change in the proof of Lemma 4.2.15. This lemma also tells us that for each integer i in the range 1 to *intmax* there is a nonterminal A with $int(A) = i$.

Example 11.3.3 Let G be the *CFG* of Example 4.2.8, that is,

$S \rightarrow TA \mid T$	$A \rightarrow +S$	$D \rightarrow)$
$T \rightarrow FB \mid F$	$B \rightarrow *T$	
$F \rightarrow a \mid (C$	$C \rightarrow SD$	

We first compute $int(A)$, for each nonterminal A in G; see Example 4.2.8. Thus, $int(F) = int(A) = int(B) = int(D) = 1$, $int(T) = 2$, $int(S) = 3$, and, finally, $int(C) = 4$. We now construct Table 11.3.2,the *left set table* of G.

Table 11.3.2 A left set table.

A	+
B	*
D	)
F	a, (
T	F
S	T
C	S

There is one row for each nonterminal and the entry for each nonterminal is the set of leftmost appearances in its productions. For example, $F \rightarrow a$ and $F \rightarrow (C$ implies the left set of F is $\{a, (\ \}$. From the left set table we construct the $FIRST_1$ table in a single pass. For A, B, D, and F their lefts sets are their $FIRST_1$ sets. For T its left set is the $FIRST_1$ set of F. Similarly, $FIRST_1(S) = FIRST_1(T)$ and $FIRST_1(C) = FIRST_1(S)$; see Table 11.3.3. In

Table 11.3.3 A $FIRST_1$ set table.

A	$+$
B	$*$
D	$)$
F	$a, ($
T	$a, ($
S	$a, ($
C	$a, ($

general for a nonterminal A

$$FIRST_1(A) = \bigcup FIRST_1(X)$$

where the union is taken over all X in the left set of A. But by construction $int(X) < int(A)$, for all such X, and, therefore, $FIRST_1(X)$ will have been computed already.

Once the $FIRST_1$ sets have been computed we can test whether the condition given in Corollary 11.3.2 holds or not. In this case it does not hold for the trivial reason that $S \to TA$ and $S \to T$ are both in G and therefore give rise to the same $FIRST_1$ sets. Hence, G is not $LL(1)$. We did not even have to compute $FIRST_1$ sets here, but in general we do.

11.3.3 Deterministic Bottom-Up Parsing

Top-down parsing is, perhaps, the more natural approach to parsing. This is, perhaps, because the goal at each stage is: can a leftmost derivation for a given nonterminal produce the given portion of the input word? This process is close to the way that words are generated in the first place.

Unfortunately, deterministic top-down parsing does have a severe limitation. The class of *LL* grammars is a highly restricted subclass of the *CFGs*. This is enough to ensure that most grammars for programming languages are not *LL*. There are two approaches to this problem. Either we squeeze the given grammar to fit into the *LL* scheme or we turn to other methods. The first approach was used in the ETH PASCAL compiler, while the second approach has been followed for other programming languages.

In this subsection we introduce an alternative scheme, the *LR* scheme, which is a deterministic bottom-up parsing scheme. By bottom-up we simply mean that we attempt to build a derivation backwards from the input word. This is in contrast to top-down methods which attempt to build a derivation from the sentence symbol. When a bottom-up scheme has a sentential form β, it attempts to find an α such that $\alpha \Rightarrow \beta$ and $S \Rightarrow^* \alpha$. Top-down schemes attempt to find a sentential form γ such that $\beta \Rightarrow \gamma$ and $\gamma \Rightarrow^* x$, the input word. Note the symmetry in these approaches. In bottom-up schemes if $\alpha \Rightarrow \beta$, then we know, by construction, that $\alpha \Rightarrow^+ x$. However, it is possible that α is not a sentential form. In top-down schemes we know, by construction, that $S \Rightarrow^* \gamma$, but it is possible that γ does not give rise to x.

On the one hand, in top-down schemes we have some nonterminal A and we have only to decide which A-production to apply. On the other hand, in bottom-up schemes we have some word α and we have to decide which nonterminal and sentential form produced it in one step. This is a more difficult decision than the top-down case. First, we have to decide which appearance of a right-hand side β of a production in the sentential form α is to be replaced. Second, we have to decide which production $A \rightarrow \beta$ should be used to replace β. Third, we have to ensure that this replacement leads to a sentential form.

If $\alpha = \alpha_1\beta\alpha_2$ is a sentential form and $A \rightarrow \beta$ is a production, then replacing α by $\alpha_1 A \alpha_2$ is said to be a *reduction*. A reduction is an un-derivation step.

Example 11.3.4A Let G be

$$S \rightarrow aSb \mid aa$$

Then with input word $aaab$ there are two appearances of aa. If we choose the first appearance we obtain Sab which is not a sentential form. In general, with input $a^{i+2}b$ we must carry out a reduction with $S \rightarrow aa$ for the rightmost subword aa. This yields a^iSb^i which is a sentential form. All other reductions with $S \rightarrow aa$ do not yield a sentential form.

It is tempting to consider *leftmost reductions* in analogy to leftmost derivations. However, Example 11.3.4 demonstrates that a leftmost reduction does not always yield a sentential form in contrast with a leftmost derivation. For this reason we introduce *quasi-leftmost reductions.* A reduction is a quasi-leftmost reduction if it yields a sentential form and, moreover, no reduction to its left yields a sentential form. If the grammar is unambiguous there will be only one such reduction. Furthermore, quasi-leftmost reductions correspond to rightmost derivations. Hence, we are dealing with right sentential forms rather than sentential forms.

Example 11.3.4B Given a^4b^2; Sa^2b^2, $aSab^2$, and a^2Sb^2 are the only possible reductions. Of these Sa^2b^2 is the only leftmost reduction and it is, clearly, not a sentential form. However, a^2Sb^2 is a quasi-leftmost reduction, since a^2Sb^2 is a sentential form and Sa^2b^2 and $aSab^2$ are not.

Example 11.3.5 Let G be

$$S \rightarrow S + T \mid T$$
$$T \rightarrow T * a \mid a$$

Given input word $a + a * a$ we obtain the quasi-leftmost reduction sequence

$T + a * a$

$S + a * a$

$S + T * a$

$S + T$

S

which corresponds to

$$S \overset{R}{\Rightarrow} S + T \overset{R}{\Rightarrow} S + T * a \overset{R}{\Rightarrow} S + a * a \overset{R}{\Rightarrow} T + a * a \overset{R}{\Rightarrow} a + a * a$$

To perform the quasi-leftmost reduction we need to determine the position of the appropriate right-hand side of a production. Such a right-hand side is called a *handle* of a right sentential form. Once a handle β is found we need only determine which production $A \rightarrow \beta$ should be used in the reduction. We can develop a general parsing algorithm for *CFGs* based on this approach, but instead we concentrate on deterministic parsing.

We can enforce determinism in the reduction process in many different ways. We consider one method; the *LR* method.

Carry out two quasi-leftmost reductions on a right-sentential form α, that is, one after the other. In other words, we have $\alpha_2 \overset{R}{\Rightarrow} \alpha_1 \overset{R}{\Rightarrow} \alpha$. Now, the handle of α_1 cannot occur to the left of the handle of α. This implies that we should scan α from left to right until its handle is determined. After a reduction has taken place we continue to scan from this position to determine the next handle.

The *LR* method solves the handle determination problem by using a fixed distance look-ahead from the current position in the right sentential form. The look-ahead together with knowledge of the right sentential form to the left of the current position uniquely determines the action of the parser. The possible actions of the parser are: report error, move to next symbol, or reduce handle. When a handle is to be reduced there is exactly one production that can be used.

We define *LR* grammars more formally before exploring one example in detail. We only deal with λ-free *CFGs* in the following development. This is for simplicity of presentation.

Definition Let $G = (N, \Sigma, P, S)$ be a λ-free *CFG* such that S does not appear in any right-hand side. Let $k \geq 0$ be an integer.

G is an *LR*(k) *grammar* if

(i) $S \overset{R}{\Rightarrow}{}^{*} \alpha A x \overset{R}{\Rightarrow} \alpha\beta x$;

(ii) $S \overset{R}{\Rightarrow}{}^{*} \gamma B y \overset{R}{\Rightarrow} \alpha\beta z$; and

(iii) $FIRST_k(x) = FIRST_k(z)$ imply that $\alpha = \gamma$, $A = B$, and $y = z$.

The k look-ahead determines $A \rightarrow \beta$ is the only production that, applied rightmost, produced $\alpha\beta$. From the reduction point of view $\alpha\beta$ together with $FIRST_k(x)$ determines at most one production $A \rightarrow \beta$ to be used in the reduction of $\alpha\beta$ to αA.

To construct a parser for an $LR(k)$ grammar G we use a technique due to Knuth. Recall that the parser must keep track of the portion of the right sentential form to the left of the current position. There are infinitely many of these. Fortunately, Knuth's technique shows these can be partitioned into a finite number of equivalence classes. The same technique also provides a simple means of obtaining the class for αA from that of $\alpha\beta$, after a reduction has been carried out. We illustrate the technique for $LR(1)$ grammars.

Definition Let $G = (N,\Sigma,P,S)$ be a λ-free CFG.
Given a derivation

$$S \overset{R}{\Rightarrow}{}^{*} \alpha A x \overset{R}{\Rightarrow} \alpha\beta x$$

in G, a prefix γ of $\alpha\beta$ is said to be a *viable prefix* of G. A viable prefix of a right sentential form is a prefix which does not extend beyond the right end of its handle. We say that $[A \rightarrow \beta_1 \cdot \beta_2, u]$ is an *item* for G, if $A \rightarrow \beta_1\beta_2$ is in P and u is in $\Sigma \cup \{\lambda\}$. An item $[A \rightarrow \beta_1 \cdot \beta_2, u]$ is *valid* for $\alpha\beta_1$, a viable prefix of G, if there is a derivation

$$S \overset{R}{\Rightarrow}{}^{*} \alpha A x \overset{R}{\Rightarrow} \alpha\beta_1\beta_2 x$$

such that $u = FIRST(x)$.

The importance of valid items stems from the following

Theorem 11.3.2 *Let $G = (N,\Sigma,P,S)$ be a λ-free CFG such that S does not appear in any right-hand side. Then G is $LR(1)$ if and only if the following condition holds for each u in $\Sigma \cup \{\lambda\}$*

Let $\alpha\beta$ be a viable prefix of a right sentential form of G. If $[A \rightarrow \beta\cdot, u]$ is valid for $\alpha\beta$, then there is no other item $[B \rightarrow \beta_1 \cdot \beta_2, v]$ that is valid for $\alpha\beta$ and has u in $FIRST(\beta_2 v)$.

Proof: This is left to Exercises. ■

This implies that parsing can be carried out by maintaining the set of valid items for the current prefix of the current right sentential form. If such a set contains an item of the form $[A \rightarrow \beta\cdot, u]$, then a reduction is carried out and the set of items is recomputed. Otherwise the current position is advanced one symbol and the next set of items is computed. The sets of items and the transitions from one set to another can be computed in advance, for all possible inputs.

Before demonstrating how to compute the set $\boldsymbol{I}_y$ of valid items for a prefix y we illustrate by example their use in the parsing process.

Example 11.3.6A Let G be

$S \to E$

$E \to E+T \mid T$

$T \to T*a \mid a$

G has eight distinct item sets $\boldsymbol{I}_\lambda$, $\boldsymbol{I}_E$, $\boldsymbol{I}_T$, $\boldsymbol{I}_a$, $\boldsymbol{I}_{E+}$, $\boldsymbol{I}_{T*}$, $\boldsymbol{I}_{E+T}$, and $\boldsymbol{I}_{T*a}$. We construct two tables from them, the *(item set) transition table* and the *action table.* A transition table has rows indexed by item sets and columns indexed by symbols of the given grammar. S is conspicuous by its absence, since it is used only at the final reduction step; it never appears in a viable prefix. The transition table for G is displayed as Table 11.3.4. Note that we use y rather than $\boldsymbol{I}_y$ to index the rows. The entries are also item sets; $\varnothing$ indicates an invalid transition. The transition table gives the next item set when given an item set and the next symbol.

An action table has rows indexed by item sets and columns indexed by all possible look-aheads. The entries are of four kinds: accept, error, reduce, and shift. The action table for G is given in Table 11.3.5. An action table provides the actions of the parser when given the current item set and the current look-ahead. Shift occurs if $[A \to \beta_1 \cdot \beta_2, u]$ is in the current item set and the current look-ahead is in $FIRST(\beta_2)$. Reduce occurs if $[A \to \beta\cdot, u]$ is in the current item set and u is the current look-ahead. Accept is a special case of reduce when $[S \to \beta\cdot, u]$ is in the current item set. An error occurs when none of the other three actions is possible.

Table 11.3.4 Transition Table

	E	T	a	$+$	$*$
λ	E	T	a	$\varnothing$	$\varnothing$
E	$\varnothing$	$\varnothing$	$\varnothing$	$E+$	$\varnothing$
T	$\varnothing$	$\varnothing$	$\varnothing$	$\varnothing$	$T*$
a	$\varnothing$	$\varnothing$	$\varnothing$	$\varnothing$	$\varnothing$
$E+$	$\varnothing$	$E+T$	a	$\varnothing$	$\varnothing$
$T*$	$\varnothing$	$\varnothing$	$T*a$	$\varnothing$	$\varnothing$
$E+T$	$\varnothing$	$\varnothing$	$\varnothing$	$\varnothing$	$T*$
$T*a$	$\varnothing$	$\varnothing$	$\varnothing$	$\varnothing$	$\varnothing$

Table 11.3.5 Action Table

	λ	a	$+$	$*$	
x	error	shift	error	error	
E	accept	error	shift	error	
T	reduce	error	reduce	shift	$E \to T$
a	reduce	error	reduce	reduce	$T \to a$
$E+$	error	shift	error	error	
$T*$	error	shift	error	error	
$E+T$	reduce	error	reduce	shift	$E \to E+T$
$T*a$	reduce	error	reduce	reduce	$T \to T*a$

When performing a shift, the next item set is found from the transition table using the current item set and the current look-ahead. When performing the reduction $\alpha\beta$ becomes αA, the next item set is found from the transition table using the item set for α and look-ahead A. This implies we keep the item sets for all viable prefixes during the parsing process. Usually a stack is used to keep (item set, symbol) pairs of the shifted symbols.

We illustrate the parsing process on the input word $a+a*a+a$

Initially we have

$$\boldsymbol{I}_\lambda \underset{\uparrow}{} a+a*a+a$$

where $\uparrow$ indicates the current position in the right sentential form.

The action table indicates a shift and the transition table gives $\boldsymbol{I}_a$ as the next item set, that is,

$$\boldsymbol{I}_\lambda a \boldsymbol{I}_a \underset{\uparrow}{} +a*a+a$$

$\boldsymbol{I}_a$ and $+$ yield a reduction step using $T \to a$, from which $\boldsymbol{I}_\lambda$ and T give $\boldsymbol{I}_T$

$$\boldsymbol{I}_\lambda T \boldsymbol{I}_T \underset{\uparrow}{} +a*a+a$$

$\boldsymbol{I}_T$ and $+$ yield another reduction $E \to T$ and $\boldsymbol{I}_\lambda$ and E yield $\boldsymbol{I}_E$

$$\boldsymbol{I}_\lambda E \boldsymbol{I}_E \underset{\uparrow}{} +a*a+a$$

$\boldsymbol{I}_E$ and $+$ cause a shift and produce $\boldsymbol{I}_{E+}$

$$\boldsymbol{I}_\lambda E \boldsymbol{I}_E + \boldsymbol{I}_{E+} \underset{\uparrow}{} a*a+a$$

Again we have a shift

$$\boldsymbol{I}_\lambda E \boldsymbol{I}_E + \boldsymbol{I}_{E+} a \boldsymbol{I}_a \underset{\uparrow}{} *a+a$$

This in turns yields the sequence

$$\boldsymbol{I}_\lambda E \boldsymbol{I}_E + \boldsymbol{I}_{E+} T \boldsymbol{I}_{E+T} \underset{\uparrow}{} *a+a$$

$$\boldsymbol{I}_\lambda E \boldsymbol{I}_E + \boldsymbol{I}_{E+} T \boldsymbol{I}_{E+T} * \boldsymbol{I}_{T*} \underset{\uparrow}{} a+a$$

$\boldsymbol{I}_\lambda E \boldsymbol{I}_E + \boldsymbol{I}_{e+} T \boldsymbol{I}_{E+T} * \boldsymbol{I}_{T*} a T_{T*a} \uparrow + a$	reduce $T \rightarrow T * a$
$\boldsymbol{I}_\lambda E \boldsymbol{I}_E + \boldsymbol{I}_{E+} T \boldsymbol{I}_{E+T} \uparrow + a$	reduce $E \rightarrow E + T$
$\boldsymbol{I}_\lambda E \boldsymbol{I}_E \uparrow + a$	
$\boldsymbol{I}_\lambda E \boldsymbol{I}_E + \boldsymbol{I}_{E+} \uparrow a$	
$\boldsymbol{I}_\lambda E \boldsymbol{I}_E + \boldsymbol{I}_{E+} a \boldsymbol{I}_a \uparrow$	reduce $T \rightarrow a$
$\boldsymbol{I}_\lambda E \boldsymbol{I}_E + \boldsymbol{I}_{E+} T \boldsymbol{I}_{E+T} \uparrow$	reduce $E \rightarrow E + T$
$\boldsymbol{I}_\lambda E \boldsymbol{I}_E \uparrow$	accept

To complete our exposition we turn to the computation of item sets.

Example 11.3.6B We begin by computing $\boldsymbol{I}_\lambda$ for the grammar G. For the viable prefix λ the associated set $\boldsymbol{I}_\lambda$ of valid items contains $[S \rightarrow \cdot E, \lambda]$ since

$$S \overset{R}{\Rightarrow}{}^* S \Rightarrow E$$

This implies $[E \rightarrow \cdot E + T, \lambda]$ and $[E \rightarrow \cdot T, \lambda]$ are also in $\boldsymbol{I}_\lambda$, since

$$S \overset{R}{\Rightarrow}{}^* E \overset{R}{\Rightarrow} E + T$$

and

$$S \overset{R}{\Rightarrow}{}^* E \overset{R}{\Rightarrow} T$$

Again this implies

$$[E \rightarrow \cdot E + T, +], [E \rightarrow \cdot T, +]$$
$$[T \rightarrow \cdot T * a, \lambda], [T \rightarrow \cdot a, \lambda]$$

are in $\boldsymbol{I}_\lambda$, by similar arguments.

Finally, we obtain

$$[T \rightarrow \cdot T * a, +], [T \rightarrow \cdot a, +]$$
$$[T \rightarrow \cdot T * a, *], \text{ and } [T \rightarrow \cdot a, *]$$

Letting $\lambda +$ denote λ or $+$, and $\lambda + *$ denote λ or $+$ or $*$ we have

$$\boldsymbol{I}_\lambda = \{[S \rightarrow \cdot E, \lambda], [E \rightarrow \cdot E + T, \lambda +], [E \rightarrow \cdot T, \lambda +], [T \rightarrow \cdot T * a, \lambda + *], [T \rightarrow \cdot a, \lambda + *]\}$$

The computational rules which we have followed can be stated as

(a) *If $S \to \alpha$ is in P, then add $[S \to \cdot\alpha,\lambda]$ to $\boldsymbol{I}_\lambda$.*
(b) *If $[A \to \cdot\beta\alpha,u]$ is in $\boldsymbol{I}_\lambda$ and $B \to \beta$ is in P, then add $[B \to \cdot\beta,x]$ to $\boldsymbol{I}_\lambda$, for all x in $FIRST(\alpha u)$. Repeat this step until no new items are added to $\boldsymbol{I}_\lambda$.*

We now compute $\boldsymbol{I}_X$ for X in $\{E,T,a,+,*\}$. Consider $\boldsymbol{I}_E$. Since $[S \to \cdot E,\lambda]$ is in $\boldsymbol{I}_\lambda$, $[S \to E\cdot,\lambda]$ is in $\boldsymbol{I}_E$. Why? Because

$$S \overset{R}{\Rightarrow}{}^* S \overset{R}{\Rightarrow} E$$

is in G. Similarly, because $[E \to \cdot E+T,\lambda+]$ are in $\boldsymbol{I}_\lambda$, $[E \to E\cdot+T,\lambda+]$ are in $\boldsymbol{I}_E$. In this case there are no more items in $\boldsymbol{I}_E$, that is,

$$\boldsymbol{I}_E = \{[S \to E\cdot,\lambda],[E \to E\cdot+T,\lambda+]\}$$

In a similar manner we obtain

$$\boldsymbol{I}_T = \{[E \to T\cdot,\lambda+],[T \to T\cdot * a,\lambda+*]\}$$

$$\boldsymbol{I}_a = \{[T \to a\cdot,\lambda+*]\},$$

and $\boldsymbol{I}_+ = \boldsymbol{I}_* = \varnothing$. Turning to $\boldsymbol{I}_{EX}$, for X in $\{E,T,a,+,*\}$, we find that $[E \to E+\cdot T,\lambda+]$ are in $\boldsymbol{I}_{E+}$. However, in this case, we also find that $[T \to \cdot T * a,\lambda+]$ and $[T \to \cdot a,\lambda+]$ are in $\boldsymbol{I}_{E+}$, since

$$S \overset{R}{\Rightarrow}{}^* E+Tx \overset{R}{\Rightarrow} E+ax$$

for $x = \lambda$ or $x = +y$. This implies that $[T \to \cdot T * a, *]$ and $[T \to \cdot a, *]$ must also be added to $\boldsymbol{I}_{E+}$, but this is all.

Given $\boldsymbol{I}_\gamma$ and X in $N \cup \Sigma$ to compute $\boldsymbol{I}_{\gamma X}$ we have used the following computational rules

(a) *If $[A \to \alpha\cdot X\beta,u]$ is in $\boldsymbol{I}_\gamma$, then add $[A \to \alpha X\cdot\beta,u]$ to $\boldsymbol{I}_{\gamma X}$.*
(b) *If $[A \to \alpha\cdot B\beta,u]$ is in $\boldsymbol{I}_{\gamma X}$ and $B \to \delta$ is in P, then add $[B \to \cdot\delta,x]$ to $\boldsymbol{I}_{\gamma X}$, for all x in $FIRST(\beta u)$. Repeat this step until no new items are added to $\boldsymbol{I}_{\gamma X}$.*

Continuing we find

$$\boldsymbol{I}_{T*} = \{[T \to T*\cdot a,\lambda+*]\}$$

$$\boldsymbol{I}_{E+T} = \{[E \to E+T\cdot,\lambda+],[T \to T\cdot * a,\lambda+*]\}$$

$$\boldsymbol{I}_{T*a} = \{[T \to T*a\cdot,\lambda+*]\}$$

and all other $\boldsymbol{I}_\gamma$ either equal one of the above items sets or are empty. Indeed, $\boldsymbol{I}_{E+a} = \boldsymbol{I}_a$ and $\boldsymbol{I}_{E+T*} = \boldsymbol{I}_{T*}$.

Given the item sets we can construct the action table from them. The transition table was constructed implicitly during the construction of the item sets when forming $\boldsymbol{I}_{yX}$ from $\boldsymbol{I}_y$ and X. For example, $(\boldsymbol{I}_\lambda,E)$ gives $\boldsymbol{I}_E$ and $(\boldsymbol{I}_{E+T},*)$ gives $\boldsymbol{I}_{T*}$.

11.4 ADDITIONAL REMARKS

11.4.1 Summary

We have proved that the following decision problems are *NP*-complete, that is, they are some of the hardest problems in ***NP***: *NTM* TIME-BOUNDED HALTING, SATISFIABILITY, 3-SATISFIABILITY, VERTEX COVER, HAMILTONIAN CIRCUIT, TRAVELING SALESMAN CHECKING, 3-DIMENSIONAL MATCHING, SUBSET SUM, and RECTANGLE INTERSECTION. The important and fundamental tool used in proving these results is the POLYNOMIAL TRANSFORMATION.

Additional evidence for the CHURCH(-TURING) THESIS has been presented by proving that both both PHRASE STRUCTURE GRAMMARS and INTERACTIVE LINDENMAYER GRAMMARS have the same power as *DTMs*.

Finally, we have presented three parsing techniques for *CFGs*. These are COCKE-KASAMI-YOUNGER, $LL(k)$, and $LR(k)$.

11.4.2 History

Cook (1971a) laid the foundations for the theory of *NP*-completeness. He concentrated on decision problems, demonstrated the importance of polynomial transformations, and proved SATISFIABILITY is *NP*-complete. Karp (1972) followed this up by providing the first collection of *NP*-complete problems. The crucial notion of a transformation has already been used in Chapter 10 for proving undecidability results. However, its use for decidable problems was foreshadowed in the work of Danzig (1960), Edmonds (1962, 1965), and Gimpel (1965), for example.

Readable introductory articles on this topic are Gardner (1985), Graham (1978), Pippenger (1978), Stockmeyer and Chandra (1979), and Karp (1986). The compendium of Garey and Johnson (1979), which we have followed, is an excellent text on the topic. Papadimitriou and Steiglitz (1982) is a more recent text that concentrates on combinatorial problems.

The formal study of hierarchical layout languages was initiated in Bentley and Ottmann (1981) and continued in Bentley et al. (1983). Examples of such languages are *CIF* and *SLL* described in Mead and Conway (1980), and *Box*, described in Newman and Sproull (1979).

Phrase structure grammars were introduced by Chomsky (1956) who proved that they have the same power as Turing machines. Interactive Lindenmayer grammars were introduced in Lindenmayer (1968) and their generative power was studied in Herman (1969) and van Dalen (1971). The books of Révész (1983) and Salomaa (1973) provide excellent introductions to phrase structure grammars, while Herman and Rozenberg (1975) and Rozenberg and Salomaa (1980) are good sources for *L* grammars. Fischer (1968) introduced macro grammars and Rosenkrantz (1969) programmed grammars.

Parsing has been a topic of major interest in formal language theory, because it is needed in compiler writing, pattern recognition, artificial intelligence, word processing, operating systems, and, many other areas.

The Cocke-Kasami-Younger algorithm was discovered independently by the three originators, see Hays (1967), Kasami (1965), and Younger (1967). Dynamic programming was introduced by Richard Bellman (see Bellman and Dreyfus, 1962 and Horowitz and Sahni, 1978). Earley (1970) gives an efficient algorithm based on the approach of Knuth (1965) for $LR(k)$ parsing. Valiant (1975) provided the first general parsing algorithm to run faster than n^3. Indeed, it requires $O(n^{\log_2 7})$ time; see Harrison (1978).

$LL(1)$ grammars were defined by Lewis and Stearns (1968) who were the first to investigate them in depth. However, Knuth gave a series of lectures on deterministic top-down parsing in 1967: these subsequently appeared as Knuth (1971). Wood (1969b) introduced left-factored grammars, which are also known as strong $LL(1)$ grammars, based on the approach of Foster (1968). The extension to a look-ahead of k symbols yielding $LL(k)$ grammars is investigated in detail in Rosenkrantz and Stearns (1970). Kurki-Suonio (1969) was the first to show that the $LL(k)$ languages form a proper hierarchy.

$LR(1)$ grammars and, in general, $LR(k)$ grammars were introduced by Knuth (1965). Their associated language families do not form a proper hierarchy: for $k \geq 1$, the $LR(k)$ languages are $LR(1)$ languages and the family of $LR(1)$ languages is exactly $\boldsymbol{L}_{DCF}$.

Aho and Ullman (1972), which formed the basis for our presentation of $LR(k)$ parsing, contains an excellent coverage of parsing. Harrison (1978) includes a thorough treatment of $LR(k)$ parsing, while Sippu and Soisalon-Soininen (1987) is a state-of-the-art parsing monograph.

11.5 SPRINGBOARD

11.5.1 Dealing with *NP*-Completeness

TRAVELING SALESMAN CHECKING (*TSC*) has been shown to be *NP*-complete. This implies that any algorithm which solves *TSC* requires exponential time, in the worst case, under the assumption that $\boldsymbol{P} \neq \boldsymbol{NP}$. By the results of Section 6.6 this, in turn, implies that TRAVELING SALESMAN (*TS*) requires exponential time, in the worst case, under the same assumption. But *TS* is not just of theoretical interest, it also has practical import. Indeed it is a problem that crops up in many day-to-day situations and, therefore, must be solved.

One approach to *TS* is to argue that worst case instances are pathological, by deriving an average case analysis of algorithms for them. This is difficult, but there has been some success; for example, see Karp (1977) who treats a geometric version of *TS* and his more recent Turing lecture, Karp(1986).

A second approach is to drop the requirement that an optimal solution be found and provide algorithms which give nearly optimal solutions. For example, assume *TS* satisfies the triangle inequality, in other words, $d(C_i, C_j) \leq d(C_i, C_k) + d(C_k, C_j)$, for all i, j and k. Then a solution which is at most twice as costly as the optimal solution can be found in deterministic polynomial time; see Rosenkrantz et al. (1977). Christofides (1976) suggested a refinement of the technique which bounds the cost of the nearly optimal solution by 3/2 times the cost of the optimal.

Consult Garey and Johnson (1979) for approaches to solving other *NP*-complete problems.

11.5.2 *LALR* Grammars

If the algorithm for generating *LR*(1) item sets is the one sketched in Section 11.3.3, then a grammar for ALGOL 60, PASCAL, or ADA will yield thousands of item sets. Since the item sets are the basis for the transition and action tables, this implies that these tables will be very large.

One successful approach to improving the situation is the merging of item sets which only differ by their look-ahead. In general, this may produce two or more distinct actions associated with each item set and input symbol. This is called an *action conflict*. *LR*(1) grammars which yield no action conflicts are called *look-ahead LR*(1) grammars or *LALR*(1) grammars.

Investigate *LALR*(1) grammars as the basis for practical compiler-writing tools, for example, *YACC*; see Johnson (1974) and Aho and Ullman (1977), and from the theoretical point of view, see Sippu et al. (1983).

11.5.3 Other Kinds of Completeness

The concept of completeness with respect to a given class or family is fundamental to the study of the complexity of languages and problems. In particular, since $\mathbf{L}_{DPSPACE} = \mathbf{L}_{NPSPACE}$ we speak of languages which are hardest in $\mathbf{L}_{DPSPACE}$ as *PSPACE-complete* languages. The collection of all decision problems whose encodings are in $\mathbf{L}_{DPSPACE}$ is denoted by $\mathbf{PS}$. A hardest problem in $\mathbf{PS}$ is also said to be *PSPACE*-complete.

Immediately, we have $\mathbf{P} \subseteq \mathbf{NP} \subseteq \mathbf{PS}$, and $\mathbf{P} = \mathbf{PS}$ implies $\mathbf{P} = \mathbf{NP}$. However, if $\mathbf{P} \neq \mathbf{PS}$, then we may still have $\mathbf{P} = \mathbf{NP}$, so *PSPACE*-complete problems are even more intractable than *NP*-complete problems. As with *NP*-complete problems, *PSPACE*-complete problems provide evidence that $\mathbf{P} \neq \mathbf{PS}$ and, indeed, that $\mathbf{NP} \neq \mathbf{PS}$.

One of the first problems to be proved *PSPACE*-complete was

REGULAR EXPRESSION NON-UNIVERSALITY
INSTANCE: A regular expression E over an alphabet Σ.
QUESTION: Is $L(E) \neq \Sigma^*$?

See Meyer and Stockmeyer (1972) for the proof of this result. Consult Garey and Johnson (1979) for an introduction to the topic.

The notions of *PTIME*-completeness (or simply *P*-completeness) and *NLOGSPACE*-completeness (an offline *NTM* which computes within logarithmic space defines a language in $\mathbf{L}_{NLOGSPACE}$) have also been studied. Again see Garey and Johnson (1979) for details.

EXERCISES

1.1 When proving, in Section 11.1.2, that *TBH* is *NP*-complete, it was essential to the proof that the time bound be encoded in unary. Explain why encoding the time bound in binary does not yield an *NP*-completeness result.

1.2 *NTM* SPACE-BOUNDED HALTING (*SBH*)
INSTANCE: An *NTM* $M = (Q, \{a,b\}, \{a,b,\mathbf{B}\}, \delta, s, f)$, a word x in $\{a,b\}^*$, and a space bound $B \geq 1$.
QUESTION: Is x accepted by M using at most B distinct cells?

Do you think *SBH* is *NP*-complete? Give the intuition behind your answer.

1.3 3-*SAT* is a simpler problem than *SAT*. This raises the question: Why not use 2-*SAT*? Unfortunately 2-*SAT* is not *NP*-complete.

(i) Prove that 1-*SAT* is in ***P***.
(ii) Prove that 2-*SAT* is in ***P***.

1.4 A collection of clauses is a tautology if, under every truth assignment, it is satisfied. Letting TAUTOLOGY be the corresponding decision problem, what results can you obtain for it?

1.5 Let *d*-*HC*, $d \geq 1$, be *HC* in which each node in the given graph has at most d edges. What can you say about 1-*HC*, 2-*HC*, 3-*HC*, etc.

1.6 Prove that the following problems are *NP*-complete.

(i) CLIQUE
INSTANCE: A graph $G = (V,E)$ and a positive integer $K \leq \#V$.
QUESTION: Does G contain a complete subgraph or clique of size at most K?

(ii) INDEPENDENT SET
INSTANCE: A graph $G = (V,E)$ and a positive integer $K \leq \#V$.
QUESTION: Is there a subset $V' \subseteq V$ such that $\#V' \leq K$ and for all u, v in V', $\{u,v\}$ is not in E? In other words, is there an independent set of size at most K.

(iii) GENERALIZED *TSC*
INSTANCE: An integer $n \geq 1$, n distinct "cities" $C_1, \ldots, C_n$, integer distances between every pair C_i, C_j and an integral bound B.
QUESTION: Is there a tour with cost $\leq B$?

(iv) DIRECTED *HSC*
INSTANCE: A digraph $G = (V,E)$.
QUESTION: Does G contain a directed Hamiltonian circuit?

(v) *k-HSC*
INSTANCE: A graph $G = (V,E)$.
QUESTION: Does G contain a circuit in which each node is visited exactly k times?

(vi) PARTITION
INSTANCE: A finite set A and a size function $\sigma : A \rightarrow \mathbb{N}$.
QUESTION: Is there a subset $A' \subseteq A$ such that $\sigma(A') = \sigma(A - A')$?

(vii) EXACT COVER BY 3-SETS ($X3C$)
INSTANCE: A finite set X with $\#X = 3m$, for some $m \geq 1$, and a collection C of 3-element subsets of x.
QUESTION: Is there a subcollection $C' \subseteq C$ such that every element of X appears in exactly one member of C'? This is an exact cover.

(viii) WEAK 3-*DM*
INSTANCE: Three sets X, Y, and Z, a relation $M \subseteq X \times Y \times Z$, and an integer n.
QUESTION: Is there a *weak matching* M', that is, a subset $M' \subseteq M$ such that $\#M' = n$ and each element of X, Y, and Z appears at most once.

(ix) POINT INTERSECTION
INSTANCE: An *HCFG*, $G = (N,P,S)$, and a point (x,y).
QUESTION: Does (x,y) belong to any rectangle $R(G)$?

(x) COMMON RECTANGLE
INSTANCE: Two *HCFGs*, $G_1 = (N_1,P_1,S_1)$ and $G_2 = (N_2,P_2,S_2)$.
QUESTION: Is there a rectangle that appears in both $R(G_1)$ and $R(G_2)$, that is, $R(G_1) \cap R(G_2) \neq \emptyset$?

1.7 What can you say about the following problems?

(i) $K-VC$
INSTANCE: A graph $G = (V,E)$.
QUESTION: Does there exist a vertex cover V' of G such that $\#V' \leq K$?

(ii) CHINESE POSTMAN
INSTANCE: A graph $G = (V,E)$.
QUESTION: Is there a cycle that includes every edge exactly once?

(iii) 2-DIMENSIONAL MATCHING
INSTANCE: Two sets X and Y such that $\#X = \#Y$ and a relation $M \subseteq X \times Y$.
QUESTION: Is there a relation $M' \subseteq M$ such that each element of

X and Y appears exactly once?

(iv) K-SUBSET SUM
INSTANCE: A finite set A, a size function $\sigma : A \rightarrow \mathbb{N}$, and a bound $B \geq 1$.
QUESTION: Is there a subset $A' \subseteq A$ such that $\sigma(A') = B$ and $\#A' = K$?

(v) 1-RI
INSTANCE: A one-dimensional $HCFG$, $G = (N,P,S)$, that is, $P \subseteq (N \times \mathbb{N}^2) \cup (N \times (N\mathbb{N})^*)$.
QUESTION: Are there two intervals I_1 and I_2 in $R(G)$ such that I_1 and I_2 intersect?

1.8 Prove that the transformations of 3-SAT to VC, VC to HC, and SS to RI are polynomial.

2.1 Define $PSGs$ and $ILGs$ that generate the following languages.

(i) $\{ww : w \text{ is in } \{a,b\}^*\}$.
(ii) $\{a^{n^2} : n \geq 0\}$.
(iii) $\{a^{2^n} : n \geq 0\}$.
(iv) $\{a^p : p \text{ is a } prime\}$.

2.2 Prove, constructively, that for every CFG there exists an equivalent $(0,0)LG$.

2.3 Prove that for each PSG there exists an equivalent PSG in which each left-hand side of a production contains at most two symbols.

2.4* Prove that every PSG is equivalent to a PSG in which the productions only have the forms: (i) $AB \rightarrow A\alpha$ and (ii) $AB \rightarrow BA$. The first type only uses left context, while the second permutes symbols.

2.5** Prove that every PSG is equivalent to a PSG in which productions only have the form (i) of Exercise 2.4.

2.6 A PSG, G, is *monotonic* if every production $\alpha \rightarrow \beta$ in G satisfies $|\alpha| \leq |\beta|$.

Prove that MEMBERSHIP is decidable for monotonic $PSGs$.

2.7 A PSG $G = (N,\Sigma,P,S)$ is said to be *context-sensitive*, or a CSG, if every production has the form $\beta_1 A \beta_2 \rightarrow \beta_1\alpha\beta_2$, where A is in N, β_1 and β_2 are in $(N \cup \Sigma)^*$, and α is in $(N \cup \Sigma)^+$. Prove that monotonic $PSGs$ and $CSGs$ generate the same families of languages.

2.8 An (m,n) *context-dependent grammar* G is given by a quadruple (N,Σ,P,S), where P, a finite set of productions, satisfies

$$P \subseteq (N \cup \Sigma)^{\leq m} \times N \times (N \cup \Sigma)^{\leq n} \times (N \cup \Sigma)^*$$

where $V^{\leq m} = V^0 \cup \ldots \cup V^m$. A production is usually written as $\beta, A, \gamma \rightarrow \alpha$, where (β, γ) is its context; A is rewritten as α if $\beta A \gamma$ is a subword of the given sentential form. A *CFG* is, simply, a $(0,0)CDG$.

(i) Prove that for every $(m,n)CDG$, $G = (N, \Sigma, P, S)$, there exists an equivalent $(m,n)CDG$, $G' = (N', \Sigma, P', S')$, in which the set of productions satisfies $P' \subseteq N^{\leq m} \times N \times N^{\leq n} \times (N \cup \Sigma)^*$.

(ii) Prove that *CDGs* are equivalent in power to *PSGs*.

(iii) Prove that for every $(m,n)CDG$ with $m,n \geq 1$, there is an equivalent $(1,1)CDG$.

(iv)* An $(m,n)CDG$ in which each context is in $\Sigma^{\leq m} \times \Sigma^{\leq n}$ is a *terminal context* $(m,n)CDG$. Prove that each such *CDG* generates a context-free language.

2.9 A $(0,0)ILG$ is said to be an *LG*; its productions are applied in parallel without any context.

(i) What are the closure properties of the family of *LG* languages?

(ii) Prove that there is an *LG* language which is not a *CFL*.

(iii) Prove that to every *LG* there exists an equivalent, modulo the empty word, λ-free *LG*.

3.1 We pose two problems about $LL(k)$ languages.

(i) Prove that $\{a^n x : n \geq 1 \text{ and } x \text{ is in } \{b, c, bbd\}^n\}$ is not an $LL(1)$ language, but is an $LL(2)$ language.

(ii) Prove that, for every $k \geq 1$, there are $LL(k+1)$ languages that are not $LL(k)$.

3.2 Prove that for every $k \geq 1$, each $LL(k)$ grammar has an equivalent λ-free $LL(k+1)$ grammar.

3.3 Prove that each reduced $LL(k)$ grammar is non-left-recursive. From this result show that each $LL(k)$ grammar has an equivalent $LL(k+1)$ grammar in Greibach normal form.

3.4 Prove that each $LL(k)$ grammar is unambiguous.

3.5 Prove that for every λ-free $LL(1)$ grammar there is an equivalent simple deterministic grammar (see Section 4.1).

3.6 Prove that the equivalence problem is decidable for λ-free $LL(1)$ grammars.

3.7 What can you prove about $LL(0)$ grammars and languages?

3.8 Prove that it is undecidable to determine whether a *CFG* is $LL(k)$, for some $k \geq 0$.

3.9 Complete the proof of the decidability of $LL(k)$-ness of a CFG, for a given $k \geq 0$.

3.10 Give an example of a $DPDA$ language which is not $LL(k)$, for any $k \geq 0$. Justify your answer.

3.11 Let $G = (N,\Sigma,P,S)$ be a CFG. For $k \geq 0$ and α in $(N \cup \Sigma)^*$, define $FOLLOW_k(\alpha)$ to be the set

$$\{x : S \Rightarrow^* \beta_1\alpha\beta_2 \text{ and } x \text{ is in } FIRST_k(\beta_2)\}$$

For $k \geq 0$, we say that a CFG $G = (N,\Sigma,P,S)$ is a *strong $LL(k)$ grammar* if the following condition holds. If $A \rightarrow \alpha$ and $A \rightarrow \beta$ are distinct A-productions, then

$$FIRST_k(\alpha\ FOLLOW_k(A)) \cap FIRST_k(\beta\ FOLLOW_k(A)) = \varnothing.$$

(i) Prove that every strong $LL(k)$ grammar is $LL(k)$.
(ii) Prove that for each $LL(k)$ grammar there exists an equivalent strong $LL(k)$ grammar.
(iii) Prove that a CFG is strong $LL(1)$ if and only if it is $LL(1)$.
(iv) Give an example of CFG which is $LL(2)$ but not strong $LL(2)$.
(v) Prove, for every $k > 1$, that there are $LL(k)$ grammars which are not strong $LL(k)$.

3.12 Prove that each $LL(1)$ grammar is $LR(1)$. Can you generalize your proof for $LL(k)$ and $LR(k)$?

3.13 Prove that each $LR(k)$ grammar is unambiguous.

3.14 Assume $L \subseteq \Sigma^*$ is an $LR(0)$ language.

(i) Prove that, for all x in Σ^+ and for all w and y in L, if wx is in L, then yx is in L.
(ii) Prove that there are $LR(1)$ languages which are not $LR(0)$.

3.15* Prove that for each $LR(k)$ grammar, $k \geq 2$, there exists an equivalent $LR(1)$ grammar.

3.16 Prove that it is undecidable whether an arbitrary CFG is $LR(k)$, for any $k \geq 0$.

3.17 Prove Theorem 11.3.2.

Bibliography

Aho, A.V., Indexed Grammars: An Extension of Context-Free Grammars, *Journal of the ACM 15*, (1968), 647-671.

Aho, A.V. (ed.), *Currents in the Theory of Computing*, Prentice-Hall, Englewood Cliffs, New Jersey, 1973.

Aho, A.V., Pattern Matching in Strings, in Book, R.V. (ed.), *Formal Language Theory: Perspectives and Open Problems*, Academic Press, New York (1980), 325-247.

Aho, A.V., and Corasick, M.J., Efficient String Matching: An Aid to Bibliographic Search, *Communications of the ACM 18*, (1975), 333-340.

Aho, A.V., and Ullman, J.D., Syntax Directed Translations and the Pushdown Assembler, *Journal of Computer and System Sciences 3*, (1969a), 37-56.

Aho, A.V., and Ullman, J.D., Properties of Syntax Directed Relations, *Journal of Computer and System Sciences 3*, (1969b), 319-334.

Aho, A.V., and Ullman, J.D., Translations on a Context-Free Grammar, *Information and Control 19*, (1971), 439-475.

Aho, A.V., and Ullman, J.D., *The Theory of Parsing, Translation and Compiling, Vol. I: Parsing*, Prentice Hall, Englewood Cliffs, New Jersey, 1972.

Aho, A.V., and Ullman, J.D., *The Theory of Parsing, Translation and Compiling, Vol. II: Compiling*, Prentice Hall, Englewood Cliffs, New Jersey, 1973.

Aho, A.V., and Ullman, J.D., *Principles of Compiler Design*, Addison-Wesley, Reading, Massachusetts, 1977.

Alagic, S., and Arbib, M.A., *The Design of Well-Structured and Correct Programs*, Springer-Verlag, Heidelberg, 1978.

Altman, E.B., *The Concept of Finite Representability*, Systems Research Center Report SRC 56-A-64-20, Case Institute of Technology, 1964.

Anderson, R.B., *Proving Programs Correct*, John Wiley & Sons, New York, 1979.

Arbib, M.A., Kfoury, A.J. and Moll, R.N., *A Basis for Theoretical Computer Science*, Springer-Verlag, New York, 1981.

Ashcroft, E.A., and Wadge, W.W., LUCID, a Formal System for Writing and Proving Programs, *SIAM Journal on Computing 5*, (1976), 336-354.

Backus, J.W., The Syntax and Semantics of the Proposed International Algebraic Language of the Zürich ACM-GAMM Conference, *Proceedings International Conference on Information Processing*, UNESCO, (1959), 125-132.

Banerji, R.B., Phrase Structure Languages, Finite Machines, and Channel Capacity, *Information and Control 6*, (1963), 153-162.

Bar-Hillel, Y., Perles, M., and Shamir, E., On Formal Properties of Simple Phrase Structure Grammars, *Zeitschrift für Phonetik Sprachwissenschaft und Kommunikations-forschung 14*, (1961), 143-172.

Barnes, B.H., A Programmer's View of Automata, *Computing Surveys 4*, (1972), 221-239.

Barnett, M.P., and Futrelle, R.P., Syntactic Analysis by Digital Computer, *Communications of the ACM 5*, (1962), 515-526.

Beckman, F.S., *Mathematical Foundations of Programming*, Addison-Wesley, Reading, Massachusetts, 1980.

Beizer, B., *Software Testing Techniques*, Van Nostrand Reinhold, New York, 1983.

Bellman, R.E., and Dreyfus, S.E., *Applied Dynamic Programming*, Princeton University Press, Princeton, New Jersey, 1962.

Bentley, J.L., and Ottmann, Th., The Complexity of Manipulating Hierarchically Defined Sets of Rectangles, *Mathematical Foundations of Computer Science 1981, Springer-Verlag Lecture Notes in Computer Science 118*, (1981), 1-15.

Bentley, J.L., Ottmann, Th., and Widmayer, P., The Complexity of Manipulating Hierarchically Defined Sets of Rectangles, in Preparata, F.P. (ed.), *Advances in Computing Research 1*, JAI Press, Greenwich, Connecticut, 1983, 127-158.

Berger, R., The Undecidability of the Domino Problem, *Memoirs of the American Mathematical Society 66*, (1966).

Berlekamp, E.R., Conway, J.H., and Guy, R.K., *Winning Ways, Volume 2: Games in Particular*, Academic Press, New York, 1982.

Berstel, J., *Transductions and Context-Free Languages*, Teubner, Stuttgart, West Germany, 1979.

Berstel, J., and Boasson, L., Une Suite Décroissante de Cônes Rationnels, *Springer-Verlag Lecture Notes in Computer Science 14*, (1974), 383-397.

Bobrow, L.S., and Arbib, M.A., *Discrete Mathematics: Applied Algebra for Computer and Information Science*, W.B. Saunders, Philadelphia, Pennsylvania, 1974.

Book, R.V. (ed.), *Formal Language Theory: Perspectives and Open Problems*, Academic Press, New York, 1980.

Bourne, S.R., *The UNIX System*, Addison-Wesley, Reading, Massachusetts, 1983.

Braffort, P., and Hirschberg, D. (eds.), *Computer Programming and Formal Systems*, North-Holland, Amsterdam, 1963.

Brainerd, W.S., and Landweber, L.H., *Theory of Computation*, John Wiley & Sons, New York, 1974.

Brzozowski, J.A., A Survey of Regular Expressions and Their Applications, *IEEE Transactions on Electronic Computers 11*, (1962), 324-335.

Brzozowski, J.A., Derivatives of Regular Expressions, *Journal of the ACM 11*, (1964), 481-494.

Brzozowski, J.A., Open Problems about Regular Languages, in Book, R.V. (ed.), *Formal Language Theory: Perspectives and Open Problems*, Academic Press, New York, 1980, 23-47.

Brzozowski, J.A., and McCluskey, Jr., E.J., Signal Flow Graph Techniques for Sequential Circuit State Diagrams, *IEEE Transactions on Electronic Computers EC-12*, (1963), 67-76.

Brzozowski, J.A., and Yoeli, M., *Digital Networks*, Prentice-Hall, Englewood Cliffs, New Jersey, 1976.

Bucher, W., and Maurer, H.A., *Theoretische Grundlagen der Programmiersprachen: Automaten und Sprachen*, Bibliographisches Institut, Zürich, 1984.

Büchi, J.R., Turing-Machines and the Entscheidungsproblem, *Mathematische Annalen 148*, (1962), 201-213.

Burge, W.H., *Recursive Programming Techniques*, Addison-Wesley, Reading, Massachusetts, 1975.

Burks, A.W. (ed.), *Essays in Cellular Automata*, University of Illinois Press, 1970.

Burks, A.W., Warren, D.W., and Wright, J.B., An Analysis of a Logical Machine Using Parenthesis-Free Notation, *Mathematical Tables and Other Aids to Computation 8*, (1954), 55-57.

Cantor, D.C., On the Ambiguity Problem of Backus Systems, *Journal of the ACM 9*, (1962), 477-479.

Choffrut, C., and Culik II, K., Properties of Finite and Pushdown Transducers, *SIAM Journal on Computing 12*, (1983), 300-315.

Chomsky, N., Three Models for the Description of Language, *IRE Transactions on Information Theory 2*, (1956), 113-124.

Chomsky, N., *Syntactic Structures*, Mouton, The Hague, Netherlands, 1957.

Chomsky, N., On Certain Formal Properties of Grammars, *Information and Control 2*, (1959), 137-167.

Chomsky, N., Context-Free Grammars and Pushdown Storage, *Quarterly Progress Report No. 65*, MIT Research Laboratory of Electronics, Cambridge, Massachusetts, (1962), 187-194.

Chomsky, N., Formal Properties of Grammars, *Handbook of Mathematical Psychology, Vol. 2*, John Wiley & Sons, New York, (1963), 323-418.

Chomsky, N., and Miller, G.A., Finite-State Languages, *Information and Control 1*, (1958), 91-112.

Chomsky, N., and Schutzenberger, M.P., The Algebraic Theory of Context Free Languages, in Braffort,P., and Hirschberg,D.(eds.), *Computer Programming and Formal Systems*, North-Holland, Amsterdam, 1963, 118-161.

Christofides, N., Worst-Case Analysis of a New Heuristic for the Traveling Salesman Problem, Technical Report, Graduate School of Industrial Administration, Carnegie-Mellon University, Pittsburgh, Pennsylvania, 1976.

Church, A., An Unsolvable Problem of Elementary Number Theory, *American Journal of Mathematics 58*, (1936), 345-363.

Church, A., The Calculi of Lambda-Conversion, *Annals of Mathematics Studies 6*, Princeton University Press, Princeton, New Jersey, (1941).

Cleaveland, J.C., and Uzgalis, R., *Grammars for Programming Languages*, Elsevier North-Holland, New York, 1977.

Clocksin, W.F., and Mullish, C.S., *Programming in PROLOG*, Springer-Verlag, Heidelberg, 1981.

Cobham, A., The Intrinsic Computational Difficulty of Functions, *Proceedings 1964 Congress for Logic, Mathematics, and Philosophy of Science*, North Holland, Amsterdam, (1964), 24-30.

Cohen, J., Nondeterministic Algorithms, *Computing Surveys 11*, (1979), 79-94.

Cohen, D.J., and Gotlieb, C.C., A List Structure Form of Grammars for Syntactic Analysis, *Computing Surveys 2*, (1970), 65- .

Comer, D., Heuristics for Trie Index Minimization, *ACM Transactions on Data Base Systems 4*, (1979), 383-395.

Comer, D., Analysis of a Heuristic for Full Trie Minimization, *ACM Transactions on Data Base Systems 6*, (1981), 513-537.

Conway, M.E., Design of a Separable Transition-Diagram Compiler, *Communications of the ACM 6*, (July 1963), 396-408.

Cook, S.A., The Complexity of Theorem-Proving Procedures, *Proceedings Third Annual ACM Symposium on the Theory of Computing*, (1971a), 151-158.

Cook, S.A., Linear-Time Simulation of Deterministic Two-Way Pushdown Automata, *Proceedings of the 1971 IFIP Congress*, North-Holland, Amsterdam (1971b), 75-80.

Dantzig, G.B., On the Significance of Solving Linear Programming Problems with Integer Variables, *Econometrica 28*, (1960), 30-44.

Davis, M., *Computability and Unsolvability*, McGraw-Hill, New York, 1958.

Davis, M. (ed.), *The Undecidable: Basic Papers on Undecidable Propositions, Unsolvable Problems, and Computable Functions*, Raven Press, Hewlett, New York, 1965.

Davis, M., Hilbert's Tenth Problem is Unsolvable, *American Mathematical Monthly 80*, (1973), 233-269.

Davis, M.D., and Weyuker, E.J., *Computability, Complexity, and Languages*, Academic Press, New York, 1983.

de Bakker, J.W., Semantics of Programming Languages, in Tou,J.(ed.), *Advances in Information Systems and Sciences, Vol. 2*, Plenum Press, New York, 1969, 173-227.

Dekker, J.C.E. (ed.), *Recursive Function Theory, Proceedings of Symposia in Pure Mathematics 5*, American Mathematical Society, Providence, Rhode Island, 1962.

DeMillo, R.A., Dobkin, D.P., Jones, A.K., and Lipton, R.J. (eds.), *Foundations of Secure Computation*, Academic Press, New York, 1978.

DeMillo, R.A., Lipton, R.J., and Perlis, A.J., Social Processes and Proofs of Theorems and Programs, *Communications of the ACM 22*, (1979), 271-280.

Denning, P.J., Dennis, J.B., and Qualitz, J.E., *Machines, Languages, and Computation*, Prentice-Hall, Englewood Cliffs, New Jersey, 1978.

Dewdney, A.K., Computer Recreations: A Computer Trap for the Busy Beaver, the Hardest-Working Turing Machine, *Scientific American 251*, (August 1984), 19-23.

Dewdney, A.K., Computer Recreations, *Scientific American 252*, (March 1985), 23.

Dijkstra, E.W., *A Discipline of Programming*, Prentice-Hall, Englewood Cliffs, New Jersey, 1976.

Earley, J., An Efficient Context-Free Parsing Algorithm, *Communications of the ACM 13*, (1970), 94-102.

Edmonds, J., Covers and Packings in a Family of Sets, *Bulletin of the American Mathematical Society 68*, (1962), 494-499.

Edmonds, J., Paths, Trees and Flowers, *Canadian Journal of Mathematics 17*, (1965), 449-467.

Ehrenfeucht, A., Karhumäki, J., and Rozenberg. G., The (Generalized) Post Correspondence Problem with Lists Consisting of Two Words is Decidable, *Theoretical Computer Science 21*, (1982), 119-144.

Ehrenfeucht, A., Parikh, R., and Rozenberg, G., Pumping Lemmas for Regular Sets, *SIAM Journal on Computing 10*, (1981), 536-541.

Eilenberg, S., *Automata, Languages, and Machines, Volume A*, Academic Press, New York, 1974.

Eilenberg, S., *Automata, Languages, and Machines, Volume B*, Academic Press, New York, 1976.

Elgot, C.C., and Mezei, J.E., On Relations Defined by Generalized Finite Automata, *IBM Journal of Research and Development 9*, (1965),47-68.

Elspas, B., Levitt, K., Waldinger, R., and Waksman, A., An Assessment of

Techniques for Proving Program Correctness, *Computing Surveys 4*, (1972), 97-147.

Engelfriet, J., Some Open Questions and Recent Results on Tree Transducers and Tree Languages, in Book, R.V. (ed.), *Formal Language Theory: Perspectives and Open Problems*, Academic Press, New York (1980), 241-286.

Engelfriet, J., Schmidt, E.M., and van Leeuwen, J., Stack Machines and Classes of Nonnested Macro Languages, *Journal of the ACM 27*, (1980), 96-117.

Evey, J., Application of Pushdown Store Machines, *Proceedings 1963 Fall Joint Computer Conference*, AFIPS Press, Montvale, New Jersey, (1963), 215-227.

Fischer, M.J., Grammars with Macro-like Productions, *Proceedings of the Ninth Annual IEEE Symposium on Switching and Automata Theory*, (1968), 131-142.

Floyd, R.W., On Ambiguity in Phrase Structure Languages, *Communications of the ACM 5*, (1962a), 526-534.

Floyd, R.W., On the Nonexistence of a Phrase Structure Grammar for ALGOL 60, *Communications of the ACM 5*, (1962b), 483-484.

Floyd, R.W., *New Proofs and Old Theorems in Logic and Formal Linguistics*, Computer Associates, Wakefield, Massachusetts, 1964a.

Floyd, R.W., The Syntax of Programming Languages — A Survey, *IEEE Transactions on Electronic Computers, Vol. EC-13*, (August 1964), 346-353. Reprinted in Rosen, S. (ed.), *Programming Systems and Languages*, McGraw-Hill, New York, 1967; and Pollack, B.W., *Compiler Techniques*, Auerbach Press, Philadelphia, Pennsylvania, 1972.

Floyd, R.W., Assigning Meaning to Programs, in Schwartz, J.T. (ed.), *Mathematical Aspects of Computer Science*, American Mathematical Society, Providence, Rhode Island (1967a), 19-32.

Floyd, R.W., Nondeterministic Algorithms, *Journal of the ACM 14*, (1967b), 636-644.

Floyd, R.W., and Ullman, J.D., The Compilation of Regular Expressions into Integrated Circuits, *Journal of the ACM 29*, (1984), 603-622.

Fosdick, L.D., and Osterweil, L.J., Data Flow Analysis in Software Reliability, *Computing Surveys 8*, (1976), 305-330.

Foster, J.M., A Syntax-Improving Program, *Computer Journal 11*, (1968) 31-34.

Foster, J.M. *Automatic Syntactic Analysis*, American Elsevier, New York, 1970.

Galler, B.A., and Perlis, A.J., *A View of Programming Languages*, Addison-Wesley, Reading, Massachusetts, 1970.

Gardner, M., *Wheels, Life and Other Mathematical Amusements*, W.H. Freeman, San Francisco, 1983.

Gardner, M., The Traveling Salesman's Travail, *Discover 6*, (April 1985), 87-90.

Garey, M.R., and Johnson, D.S., *Computers and Intractability: A Guide to the Theory of NP-Completeness*, W.H. Freeman, San Francisco, 1979.

Gecseg, F. and Steinby,M., *Tree Automata*, Akademia Kiado, Budapest, 1984.

Gimpel, J.F., A Method of Producing a Boolean Function Having an Arbitrarily Prescribed Prime Implicant Table, *IEEE Transactions on Computers 14*, (1965), 485-488.

Ginsburg, S., *The Mathematical Theory of Context-Free Languages*, McGraw Hill, New York, 1966.

Ginsburg, S., and Greibach, S.A., Abstract Families of Languages, in Ginsburg, Greibach, and Hopcroft (1969), 1-32.

Ginsburg, S., Greibach, S.A., and Hopcroft, J.E., Studies in Abstract Families of Languages, *Memoirs of the American Mathematical Society 87*, Providence, Rhode Island, 1969.

Ginsburg, S., and Rice, H.G., Two Families of Languages Related to ALGOL, *Journal of the ACM 9*, (1962), 350-371.

Ginsburg, S., and Rose, G.F., Some Recursively Unsolvable Problems in ALGOL-like Languages, *Journal of the ACM 10*, (1963a), 29-47.

Ginsburg, S., and Rose, G.F., Operations Which Preserve Definability in Languages, *Journal of the ACM 10*, (1963b), 175-195.

Ginsburg, S., and Spanier, E.H., Quotients of Context-Free Languages, *Journal of the ACM 10*, (1963), 487-492.

Goldstine, J., *A Simplified Proof of Parikh's Theorem*, Unpublished manuscript, c.1980.

Gonnet, G.H., *Handbook of Algorithms and Data Structures*, Addison-Wesley, Reading, Massachusetts, 1984.

Gonnet, G.H., and Tompa, F.W., A Constructive Approach to the Design of Algorithms and Data Structures, *Communications of the ACM 26*, (1983), 912-920.

Gouda, M.G., and Rosier, L.E., Priority Networks of Communicating Finite State Machines, *SIAM Journal on Computing 14*, (1985), 569-584.

Graham, R.L., The Combinatorial Mathematics of Scheduling, *Scientific American 238*, 3 (1978), 124-132.

Gray, J.N., and Harrison, M.A., The Theory of Sequential Relations, *Information and Control 9*, (1966), 435-468.

Greibach, S.A., The Undecidability of the Ambiguity Problem for Minimal Linear Grammars, *Information and Control 6*, (1963), 117-125.

Greibach, S.A., A New Normal Form Theorem for Context-Free Phrase Structure Grammars, *Journal of the ACM 12*, (1965), 42-52.

Greibach, S.A., Chains of Full AFLs, *Mathematical Systems Theory 4*, (1970), 231-242.

Greibach, S.A., A Generalization of Parikh's Theorem, *Discrete Mathematics 2*, (1972), 347-355.

Greibach, S.A, The Hardest Context-Free Language, *SIAM Journal on Computing 2*, (1973), 304-310.

Gries, D., *Compiler Construction for Digital Computers*, John Wiley & Sons, New York, 1971.

Gries, D., *The Science of Programming*, Springer-Verlag, New York, 1981.

Gruska, J., A Characterization of Context-Free Languages, *Journal of Computer and System Sciences 5*, (1971), 353-364.

Hantler, S.L., and King, J.C., An Introduction to Proving the Correctness of Programs, *Computing Surveys 8*, (1976), 331-353.

Harrison, M.A., *Introduction to Switching and Automata Theory*, McGraw-Hill, New York, 1965.

Harrison, M.A., *Introduction to Formal Language Theory*, Addison-Wesley, Reading, Massachusetts, 1978.

Harrison, M.A., Ruzzo, W.L., and Ullman, J.D., Protection in Operating Systems, *Communications of the ACM 19*, (1976), 461-471.

Hartmanis, J., Context-Free Languages and Turing Machine Computations, in Schwartz, J.T. (ed.), *Mathematical Aspects of Computer Science*, American Mathematical Society, Providence, Rhode Island (1967), 42-51.

Hartmanis, J., and Hopcroft, J.E., An Overview of the Theory of Computational Complexity, *Journal of the ACM 18*, (1971), 444-475.

Hartmanis, J., Lewis, P.M., II, and Stearns, R.E., Hierarchies of Memory Limited Computations, *Proceeedings of the Sixth Annual Symposium on Switching Circuit Theory and Logical Design*, (1965), 179-190.

Hartmanis, J., and Stearns, R.E., On the Computational Complexity of Algorithms, *Transactions of the AMS 117*, (1965), 285-306.

Hayes, B., Computer Recreations, *Scientific American 249*, (December 1983), 19-28.

Hayes, B., Computer Recreations, *Scientific American 250*, (March 1984), 10-16.

Hays, D.G., *Introduction to Computational Linguistics*, American Elsevier, New York, 1967.

Henderson, P., *Functional Programming: Application and Implementation*, Prentice-Hall, Englewood Cliffs, New Jersey, 1980.

Hennie, F.C., *Introduction to Computability*, Addison-Wesley, Reading, Massachusetts, 1977.

Hennie, F.C., and Stearns, R.E., Two-Tape Simulation of Multitape Turing Machines, *Journal of the ACM 13*, (1966), 533-546.

Herman, G.T., The Computing Ability of a Developmental Model for Filamentous Organisms, *Journal of Theoretical Biology 25*, (1969), 421-435.

Herman, G.T., A Biologically Motivated Extension of ALGOL-like Languages, *Information and Control 22*, (1973), 487-502.

Herman, G.T., and Rozenberg, G., *Developmental Systems and Languages*, American Elsevier, New York, 1975.

Hermes, H., *Enumerability, Decidability, Computability*, Springer-Verlag, New York, 1969.

Hoare, C.A.R., and Allison, D.C.S., Incomputability, *Computing Surveys 4*, (1972), 169-178.

Hoare, C.A.R., and Lauer, P., Consistent and Complementary Formal Theories of the Semantics of Programming Languages, *Acta Informatica 3*, (1974), 135-153.

Hoare, C.A.R., and Wirth, N., An Axiomatic Definition of the Programming Language PASCAL, *Acta Informatica 2*, (1973), 335-355.

Hodges, A., *Alan Turing: The Enigma*, Burnett Books Ltd., London, 1983.

Hopcroft, J.E., An $n \log n$ Algorithm for Minimizing the States in a Finite Automaton, in Kohavi, Z., and Paz, A. (eds.), *Theory of Machines and Computations*, Academic Press, New York (1971), 189-196.

Hopcroft, J.E., Turing Machines, *Scientific American 250*, (May 1984), 86-98.

Hopcroft, J.E., and Ullman, J.D., *Introduction to Automata Theory, Languages, and Computation*, Second Edition, Addison-Wesley, Reading, Massachusetts, 1979.

Horowitz, E., and Sahni, S., *Fundamentals of Computer Algorithms*, Computer Science Press, Potomac, Maryland, 1978.

Ikeno, N., A 6-Symbol, 10-State Universal Turing Machine, *Proceedings of the Institute of Electrical Communications*, Tokyo, 1958.

Irons, E.T., A Syntax Directed Compiler for ALGOL 60, *Communications of the ACM 4*, (1961), 51-55.

Johnson, J.H., *Formal Models for String Similarity*, Ph.D. Dissertation, Department of Computer Science, University of Waterloo, 1983.

Johnson, S.C., YACC — Yet Another Compiler Compiler, *Computer Science Technical Report 32*, Bell Laboratories, Murray Hill, New Jersey, 1974.

Johnson, W.L., Porter, J.H., Ackley, S.I., and Ross, D.T., Automatic Generation of Efficient Lexical Analyzers Using Finite-State Techniques, *Communications of the ACM 11*, (1968), 805-813.

Karp, R.M., Reducibility Among Combinatorial Problems, in Miller, R.E., and Thatcher, J.W. (eds.), *Complexity of Computer Computations*, Plenum Press, New York (1972), 85-103.

Karp, R.M., Probabilistic Analysis of Partitioning Algorithms for the Traveling Salesman Problem in the Plane, *Mathematics of Operations Research 2*, (1977), 209-224.

Karp, R.M., Combinatorics, Complexity, and Randomness, *Communications of the ACM 29*, (1986), 98-109.

Kasami, T., An Efficient Recognition and Syntax Algorithm for Context-Free

Languages, *Scientific Report AFCRL-65-758*, Air Force Cambridge Research Laboratory, Bedford, Massachusetts, (1965).

Kernighan, B.W., and Plauger, P.J., *Software Tools*, Addison-Wesley, Reading, Massachusetts, 1976.

Kfoury, A.J., Moll, R.N., and Arbib, M.A., *A Programming Approach to Computability*, Springer-Verlag, New York, 1982.

Kleene, S.C., General Recursive Functions of Natural Numbers, *Mathematische Annalen 112*, (1936), 727-742.

Kleene, S.C., Recursive Predicates and Quantifiers, *Transactions of the American Mathematical Society 53*, (1943), 41-73.

Kleene, S.C., *Introduction to Metamathematics*, D. Van Nostrand, Princeton, New Jersey, 1952.

Kleene, S.C., Representation of Events in Nerve Nets and Finite Automata, in Shannon, C.E., and McCarthy, J. (eds.), *Automata Studies*, Princeton University Press, Princeton, New Jersey (1956), 3-42.

Kleine Buning, H., and Ottmann, Th., Kleine Universelle Mehrdimensionale Turingmaschinen, *Elektronische Informationsverarbeitung und Kybernetik 13*, (1977), 179-201.

Knuth, D.E., On the Translation of Languages from Left to Right, *Information and Control 8*, (1965), 607-639.

Knuth, D.E., The Remaining Trouble Spots in ALGOL 60, *Communications of the ACM 10*, (1967), 611-618.

Knuth, D.E., Top-Down Syntax Analysis, *Acta Informatica 1*, (1971), 79-110.

Knuth. D.E., *The Art of Computer Programming, Vol.3: Sorting and Searching*, Addison-Wesley Publishing Co., Reading, Massachusetts, 1973.

Knuth, D.E., Big Omicron and Big Omega and Big Theta, *SIGACT News 8*, (1976), 18-23.

Knuth, D.E., Morris, J.H., Jr., and Pratt, V.R., Fast Pattern Matching in Strings, *SIAM Journal on Computing 6*, (1977), 323-350.

Korenjak, A.J., and Hopcroft. J.E., Simple Deterministic Languages, *Proceedings of the Seventh Annual IEEE Symposium on Switching and Automata Theory*, (1966), 36-46.

Kuich, W., and Salomaa, A., *Semirings, Automata, Languages*, Springer-Verlag, New York, 1985.

Kurki-Suonio, R., Note on Top-Down Languages, *BIT 9*, (1969), 225-238.

Larson, L.C., *Problem-Solving through Problems*, Springer-Verlag, New York, 1983.

Lauer, P.E., Torrigiani, P.R., and Shields, M.W., COSY: A System Specification Language Based on Paths and Processes, *Acta Informatica 12*, (1979), 109-158.

Lewis, H.R., and Papadimitriou, C., *Elements of the Theory of Computation*, Prentice-Hall, Englewood Cliffs, New Jersey, 1981.

Lewis, P.M., II, Rosenkrantz, D.J., and Stearns, R.E., *Compiler Design Theory*, Addison-Wesley, Reading, Massachusetts, 1976.

Lewis, P.M., II, and Stearns, R.E., Syntax-Directed Transduction, *Journal of the ACM 15*, (1968), 465-488.

Lewis, P.M., II, Stearns, R.E., and Hartmanis, J., Memory Bounds for Recognition of Context-Free and Context-Sensitive Languages, *Proceedings of the Sixth Annual IEEE Symposium on Switching Circuit Theory and Logical Design*, (1965), 191-202.

Lindenmayer, A., Mathematical Models for Cellular Interactions in Development, Parts I and II, *Journal of Theoretical Biology 18*, (1968), 280-315.

Lindenmayer, A., Developmental Systems without Cellular Interaction, Their Languages and Grammars, *Journal of Theoretical Biology 30*, (1971), 455-484.

Linger, R.C., Mills, H.D., and Witt, B.I., *Structured Programming: Theory and Practice*, Addison-Wesley, Reading, Massachusetts, 1979.

Loeckxx, J., and Sieber, K., *The Foundations of Program Verification*, John Wiley & Sons, New York, 1984.

Machtey, M., and Young, P.R., *An Introduction to the General Theory of Algorithms*, North-Holland, New York, 1978.

Mallozi, J.S., and De Lillo, N.J., *Computability with PASCAL*, Prentice-Hall, Englewood Cliffs, New Jersey, 1984.

Marcotty, M., Ledgard, H.F., and Bochmann, G.V., A Sampler of Formal Definitions, *Computing Surveys 8*, (1976), 191-276.

Markov, A.A., *The Theory of Algorithms* (translated from the Russian by J.J. Schorr-kon), U.S. Dept. of Commerce, Office of Technical Services, No. OTS 60-51085, 1954.

McCarthy, J., Recursive Functions of Symbolic Expressions and Their Computation by Machine, Part I, *Communications of the ACM 3*, (1960), 184-195.

McCarthy, J., A Basis for a Mathematical Theory of Computation, in Braffort, P., and Hirschberg, D. (eds.), *Programming and Formal Systems*, North-Holland, Amsterdam (1963), 33-70.

McCarthy, J., and Painter, J., Correctness of a Compiler for Arithmetic Expressions, in Schwartz, J.T. (ed.), *Mathematical Aspects of Computer Science*, American Mathematical Society, Providence, Rhode Island (1967), 33-41.

McCulloch, W.S., and Pitts, W., A Logical Calculus of the Ideas Immanent in Nervous Activity, *Bulletin of Mathematical Biophysics 5*, (1943), 115-133.

McNaughton, R., *Elementary Computability, Formal Languages, and Automata*, Prentice-Hall, Englewood Cliffs, New Jersey, 1982.

McNaughton, R., and Papert, S., *Counter-Free Automata*, The M.I.T. Press, Cambridge, Massachusetts, 1971.

McNaughton, R., and Yamada, H., Regular Expressions and State Graphs for Automata, *IEEE Transactions on Electronic Computers 9*, (1960), 39-47.

McWhirter, I.P., Substitution Expressions, *Journal of Computer and System Sciences 5*, (1971), 629-637.

Mead, C.A., and Conway, L.A., *Introduction to VLSI Systems*, Addison-Wesley, Reading, Massachusetts, 1980.

Meyer, A.R., and Stockmeyer, L.J., The Equivalence Problem for Regular Expressions with Squaring Requires Exponential Time, *Proceedings of the Thirteenth Annual IEEE Symposium on Switching and Automata Theory*, (1972), 125-129.

Minsky, M.L., A 6-Symbol, 7-State Universal Turing Machine, *MIT Laboratory Group Report 54G-0027*, August, 1960.

Minsky, M.L., Size and Structure of Universal Turing Machines Using Tag Systems, Marcel Dekker (1962), 229-238.

Minsky, M.L., *Computation: Finite and Infinite Machines*, Prentice Hall, Englewood Cliffs, New Jersey, 1967.

Moore, E.F., Gedanken Experiments on Sequential Machines, in Shannon, C.E., and McCarthy, J. (eds.), *Automata Studies*, Princeton University Press, Princeton, New Jersey (1956), 129-153.

Moore, G.B., Kuhns, J.L., Trefftzs, J.L., and Montgomery, C.A., Accessing Individual Records from Personal Data Files Using Non-Unique Identifiers. *NBS Special Publication 500-2*, U.S. Department of Commerce, National Bureau of Standards, 1977.

Myhill, J., Finite Automata and the Representation of Events, *WADD TR-57-624*, Wright Patterson AFB, Ohio, (1957), 112-137.

Naur, P. (ed.), Report on the Algorithmic Language ALGOL 60, *Communications of the ACM 3*, (1960), 299-314, revised in *Communications of the ACM 6*, (1963), 1-17.

Newell, A., and Shaw, J.C., Programming the Logic Theory Machine, *Proceedings of the Western Joint Computer Conference*, (1957), 230-240.

Newman, W., and Sproull, R., *Principles of Interactive Computer Graphics*, Second Edition, McGraw-Hill, New York, 1979.

Oettinger, A.G., Automatic Syntactic Analysis and the Pushdown Store, *Proceedings of the Symposia in Applied Mathematics 12*, American Mathematical Society, Providence, Rhode Island, (1961), 104-109.

Ogden, W., A Helpful Result for Proving Inherent Ambiguity, *Mathematical Systems Theory 2*, (1968), 191-194.

Pagan, F.G., *Formal Specification of Programming LanguagesL: A Panoramic Primer*, Prentice-Hall, Englewood Cliffs, New Jersey, 1981.

Pansiot, J.J., A Note on Post's Correspondence Problem, *Information Processing Letters 12*, (1981), 233.

Papadimitriou, C.H., and Steiglitz, K., *Combinatorial Optimization: Algorithms and Complexity*, Prentice-Hall, Englewood Cliffs, New Jersey, 1982.

Parikh, R.J., On Context-Free Languages, *Journal of the ACM 13*, (1966), 570-581.

Pavlenko, V.A., Post Combinatorial Problem with Two Pairs of Words, *Dokladi AN Ukr. SSR 33*, (1981), 9-11.

Peter, R., *Recursive Functions*, Academic Press, New York, 1967.

Pilling, D.L., Commutative Regular Equations and Parikh's Theorem, *Journal of the London Mathematical Society II, 6*, (1973), 663-666.

Pippenger, N., Complexity Theory, *Scientific American 238*, 6 (1978), 114-124.

Post, E.L., Finite Combinatory Processes-Formulation I, *Journal of Symbolic Logic 1*, (1936), 103-105.

Post, E.L., Recursive Unsolvability of a Problem of Thue, *Journal of Symbolic Logic 12*, (1947), 1-11.

Prather, R.E., *Discrete Mathematical Structures for Computer Science*, Houghton Mifflin, Boston, 1976.

Pratt, T.W., The Formal Analysis of Computer Programs, in Pollack, S.V. (ed.), *Studies in Computer Science*, Studies in Mathematics, The Mathematical Association of America (1982), 169-195.

Priese, L., Towards a Precise Characterization of the Complexity of Universal and Nonuniversal Turing Machines, *SIAM Journal on Computing 8*, (1979), 508-523.

Rabin, M.O., and Scott, D., Finite Automata and Their Decision Problems, *IBM Journal of Research and Development 3*, (1959), 115-125.

Rado, T., On Noncomputable Functions, *Bell System Technical Journal 41*, (1962), 877-884.

Révész, G.E., *Introduction to Formal Language Theory*, McGraw-Hill, New York, 1983.

Reynolds, J.C., *The Craft of Programming*, Prentice-Hall, Englewood Cliffs, New Jersey, 1981.

Robinson, R.M., Undecidability and Nonperiodicity for Tilings of the Plane, *Inventiones Mathematicae 12*, (1971), 177-209.

Rogers, Jr., H., *The Theory of Recursive Functions and Effective Computability*, McGraw-Hill, New York, 1967.

Rosenkrantz, D.J., Matrix Equations and Normal Forms for Context-Free Grammars, *Journal of the ACM 14*, (1967), 501-507.

Rosenkrantz, D.J., Programmed Grammars and Classes of Formal Languages, *Journal of the ACM 16*, (1969), 107-131.

Rosenkrantz, D.J., and Stearns, R.E., Properties of Deterministic Top-Down Grammars, *Information and Control 17*, (1970), 226-256.

Rosenkrantz, D.J., Stearns, R.E., and Lewis, P.M., An Analysis of Several Heuristics for the Traveling Salesman Problem, *SIAM Journal on Computing 6*, (1977), 563-581.

Rozenberg, G., Extension of Tabled 0L Systems and Languages, *International Journal of Computer and Information Sciences 2*, (1973), 311-334.

Rozenberg, G., and Salomaa, A., *The Mathematical Theory of L Systems*, Academic Press, New York, 1980.

Rustin, R. (ed.), *Formal Semantics of Programming Languages*, Prentice-Hall, Englewood Cliffs, New Jersey, 1972.

Salomaa, A., *Theory of Automata*, Pergamon Press, London, 1969.

Salomaa, A., *Formal Languages*, Academic Press, New York, 1973.

Salomaa, A., *Computation and Automata*, Cambridge University Press, Cambridge, England, 1985.

Savitch, W.J., *Abstract Machines and Grammars*, Little, Brown, Boston, 1982.

Savitch, W.J., Relationships between Nondeterministic and Deterministic Tape Complexities, *Journal of Computer and System Sciences 4*, (1970), 177-192.

Scheinberg, S., Note on the Boolean Properties of Context-Free Languages, *Information and Control 3*, (1960), 372-375.

Schutzenberger, M.P., On Context-Free Languages and Pushdown Automata, *Information and Control 6*, (1963), 246-264.

Scott, D., Some Definitional Suggestions for Automata Theory, *Journal of Computer and System Sciences 1*, (1967), 187-212.

Scowen, R.S., An Introduction and Handbook for the Standard Syntactic Metalanguage, *National Physical Laboratory Report DITC 19/83*, 1983.

Shannon, C.E., and McCarthy, J. (eds.), *Automata Studies*, Princeton University Press, Princeton, New Jersey, 1956.

Shepherdson, J.C., The Reduction of Two-Way Automata to One-Way Automata, *IBM Journal of Research and Development 3*, (1959), 198-200.

Shepherdson, J.C., and Sturgis, H.E., Computability of Recursive Functions, *Journal of the ACM 10*, (1963), 217-255.

Sippu, S., and Soisalon-Soininen, E., *Parsing Theory*, Springer-Verlag, New York, 1987.

Sippu, S., Soisalon-Soininen, E., and Ukkonen, E., The Complexity of LALR(k) Testing, *Journal of the ACM 30*, (1983), 259-270.

Smith, A.R., Plants, Fractals, and Formal Languages, *Computer Graphics 18*, (1984), 1-10.

Solow, D., *How to Read and Do Proofs*, John Wiley & Sons, New York, 1982.

Stanat, D.F., and McAllister, D.F., *Discrete Mathematics in Computer Science*, Prentice-Hall, Englewood Cliffs, New Jersey, 1977.

Stockmeyer, L.J., and Chandra, A.K., Intrinsically Difficult Problems, *Scientific American 240,*, 5 (1979), 140-159.

Stone, H.S., *Discrete Mathematical Structures and Their Applications*, SRA, Chicago, Illinois, 1973.

Tarjan, R.E., A Unified Approach to Path Problems, *Journal of the ACM 28*, (1981), 577-593.

Tennent, R.D., The Denotational Semantics of Programming Languages, *Communications of the ACM 19*, (1976), 437-453.

Tennent, R.D., *Principles of Programming Languages*, Prentice-Hall, Englewood Cliffs, New Jersey, 1981.

Thatcher, J.W., Characterizing Derivation Trees of a Context-Free Grammar through a Generalization of Finite-Automata Theory, *Journal of Computer and System Sciences 1*, (1967), 317-322.

Thatcher, J.W., Tree Automata: An Informal Survey, in Aho, A.V. (ed.), *Currents in the Theory of Computing*, Prentice-Hall, Englewood Cliffs, New Jersey (1973), 143-172.

Thompson, K., Regular Expression Search Algorithm, *Communications of the ACM 11*, (1968), 419-422.

Thue, A., Probleme über Veränderungen von Zeichenreihen nach gegebenen Regeln, *Skrifter utgit av Videnskapsselskapet i Kristiania*, I. Matematisk-naturvidenskabelig klasse 1914, 10 (1914).

Tremblay, R.E., and Manohar, R.P., *Discrete Mathematical Structures with Applications to Computer Science*, McGraw-Hill, New York, 1975.

Turing, A.M., On Computable Numbers with an Application to the Entscheidungs-Problem *Proceedings of the London Mathematical Society 2*, (1936), 230-265. A correction, *ibid., 43*, 544-546.

Ullman, J.D., *Computational Aspects of VLSI*, Computer Science Press, Rockville, Maryland, 1984.

Valiant, L.G., General Context-Free Recognition in Less than Cubic Time, *Journal of Computer and Systems Sciences 10*, (1975), 308-315.

van Dalen, D., A Note on some Systems of Lindenmayer, *Mathematical Systems Theory 5*, (1971), 128-140.

van Leeuwen, J., A Generalization of Parikh's Theorem in Formal Language Theory, *Proceedings of ICALP '74, Springer-Verlag Lecture Notes in Computer Science 14*, (1974a), 17-26.

van Leeuwen, J., Notes on Pre-Set Pushdown Automata, *Springer-Verlag Lecture Notes in Computer Science 15*, (1974b), 177-188.

van Wijngaarden, A. (ed.), Report on the Algorithmic Language ALGOL 68, *Numerische Mathematik 14*, (1969), 79-218.

van Wijngaarden, A., Mailloux, B.J., Peck, J.E.L., Koster, C.H.A., Sintzoff, M., Lindsey, C.H., Meertens, L.G.L.T., Fisker, R.G. (eds.), Revised Report on the Algorithmic Language ALGOL 68, *Acta Informatica 5*, (1974), 1-236.

Vere, S., Translation Equations, *Communications of the ACM 13*, (1970), 83-89.

Watanabe, S., On a Minimal Universal Turing Machine, *MCB Report*, Tokyo, 1960.

Watanabe, S., 5-Symbol 8-State and 5-Symbol 6-State Universal Turing Machines, *Journal of the ACM 8*, (1961), 476-483.

Wegner, P., The Vienna Definition Language, *Computing Surveys 4*, (1972a), 5-63.

Wegner, P., Programming Language Semantics, in Rustin, R. (ed.), *Formal Semantics of Prgramming Languages*, Prentice-Hall, Englewood Cliffs, New Jersey (1972b), 149-248.

Winston, P.H., *Artificial Intelligence*, Addison-Wesley Publishing Co., Reading, Massachusetts, 1977.

Wirth, N., *Systematic Programming: An Introduction*, Prentice-Hall, Englewood Cliffs, New Jersey, 1973.

Wirth, N., Data Structures and Algorithms, *Scientific American 251*, (September 1984), 60-69.

Wise, D.S., A Strong Pumping Lemma for Context-Free Languages, *Theoretical Computer Science 3*, (1976), 359-370.

Wolfram, S., Farmer, J.D., and Toffoli, T. (eds.), Cellular Automata: Proceedings of an Inter-Disciplinary Workshop, *Physica 10D*, Nos. 1 and 2, 1984.

Wood, D., The Normal Form Theorem — Another Proof, *Computer Journal 12*, (1969a), 139-147.

Wood, D., The Theory of Left-Factored Languages, *Computer Journal 12*, (1969b), 349-356, and *13* (1970), 55-62.

Wood, D., A Few More Trouble Spots in ALGOL 60, *Communications of the ACM 12*, (1969c), 248.

Wood, D., *Paradigms and Programming with PASCAL*, Computer Science Press, Rockville, Maryland, 1984.

Yntema, M.K., Cap Expressions for Context-Free Languages, *Information and Control 8*, (1971), 311-318.

Younger, D.H., Recognition and Parsing of Context-Free Languages in Time n^3, *Information and Control 10*, (1967), 189-208.

Index

2-*CDTM* 311
2-*DPDA* 271
2-*DTM* 312
2-normal form 263
2-*NPDA* 263
2-P*DTM* 306
2*SNF-PDA* 383
4-*CDTM* 309

a-length 64
A-production 191
abstract family
  of languages 438
  of relations 444
accept action 519
acceptance
  2*DPDA* 272
  DFA 102
  DPDA 253
  DTM 286
  NFA 115ff.; 120
  NPDA 260
  NTM 316
accepting configuration sequence
  DFA 102
  DPDA 253
  DTM 286
  NTM 316
acceptor; see automaton; recognizer.
Ackermann function 362
action conflict 525
action table 519
active nonterminal 201
adjacency matrix 12
adjacent 11
AED RWORD 364
AFL 438
Aho, A.V. 80, 143, 238, 274, 364, 365, 441, 524, 525
Alagic, S. 86
ALGOL-like languages 240
algorithm function 59
algorithm-computable function 60
algorithms
  Check 400
  Eliminate One Unit Production 220
  GNF Conversion 227
  Iterative Subset Construction 120
  λ-Nonterminals 216
  NFA to DFA 120
  *NFA to DFA*2 121
  Reachability 14
  Reachable Symbols 214
  Reduce Largest Productions 221
  Simulate 399-400
  Square 55-56
  Subset Construction 120
  Surface Configuration 385-386
  Terminating Symbols 212
Allison, D.C.S. 462
alph 68
alphabet 63, 64
Altman, E.B. 441
ambiguity 203ff., 238
Anderson, R.B. 85
antisymmetry 23

approximation
  language 374-375
  transitive closure 26, 27
Arbib, M.A. 37, 80, 86
Ashcroft, E.A. 85, 86
associativity 68
asymmetry 22
automaton
  cellular 144
  data structure 274
  deterministic finite 98ff.
  deterministic pushdown 248ff., 249-250
  deterministic Turing machine 280ff.
  extended finite 167
  finite 95ff., 114
  finite buffer pushdown 274
  finite state; see finite.
  λ-*NFA* 135ff.
  λ-transition deterministic pushdown 256
  lazy *FA* 140ff.
  lazy *PD* 274
  loop-free *DFA* 133ff.
  multihead 274
  multipushdown 274
  nondeterministic counter 270
  nondeterministic finite 114ff.
  nondeterministic pushdown 259
  preset pushdown 274
  tree 145
  tree *DFA* 133ff.
  two-way deterministic pushdown 271

Backus, J.W. 189, 237
Backus-Naur form 189
Banerji, R.B. 441
Bar-Hillel, Y. 406, 440, 441, 462
Barnes, B.H. 143
Barnett, M.P. 364
base element 5
Beckman, F.S. 80
Beizer, B. 181
Bellman, R.E. 524
Bentley, J.L. 91, 523
Berger, R. 463
Berlekamp, E.R. 144, 337
Berstel, J. 80, 364, 441
big-O notation 19
big-omega notation 19
big-theta notation 19
bijection 18
binary relation 20ff., 339
bit matrix 11
blank symbol 280
block code 345
BNF 189
Boasson, L. 441
Bobrow, L.S. 37
Boggle 177
Boolean
  algebra 413
  function 19
  operations 8
bottom-up parsing 515-523
bounded delay code 422-423
Bourne, S.R. 179
Brainerd, W.S. 80
Brzozowski, J.A. 143, 179, 180, 181
Bücher, W. 80
Burge, W.H. 81
Burks, A.W. 144, 274
busy beaver 331

Caesar cipher 4, 420, 421, 422
call digraph 10
canonical derivation 198
canonical derivation sequence 200
Cantor, D.C. 237, 463
cardinality
  finite sets 4
  infinite sets 27ff.
cartesian product 9
catenation 434
  languages 68
  words 64
cell 99
CFG 190
CFL 194
Chandra, A.K. 333, 523
characteristic function 17
checker 61
checking off. 296ff.
child 15
Choffrut, C. 364
choice 113
Chomsky, N. 54, 55, 143, 211, 212, 221, 223, 237, 274, 406, 463, 497, 523
Christofides, N. 525
Church's thesis 359
Church, A. 54, 55, 81
circular production 220
Cleaveland, J.C. 239
Clocksin, W.F. 85, 86
closed 5, 412, 413
closure 412ff., 413
  definition 5
  of language; see star.

of set 4, 5, 24ff.
properties 412ff.
CNF 221
Cobham, A. 331
Cocke, J. 212, 524
Cohen, D.J. 143, 237
collection; see set.
Comer, D. 143
complement
of languages 68
of sets 6
complete *DFA* 103
completion state; see sink state.
composition 360, 435ff.
of functions 18
of relations 21
computable function 333, 356
computation; see configuration sequence.
computing function 289, 356
computer graphics 240
concatenation; see catenation.
cone 438
configuration
2-DPDA 272
DFA 100
DPDA 250
DTM 281
DTT 355
EFA 168
FT 340
NFA 115
NPDA 260
NTM 316
PDT 353
configuration sequence
DFA 101
DPDA 252
DTM 285
DTT 355
EFA 168
FT 340
NFA 115
NPDA 260
NTM 316
PDT 353
constant function 361
constructions
2−*NPDA to CFG* 263
CFG to NPDA 261
NPDA to 2−*NPDA* 264
s−*CFG to DPDA* 256
constructor 5
containment 4, 373ff.
context-free
expression 239
grammar; see grammar.
language 413-417, 426-427
substitution 426, 427
context-sensitive language 334
contradiction, proof by 30ff.
Conway, J.H. 237, 337
Conway, L.A. 523
Cook, S.A. 406, 527
correctness 60, 85
countability 27ff.
counter tape 306
counting 35ff.
cross product 417
Culik II, K. 364
current state 100, 250
current symbol 100, 250
cycle 10
cylinder 438

Danzig, G.B. 523
data compression 344
data structure automaton; see automaton.
Davis, M.D. 50, 333, 364, 365, 462, 463
de Bakker, J.W. 84
De Lillo, N.J. 80
De Morgan's rules 7ff.
decidability 446ff.
decidable language 294
decision
algorithm 60
maker 291
problem 60, 446ff.
decision-making Turing machine 291
decoding 341
definition of languages 71ff.
definition of sets
closure 5
inductive 4
recursive 4
degree 10
DeMillo, R.A. 85, 145
Denning, P.J. 80
derivation 193
derivation sequence 193
derivation tree; see syntax tree.
derive 191
deterministic 97
CFL 259
context-free languages 382
finite acceptor; see automaton.
parsing
bottom-up 515-523

top-down 507-515
Turing transduction 438
deterministic-space complexity 327
deterministic-time complexity 327
Dewdney, A.K. 331, 332
DFA language 102
DFA; see automaton.
diagonalization 362, 36ff., 404
difference 6
digraph 9ff., 20
Dijkstra, E.W. 86
direct proof 28ff.
directed graph; see digraph.
directed tree; see tree.
distinguishability 128
domino 463
DPDA 249
Dreyfus, S.E. 524
DTM 280
DTM languages 403ff., 417-420, 427-
DTM substitution 427-428
DTML 286
DTT 355
dual digraph 14
Dyck language 252

Earley, J. 524
ECFG 233
edge 10
edge label 15
Edmonds, J. 523
EFA; see automaton.
Ehrenfeucht, A. 407, 463
Eilenberg, S. 143
element
of language; see word.
of sequence 8
of set 3
Elgot, C.C. 364
Elspas, B. 85
empty
language, ∅
path 10
production 215
pushdown 249
sequence 9
set, ∅ 3
tree 15
word, λ 64
word set, {λ} 68
encoding 77ff., 321-322, 325-326, 341, 475
Engelfriet, J. 145, 274
enumerability 27ff.
equality of
cardinality 27
functions 18
sets 4
words 66
equations 240
equivalence
class 23
relation 23
equivalence of
algorithms 60
CFGs 194
DFAs 107
DTMs 286
NFAs 116
regular expressions 161
error action 519
Evey, J. 274, 364
execute cycle 100
extended
BNF 234
CFG 233
finite automaton; see automaton.
regular expression 175ff.
transition function 167
extension 375
external node 15

FA; see automaton.
false instance 62
family of
CF languages 194
DCF languages 259
DFA languages 102
DTM languages 286
finite languages 71, 373-375
finite transductions 341
languages 69
linear languages 232
NFA languages 116, 117
NPDA languages 260
pushdown transductions 353
recursive languages 294
REG languages 160
SDTs 351
SSDTs 350
family relationships 373ff.
file compaction 365
final configuration
DTT 355
FT 340
PDT 353
final state 96
final surface configuration 384
finite

control 99
digraph 10
index 24
language 71, 373-375
sequence 8
set 3
substitution 424
transducer 339ff., 340
transduction 341, 424, 431-433, 435-436
transition relation 259, 316, 340, 353
finite-buffer *PDA* 274
finite-degree ambiguity 238
finite specification of
Boolean functions 79
languages 79
Fischer, M.J. 441, 497, 524
Floyd, R.W. 85, 143, 181, 237, 238, 463
Forkes, D. 89
formal semantics 84
Fosdick, L.D. 182
Foster, J.M. 524
free monoid 65
FT 340
function 17ff., 339, 356
of language 71
Futrelle, R.P. 364

Galler, B.A. 82
Gardner, M. 144, 523
Garey, M.R. 333, 523, 525, 526
Gecseg, F. 145
general parsing 504-507
generation 191
generator 73, 439
generic
element 8
specification 55
Gimpel, J.F. 523
Ginsburg, S. 80, 240, 406, 441, 462
GNF 224
Goldstine, J. 406
Gonnet, G.H. 143, 239
Gotlieb, C.C. 237
Gouda, M.G. 144
Graham, R.L. 333, 523
grammar
context-free 187ff., 190
derivation bounded 410
E0L 239
extended context-free 232ff.
hierarchical context-free 492-497
interactive Lindenmayer 500-503
LALR 525
linear 231
left-derivation bounded 410
left linear 231
left ultralinear 410
LL 513
LR 518
meta-linear 409
phrase structure 497-500
pure 409
regular 231
right linear 231
SDT 350
sequential 409
simple deterministic 254
SSDT 349
ultralinear 410
unambiguous 409
W- 238-239
graph
directed; see digraph.
undirected 48
Gray, J.N. 365
Greibach normal form; see normal form.
Greibach, S.A. 208, 211, 212, 224, 237, 406, 441
Gries, D. 86, 274
Gruska, J. 239
Guy, R.K. 337

halting 281
handle 517
hanging 282
Hantler, S.L. 85
hardest *CFL* 439
Harrison, M.A. 80, 143, 144, 145, 238, 239, 365, 407, 524
Hartmanis, J. 331, 463
Hayes, B. 143, 144
Hays, D.G. 524
height 41, 392
Henderson, P. 81
Hennie, F.C. 80
Herman, G.T. 240
Hermes, H. 80, 364, 365, 523, 524
hierarchy 373ff.
Hilbert, D. 463
Hoare, C.A.R. 84, 85, 462
Hodges, A. 331
homomorphism; see morphism.
Hopcroft, J.E. 80, 143, 237, 274, 331, 333, 334, 407
Horowitz, E. 524

identity
function 17
language 69

word 64
Ikeno, N. 332
in-degree 10
inaccessibility 127
incomparability 4
incomplete *DFA* 103
index of equivalence relation 24
indistinguishability 128
induction, proof by 31ff.
inductive definition 4
inequality 4
inequivalence
of *DFA* 107
of *NFA* 116
inference problem 375
infinite
degree ambiguity 238
index 24
sequence 8
set 3
inherent ambiguity 203
initial
configuration
of *DTT* 355
of *FT* 340
of *PDT* 353
element 5
function 361
pushdown symbol 249
surface configuration 384
injective function 17
input
file 99
symbol 98
tape 99
instance 330
internal node 15
intersection 6
intractable language 328
intransitivity; see nontransitivity.
inverse morphism 421, 429-431
Irons, E.T. 364
irreflexivity 22
is in; see member of.
is not in; see not a member of.
isomorphism
minimal *DFA* 133
renaming 107
item 518

Johnson, D.S. 333, 523, 525, 526
Johnson, J.H. 365
Johnson, S.C. 364, 525
Johnson, W.L. 364
Joyce, J. 63

Karp, R.M. 523, 524
Kasami, T. 212, 524
Kernighan, B.W. 179
Kfoury, A.J. 80
King, J.C. 85
Kleene, S.C. 54, 55, 80, 83, 179, 364
Kleine Buning, H. 333
Knuth, D.E. 84, 143, 518, 524
Korenjak, A.J. 274
Kuich, W. 237
Kurki-Suonio, R. 524

label 15
λ-calculus 81
λ-*DPDA* 256
λ-equivalence 217
λ-free *CFG* 216
λ-free morphism 72
λ-*NFA*; see λ-transition *NFA*.
λ-nonterminal 216
λ-production; see empty production.
λ-transition 135, 259
λ-transition *NFA* 135
λ-transition *DPDA* 256
Landin, P.J. 81
Landweber, L.H. 80
language 63, 68ff.
CF 194
deterministic context-free 259
DFA
DTM 286
finite 71, 373-375
linear 231
NFA
recursive 294
regular 160
language of
CFG 194
decision-making Turing machine 294
DFA 102
DPDA 253
NFA 116
NPDA 260
regular expression 160
yes instances 79
language expression 74ff.
Larson, L.C. 37
Lauer, P.E. 84, 182
lazy
finite automaton; see automaton.
PDA 274
left

endmarker 271
movement 271
quotient 434
set 514
left-recursion 224
leftmost
derivation 201
reduction 516
length of
path 10
sequence 9
word 64
length set 379
letter 64
Lewis, H.R. 80
Lewis, P.M., II 275, 331, 364, 365, 524
lexical analysis 343, 364
life, game of 144
Lindenmayer, A. 54, 55, 239, 240, 497, 523
linear order 24
linear-bounded automaton 334
Linger, R.C. 86
Loeckxx, J. 85
look-ahead 508
look-ahead LR 525
loop 133
loop-free *DFA* 133ff.
looping state 133
lower bound 19

m,*n*-place function 18
m-place function 18
machine; see automaton; recognizer; transducer.
Machtey, M. 83, 364
Mallozi, J.S. 80
Manohar, R.P
mapping 71, 420
Marcotty, M. 84
marking 212, 214, 216
Markov algorithm 81
Markov, A.A. 54, 55, 81, 359
Maurer, H.A. 80
McAllister, D.F. 37
McCarthy, J. 81, 84, 143
McCluskey, E.J. 179
McCulloch, W.S. 143
McNaughton, R. 80, 143, 179
McWhirter, I.P. 239
Mead, C.A. 523
member of 4
member; see element
metalanguage 189
Meyer, A.R. 525
Mezei, J.E. 364
Miller, G.A. 143
minimal
cycle 10
DFA 127ff.
path 10
minimalization 363
minimization 127ff.
Minsky, M.L. 80, 332, 333
mirror image; see reversal.
modulo counting 107
monoid 65
Moore, E.F. 143, 365, 462
Moore, G.B.
morphism 71, 420
Morse code 346, 421
Mullish, C.S. 85, 86
multihead automaton; see automaton.
multihead *DTM* 318ff.
multiplicity 8
multipushdown automaton; see automaton.
multiset digraph 14
multisets 8
multitape *DTM* 320
Myhill, J. 406

n-ary relation 20
n-step derivation 193
Naur, P. 84, 189
NCA 270
Newell, A. 274
Newman, W. 523
next state 100, 250
NFA; see automaton.
no instance 62, 330
no instance language 294
node 10
node connectivity 13
node label 15, 16
non-left-recursion 224-225
non-right-recursion 224-225
nondeterminism 113
nondeterministic
counter automaton; see automaton.
finite automaton; see automaton.
pushdown automaton; see automaton.
Turing machine 316ff.
nondeterministic-space complexity 327
nondeterministic-time complexity 327
nonenumerability 28
nonlanguage 375, 381, 390, 394, 397, 440
nonterminal
alphabet 190

symbol 190
nonterminating symbol 212
nontotal function 17
nontransitivity 23
normal form
binary 221, 223
Chomsky 221, 223
Greibach 208, 224, 228, 253, 261, 262, 353
one-standard 230, 231
two-standard 224, 229
not a member of 4
NP-complete problems 472ff.
NP-completeness 333, 471ff.
NP-hardness 474
NPDA 259
NTM 316
null production; see empty production.

Oettinger, A.G. 274
off-line *DTM* 305
Ogden, W. 407
one-to-one into function; see injective function.
one-to-one onto function; see bijection.
onto function; see surjective function.
operations 4, 6ff.
ordered
pair 9
tree; see tree.
tuple; see tuple.
ordering problems 472-473
Osterweil, L.J. 182
Ottmann, Th. 333, 523
out-degree 10
output
alphabet 339, 349, 353, 355
of *FT* 340
of *PDT* 353
tape 339, 352
word 339

P = *NP* problem 328
Pagan, P.G. 84
Painter, J. 84
Pansiot, J.J. 463
Papadimitriou, C.H. 80, 523
Papert, S. 143
parent 15
Parikh, R.H. 394, 395, 406
Parikh's theorem 395
parser 73
parsing 237-238, 504-523
partial
function 17
order 24
recursive function 363
passive nonterminal 201
path 10
pattern matching 117ff., 125ff., 231-232, 272-273
Paull, M.C. 237
Pavlenko, V.A. 463
PCP 463
PDT 352
period 380
periodic 380
Perles, M. 462
Perlis, A.J. 82
Peter, R. 80, 364
pigeonhole principle 33ff., 225, 377, 392, 393, 395
Pilling, D.L. 406
Pippenger, N. 333, 523
Pitts, W. 143
Plauger. P.J. 179
plus
of λ-*NFA* 136
of language 69
polynomial
equivalence 474
time-bounded *DTM* 472, 474
time-bounded *NTM* 472
transformation 473
polynomial-space-bounded language 327
polynomial-time-bounded language 327
pop 383
positive closure; see plus.
Post system 82
Post's correspondence problem 463
Post, E.L. 54, 55, 83, 237, 359, 463
power set 8
power of
language 69
relation 22
word 65
Prather, R.E. 37
Pratt, T.W. 84
pre-leftmost derivation 201
pre-rightmost derivation 201
predecessor 10
prefix 66
preset pushdown automaton; see automaton.
Priese, L. 333
primitive recursion 360
primitive recursive function 361

problem
  function 58
  instance 55
  specification 55
  uncomputable 446
problems
  3-DIMENSIONAL MATCHING 489, 490-491
  3-DIMENSIONAL RELAXED MATCHING 490-492
  3-SATISFIABILITY 480-481, 482-484, 491-492
  CFG AMBIGUITY 205
  CFG CONTAINMENT 455, 458-461
  CFG EMPTINESS 451
  CFG EQUIVALENCE 455, 458-461
  CFG FINITENESS 451
  CFG INTERSECTION 455-458
  CFG MEMBERSHIP 207, 447
  CFG PARSING 208
  CFG UNIVERSALITY 461
  CFL AMBIGUITY 206, 462
  CHESS QUEENS 112
  CHESS QUEENS CHECKING 112
  DFA CONTAINMENT 454
  DFA EMPTINESS 450
  DFA EQUIVALENCE 108, 133, 455
  DFA FINITENESS 134, 451
  DFA INTERSECTION 455
  DFA MEMBERSHIP 103, 447
  DFA UNIVERSALITY 450
  DTM EMPTINESS 452, 453-454
  DTM EMPTYWORD MEMBERSHIP 453
  DTM FINITENESS 452, 454
  FERMAT TESTING 60
  HALTING 446, 448
  HAMILTONIAN CIRCUIT 481-482
  NTM TIME BOUNDED HALTING 475-476
  RECTANGLE INTERSECTION 495-497
  REGULAR EXPRESSION NON-UNIVERSALITY 525
  RESTRICTED HALTING 449
  S-CFG MEMBERSHIP 254
  SATISFIABILITY 477-479
  SQUARE 55
  SQUARE TESTING 60
  SUBSET SUM 489, 490, 495-497
  TRAVELING SALESMAN 328
  TRAVELING SALESMAN CHECKING 328, 481-482, 77
  UNAMBIGUOUS *CFG* CONSTRUCTION 206
  VERTEX COVER 481-488
product
  of λ-*NFA* 162
  of languages; see catenation.
production 190
production system 82
program
  construction 86
  correctness 85
projection function 361
proof methods 28ff.
proof techniques 33ff.
proper
  containment 4, 373ff.
  prefix 66
  subword 66
  suffix 66
properties of operations
  associativity 6
  commutativity 6
  distributivity 6
  idempotence 6
  involution 6
protocols 144
PSPACE-completeness 333
pumping 376ff., 387ff.
pumping argument 378
pumping lemma 376, 389-390, 395
push 383
pushdown
  alphabet 249
  state diagram 251
  symbol 249
  tape 306
  transducer 352, 509-512, 519-521
  transduction 436-437

QED 364
quasi-leftmost reduction 516
quotient 434, 443

Rabin, M.O. 143, 462
Rado, T. 331
reachability
  in *CFGs* 212
  in *DFAs* 121
  in digraphs 73
read-only input tape 305
realizability 384
realizable pair 385
recognizer 73
recursion
  direct 11
  mutual 11

recursive
  definition 4
  function 83, 363
  function theory 365
  languages 401ff.
  subprogram 10
recursively enumerable 334
recursively enumerable language 334
reduce action 519
reduction 516
  no-move 301
  state 303
  tape symbol 301
redundant symbol 212
refinement 12
reflexive transitive closure 22
reflexivity
  of digraph 12
  of relation 22
REG language; see language, regular.
REGL; see language, regular.
regular
  expression 156, 155ff.
  intersection 414
  language 160, 375, 413, 424-425
  left quotient 434
  relation 443
  right quotient 434
  substitution 426
rejection
  for *DFA* 102
  for *NFA* 116, 120
relation 19ff.
relative power 373ff.
renaming 106, 421
replace; see substitute.
resource-bounded Turing machines 326ff.
restricted *NPDA* 384
reversal
  of language 70, 419
  of word 67
Révész, G.E. 80, 406, 334, 523
rewrite 191
rewriting sequence; see derivation sequence.
Reynolds, J.C. 86
Rice, H.G. 240, 406
right
  endmarker 271
  movement 271
  quotient 434
  sentential form 516
right-hand side 191
right-recursion 224
rightmost derivation 201
rNPDA 384
Robinson, R.M. 463
Rogers, Jr., H. 83, 364, 365
Rose, G.F. 441, 463
Rosenkrantz, D.J. 237, 275, 365, 441, 497, 524, 525
Rosier, L.E. 144
Rozenberg, G. 239, 441, 524
rule 190, 349
rule 349
Rustin, R. 84

s-CFG 254
Sahni, S. 524
Salomaa, A. 80, 181, 237, 239, 334, 406, 523, 524
Savitch, W.J. 80, 407
Scheinberg, S. 440
Schutzenberger, M.P. 237, 274, 463
Scott, D. 80, 143, 462
Scowen, R.S. 237
SDTG 350
search tree 208
security 144-145
semantics 84
semi-*AFL* 438
sentence 194
sentence symbol 190
sentential derivation 194
sentential form 194
sequences 8ff.
set 3ff.
  of English words 3
  of integers, $\mathbb{Z}$ 3
  of nonnegative integers, $\mathbb{N}_0$ 3
  of positive integers, $\mathbb{N}$ 3
  of reals, $\mathbb{R}$ 3
Shamir, E. 462
Shannon, C.E. 143, 331
Shaw, J.C. 274
Shepherdson, J.C. 143
shift action 519
shuffle 443
sibling 15
Sieber, K. 85
simple syntax-directed translation grammar 349
simplification
  of *CFGs*, 211ff.
  of *DTMs*; 305, see also reduction.
  of *FAs* 127
sink state 109
Sippu, S. 524, 525

size
  of sequence; see length.
  of set; see cardinality.
small *PDAs* 275
Smith, A.R. 240
Soisalen-Soininen, E. 524
Solow, D. 37
space-bounded *DTM* 327
spaghetti programming 145-146
Spanier, E.H. 441
specification 3ff.
Sproull, R. 523
SSDTG 349
Stanat, D.F. 37
star
  of λ-*NFA* 162
  of language 69, 434
start state 96
state 95, 98
  diagram 95ff.
  symbol 98
  table 98
  transition; see transition.
Stearns, R.E. 275, 331, 364, 365, 524
Steiglitz, K. 523
Steinby, M. 145
Stockmeyer, L.J. 333, 523, 525
Stone, H.S. 37
string similarity 365
strong connectivity 13
subset 4
substitute 191
substitution 423-424
subtree 15
subword 66
succ 361
successor 10
successor function 361
suffix 66
super-configuration 118
super-configuration sequence 118
superset 4
surface configuration 383, 384
surjective function 17
symbol 64
symbol reachability 127
symmetry
  of digraph 12
  of relation 22
synchronization 502
syntactic monoid 143
syntax
  diagram 138, 187ff., 234ff.
  tree 197ff., 387ff.
syntax-directed translation 364
system 96

tail-end recursion 231
tape
  alphabet 281
  symbol 281
Tarjan, R.E. 182
Tennent, R.D. 84
terminal
  alphabet 190
  left appearance 225
  sentential form 194
  symbol 190
terminating symbol 212
Thatcher, J.W. 145
Thompson, K. 364
threshold 380
Thue, A. 237
tiling 463
time-bounded *DTM* 327
Tompa, F.W. 239
top-down parsing 504-515
topmost pushdown symbol 250, 385
total
  function 17
  order 14, 24
tour 77
tractable language 328, 383
transducer 339
  deterministic Turing 355
  finite 340
  nondeterministic pushdown 352
transduction 339
  finite 341
  pushdown 353
transformation 472
transition 96, 97
  diagram 96
  function 98, 250, 281, 355
  relation 114
  sequence; see configuration sequence.
  table 98, 519
transitive closure
  of digraph 13
  of relation 22
transitivity
  of digraph 12
  of relation 23
translation 339
  expression 365
  finite 341
  form 349

grammar 339, 348ff., 349
pushdown 353
simple syntax-directed 350
syntax-directed 351
tree 15ff.
tree *DFA* 133ff.
Tremblay, R.E. 37
trie; see tree *DFA*.
trio 438
true instance 62
tuple 9
Turing, A.M. 54, 55, 80, 331, 333, 364, 407, 462
Turing
machine; see automaton.
micro 332
state diagram 283
Turing-computable function 356
two-pushdown off-line *DTM* 306
two-way
deterministic pushdown automaton; see automaton.
infinite tape *DTM* 312

u-connectivity 13
u-reachability
in *DFA* 121
in digraph 13
Ullman, J.D. 80, 181, 237, 238, 274, 333, 334, 364, 365, 407, 524, 525
ultimately periodic 380
un-derivation step 516
unary language 379
unary regular language 379ff.
uncountability 28
undecidability 446ff.
undef 17
undefined value 17
union
of λ-*NFA* 136, 162
of languages
of sets 6ff.
uniquely decodable 422
unit production 220
unit-free *CFG* 220
universal
language 64
Turing machine 320ff.
universe 3
of English words 3
of integers $\mathbb{Z}$ 3
unreachability
in *DFAs* 127
in *CFGs* 212
upper bound 19
Uzgalis, R. 239

Valiant, L.G. 524
valid item 518
validator 61
van Dalen, D. 523
van Leeuwen, J. 237, 274, 406
van Wijngaarden, A. 238, 239
variable 190
variable-length code 345
vectors; see tuples.
Venn diagram 6
Vere, S. 365
verifier 61
vertex 10
viable prefix 518
VLSI 492
vocabulary 191

W-grammar 238-239
Wadge, W.W. 85, 86
Watanabe, S. 332
Wegner, P. 84
weight of tree 41
Weyuker, E.J. 80
Winston, P.H. 83
Wirth, N. 84, 143, 237
Wolfram, S. 144
Wood, D. 84, 86, 237, 524
word 63, 64
work tape 305
worst-case measure 326

YACC 364, 525
Yamada, H. 179
yes instance 62, 330
yes instance language 330, 294
yield of tree 200, 392
yields 101
Yntema, M.K. 239
Yoeli, M. 143
Young, P.R. 83, 364
Younger, D.H. 212, 524

zero
function 361
language 69

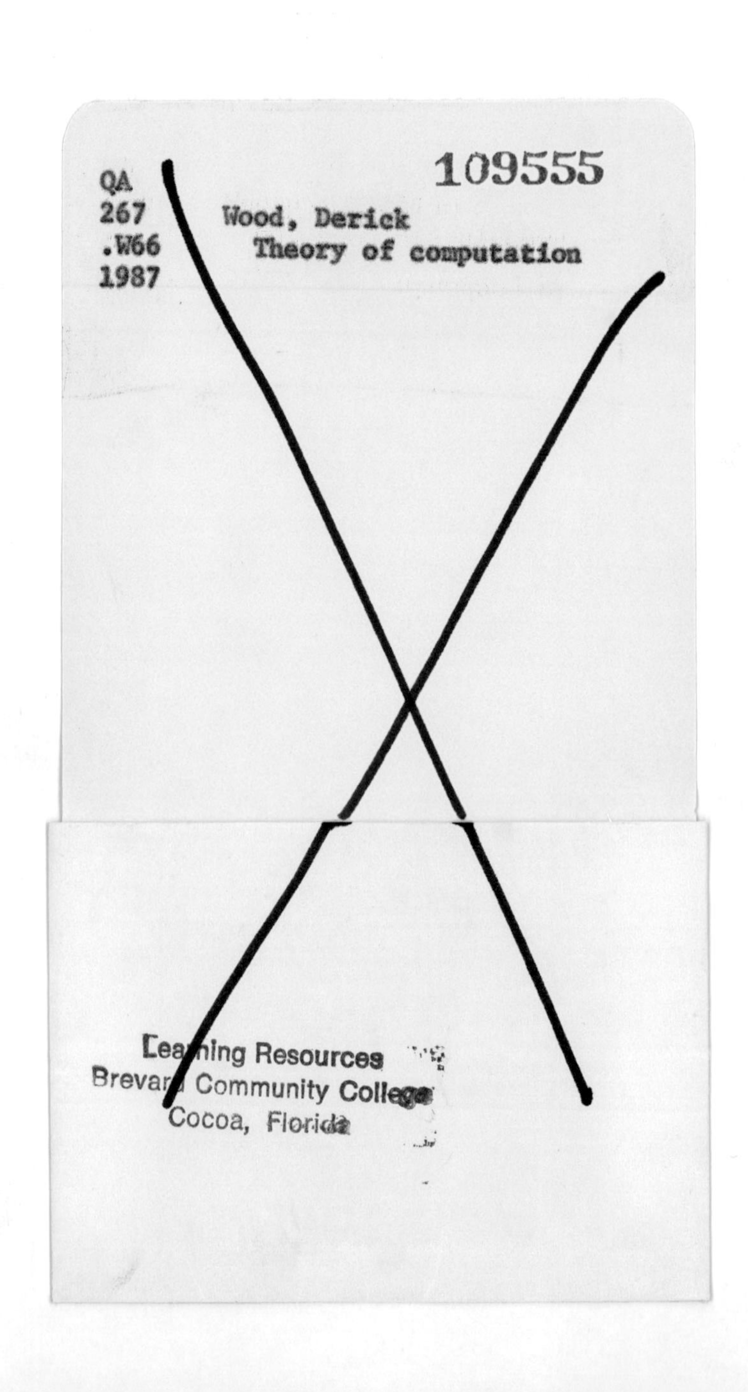
QA
267
.W66
1987
109555
Wood, Derick
Theory of computation
Learning Resources
Brevard Community College
Cocoa, Florida